Manufacturing Engineering

Manufacturing Engineering

Edited by **Jeff Hansen**

Clanrye International
New Jersey

Published by Clanrye International,
55 Van Reypen Street,
Jersey City, NJ 07306, USA
www.clanryeinternational.com

Manufacturing Engineering
Edited by Jeff Hansen

International Standard Book Number: 978-1-63240-541-8 (Hardback)

Printed in China.

Contents

Preface

Manufacturing engineering involves planning, designing, and packaging i.e. conversion of raw materials into a finished products. This field aids the manufacturing processes and related industries by reducing their processing time and increasing profits. Many technological advancements have been aided by applied research in this field and the most important ones being automation and robotics. This book presents the complex subject of manufacturing in the most comprehensible and easy to understand language. From theories to research to practical applications, case studies related to all contemporary topics of relevance to this field have been included in this book. It includes contributions of international veterans which makes it a beneficial guide for engineers, manufacturers, professionals, academicians and students associated with this field.

After months of intensive research and writing, this book is the end result of all who devoted their time and efforts in the initiation and progress of this book. It will surely be a source of reference in enhancing the required knowledge of the new developments in the area. During the course of developing this book, certain measures such as accuracy, authenticity and research focused analytical studies were given preference in order to produce a comprehensive book in the area of study.

This book would not have been possible without the efforts of the authors and the publisher. I extend my sincere thanks to them. Secondly, I express my gratitude to my family and well-wishers. And most importantly, I thank my students for constantly expressing their willingness and curiosity in enhancing their knowledge in the field, which encourages me to take up further research projects for the advancement of the area.

Editor

1

Design and Fabrication of Diffractive Light-Collecting Microoptical Device with 1D and 2D Lamellar Grating Structures

ChaBum Lee

Department of Mechanical Engineering, University of South Carolina, 300 Main Street, Columbia, SC 29208, USA

Correspondence should be addressed to ChaBum Lee; chabum@mailbox.sc.edu

Academic Editor: Konstantinos Salonitis

This paper presents the optimal design method of diffractive light-collecting microoptical device and its fabrication method by E-beam lithography, fast atom beam etching, and hot-embossing processes. The light-collecting device proposed in the paper is comprised of 9 (3×3) blocks of optical elements: 4 blocks of 1D lamellar grating structures, 4 blocks of 2D lamellar grating structures, and a single block of nonpatterned element at the center, which acts for lens to be able to collect the diffracted and transmitted lights from the lamellar grating structures into the focus area. The overall size of the light-collecting device is $300 \times 300\ \mu m^2$, and the size of each block was practically designed as $100 \times 100\ \mu m^2$. The performance of 1D and 2D lamellar grating structures was characterized in terms of diffraction efficiency and diffraction angle using a rigorous coupled-wave analysis (RCWA) method, and those geometric parameters, depth, pitch, and orientation, were optimized to achieve a high light-collecting efficiency. The master molds for the optimized structures were fabricated on Si substrate by E-beam lithography and fast atom beam etching processes. The 100 μm thick patterned polymethyl methacrylate (PMMA) film was then replicated by a hot-embossing process. As a result, the patterned PMMA film collected 63.0% more incident light than a nonpatterned one.

1. Introduction

Microoptical devices or hybrid integrated optical devices have been very important in the field of optical application systems such as optical communication systems, optical information processing systems, and optical sensing systems to achieve compactness and high performance [1–4]. Laser diodes (LD) and light emitting diodes (LEDs) have been widely used as light sources in these systems given their compactness, low driving current, and capability of high speed modulation [4]. The light source transmits through optical fibers in many cases; thus the output beam emitting from the source has far-field radiation angles that need an external lens such as a collimating lens or a focusing lens. But these lenses are so bulky that it is difficult to compact collimated or focus light sources for microoptical applications [3–5].

The diffractive optical elements (DOEs) can overcome such problems. Recently, a great deal of research in DOEs has been performed [6–13]. The DOEs play an important role in many optical applications, including optical telecommunications components, multiple imaging, light-collecting, and spectroscopy applications, because of their high uniformity, light weight, and miniaturization in size [1–4]. The most popular light-collecting device is a Fresnel lens as seen in Figure 1. A Fresnel lens is an optical component which can be used as a cost-effective, light-weight alternative to conventional continuous surface optics. Many fabrication methods have been proposed to produce Fresnel lenses. Fugita et al. fabricated the stepwise Fresnel lens on a Si substrate by using E-beam lithography [9, 10]. Yan et al. fabricated the continuous Fresnel lens on a germanium (Ge) substrate by using ultraprecision machining [11], and Joo et al. fabricated the continuous Fresnel lens on polymeric substrate by using ultraprecision machining [12]. However, MEMS-based fabrication methods such as E-beam lithography, photolithography, and holography are limited to machining circularly curved facets and low material removal efficiency [6–10]. On the other hand, ultraprecision machining-based fabrication methods rely on more complicated mechanisms depending on the degree of the size effect due to small ratio of depth of cut to the tool edge radius. In addition, both MEMS-based and ultraprecision machining-based machining methods are expensive [11, 14]. An evident disadvantage of using a lens with grooves is the possibility of losing light due to incidence on the draft facet. Making the facet perfectly vertical (i.e., perpendicular to the incident plane of light) works to minimize the optical loss (draft loss) [9–12]. However there is little work regarding thin film-type Fresnel lens for microoptics light-collecting applications.

Keeping pace with these interests in DOEs, many analysis methods have been introduced [6, 12, 13, 15–17]. Scalar diffraction theory is one of the widely used methods for the design and analysis of DOEs. The Fresnel or Fraunhofer diffraction integrals are commonly employed, and these integrals are generally calculated with the fast Fourier transform (FFT). Although this approach is relatively simple and agrees with experimental results to some extent, its applicability is limited due to diffraction limit [15–17]. When the grating size becomes so small that its scale is comparable to or less than the wavelength, the simplified approximations made in the scalar method are not valid and the polarizing nature of the light cannot be ignored in this regime. Therefore a more accurate fully electromagnetic analysis is required. The rigorous coupled-wave analysis (RCWA) provides means to solve these problems and enables characterizing the diffraction phenomenon of the DOEs with respect to TE (electric field perpendicular to the incident plane) and TM (magnetic field perpendicular to the incident plane) polarization directions.

Here thin film-type and polymethyl-methacrylate (PMMA) diffractive light-collecting microoptical device were proposed. To minimize the possibility of lost light due to incidence on the draft facet, the facet of the Fresnel lens was designed perfectly vertical. The proposed device was designed by RCWA and fabricated by E-beam lithography, fast atom beam (FAB) etching, and hot-embossing processes. The optical testing method and its performance are discussed in the following.

2. Principles

The configuration of the light-collecting device is presented in Figure 2, where L is length of a single block, F is the filling factor, P_1, P_2, and P_3 are periods of each block, d is the depth of lens, f_d is the focal length, and n_1 and n_0 are refractive index of PMMA and air. It is comprised of 9 (3×3) blocks of optical elements: 4 blocks of 1D lamellar grating structures, 4 blocks of 2D lamellar grating structures, and a single block of nonpatterned element at the center, which acts for lens to be able to collect the diffractive and transmitted lights from the lamellar grating structures into the center of the structure. The overall size of the light-collecting device is 300 $\times$ 300 μm^2, and the size of each block was practically designed as 100 $\times$ 100 μm^2. As seen in Figure 2(b), the incident light is diffracted from 8 blocks of 1D and 2D lamellar grating structures, and each diffracted light will be focused on a projected nonpatterned area with f_d.

Table 1: Optical and geometrical conditions of 1D and 2D DOEs for optical analysis.

Type	λ [μm]	n_1/n_0	P [μm]	F	D [μm]
1D	0.65	1.49	2.0	0.5	0.65
2D			1.428		

The pitches of 1D and 2D grating structures were determined by [5]

$$n_0 \sin(\theta_0) + n_1 \sin(\theta_1) = m\frac{\lambda}{P}, \tag{1}$$

where m is the number of diffraction order. The pitches of 1D grating structures, P_2 and P_3, were practically set to 2.0 μm, which corresponded to diffraction angle, $\pm 12.6^\circ$ in degree. In accordance with the pitch of 1D grating structures, the pitch of 2D grating structures was determined to make the incident light diffract into the center area, which was $\pm 17.7^\circ$ in degree. The phase-matched depth, d, was optimized using RCWA under the optical and geometrical properties as summarized in Table 1. Diffraction efficiencies were calculated by RCWA with respect to the grating depth as seen in Figure 3. The 0th order diffraction efficiencies reached their minimum at depths of 0.63 μm and 0.66 μm for TE and TM waves, respectively, while the $\pm$1st order diffraction efficiencies reached their maximum at these depths. The $\pm$1st order diffraction efficiencies with respect to polarization directions, $\eta_{1,\mathrm{TE}}$ and $\eta_{1,\mathrm{TM}}$, were estimated to be 38.3% and 38.7%, at a depth of 0.63 μm and 0.66 μm, respectively. Although small difference existed between diffraction efficiencies with respect to TE and TM waves, it is small enough that the depth was set to 0.65 μm. Diffraction angles were calculated to confirm the estimated angles using the fast Fourier transform as seen in Figure 4, which showed good agreement with the estimated ones. From the pitches of grating structures, the focal length was calculated, 448 μm.

3. Fabrication

The diffractive light-collecting microoptical device as a mold master was fabricated on a Si substrate by E-beam lithography (ELS 3700, Elionix) and FAB (FAB 60ML, Abara) processes, and the PMMA replica was duplicated by hot-embossing process. E-beam lithography and FAB processes were shown in Figure 5. An n-type of 500 μm thick Si substrate (20 mm by 20 mm) was used for E-beam direct writing. For the fabrication, 1000 nm thick PR (photoresistor, ZEP 520, Nippon Zeon Co., Ltd.) was coated on the silicon substrate and prebaked at 180 degree for 3 min as seen in Figure 5(b). After that, E-beam direct writing was used to fabricate the PR patterned 1D and 2D DOEs; the acceleration voltage, beam current, and dose time were 30 KV, 10 pA, and 35 μs, respectively, which corresponds to a dose of 56 μC/cm^2. Next, E-beam exposed silicon substrate was developed in ZEP 520 developer for 5 min and washed in the rinse-filled beaker (methyl isobutyl ketone) for 30 sec in the water pool at

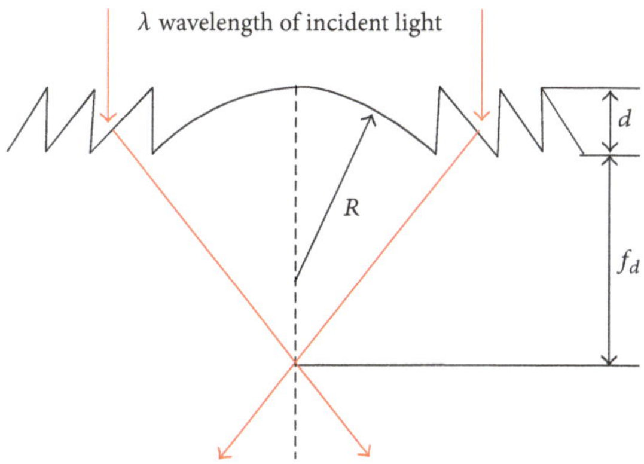

FIGURE 1: Schematics of general Fresnel lens: d depth of lens, f_d focal length, and R radius of curvature.

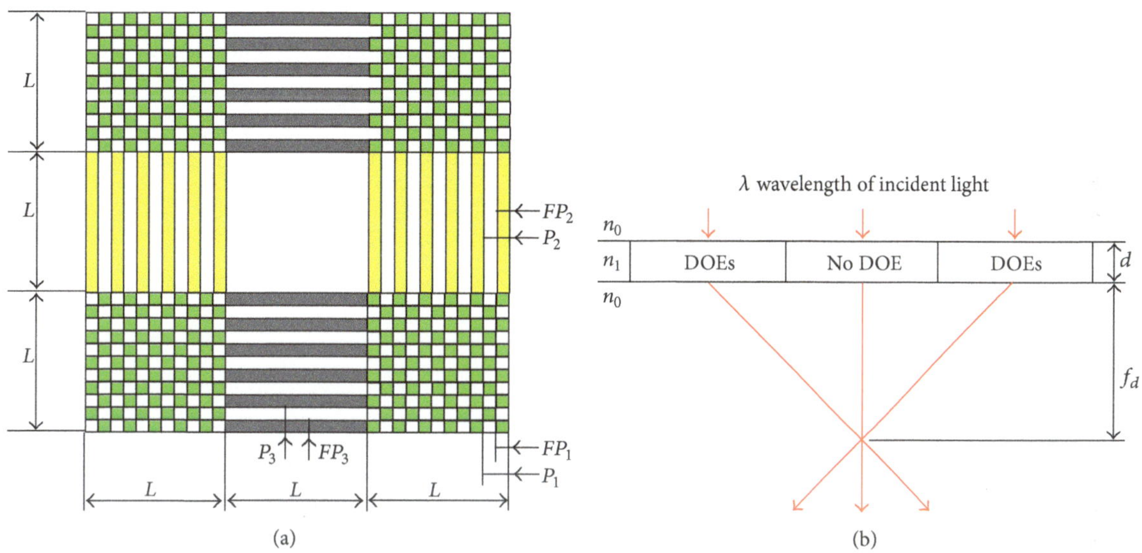

FIGURE 2: Schematics of the proposed light-collecting device: (a) overall layout: 9 (3 × 3) blocks of optical elements: 4 blocks of 1D lamellar grating structures, 4 blocks of 2D lamellar grating structures, and a single block of nonpatterned element at the center and (b) cross-sectional view: d depth of lens and f_d focal length.

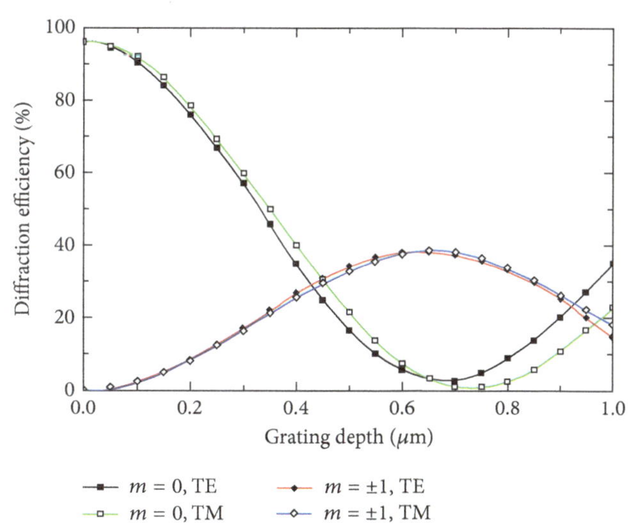

FIGURE 3: Diffraction efficiency curve with respect to the grating depth.

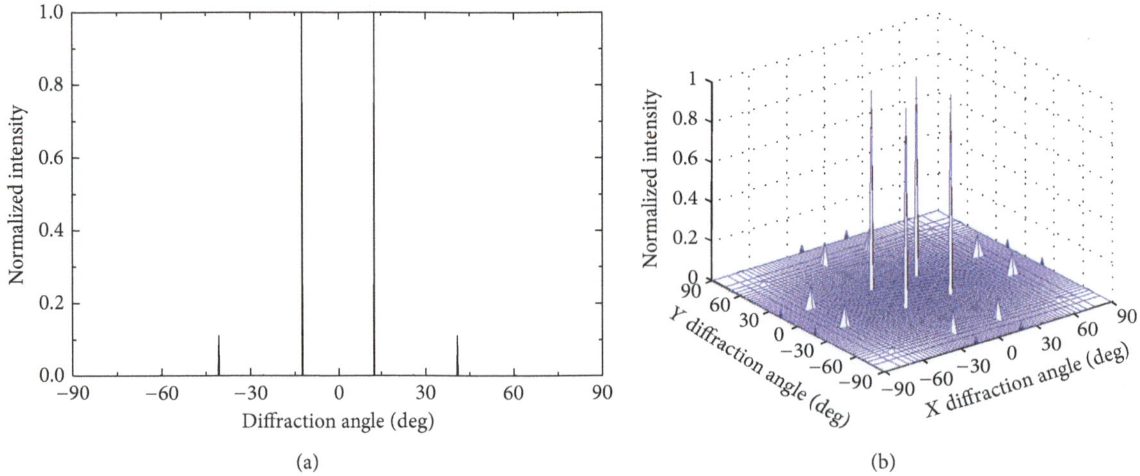

FIGURE 4: The estimated diffraction angles of 1D and 2D DOEs; (a) $\theta \pm 1\text{st} = \pm 12.6°$ for 1D, (b) $\theta \pm 1\text{st} = \pm 17.7°$ for 2D.

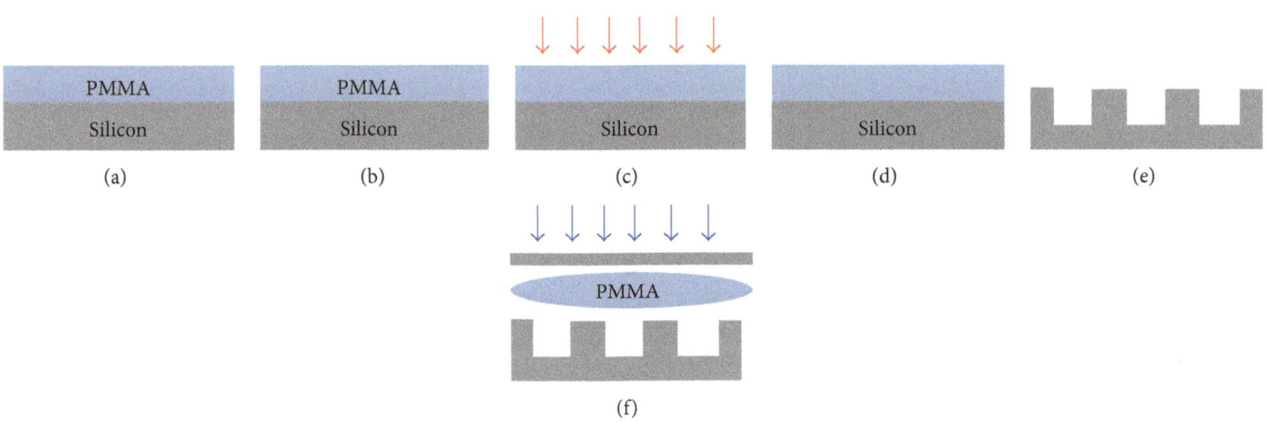

FIGURE 5: Fabrication process of DOEs; (a) silicon substrate cleaning, (b) PR coating, (c) E-beam lithography, (d) developing, (e) FAB etching, and (f) hot-embossing molding.

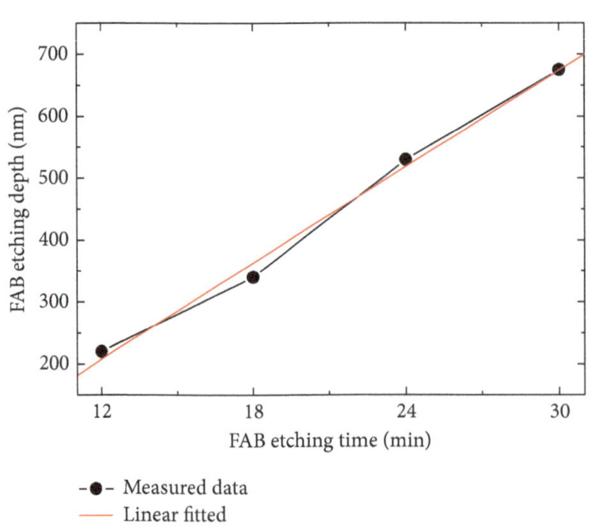

FIGURE 6: Calibration of FAB etching ratio: 21.0 nm/min.

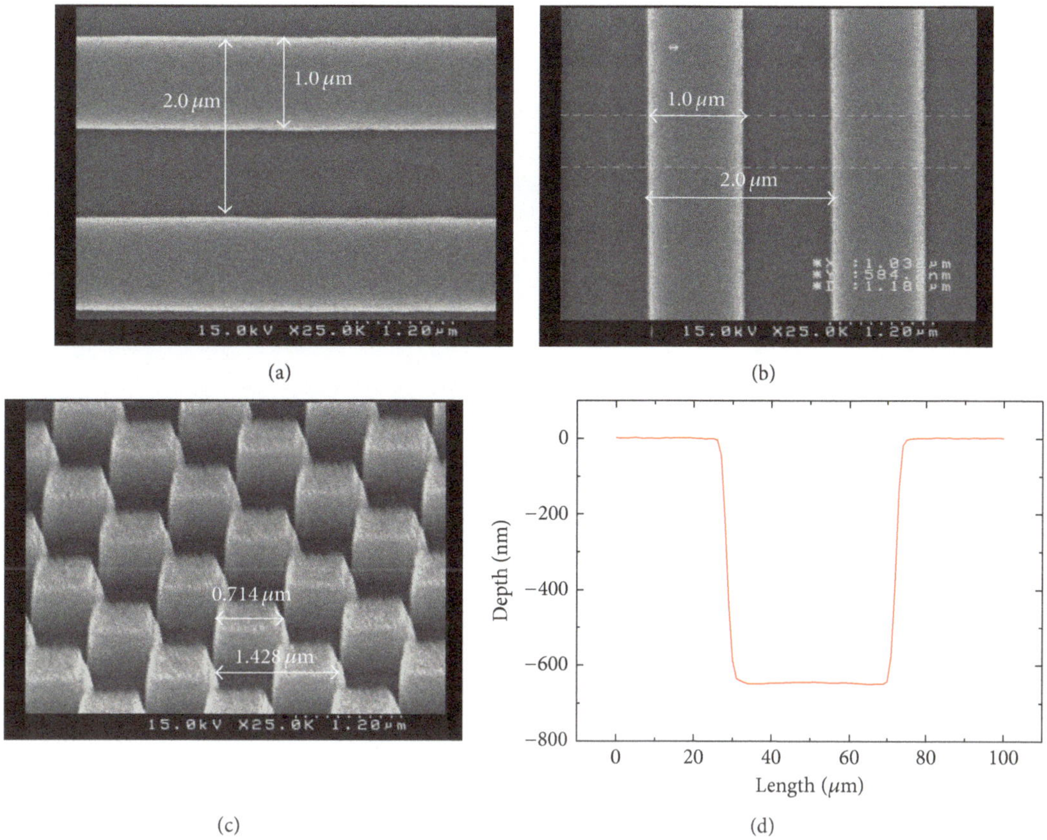

Figure 7: Scanning electron micrograph of the fabricated 1D and 2D DOEs and the depth measured by surface profiler; (a) horizontal 1D DOEs (yellow structures in Figure 2), (b) vertical 1D DOEs (gray structures in Figure 2), (c) 2D DOEs, and (d) the depth of each DOE.

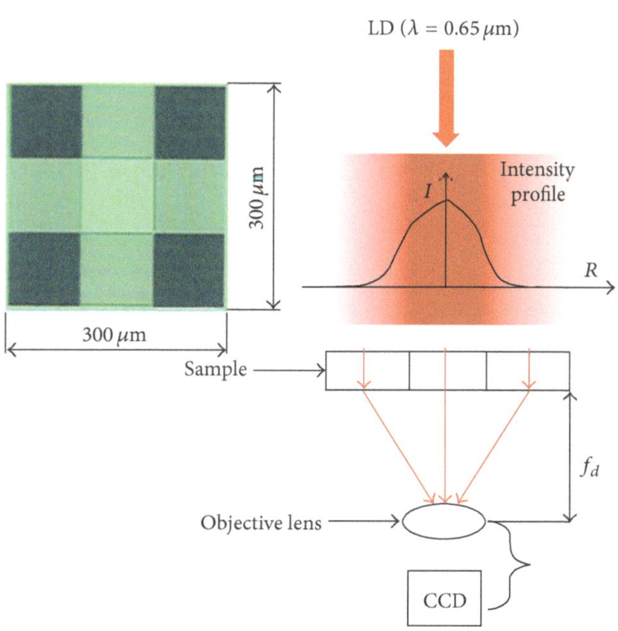

Figure 8: Experimental setup for focal length measurement.

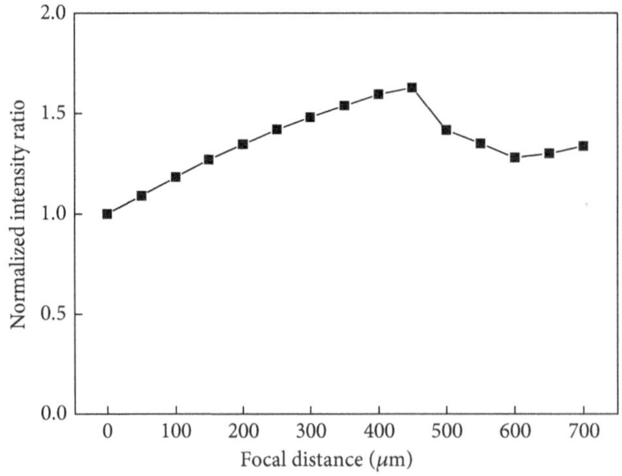

Figure 9: Measurement result of the intensity collected at each focal distance.

a temperature of 23°C and then postbaked for 3 min at 110°C. As seen in Figure 5(d), the PR patterned 1D and 2D DOEs (300 μm by 300 μm) on the silicon substrate were fabricated. FAB etching has a good directionality and a constant etching ratio (1 : 1) between the silicon and PR [10, 13, 18, 19]. The FAB etching ratio, 21.0 nm/min, was calibrated under an etching condition as shown in Figure 6. The DOEs with a depth of 0.65 μm and a period of 2.0 μm and 1.43 μm were successfully fabricated in Figure 5(e) and both fabricated 1D and 2D DOEs were taken by scanning electron microscope and etched depth was measured by surface profiler (Tencor). These results were shown in Figure 7. The etched depth of Si substrate was approximately 0.65 μm the same as design value. Last, the patterned Si substrate was hydrophobically coated to make the mold release easier. The 100 μm thick PMMA patterned film was replicated from the patterned Si substrate as a master mold by hot-embossing operation. The optimal hot-embossing condition was investigated by its process sequence: preheating, pressing, cooling, and demolding [10, 12, 13]. The hot-embossing condition was set: molding temperature 150°C, applied pressure 0.6 MPa, pressing time 5 min, and demolding temperature 25°C.

4. Results

The measurement setup was shown in Figure 8. The LD (λ = 0.65 μm, Neoark) was used as the light source with a Gaussian intensity distribution, and the optical fiber was used to deliver the light from the LD. The fabricated device was placed after the optical fiber, and the light-collecting result was measured by the CCD while moving the objective lens upward and downward manually. The digital dial gauge was used to measure the moving height. The measurement results were observed at every 50 μm from 0 to 700 μm and these images were photo-taken by CCD. The intensities of each image were estimated by gray scale, and the normalized intensity ratio was calculated with respect to focal distances as shown in Figure 9. As a result, the maximum intensity was measured at a focal distance with 450 μm and increased by 63.0% compared to the intensity of nonpatterned area only, which has good agreement with estimated focal length, 448 μm.

5. Conclusion

The new microoptical device for the purpose of the light-collecting was proposed and fabricated. Grating structures were optimized by RCWA and fabricated by E-beam lithography and fast atom beam etching processes. The mold master fabricated on Si substrate was replicated into thin film-type (100 μm) PMMA replica by hot embossing. From the measurement results, the device was successfully fabricated, and the shape of grating structures, pitch, filling factor, and depth were the same as designed. The maximum intensity was measured at a focal distance of 450 μm and increased by 63.0% compared to the intensity of nonpatterned area only, which has good agreement with estimated focal length, 448 μm. As a result, the proposed light-collecting lens system showed high performance including high uniformity, light weight, and miniaturization in size for the purpose of the LD light focusing applications. This design approach is expected to be applied to various microoptical applications such as color mixing/filtering, light collimation, and light focusing areas.

Conflict of Interests

The author declares that there is no conflict of interests regarding the publication of this paper.

References

[1] N. F. Borrelli, *Microoptics Technology*, Marcel Dekker, 2nd edition, 2004.

[2] G. I. Hatakoshi and S. I. Tanaka, "Grating lenses for integrated optics," *Optics Letters*, vol. 2, no. 6, pp. 142–144, 1978.

[3] J. Y. Joo, D.-K. Woo, S. S. Park, and S.-K. Lee, "Design and fabrication of a fingerprint imager with compact LED illumination

and compact imaging optics," *Optics Express*, vol. 18, no. 18, pp. 18932–18944, 2010.

[4] H. Goto, Y. Ito, H. Sekii, S. Ogata, and K. Imanaka, "Micro collimated light source using micro Fresnel lens," in *Micro System Technologies 90*, pp. 471–476, 1990.

[5] J. W. Goodman, *Introduction to Fourier Optics*, McGraw-Hill, San Francisco, Calif, USA, 1968.

[6] I. Kallioniemi, T. Ammer, and M. Rossi, "Optimization of continuous-profile blazed gratings using rigorous diffraction theory," *Optics Communications*, vol. 177, no. 1, pp. 15–24, 2000.

[7] C. David, J. Wei, T. Lippert, and A. Wokaun, "Diffractive grey-tone phase masks for laser ablation lithography," *Microelectronic Engineering*, vol. 57-58, pp. 453–460, 2001.

[8] M. Okano, H. Kikuta, Y. Hirai, K. Yamamoto, and T. Yotsuya, "Optimization of diffraction grating profiles in fabrication by electron-beam lithography," *Applied Optics*, vol. 43, no. 27, pp. 5137–5142, 2004.

[9] T. Fugita, H. Nishihara, and J. Koyama, "Blazed gratings and Fresnel lenses fabricated by electron-beam lithography," *Optics Letters*, vol. 7, no. 12, pp. 578–580, 1982.

[10] D.-K. Woo, K. Hane, and S.-K. Lee, "Fabrication of a multi-level lens using independent-exposure lithography and FAB plasma etching," *Journal of Optics A: Pure and Applied Optics*, vol. 10, no. 4, Article ID 044001, 2008.

[11] J. Yan, K. Maekawa, J. Tamaki, and T. Kuriyagawa, "Micro grooving on single-crystal germanium for infrared Fresnel lenses," *Journal of Micromechanics and Microengineering*, vol. 15, no. 10, pp. 1925–1931, 2005.

[12] J. Y. Joo, C. S. Kang, S. S. Park, and S.-K. Lee, "LED beam shaping lens based on the near-field illumination," *Optics Express*, vol. 17, no. 26, pp. 23449–23458, 2009.

[13] C. B. Lee, T. Kuriyagawa, D.-K. Woo, and S.-K. Lee, "Optimizing the fabrication process of a high-efficiency blazed grating through diamond scribing and molding," *Journal of Micromechanics and Microengineering*, vol. 20, no. 5, Article ID 055028, 2010.

[14] C. B. Lee, K. Hane, and S.-K. Lee, "The optimization of sawtooth gratings using RCWA and its fabrication on a slanted silicon substrate by fast atom beam etching," *Journal of Micromechanics and Microengineering*, vol. 18, no. 4, Article ID 045014, 2008.

[15] D. Dornfeld, S. Min, and Y. Takeuchi, "Recent advances in mechanical micromachining," *CIRP Annals—Manufacturing Technology*, vol. 55, no. 2, pp. 745–768, 2006.

[16] M. G. Moharam and T. K. Gaylord, "Rigorous coupled-wave analysis of planar-grating diffraction," *Journal of the Optical Society of America*, vol. 71, no. 7, pp. 811–818, 1981.

[17] M. G. Moharam, D. A. Pommet, E. B. Grann, and T. K. Gaylord, "Stable implementation of the rigorous coupled-wave analysis for surface-relief gratings: enhanced transmittance matrix approach," *Journal of the Optical Society of America A*, vol. 12, no. 5, pp. 1077–1086, 1995.

[18] M. G. Moharam and T. K. Gaylord, "Diffraction analysis of dielectric surface-relief gratings," *Journal of the Optical Society of America*, vol. 72, no. 10, pp. 1385–1392, 1982.

[19] F. Shimokawa and H. Kuwano, "New high-power fast atom beam source," *Journal of Vacuum Science & Technology A*, vol. 12, no. 5, pp. 2739–2744, 1994.

2

Finite Element Analysis of PMMA Stretch Blow Molding

Afef Bougharriou, Mohieddine Jeridi, Mohamed Hdiji, Anoir Boughrira, and Kacem Saï

UGPMM, National Engineering School of Sfax, Route Soukra Km 2.5, BP 1173, Sfax, Tunisia

Correspondence should be addressed to Kacem Saï; kacemsai@yahoo.fr

Academic Editor: Hailiang Yu

The electric bubbles are a useful product made of PMMA material. They are produced by the stretch blow molding process. Thickness, which reflects the quality of the electric bubble, is a crucial parameter that deserves special attention for the molding process. In this work, finite element simulations of the stretch blow molding process are performed aiming at the determination of the preform geometry to ensure homogeneous thickness of the finished product. The geometrical parameters of the preform are optimized allowing a better homogeneity thickness compared to existing data. The predicted parameters allow the improvement of the thickness distribution. The standard deviation of the thickness is reduced to about 95% compared to the existing bubble.

1. Introduction

Stretch blow molding is one of the most important manufacturing processes of polymer via injection process [1–3] or extrusion process [4, 5] to produce hollow plastic parts. The majority of clear bottles are produced using injection stretch blow molding process. Accordingly, several works were devoted to the simulation of the stretch blow molding in order to optimise the process and improve the quality of these containers. Chung [6] proposed a finite element (FE) model for injection stretch blow molding of PET bottles using the software ABAQUS. These simulations focused on the optimisation of the side wall thickness of the bottle by adjusting the variations of histories of plunger movement and gas pressure. Cosson et al. [7] used the so-called constrained natural elements method to simulate and optimize the stretch blowing of a PET preform by assuming an axisymmetrical and viscoplastic preform model under an internal pressure. The thickness of the bottle is linked to the variation of the stretch rod speed and time where the blow pressure is applied. Cosson et al. [8] simulated the stretch blow molding process of PET bottles at the usual process temperature. The simulation results modeled the crystalline microstructure evolution and predicted the elastic behaviour of the final bottle. It also provides Young's modulus distributions in the bottles. Tan et al. [9] developed a 2D isothermal FE simulation of the injection stretch blow molding process for PET containers. To inflate the PET preform, two approaches were used in the simulation: (i) a direct pressure input and (ii) a constant mass flow rate of air. The results show that the simulation with the fluid flow method gives good prediction of the volume versus time curve and the preform shape evolution. Tan et al. [9] concluded that applying the pressure via a mass flow rate of air is the most appropriate method to model the pressure inside the preform. Salomeia et al. [10] simulated the stretch blow molding process for PET bottle by the means of thermodynamic model implemented in an ISBM. The simulations provide the evolution of the pressure and the temperature profiles on both inside and outside the surfaces. Yang et al. [11] proposed a FE model for the injection stretch blow molding process of PET bottles aiming at optimizing the process conditions (i.e., the bottle wall thickness, the distributions of the material, the temperature, and the strain). Gupta et al. [12] simulated the blow molding process for PET bottle in order to study the thickness and the stress distribution. Zimmer and Stommel [13] presented FE simulations for stretch blow molding with free blow trials of a PET bottle. The model provides the temperature distribution and the stress-strain fields. The poly(methyl methacrylate) (PMMA), material of this study, is an economical, versatile general purpose material. Various types of PMMA presenting high mechanical properties are used in different fields and applications. Numerous studies were devoted to the modeling of the mechanical behavior of this

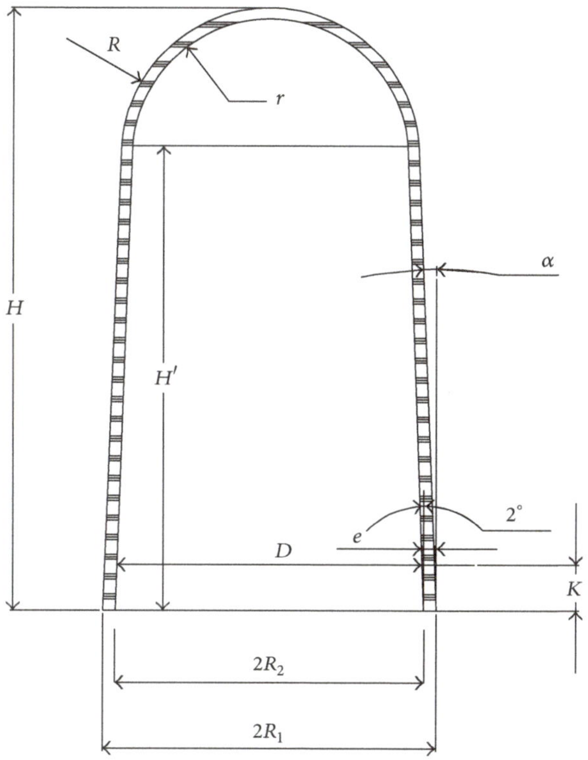

(a) Geometrical parameterization

(b) Typical mesh (zoom)

FIGURE 1: Geometry of the preform.

material; see, for instance, [14–17]. The FE studies related to the simulations of stretch blow molding process for PMMA material are scarce. Dong et al. [17] presented a FE simulation of thermoforming process of rectangular bubble inflation. The obtained critical material parameters of PMMA have been used in the simulation of free inflation of a bubble profile. The light bubbles are highly demanded in the Tunisian market. The Tunisian market imports these bubbles or the manufactured molds from abroad. IMR, a Tunisian company specialized in design and manufacture of injection tools, aims at producing these bubbles at lower cost and good quality. IMR needs to optimize the preform geometry leading to quasihomogeneous distribution of the thickness along the bubbles. In this work, FE simulations of the stretch blow molding process are performed aimed at the determination of the preform geometry to ensure homogeneous thickness of the finished product. The paper is organized in the following manner: in Section 2 the preform geometry (i.e., mesh), the boundaries conditions, the loading, and the material behavior are detailed. Section 3 is devoted to the discussion of the FE results in terms of stress and bubble thickness distributions. Finally, recommendations are given on the geometric design of the preform.

2. Finite Element Modeling

The simulations are performed using the FE software ABAQUS. The material behavior may affect in a considerable way the obtained results. To this end, the FE model of the molding process is investigated taking into account the hyperelastic behavior of the PMMA at 150°C by the means of Mooney-Rivlin model.

TABLE 1: Number of elements in the FE model.

α/e	3.6	3.8	4	4.2
2.1	4221/1872	4005/1776	3825/1696	3609/1600
2.3	—	4401/1952	4149/1840	3879/1720
2.5	—	—	—	4329/1920

2.1. Geometry and Meshing. The design parameters of the preform are shown in Figure 1(a). The geometry is controlled by the three independent parameters: e, α, and H. Assume that the deformation of the preform occuring under constant volume ($V = V_0$) reduces the number of parameters from 3 to 2 as will be shown below. The FE analysis is performed using six-noded axisymmetrical elements with a full integration scheme. The preform is designed according to the pairs (α, e), Table 1. The mesh size is chosen in such a way that subsequent refinement does not affect the stress-strain behavior much. A typical FE mesh is shown in Figure 1(b).

The volume of the preform is calculated as the sum of a semisphere and a conical cylinder as shown in Figure 1. After some algebraic manipulations, the expression of the volume of the preform is given by the three terms A, B, and S:

$$V = A + S - B \tag{1}$$

with

$$A = \frac{\pi}{3}H'\left(R_1^2 + R^2 + R_1 R\right),$$
$$B = \frac{\pi}{3}H'\left(R_2^2 + r^2 + R_2 r\right), \quad (2)$$
$$S = \frac{4\pi}{6}\left(R^3 - r^3\right).$$

A, B, and S depend on the geometrical characteristics of the preform:

$$R = D - \left(H' - K\right)\tan\alpha,$$
$$r = (D - e) - \left(H' - K\right)\tan 2,$$
$$R_1 = (D + K\tan\alpha), \quad (3)$$
$$R_2 = (D - e + K\tan 2),$$
$$H = R + H'.$$

It is assumed that the deformation of the preform occurs under constant volume during the mold flow process ($V = V_0$). This leads to a third degree equation in e:

$$V = V_0 \Longrightarrow e^3 + ae^2 + be + c = 0, \quad (4)$$

where

$$a = -3\left(D - \left(H' - K\right)\tan 2\right) - \frac{3}{2}H',$$
$$b = 3\left(D - \left(H' - K\right)\tan 2\right)^2 + \frac{3}{2}H'\left(2D + \left(2K - H'\right)\tan 2\right),$$
$$c = \frac{3}{2\pi}\left(A - V_0\right) + \frac{1}{\pi}\left(R^3 - \left(D - (H' - K)\tan 2\right)^3\right) - \frac{1}{2}\left((D + K\tan 2)^2 + \left(D - \left(H' - K\right)\tan 2\right)^2 + (D + K\tan 2)\left(D - \left(H' - K\right)\tan 2\right)\right). \quad (5)$$

Analytical solution can easily be obtained for the third degree (4). Two cases are distinguished depending on the value of the determinant $\Delta = q^2 + 4p^3/27$ ($q = 2a^3/27 - ab/3 + c$ and $p = b - a^2/3$). For the studied problem Δ is negative, so that 3 real zeros verify (4). Each solution k ($k = 0$, $k = 1$, and $k = 3$) is given by the means of polar coordinates:

$$e = 2r^{1/3}\cos\left(\frac{\theta}{3} + \frac{2k\pi}{3}\right) \quad (6)$$

with $r = 0.5\sqrt{q^2 - \Delta}$ and $\theta = \arctan(-\sqrt{-\Delta/q})$. Only one solution lies on the allowed range. The diagram in Figure 2 can be divided into regions. The divisions between these regions are bounded by the curve with the equation $H = H_{max}$. The feasible region is above this curve, whereas the infeasible region below the curve leads to preform geometry in which the height H is greater than the height of the mold.

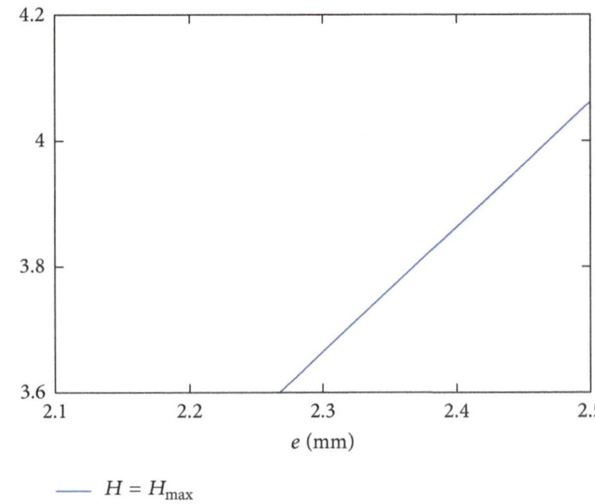

FIGURE 2: Limit curve.

2.2. Boundary Conditions and Loading. The following boundary conditions are considered for the FE analysis.

(i) The "container" is fully constrained and the preform could move only along the axial direction.

(ii) A displacement along the z-axis (Figure 3) to the bottom of the mold is applied first to the preform. This displacement is preconized by the manufacturer. Accordingly, the preform moves down to be in contact with the mold.

(iii) A pressure of 0.3 MPa was applied in the second step to the inside of the preform with a gradual increase.

(iv) A friction law, corresponding to Coulomb law with a constant friction coefficient $\mu = 0.1$, is considered between the preform and the "container." It is worth noting that some simulations (see Section 3) in which μ was taken in the range of 0.1 to 0.5 have shown that the results were not affected by the value of the friction coefficient.

2.3. Material Behavior Law. To describe the mechanical behavior of the PMMA, hyperelastic and viscoelastic models are the two commonly used approaches. The emphasis of current study is a phenomenological hyperelastic approach due to its better suitability for the PMMA which is isotropic, highly elastic, and incompressible for the forming conditions [17]. It is already established that the Mooney-Rivlin model is accurate in simulating polymer sheet forming; see, for instance, [17, 18]. It should be noted that this approach fails to represent the speed influence. The material behavior is described with a specific free energy density W which is a function of strain and temperature. For isotropic behavior, W depends on the invariants of the transformation tensor F which is given by

$$dx = F dX,$$
$$B = F^T F, \quad (7)$$

where dX is the elementary vector in the initial configuration, dx is the transformation of dX in the actual configuration,

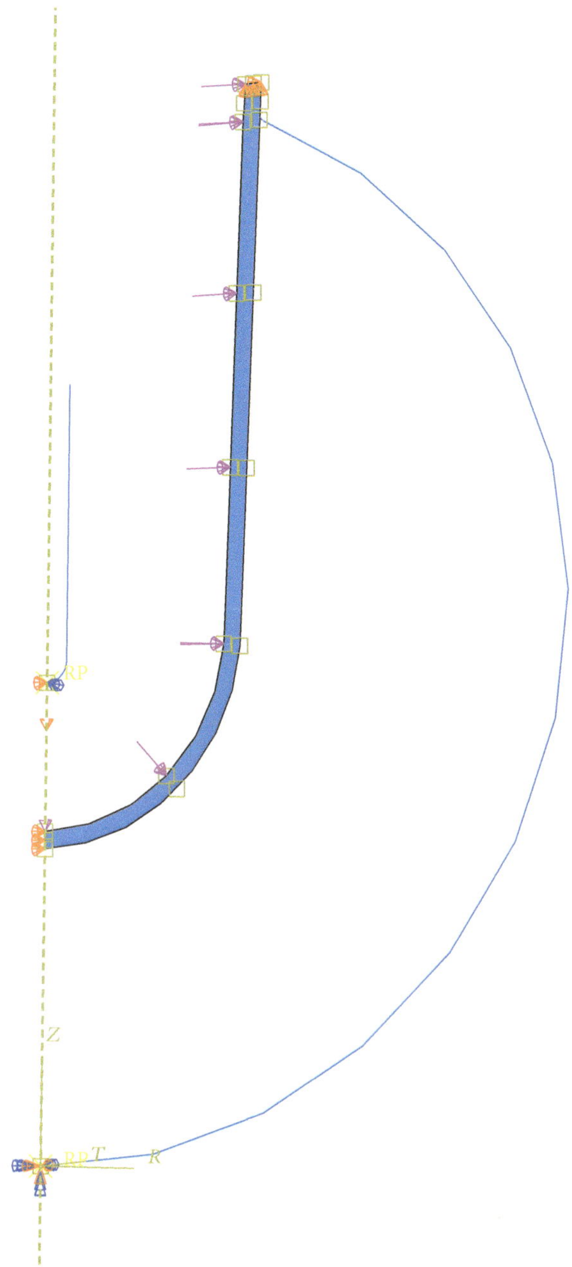

FIGURE 3: Loading.

F is the transformation gradient tensor, and B is the Cauchy-Green Gauche tensor. The stresses can be obtained as the partial derivatives of the strain energy density function with respect to invariants I_1, I_2, and I_3. For isotropic materials, W is a linear combination of the invariants I_1, I_2, and I_3 of the strain tensor $W = W(I_1, I_2, I_3)$, where $I_1 = \text{trace}B$, $I_2 = (1/2)((\text{trace}B^2) + (\text{trace}B)^2)$, and $I_3 = \det(B)$. These invariants can be expressed with the elongations (λ_i, $i = 1, 2$, and 3) as follows:

$$\begin{aligned} I_1 &= \lambda_1^2 + \lambda_2^2 + \lambda_3^2, \\ I_2 &= \lambda_1^2\lambda_2^2 + \lambda_1^2\lambda_3^2 + \lambda_2^2\lambda_3^2, \\ I_3 &= \lambda_1^2\lambda_2^2\lambda_3^2. \end{aligned} \quad (8)$$

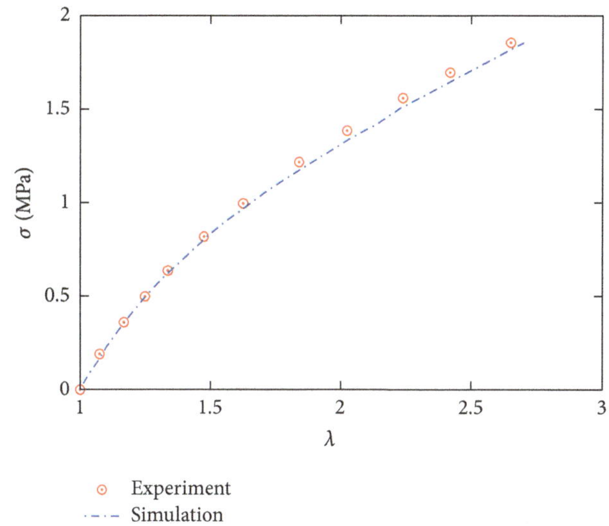

FIGURE 4: Stress strain curve.

The principal stretch ratio λ is the ratio of the gauge lengths after and before the test ($\lambda = L/L_0$). The hyperelastic theory (Mooney-Rivlin model) is adopted in this study for the simulation of the nonlinear material behavior. The strain energy density W is expressed by Mooney-Rivlin model as follows:

$$W = C_{10}(I_1 - 3) + C_{20}(I_1 - 3), \quad (9)$$

where C_{10} and C_{20} are two material parameters. For the case of incompressible material under uniaxial elongation ($\lambda_1 = \lambda$, $\lambda_2 = \lambda_3 = 1/\sqrt{\lambda}$), the true Cauchy stress differences can be calculated as

$$\begin{aligned} \sigma_{11} - \sigma_{33} &= 2C_{10}\left(\lambda^2 - \frac{1}{\lambda}\right) + 2C_{20}\left(\lambda - \frac{1}{\lambda^2}\right), \\ \sigma_{22} - \sigma_{33} &= 0. \end{aligned} \quad (10)$$

In the case of simple tension, $\sigma_{22} = \sigma_{33} = 0$, the uniaxial behavior law writes

$$\sigma_{11} = \left(2C_{10} + \frac{2C_{20}}{\lambda}\right)\left(\lambda^2 - \frac{1}{\lambda}\right). \quad (11)$$

The Mooney-Rivlin parameters of PMMA material are taken from the literature. One element test subjected to monotonic tensile loading is carried out in order to check the accuracy of the material parameters in simulating the experimental stress/elongation curve (Figure 4). It can be seen that the experimental data are well described by the Mooney-Rivlin model for the parameters values of $C_{10} = -0.0303$ MPa and $C_{20} = 0.4503$ MPa at the temperature of 150°C.

3. Results and Discussions

In this section, the FE results are discussed in terms of thickness distribution. Since the friction behaviour needed for the simulation of the stretch blow molding process is unknown, parametric study related to the effect of friction coefficient between the mold and the preform is preliminarily

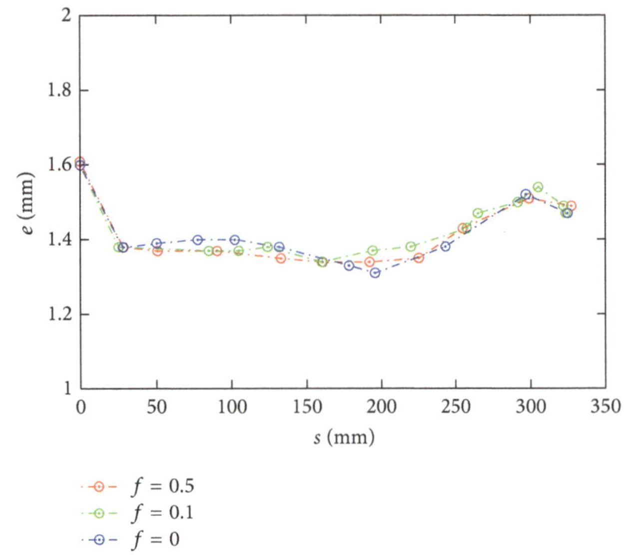

FIGURE 5: Effect of the friction coefficient.

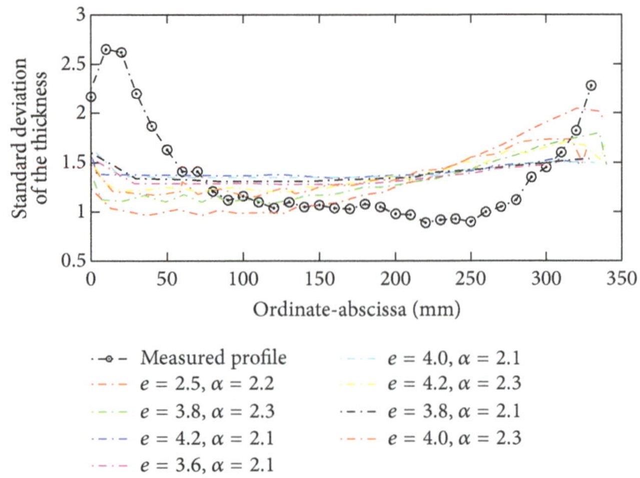

FIGURE 6: Bubble thickness evolution ordinate-abscissa.

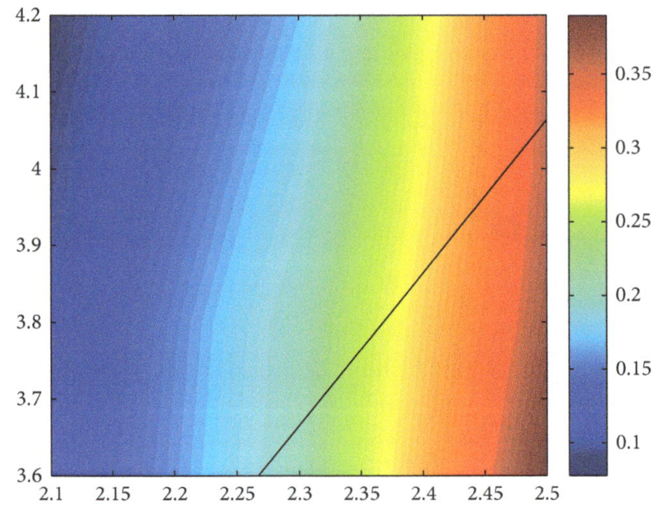

FIGURE 7: Contour plot of the standard deviation of the thickness (α versus e).

conducted. The prediction using three different friction coefficients, namely, 0.0 (frictionless), 0.1, and 0.5, is investigated. Figure 5 shows that the variation of the friction coefficient does not affect the bubble thickness. Further simulations will be carried out with friction coefficients of 0.1. Figure 6 shows the distribution of the final product thickness versus the ordinate-abscissa for several geometries of the preform. Each geometry is characterized by a given pair (α, e) (given in Table 1). The thicknesses are in agreement with the results obtained by Cosson et al. [7] and Yang et al. [11]. The simulation results are superimposed to the measured data of an existing bubble. Figure 6 shows how the scattering parameters (e and α) can greatly improve the thickness uniformity. The standard deviation of the thickness is reduced to about 95% compared to the existing bubble. For a better evaluation of the effect of the geometrical parameters e and α of the preform on the final results, a contour plot of the standard deviation of the thickness is shown in Figure 7. Classically, an appropriate cost function is mathematically formulated in which the objective function depends on the standard deviation of the computed thicknesses. In our work, the number of driving parameters is reduced to 2 (i.e., e and α). It is then possible, thanks to the contour plot of the standard deviation, to optimize manually the thickness distribution. It can be seen that the thickness is widely spread for high values of e and low values of α. Low standard deviation showing that the thickness is clustered closely around a mean value is noticed for high values of e and low values of α. The increase of the preform thickness improves the thickness distribution of the bubble and decreases the height. Accordingly, the stretching increases and improves the mechanical properties of material of the finished product as demonstrated by Cosson et al. [8]. The "optimal" distribution of the thickness should be obtained for $\alpha \simeq 2.1^{\circ}$C and $e \simeq 4.2$ mm. It gives a light bubble with a uniform distribution of the thickness in addition to a better mechanical quality [7, 8].

4. Conclusion

In this study, a finite element simulation is carried out to simulate the stretch blow molding. The geometrical parameters of the preform, leading to a homogeneous thickness distribution of the electric bubbles, are predicted. The simulation results show an improvement compared to an existing bubble in terms of thickness distribution and mechanical properties. Several possible extensions can be performed in the near future:

(i) the determination of the blow injection mold geometry allowing the realization of the electric bubbles,

(ii) the choice of the mold cavity disposition, the alimentation system, and the cooling mechanism,

(iii) the simulation of the process using the designed mold to optimise the implementation parameters such as cooling time, filling time, and closing force.

Conflict of Interests

The authors declare that there is no conflict of interests regarding the publication of this paper.

References

[1] A. Bendada, F. Erchiqui, and A. Kipping, "Understanding heat transfer mechanisms during the cooling phase of blow molding using infrared thermography," *NDT and E International*, vol. 38, no. 6, pp. 433–441, 2005.

[2] F. Daver and B. Demirel, "A simulation study of the effect of preform cooling time in injection stretch blow molding," *Journal of Materials Processing Technology*, vol. 212, no. 11, pp. 2400–2405, 2012.

[3] G. H. Menary, C. G. Armstrong, R. J. Crawford, and J. P. McEvoy, "Modelling of poly(ethylene terephthalate) in injection stretch-blow moulding," *Plastics, Rubber and Composites Processing and Applications*, vol. 29, no. 7, pp. 360–370, 2000.

[4] K. Peplinski and A. Mozer, "Ansys-polyflow software use to select the parison diameter and its thicknes distribution in blowing extrusion," *Journal of Polish CIMAC*, 2010.

[5] K. Peplinski and A. Mozer, "Comparison of bottle wall thickness distribution obtain in real manufacturing conditions and in ansys polyflow simulation environment," *Journal of Polish CIMAC*, vol. 7, pp. 231–235, 2012.

[6] K. Chung, "Finite element simulation of pet stretch/blowmolding process," *Journal of Materials Shaping Technology*, vol. 7, no. 4, pp. 229–239, 1989.

[7] B. Cosson, L. Chevalier, and J. Yvonnet, "Optimization by the C-NEM method of the stretch-blow molding process of a pet bottle near Tg," *International Journal of Material Forming*, vol. 1, no. 1, pp. 707–710, 2008.

[8] B. Cosson, L. Chevalier, and G. Régnier, "Simulation of the stretch blow moulding process: from the modelling of the microstructure evolution to the end-use elastic properties of polyethylene terephthalate bottles," *International Journal of Material Forming*, vol. 5, no. 1, pp. 39–53, 2012.

[9] C. W. Tan, G. H. Menary, Y. Salomeia et al., "Modelling of the injection stretch blow moulding of PET containers via a Pressure-Volume-time (PV-t) thermodynamic relationship," *International Journal of Material Forming*, vol. 1, no. 1, pp. 799–802, 2008.

[10] Y. M. Salomeia, G. H. Menary, and C. G. Armstrong, "Instrumentation and modelling of the stretch blow moulding process," *International Journal of Material Forming*, vol. 3, no. 1, pp. 591–594, 2010.

[11] Z. J. Yang, E. Harkin-Jones, G. H. Menary, and C. G. Armstrong, "A non-isothermal finite element model for injection stretch-blow molding of PET bottles with parametric studies," *Polymer Engineering and Science*, vol. 44, no. 7, pp. 1379–1390, 2004.

[12] S. Gupta, V. Uday, A. S. Raghuwanshi, S. Chowkshey, S. N. Das, and S. Suresh, "Simulation of blow molding using ansys polyflow," *APCBEE Procedia*, vol. 5, pp. 468–473, 2013.

[13] J. Zimmer and M. Stommel, "Method for the evaluation of stretch blow molding simulations with free blow trials," *IOP Conference Series: Materials Science and Engineering*, vol. 48, Article ID 012004, 2013.

[14] N. Saad-Gouider, R. Estevez, C. Olagnon, and R. Séguéla, "Calibration of a viscoplastic cohesive zone for crazing in PMMA," *Engineering Fracture Mechanics*, vol. 73, no. 16, pp. 2503–2522, 2006.

[15] Z. H. Stachurski, "Strength and deformation of rigid polymers:the stress-strain curve in amorphous PMMA," *Polymer*, vol. 44, no. 19, pp. 6067–6076, 2003.

[16] M. Nasraoui, P. Forquin, L. Siad, and A. Rusinek, "Influence of strain rate, temperature and adiabatic heating on the mechanical behaviour of poly-methyl-methacrylate: experimental and modelling analyses," *Materials and Design*, vol. 37, pp. 500–509, 2012.

[17] Y. Dong, R. J. T. Lin, and D. Bhattacharyya, "Determination of critical material parameters for numerical simulation of acrylic sheet forming," *Journal of Materials Science*, vol. 40, no. 2, pp. 399–410, 2005.

[18] G. Marckmann, E. Verron, and B. Peseux, "Finite element analysis of blow molding and thermoforming using a dynamic explicit procedure," *Polymer Engineering and Science*, vol. 41, no. 3, pp. 426–439, 2001.

3

Modeling of Tool Wear in Vibration Assisted Nano Impact-Machining by Loose Abrasives

Sagil James and Murali M. Sundaram

Department of Mechanical and Materials Engineering, University of Cincinnati, Cincinnati, OH 45221, USA

Correspondence should be addressed to Murali M. Sundaram; murali.sundaram@uc.edu

Academic Editor: Tuğrul Özel

Vibration assisted nano impact-machining by loose abrasives (VANILA) is a novel nanomachining process that combines the principles of vibration assisted abrasive machining and tip-based nanomachining, to perform target specific nanoabrasive machining of hard and brittle materials. An atomic force microscope (AFM) is used as a platform in this process wherein nanoabrasives, injected in slurry between the workpiece and the vibrating AFM probe which is the tool, impact the workpiece and cause nanoscale material removal. The VANILA process are conducted such that the tool tip does not directly contact the workpiece. The level of precision and quality of the machined features in a nanomachining process is contingent on the tool wear which is inevitable. Initial experimental studies have demonstrated reduced tool wear in the VANILA process as compared to indentation process in which the tool directly contacts the workpiece surface. In this study, the tool wear rate during the VANILA process is analytically modeled considering impacts of abrasive grains on the tool tip surface. Experiments are conducted using several tools in order to validate the predictions of the theoretical model. It is seen that the model is capable of accurately predicting the tool wear rate within 10% deviation.

1. Introduction

Well-known tip-based nanomachining processes such as nanoscratching and nanoindentation involve direct tool contact with the workpiece and result in high tool wear. Vibration assisted nano impact-machining by loose abrasives (VANILA) process, a noncontact vibration tip-assisted nanomachining process that uses loose abrasives, has been investigated and can be used to perform target specific impact-based machining of nanoscale features on hard and brittle materials, such as glass and ceramics materials [1, 2]. This process uses an AFM platform in which a slurry of nanodiamond abrasive particles is introduced between the tool and the workpiece. The machining is conducted in tapping mode where the tool vibrates at resonance within the slurry of abrasive nanoparticles. The impact of the tool along with the acoustic field generated moves the abrasive grain towards the workpiece surface causing a high velocity impact and material removal. A schematic of the VANILA process is shown in Figure 1.

The feasibility of nanoscale machining using the VANILA process was successfully demonstrated on hard and brittle materials such as borosilicate glass and silicon. Patterns of nanocavities were successfully machined to demonstrate the controllability and repeatability of the process as shown in Figure 2 [2].

The VANILA process uses an AFM probe as the vibrating tool to impact the nanoabrasive grains which causes material removal at nanoscale. Earlier studies using scanning electron microscopy (SEM) imaging have shown that tool wear in the VANILA process is comparatively lower than that in other nanomachining processes such as nanoscratching and nanoindentation involving direct tool contact with the workpiece (Figure 3) [2]. The accuracy and precision of any tip-based nanomachining process is limited by the tip's sharpness, material, and geometry [3]. The level of precision of the nanomachined features continuously degrades during the machining process as the probe tip inevitably wears resulting in an increase of contact diameter and forces acting on it [3]. Establishing the point at which the tool is considered

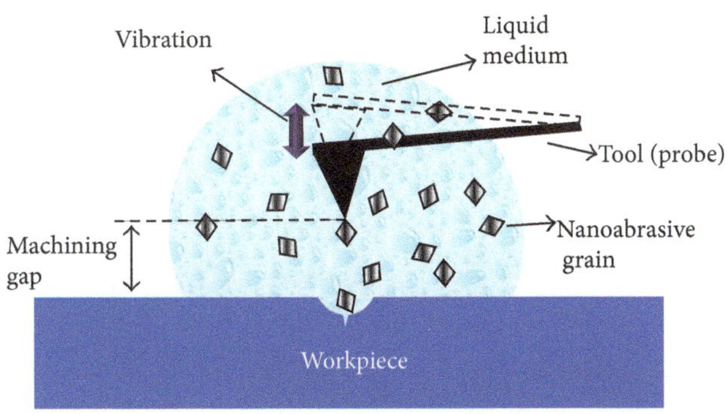

FIGURE 1: Schematic diagram of the VANILA process.

worn is important, since, after this point, machining results are no longer acceptable [4]. Thus quantification of tool wear to predict the tool life is of great significance. This study aims to develop a theoretical model to predict the tool wear in the VANILA process.

2. Mechanism of Tool Wear during VANILA Process

The nanoscale wear of AFM tip often refers to the volumetric wear of the tip due to the height loss and varying tip radius. The AFM probes, in general, have a sharp pyramidal tip of the order of few nanometers. The wear during operation causes tip to become blunt with larger tip. Studies have found that the tip shape and geometry significantly influence its deformation behaviors at nanoscales [5–7]. A review of the available literature reveals that that AFM tip wear occurs through multiple wear mechanisms [8, 9] such as abrasive and adhesive wear [10], low-cycle fatigue and material fracture [10], plastic deformation [11], coating failure, corrosive wear, and tribochemical wear [11], and thermochemical wear [12]. Recently, abrasive wear of AFM tips has been studied at atomic scales and it is found that the damaged surface is smoothly modified [13]. This basic abrasive-wear mechanism is often termed as gradual wear due to atom-by-atom attrition [14], whereby single atoms or cluster makes up the wear debris.

During the VANILA process, the resonating tool tip repeatedly impacts the nanoabrasive grains suspended in the slurry. The tool tip wear caused by these nanoscale impacts in rapid succession would lead to considerable size effects both in the length and in the time scales. Decohesion forces or pull-out forces become significant which may lead to gradual removal of atoms from the tool tip. This phenomenon is further supported by the fact that, at such lesser tool dimensions, the energy required to break the atomic bonds decreases due to bond stretching [14]. Based on the literature study, it is understood that, although several mechanisms could be involved, wear due to gradual atom-by-atom loss would be the most relevant wear mechanism causing the tool wear in the VANILA process [14]. Thus, based on this information, a wear model is developed to predict and quantify the tool wear during the VANILA process. Other wear mechanisms that could affect the tool wear process are ductile mode plastic deformation and nanoscale cracking [15]. However, these mechanisms would be more relevant when the tool tip size is considerably larger than the abrasive grain size.

3. Modeling of Tool Wear in VANILA Process

During the VANILA process, the tool tip and workpiece surface are maintained at a distance. The tool wears due to the impact of nanoabrasive grains. For the purpose of analytical modeling of tool wear, the following simplification assumptions are made.

3.1. Simplification Assumptions. The simplification assumptions made in developing the theoretical model are listed as follows.

(1) The abrasive grains are assumed to be spherical and rigid and remain intact throughout the machining process.

(2) Abrasive nanoparticle and tool material are electrically and chemically neutral and the electrostatic effects during impact are neglected.

(3) The effects of surface adhesion and interactions outside the region of contact are not considered in this study.

(4) For the theoretical modeling, the tool tip is assumed to be flat. The change in profile of the tool tip during the machining process is not considered in this study.

(5) The effects of abrasive grain impact on the sides of the tool tip are not considered in this study.

(6) The collisions between the abrasives grains are assumed to be perfectly elastic.

3.2. Impact Velocity and Forces. During the VANILA process, the abrasive grains moving in the slurry medium impact the tool which is excited at resonance. The resultant velocity of impact ($\mathbf{V_I}$) has two components: (a) velocity of vibrating tool ($\mathbf{V_t}$) and (b) velocity of abrasive grain ($\mathbf{V_a}$). The motion of tool tip can be expressed as $A_t e^{i(\omega t+\pi/2)}$. The phase associated with the total force can be considered as $\pi/2$ due to the fact that machining is conducted at resonance frequency and since on resonance the oscillations of the cantilever follow the total force with a phase delay of $\pi/2$ [16]. Average velocity of the tool tip ($\mathbf{V_t}$) can be expressed as

$$\mathbf{V_t} = 0.5\omega A_t = \pi f_t A_t. \quad (1)$$

The velocity of the nanoabrasive grain within the fluid medium can be considered to be equal to the velocity of the fluid molecule. The acoustic streaming motion generated by the acoustic field near the tool tip results in the motion of the fluid slurry between the tool tip and the workpiece surface. The nanoparticle within the slurry is transported along with the streaming fluid molecules and causes impact with the

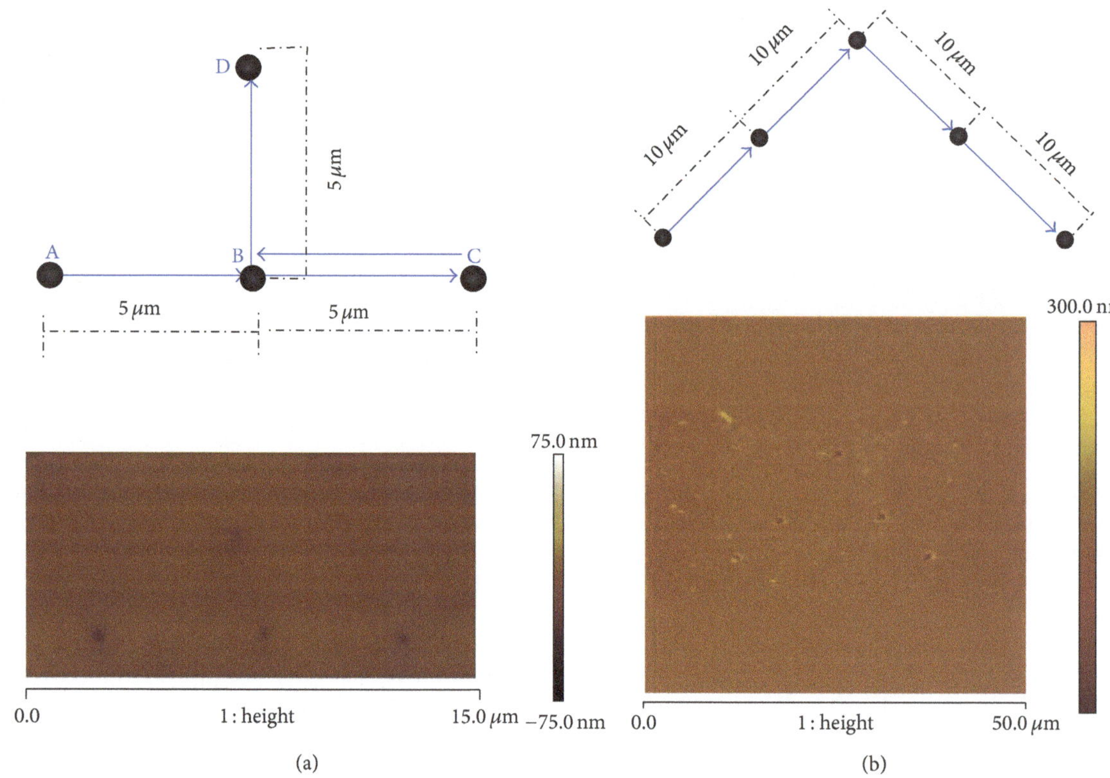

FIGURE 2: Pattern design and AFM image of nanocavity pattern machined using the VANILA process (a) on silicon substrate [2] and (b) on borosilicate glass substrate.

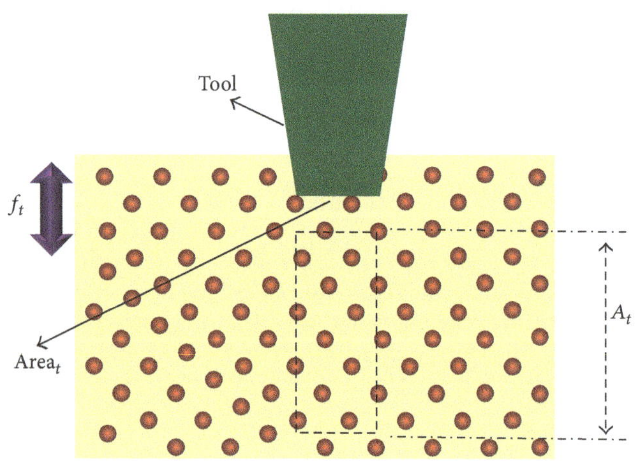

FIGURE 3: Schematic showing machining zone in VANILA process.

tool tip surface. The nanoabrasive grain velocity ($\mathbf{V_a}$) can be estimated using the relation [17]

$$\mathbf{V_a} = \left[\frac{I_t}{\rho_{fl} c_{fl}}\right]^{1/2}, \tag{2}$$

where I_t is the vibration intensity of the acoustic waves generated at the tool tip and can be calculated by dividing the acoustic power by the area of the radiating surface which is the tool tip surface area [18]:

$$I_t = \frac{P_t}{(\pi R_t^2)}. \tag{3}$$

The acoustic power generated at the tip which is excited at its resonant frequency can be found as [19]

$$P_t = \frac{\pi k_t A_t^2 f_t}{Q_t}. \tag{4}$$

The impact velocity $\mathbf{V_I}$ can be expressed as

$$\mathbf{V_I} = \mathbf{V_t} + \mathbf{V_a}. \tag{5}$$

According to the law of conservation of momentum, the impact force (F_I) on the tool tip surface due to the impact with the abrasive nanoparticles can be expressed as [20]

$$\mathbf{F_I} = \frac{m_a \mathbf{V_I} \sin\theta}{\Delta t}, \tag{6}$$

where θ is the angle of impact and Δt is the response time defined as

$$\Delta t = \frac{2\rho_{fl} R_a^2}{9\rho_a \nu_{fl}}. \tag{7}$$

The response time can be considered as the time scale over which the particles respond to changes in flow conditions [20]. Also, the mass term in the expression for F_I only considers the abrasive grain mass since the tool tip mass is much larger than the grain mass.

3.3. Establishing Threshold Criteria. When the impact force F_I is larger than the critical pull-out force F_c, the tool atoms are able to pass through a critical energy barrier resulting in gradual atom-by-atom attrition [14]. The critical pull-out force can be expressed as [21]

$$\mathbf{F_c} = \frac{3}{2}\pi w R^*, \tag{8}$$

where w is the Dupré energy of adhesion and R^* is effective tool tip radius which is given by [21]

$$R^* = \left(\frac{1}{R_a} + \frac{1}{R_t}\right)^{-1}. \tag{9}$$

The effective radius term would enable modeling the tool wear in the VANILA process at various instants of wear. This formulation of R^* is commonly used in classical Hertzian contact theory to determine the contact radius between the impacting bodies.

3.4. Estimation of Tool Wear Rate. Tool wear rate TWR is proportional to the product of volume of tool removed τ_t by single impact, the total number of active abrasive grains n_a, and the frequency of vibration f_t:

$$\text{TWR} \propto \tau_t * n_a * f_t. \tag{10}$$

Introducing a proportionality factor λ termed as wear coefficient, the tool wear rate can be determined as

$$\text{TWR} = \lambda * \tau_t * n_a * f_t. \tag{11}$$

3.5. Volume Removed by Single Impact. Since the wear happens at atomistic scales, the classical theories of mechanics may not be applicable. A quantum mechanism based approach is thus used to estimate the volume τ_t removed by single impact of the abrasive grain. The tool volume removed τ_t due to interaction with single nanoabrasive grain can be deduced by the D'Acunto model [13, 22, 23]:

$$\tau_t = \Delta t * n_{\text{atom}} * w_{\text{atom}} * \text{vol}_{\text{atom}}, \tag{12}$$

where n_{atom} is the number of atoms passing through the energy barrier, w_{atom} is the transition rate, and vol_{atom} is the individual atomic volume. Δt corresponds to the time of interaction between the abrasive grain and the tool which is described in earlier section of this paper. The term $(n_{\text{atom}} * w_{\text{atom}})$ can be calculated using the well-known Arrhenius kinetics [24] as follows [22]:

$$w_{\text{atom}} * n_{\text{atom}} = f_{\text{atom}} * \overline{\rho_t} * A_c * e^{-\Delta E/k_B T}, \tag{13}$$

where ΔE is the effective activation energy barrier, k_B is the Boltzmann constant, $\overline{\rho_t}$ is the surface atomic density of the tool tip, A_c is the area of contact between the tool and abrasive grain which is circular region having diameter R^*; that is, $A_c = \pi R^{*2}$. The term f_{atom} corresponds to the attempt frequency or oscillation frequency.

The activation energy barrier for the atoms to be removed from the tool is considerably reduced by the bond stretching due to impact of the abrasive grain. Thus the effective activation energy barrier ΔE is the difference of the activation energy E_{act} of the tool material and the energy of impact $\mathbf{E_i} = (1/2)m_a\mathbf{V_I^2}$ of the abrasive grain ($\Delta E = \mathbf{E_{act}} - \mathbf{E_i}$).

3.6. Number of Active Abrasives. n_a is the number of active abrasive grains in the machining zone which can be derived as follows. The vibrating tool impacts the abrasive grains within the volume of an imaginary cylinder having diameter equal to tool tip diameter and height equal to the amplitude of vibration of the tool as shown in dotted line in Figure 3. It is to be noted that this figure is not drawn to the actual scale and its purpose is to show the method of determining the active abrasive grains.

The volume of active abrasive particles within this imaginary cylinder can be expressed as

$$\text{Volume} = \text{Area}_t A_t c_z, \tag{14}$$

where Area_t is the cross-sectional area of tool tip given by $\text{Area}_t = \pi R_t^2$, c_z is the slurry volume concentration, and A_t is the amplitude of tool vibration.

The volume of active abrasive particles within the imaginary cylinder can also be expressed as the product of the number of abrasive nanoparticles (n_a) within the imaginary cylinder and volume of a single grain:

$$\text{Volume} = n_a * \frac{4\pi R_a^3}{3}. \tag{15}$$

Using the above equalities, n_a can be calculated as

$$n_a = \frac{\left(3R_t^2 A_t c_z\right)}{\left(4R_a^3\right)}. \tag{16}$$

3.7. Tool Wear Rate Model. Substituting the expressions for n_a and τ_t in (11), the tool wear rate TWR can be expressed as

$$\text{TWR} = \frac{\pi\lambda}{6} * f_t A_t * \frac{\left(R_t R^*\right)^2}{R_a} c_z * \frac{\rho_{fl}}{\rho_a \nu_{fl}} * \overline{\rho_t} * f_{\text{atom}} * \text{vol}_{\text{atom}} * e^{-\Delta E/k_B T}. \tag{17}$$

4. Experimental Verification

The VANILA process experiments are performed using several commercially available tapping mode AFM probes and 10 nm sized nanodiamond abrasive grains suspended in deionized water-based slurry. The tools used are made of commercially available tapping mode probes (NSC16 series MikroMasch make) made of n-type silicon having a tip diameter in the range of 8–10 nm and is excited at its resonance frequency (5–10 KHz range). The tool tip is vibrated at its resonance frequency using the cantilever tuning operation

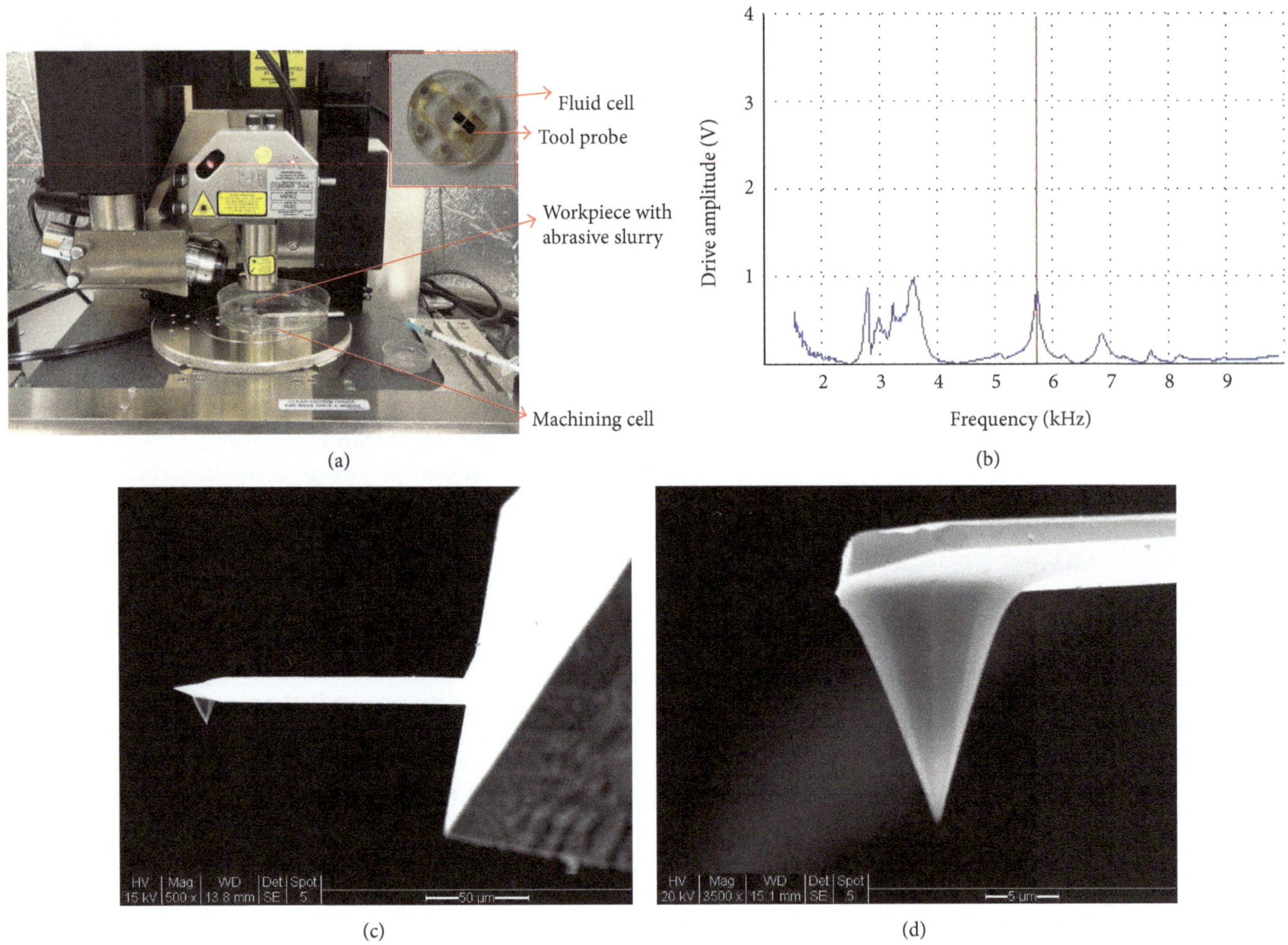

FIGURE 4: (a) Experimental setup (Inset: fluid cell), (b) cantilever tuning curve, (c) tool probe used for machining, and (d) zoomed view of tool tip.

in the Nanoscope IIIa software of the AFM. Representative drive amplitude versus frequency curve during the cantilever tuning for VANILA process is shown in Figure 4(b). The resonance frequency is chosen from the peaks shown during the cantilever tuning.

The maximum vibration amplitude of the tool (A_t) is approximately 200 nm with a quality factor (Q_t) ranging from 1to 50 and spring constant (k_t) 0.5–45 N/m. Properties of deionized water that are considered in this study are dynamic viscosity (ν_{fl}) = 0.00078 Kg/m·s, density (ρ_{fl}) = 1000 Kg/m^3, and speed of sound in water (c) as 1500 m/s. AFM (Veeco Dimension 3100 with Nanoscope III) is used to perform experiments along with a direct drive fluid cantilever holder (DDFCH) for holding the tool. The tool tip is vibrated at its resonant frequency. Polished silicon wafers are used as the workpiece substrate and the abrasive slurry is introduced on the workpiece surface using a syringe. The tool tip is maintained at a distance of 100–200 nm above the substrate surface to avoid contact, while the amplitude of vibration is approximately 100 nm. The machining set-up used for the VANILA process is shown in Figure 4(a). Figures 4(c) and 4(d) show the tool tip (and enlarged view) used for the VANILA process. The common experimental conditions are listed in Table 1.

TABLE 1: Experimental conditions for VANILA process.

Machine	AFM dimension 3100
Abrasive	10 nm sized diamond nanoparticles
Tool	Silicon with ~8 nm tip radius
Liquid medium	Deionized water
Slurry concentration	0.5 vol. conc %
Temperature	300 K

Each tip is used to machine several nanocavities before it is analyzed using scanning electron microscope (SEM) images (make - FEI XL30 ESEM). In order to determine the wear volume with nanoscale precision, SEM images are analyzed and relevant tool dimensions such as its tip length and radius are estimated as shown in Figure 5. Using this information, three-dimensional computer models of the tool tips are created using 3D modeling software SolidWorks in Figure 5(c). The model is sliced at a specified length which is obtained from the SEM images of the worn tip as

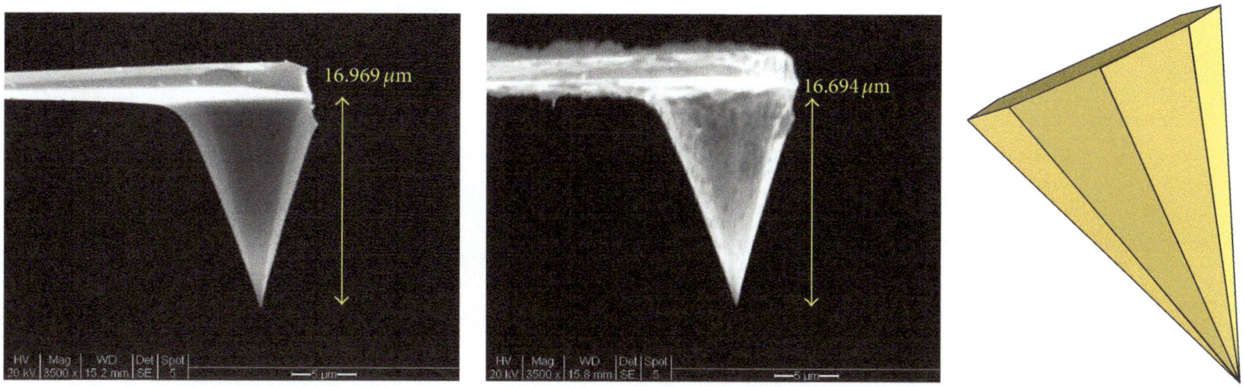

(a) Unused tool tip (b) Worn tool tip after 300 s (c) CAD model of tool tip

FIGURE 5: SEM images and solid model of tool tip used for VANILA process.

shown in Figure 5(b). The volume of the slice of the tool is geometrically determined using the mass properties tool in SolidWorks. Experimental tool wear rate (TWR) is estimated by dividing the calculated wear volume by the total time spent by the resonating tip in the abrasive slurry. During experimentation, the time is measured using a stopwatch.

In order to calculate the tool wear rate TWR, the wear coefficient λ needs to be determined empirically. For this purpose, a series of experiments are conducted and the experimental wear rates are calculated. Individual wear coefficient values are then determined by dividing the experimental wear rate values by the theoretical predictions (assuming λ as unity). An average of the individual λ values (9601, 11034, and 8034) is then considered as the final λ value for the tool wear rate model. During this process, the value of the wear coefficient λ is estimated to be approximately 9500.

In these calculations, the surface atomic density of the tool tip $\overline{\rho_t}$ is taken as 10^{16} atoms/cm^2 [25], the response time is approximately 10^{-12} seconds, and the attempt frequency f_{atom} has typical values of 10^{13} cps for silicon [26]. The atomic volume of silicon (vol_{atom}) is taken as 0.02 nm^3 with an activation energy E_{act} value of 0.9 eV [14]. Based on the experimental conditions, the impact velocity is approximately in the range of 2-3 m/s and the impact force is estimated in the range of 20–50 nN assuming the angle of impact to be normal. The Dupré energy of adhesion for silicon is taken as 4 mJ/m^2 [27] and the critical pull-out force F_c is estimated to be 0.19 nN which is lower than the force of impact and thus the wear could progress. The duration of machining for the tools is varied between 100 s and 1200 s where the tool is used to machine multiple nanoscale cavities on the workpiece. Figure 6 shows the plot between the theoretically predicted TWR and the experimental TWR for machining durations ranging from 100 s to 1200 s for a λ value of 9500.

From the study, it is seen that during the VANILA process, the typical tool wear rates are in the order of 10^3 nm^3/s compared to the material removal rates (MRR) from the workpiece which is in the order of 10^5 nm^3/s [28]. From the plot, it is seen that the TWR model is capable of predicting the tool wear rates within 10% accuracy. It is seen that the

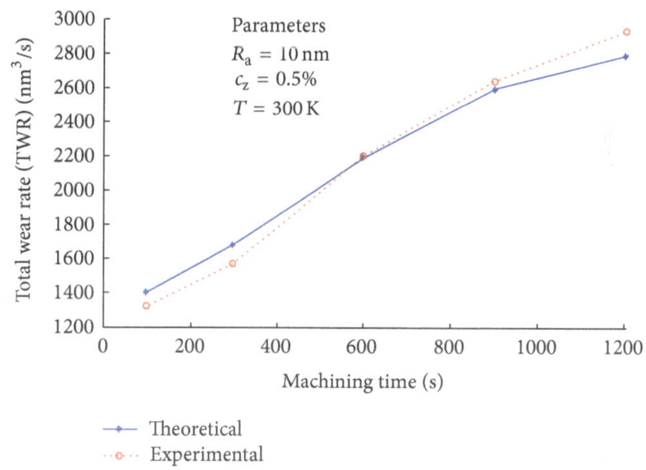

FIGURE 6: Plot showing tool wear rates from experiments versus theoretical predictions for varying machining duration.

tool wear increases gradually with the machining duration. This could be due to the fact that, as the machining duration increases, the number of active abrasive grains which affect the tool wear process increases resulting in increased tool wear rates. For the theoretical calculation of TWR, final tool tip radius value (from the SEM images of worn tool tips) is used to calculate the R^* value.

5. Conclusion

In this study, the tool wear involved in the VANILA process is studied using analytical modeling and experimental methods. Tool wear rates are estimated to be in the order of 10^3 nm^3/s during the machining process for tool tips made of silicon. In order to validate the model, experiments are performed with different tools for varying machining durations. Reasonably good quantitative correlations (within 10%) are obtained between the model predictions and the corresponding experimental observations.

Nomenclature

Tool Variables

R_t: Radius of tool tip (nm)
Area_t: Cross-sectional area of tool tip (nm^2)
f_t: Frequency of vibration of tool (KHz)
A_t: Amplitude of vibration of tool (nm)
$\mathbf{V_t}$: Velocity of tool tip (m/s)
Q_t: Quality factor of tool
k_t: Spring constant of tool
P_t: Power generated at tool tip
I_t: Acoustic vibration intensity.

Fluid Variables

ρ_{fl}: Density of fluid (kg/m^3)
ν_{fl}: Kinematic viscosity of fluid (m^2/s)
c_{fl}: Acoustic velocity in liquid medium (kg/m^3).

Abrasive Grain Variables

R_a: Radius of abrasive grain (nm)
R^*: Effective radius (nm)
ρ_a: Density of abrasive grain (kg/m^3)
m_a: Mass of abrasive grain (kg)
c_z: Volume concentration of abrasive grain slurry (%)
$\mathbf{V_t}$: Impact velocity of abrasive grain (m/s).

Process Parameters

w: Dupré energy of adhesion (J/m^2)
Δt: Response time (s)
$\mathbf{V_t}$: Impact velocity (m/s)
$\mathbf{F_t}$: Impact force (nN)
$\mathbf{F_c}$: Critical pull-out force (nN)
λ: Wear coefficient
ΔE: Effective activation energy barrier (J)
E_{act}: Activation energy (J)
E_i: Kinetic energy of impact (J)
$\overline{\rho_t}$: Surface atomic density of the tool tip
A_c: Area of contact
f_{atom}: Attempt frequency or oscillation frequency
vol_{atom}: Volume of tool atom
τ_t: Volume removed in a single impact (nm^3)
θ: Angle of impact ($\circ$)
k_B: Boltzmann constant (J/K)
T: Temperature (K)
n_{atom}: Number of atoms passing the energy barrier
w_{atom}: Transition rate
n_a: Number of active abrasive grains.

Conflict of Interests

The authors declare that there is no conflict of interests regarding the publication of this paper.

Acknowledgments

This material is based upon work supported by the National Science Foundation under Grant nos. CMMI-1137968, CMMI-1120382, and CBET-1239779. The research facilities provided by the Advanced Material Characterization Center at the University of Cincinnati are acknowledged.

References

[1] M. M. Sundaram and S. James, "Vibration assisted nano abrasive machining," in *Proceedings of the International Conference on Precision, Meso, Micro, and Nano Engineering*, College of Engineering, Pune, India, 2011.

[2] S. James and M. M. Sundaram, "A feasibility study of vibration-assisted nano-impact machining by loose abrasives using atomic force microscope," *Journal of Manufacturing Science and Engineering*, vol. 134, no. 6, Article ID 61014, 2012.

[3] R. Agrawal, N. Moldovan, and H. D. Espinosa, "An energy-based model to predict wear in nanocrystalline diamond atomic force microscopy tips," *Journal of Applied Physics*, vol. 106, no. 6, Article ID 064311, 2009.

[4] G. M. Robinson, M. J. Jackson, and M. D. Whitfield, "A review of machining theory and tool wear with a view to developing micro and nano machining processes," *Journal of Materials Science*, vol. 42, no. 6, pp. 2002–2015, 2007.

[5] C. Lu, Y. Gao, G. Michal, N. N. Huynh, H. T. Zhu, and A. K. Tieu, "Atomistic simulation of nanoindentation of iron with different indenter shapes," *Proceedings of the Institution of Mechanical Engineers Part J: Journal of Engineering Tribology*, vol. 223, no. 7, pp. 977–984, 2009.

[6] K. Komvopoulos and W. Yan, "Molecular dynamics simulation of single and repeated indentation," *Journal of Applied Physics*, vol. 82, no. 10, pp. 4823–4830, 1997.

[7] P.-Z. Zhu, Y.-Z. Hu, H. Wang, and T.-B. Ma, "Study of effect of indenter shape in nanometric scratching process using molecular dynamics," *Materials Science and Engineering A*, vol. 528, no. 13-14, pp. 4522–4527, 2011.

[8] H.-J. Kim, S.-S. Yoo, and D.-E. Kim, "Nano-scale wear: a review," *International Journal of Precision Engineering and Manufacturing*, vol. 13, no. 9, pp. 1709–1718, 2012.

[9] J. Liu, J. K. Notbohm, R. W. Carpick, and K. T. Turner, "Method for characterizing nanoscale wear of atomic force microscope tips," *ACS Nano*, vol. 4, no. 7, pp. 3763–3772, 2010.

[10] A. Khurshudov and K. Kato, "Wear of the atomic force microscope tip under light load, studied by atomic force microscopy," *Ultramicroscopy*, vol. 60, no. 1, pp. 11–16, 1995.

[11] M. L. Bloo, H. Haitjema, and W. O. Pril, "Deformation and wear of pyramidal, silicon-nitride AFM tips scanning micrometre-size features in contact mode," *Measurement: Journal of the International Measurement Confederation*, vol. 25, no. 3, pp. 203–211, 1999.

[12] K. Cheng, X. Luo, R. Ward, and R. Holt, "Modeling and simulation of the tool wear in nanometric cutting," *Wear*, vol. 255, no. 7-12, pp. 1427–1432, 2003.

[13] M. D'Acunto, "Theoretical approach for the quantification of wear mechanisms on the nanoscale," *Nanotechnology*, vol. 15, no. 7, pp. 795–801, 2004.

[14] B. Gotsmann and M. A. Lantz, "Atomistic wear in a single asperity sliding contact," *Physical Review Letters*, vol. 101, no. 12, Article ID 125501, 2008.

[15] S. James and M. M. Sundaram, "A molecular dynamics study of the effect of impact velocity, particle size and angle of impact of abrasive grain in the Vibration Assisted Nano Impact-machining by Loose Abrasives," *Wear*, vol. 303, no. 1-2, pp. 510–518, 2013.

[16] O. Sahin, C. F. Quate, O. Solgaard, and A. Atalar, "Resonant harmonic response in tapping-mode atomic force microscopy," *Physical Review B: Condensed Matter and Materials Physics*, vol. 69, no. 16, Article ID 165416, 2004.

[17] Q. Qi and G. J. Brereton, "Mechanisms of removal of micron-sized particles by high-frequency ultrasonic waves," *IEEE Transactions on Ultrasonics, Ferroelectrics, and Frequency Control*, vol. 42, no. 4, pp. 619–629, 1995.

[18] D. M. VandenBurg, *The Sonochemical Remediation of Phthalate Esters: An Investigation into Products and Kinetics*, University of Bath, Bath, UK, 2011.

[19] J. P. Cleveland, B. Anczykowski, A. E. Schmid, and V. B. Elings, "Energy dissipation in tapping-mode atomic force microscopy," *Applied Physics Letters*, vol. 72, no. 20, pp. 2613–2615, 1998.

[20] S. M. Booij, *Fluid Jet Polishing: Possibilities and Limitations of a New Fabrication Technique*, 2003.

[21] D. Maugis, "Adhesion of spheres: the JKR-DMT transition using a Dugdale model," *Journal of Colloid And Interface Science*, vol. 150, no. 1, pp. 243–269, 1992.

[22] R. Bassani and M. D'Acunto, "Nanotribology: tip-sample wear under adhesive contact," *Tribology International*, vol. 33, no. 7, pp. 443–452, 2000.

[23] T. D. B. Jacobs, B. Gotsmann, M. A. Lantz, and R. W. Carpick, "On the application of transition state theory to atomic-scale wear," *Tribology Letters*, vol. 39, no. 3, pp. 257–271, 2010.

[24] K. J. Laidler, "The development of the arrhenius equation," *Journal of Chemical Education*, vol. 61, no. 6, pp. 494–498, 1984.

[25] M. D'Acunto, "Controlling wear on nanoscale," in *Scanning Probe Microscopy in Nanoscience and Nanotechnology*, pp. 647–686, Springer, 2010.

[26] C. M. Bertoni, V. Bortolani, C. Calandra, and E. Tosatti, "Dielectric matrix and phonon frequencies in silicon," *Physical Review Letters*, vol. 28, no. 24, pp. 1578–1581, 1972.

[27] V. V. Tsukruk and V. N. Bliznyuk, "Adhesive and friction forces between chemically modified silicon and silicon nitride surfaces," *Langmuir*, vol. 14, no. 2, pp. 446–455, 1998.

[28] S. James and M. Sundaram, "Modeling of material removal rate in vibration assisted nano impact-machining by loose abrasives," *Journal of Manufacturing Science and Engineering*, 2014.

4

Resistance Spot Welding Optimization Based on Artificial Neural Network

Thongchai Arunchai,[1] Kawin Sonthipermpoon,[1] Phisut Apichayakul,[1] and Kreangsak Tamee[2]

[1] *Department of Industrial Engineering, Faculty of Engineering, Naresuan University, Phitsanulok 65000, Thailand*
[2] *Department of Computer Science and Information Technology, Faculty of Science, Naresuan University, Phitsanulok 65000, Thailand*

Correspondence should be addressed to Thongchai Arunchai; t_ch@hotmail.com

Academic Editor: Quanfang Chen

Resistance Spot Welding (RSW) is processed by using aluminum alloy used in the automotive industry. The difficulty of RSW parameter setting leads to inconsistent quality between welds. The important RSW parameters are the welding current, electrode force, and welding time. An additional RSW parameter, that is, the electrical resistance of the aluminum alloy, which varies depending on the thickness of the material, is considered to be a necessary parameter. The parameters applied to the RSW process, with aluminum alloy, are sensitive to exact measurement. Parameter prediction by the use of an artificial neural network (ANN) as a tool in finding the parameter optimization was investigated. The ANN was designed and tested for predictive weld quality by using the input and output data in parameters and tensile shear strength of the aluminum alloy, respectively. The results of the tensile shear strength testing and the estimated parameter optimization are applied to the RSW process. The achieved results of the tensile shear strength output were mean squared error (MSE) and accuracy equal to 0.054 and 95%, respectively. This indicates that that the application of the ANN in welding machine control is highly successful in setting the welding parameters.

1. Introduction

In automotive production, each automobile has approximately 7,000 to 12,000 spot welds. The welds are done using the Resistance Spot Welding (RSW) process which is done by a computer controlled robotic welder. The use of RSW on lightweight aluminum alloy is increasing [1, 2]. RSW is a rapid joining technique extensively used to join thin shell assemblies in automotive manufacturing operations. It is an important process to ensure strong structural car bodies using lightweight materials to save both energy and natural resources [3]. The aluminum alloy material is a low density material with significant mechanical properties which is expected to be extensively used in the future to partially replace steel which is currently the primary production material in automobiles [4]. The 6061-T6 aluminum alloy is of light weight and has significant mechanical properties which are of interest in this research.

The weld quality of the RSW process has been a significant problem for the automotive industry. Manual calculation of welding parameters, operator experience, and technician expertise in adjusting the parameter settings have not been consistently accurate or correct. The calculations have previously been unable to be confirmed against optimal parameters [5]. The parameter settings of each welding machine have been difficult because there are many sensitive factors. This has led to the high cost of requiring many specimens of the construction material to be tested to achieve adequate experimental results to derive optimal parameter values. Each car body factory has more than 200 welding machines. In order to get standard weld quality overall has led to high costs of adjusting the parameter settings for each different model of welding machine, for different materials (e.g., thickness), for replacement electrodes, and so on [6–8]. Therefore, it is important to understand the parameter relationships in the RSW process, quality improvement measures, assessment

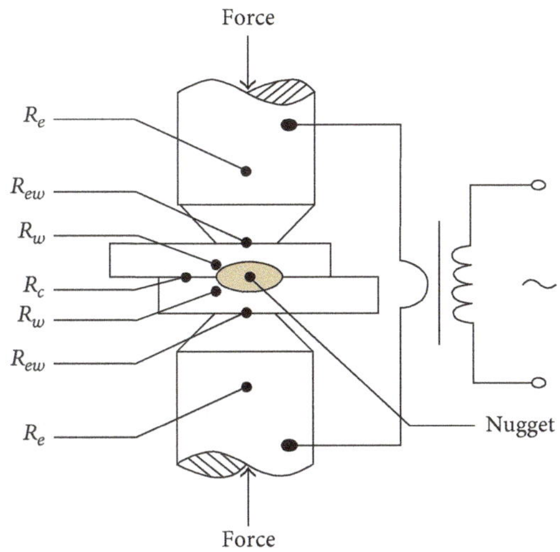

FIGURE 1: The resistance in RSW process [10].

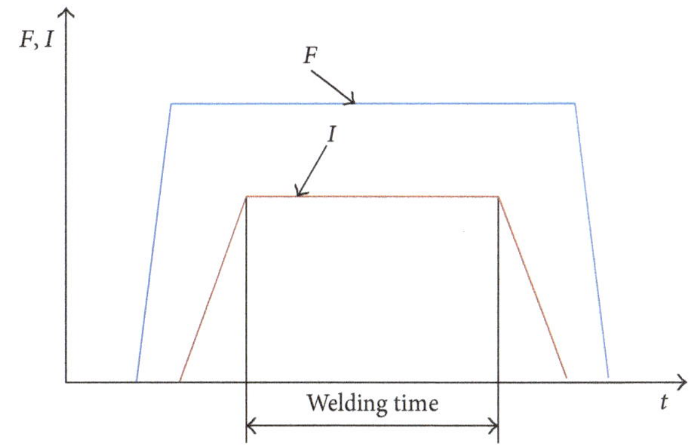

FIGURE 2: RSW schematic diagram.

and efficiency prediction, and appropriate parameter optimization.

The problems mentioned have motivated this study, formally titled "parameter optimization for resistance spot welding of 6061-T6 aluminum alloy based on artificial neural network." The estimation of RSW parameters using ANN will be more efficiently and correctly optimized. The ANN proved to be effective in resolving both linear and nonlinear functions required for adjusting the RSW parameter settings of computer controlled robotic welders, especially in the automotive industry.

2. RSW Process

The objective of the RSW process is to generate heat rapidly in the joints of the material being welded while minimizing conduction of heat to cooler adjacent material. This heat generation can be expressed by

$$Q = I^2 Rt, \tag{1}$$

where Q is the heat energy in joules, I is the current in amperes, R is the resistance in ohms, and t is the time in seconds [9]. The series of resistances contributed by the secondary circuit resistance welding machine are shown in Figure 1.

From Figure 1, the resistance in the RSW process depends on the total resistance which can be expressed by

$$R = R_c + 2R_w + 2R_{ew} + 2R_e, \tag{2}$$

where R is the total resistance, R_e is the electrode resistance, R_{ew} is the specimen electrode contact resistance, R_w is the specimen resistance, and R_c is the specimen's contact resistance.

The resistance transformation of the material work piece affects the melting temperature in the spot weld in the RSW process and the weld ability of the material [11, 12]. Optimal electrode force is needed for the explosive solutions during material smelting [13, 14], while the intensity of the electric welding current is an important factor that affects heat generation leading to smelting. When the electric intensity is high the spot weld will also have intensive heat. This will affect the nugget size and strength of the weld joint [15]. The nugget size and weld joint strength depend on the resistance and welding time, so the welding duration influences the mechanical properties of the weld joints [16, 17]. Melting occurs as a result of the relationship between welding current, welding time, and electrode force, as shown in schematic diagram RSW cycle Figure 2, where I is welding current, t is welding time, and F is electrode force.

Researchers have studied the relationship of the RSW parameters and confirm that the welding process parameters have a great influence on the weld quality [18, 19]. In addition, the resistance of the material due to the thickness of the work piece is also important, as can be seen in greater nugget growth in thicker pieces than thin pieces. In different materials, a satisfactory nugget will be generated in high resistance materials or materials with lower thermal conductivity [20]. Therefore, this study adds the resistance as a factor.

3. Material and Methods

This research studied the RSW process under actual conditions in a car body factory which used gun welds controlled by a MFDC Rexroth Bosch welder. Aluminum alloy specimens (6061-T6) of both 1 and 2 mm thickness were welded at three resistance levels. Figure 3 shows the three resistance levels of the high, medium, and low thicknesses of 2-2, 2-1, and 1-1 mm material, respectively.

The mechanical and chemical properties based on KAISER ALUMINUM [21] are shown in Tables 1 and 2, respectively.

This research used 6 mm copper alloy electrode tips. The experimental design included full factorial analysis based on low, medium, and high thickness of material, with all parameter settings varied at each level of thickness, giving $3^4 \times 3 = 243$ runs. The parameter settings are welding current,

TABLE 1: 6061-T6 aluminum alloy chemical properties.

6061	Si	Fe	Cu	Mn	Mg	Cr	Zn	Ti	Zr	Other	Max.
Min. (wt%)	0.40	0.0	0.15	0.00	0.8	0.04	0.00	0.00	0.00	Each	0.05
Max. (wt%)	0.8	0.7	0.40	0.15	1.2	0.35	0.25	0.15	0.05	Total	0.15

TABLE 2: 6061-T6 aluminum alloy mechanical properties.

Tensile tests	Temper T6 Ultimate (MPA)	Yield (MPA)	Elongation (%)
Min.: Max.	337 : 340	286 : 288	13.6 : 13.9

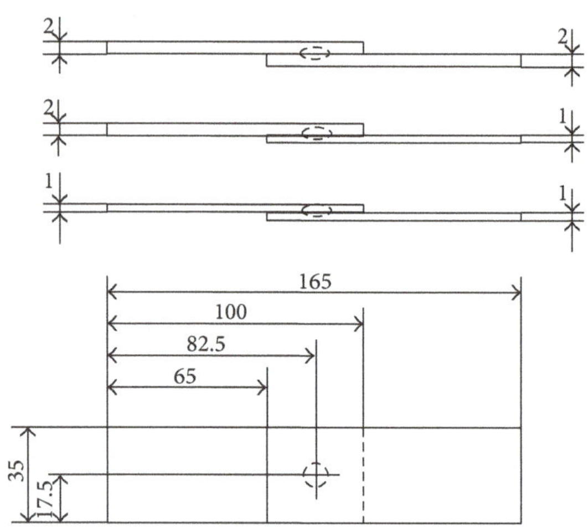

FIGURE 3: The specimen dimensions (mm).

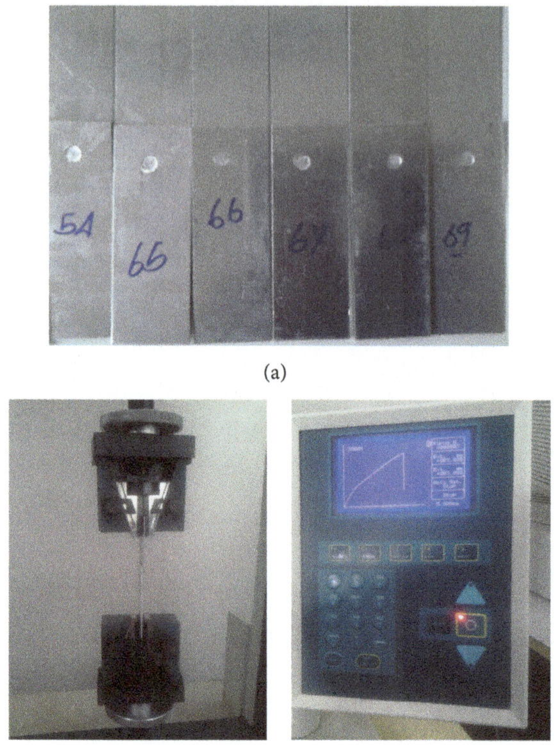

(a)

(b) (c)

FIGURE 4: (a) is the weld joint, (b) is the tested clamping, and (c) is the graphic monitor.

electrode force, welding time, and resistance of 20, 28, and 36 kA, 2, 4, and 6 kN, 100, 150, and 200 ms and 2-2, 2-1, and 1-1 mm, respectively. Subsequently the welding parameters were input according to the experimental design and welding tests were done. The finished weld joints and the maximum tensile shear strength were tested and recorded. The tests were done using the HOUNSFIELD 50 kN at maximum load as shown in Figure 4.

In this research, 75% of the 243 experimental results of the ANN were used to train the robotic welders. 25% of the experimental results were selected randomly for testing, for comparison between the predicted shear strength and the experimental shear strength results. The ANN models were configured to establish the relationship between the welding parameters and the shear strength. The ANN model is a multilayered feed-forward algorithm, as shown in Figure 5. In that algorithm, there are three layers: the input layer, the output layer, and the hidden layer between the input and output layers. In the input layer, the number of the input neurons is set to the level of welding parameters in each run order, and the output layer is the shear strength.

The ANN was designed and tested for estimating the shear strength by using the welding parameters as input data and the shear strength as output data, using the multilayered feed-forward algorithm, and the ANN was trained by the backpropagation algorithm. The final transfer function was a sigmoidal function and was a pure linear function.

4. Results and Discussion

This work studied the application of the ANN to calculate optimized parameters for controlling welding robots. Table 3 shows the tensile shear strength achieved from actual application of the experimental results as calculated by the ANN; this variable is the calculation of the mean squared error (MSE) of the shear strength. The MSE is equal to 0.05. According to these results, it can be seen that the ANN determined appropriate parameter optimization with satisfactory results.

The shear strength is obtained when applying the experimental results from the ANN. Applying the ANN calculations results in good parameter optimization as shown in Figure 6.

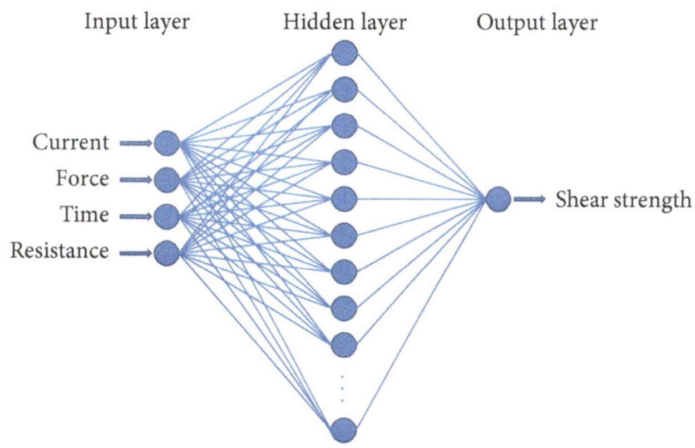

FIGURE 5: The ANN structure for shear strength prediction.

TABLE 3: Show calculation of the mean squared error (MSE).

Number	Shear strength (kN)	ANN prediction	Squared error	Number	Shear strength (kN)	ANN prediction	Squared error	Number	Shear strength (kN)	ANN prediction	Squared error
1	1.92	2.0571	0.018796	21	3.585	3.669	0.007056	41	3.271	3.3765	0.01113
2	3.11	3.1568	0.00219	22	1.985	2.0526	0.00457	42	3.623	3.76	0.018769
3	2.695	2.293	0.161604	23	3.611	3.6838	0.0053	43	3.447	3.6838	0.056074
4	2.505	2.3748	0.016952	24	2.892	3.1432	0.063101	44	1.92	2.0571	0.018796
5	3.598	3.8369	0.057073	25	3.585	3.669	0.007056	45	1.842	1.9775	0.01836
6	2.558	2.4952	0.003944	26	2.84	3.1568	0.100362	46	3.802	3.9422	0.019656
7	3.271	3.3765	0.01113	27	3.623	3.76	0.018769	47	3.836	3.8369	$8.1E-07$
8	3.485	3.0774	0.166138	28	2.662	2.6579	$1.68E-05$	48	1.963	2.1044	0.019994
9	3.505	3.592	0.007569	29	3.572	2.9412	0.397909	49	2.816	3.1432	0.10706
10	1.882	2.3376	0.207571	30	1.89	1.9507	0.003684	50	2.866	2.9608	0.008987
11	2.405	2.9412	0.28751	31	3.566	3.592	0.000676	51	1.884	2.0143	0.016978
12	3.458	3.669	0.044521	32	3.185	3.1284	0.003204	52	2.843	2.6913	0.023013
13	3.848	3.8369	0.000123	33	3.84	3.9422	0.010445	53	2.708	2.6617	0.002144
14	3.878	3.9422	0.004122	34	2.852	2.4274	0.180285	54	3.314	3.3765	0.003906
15	2.468	2.5167	0.002372	35	3.525	3.592	0.004489	55	3.235	2.6829	0.304814
16	3.316	3.1432	0.02986	36	3.626	3.6838	0.003341	56	3.285	3.1568	0.016435
17	2.865	2.5445	0.10272	37	1.942	2.1044	0.026374	57	2.837	2.6913	0.021228
18	1.99	2.1174	0.016231	38	3.066	3.1284	0.003894	58	2.702	2.8238	0.014835
19	2.883	2.8238	0.003505	39	2.861	3.1284	0.071503	59	3.485	3.0774	0.166138
20	2.685	2.5167	0.028325	40	3.623	3.76	0.018769	60	2.84	2.304	0.287296
					MSE = 0.053978						

ANN is therefore seen as an appropriate method optimization for finding an answer that is both linear and nonlinear.

The ANN model is composed of three different layers: input layer, hidden layer, and output layer. The ANN was used to calculate the shear strength at the output layer and the RSW parameters at the input layer while training and using the backpropagation algorithm. 75% of the 243 data values from the full factorial experimental results were used for training and 25% of the data were used for testing the accuracy of the ANN's calculated values. The results show that hidden nodes, learning rate, MSE goal, and maximum learning epoch are 30, 0.001, 0.04, and 3,000,000, respectively.

When tested, 20 samples were found to have a mean squared error (MSE) of 0.047896 kN. This indicates that the ANN model is capable of predicting the shear strength for adjusting the RSW parameters under test with a small sample which are shown in Table 4 and Figure 7, respectively.

From this study, it is suggested that the RSW operator can use a small sample set appropriate as input into the ANN model which would be sufficient for setting the optimization

Table 4: Calculation of the mean squared error (MSE) case of 20-unit experiment.

Number	Shear strength (kN)	ANN prediction	Squared error	Number	Shear strength (kN)	ANN prediction	Squared error	Number	Shear strength (kN)	ANN prediction	Squared error
1	3.598	3.8369	0.057073	8	3.611	3.6838	0.0053	15	3.626	3.6838	0.003341
2	3.485	3.0774	0.166138	9	3.585	3.669	0.007056	16	3.623	3.76	0.018769
3	3.505	3.592	0.007569	10	3.623	3.76	0.018769	17	3.623	3.76	0.018769
4	3.458	3.669	0.044521	11	3.572	2.9412	0.397909	18	3.802	3.9422	0.019656
5	3.848	3.8369	0.000123	12	3.566	3.592	0.000676	19	3.836	3.8369	0.000000
6	3.878	3.9422	0.004122	13	3.84	3.9422	0.010445	20	3.485	3.0774	0.166138
7	3.585	3.669	0.007056	14	3.525	3.592	0.004489				
MSE = 0.047896											

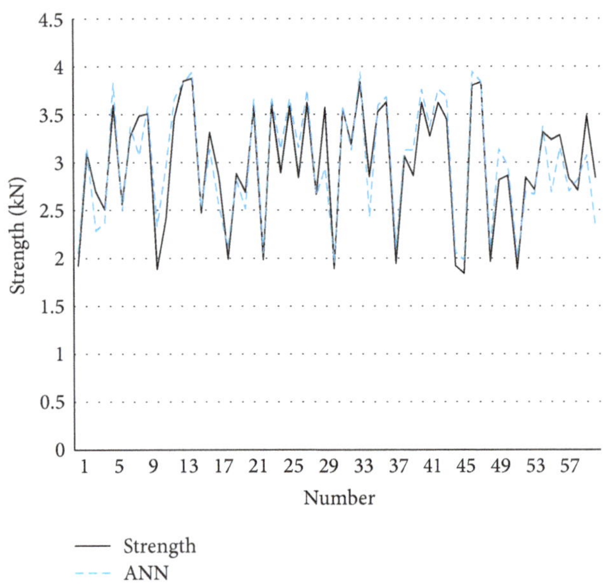

Figure 6: ANN prediction.

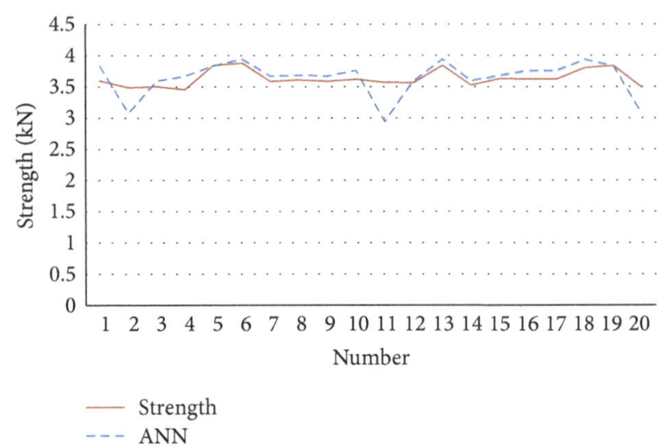

Figure 7: ANN prediction case of 20-unit experiment.

parameters. However, the accuracy of the parameters may be affected by other factors outside the scope of the ANN, such as the particular welding machine model, the surface condition of the specimen material, the cooling effect of the environment, and electrode wear.

5. Conclusion

The full factorial experimental results and ANN were developed and successfully tested in an auto body industry plant in Thailand. The results were MSE and accuracy equal to 0.054 and 95%, respectively. Calculating a reliable estimation of shear strength enables parameter settings to achieve high weld quality and reduces both the setting time process and the sample testing. Future work should test the accuracy of other algorithms used to calculate these parameters to enable comparison of the accuracy of other algorithms.

Conflict of Interests

The authors declare that there is no conflict of interests regarding the publication of this paper.

Acknowledgments

The authors would like to express their profound gratitude to the UK Engineering and Supply Co. Ltd. and Thonburi Automotive Assembly Plant Co., Ltd., Thailand, for allowing use of their premises and providing research equipment to support the project.

References

[1] H. Huh and W. Kang, "Electrothermal analysis of electric resistance spot welding processes by a 3-D finite element method," *Journal of Materials Processing Technology*, vol. 63, no. 1–3, pp. 672–677, 1997.

[2] W. Li, S. Cheng, S. J. Hu, and J. Shriver, "Statistical investigation on resistance spot welding quality using a two-state, sliding-level experiment," *Journal of Manufacturing Science and Engineering, Transactions of the ASME*, vol. 123, no. 3, pp. 513–520, 2001.

[3] J. A. Khan, L. Xu, Y.-J. Chao, and K. Broach, "Numerical simulation of resistance spot welding process," *Numerical Heat Transfer Part A: Applications*, vol. 37, no. 5, pp. 425–446, 2000.

[4] J. Zhu, L. Li, and Z. Liu, "CO_2 and diode laser welding of AZ31 magnesium alloy," *Applied Surface Science*, vol. 247, no. 1–4, pp. 300–306, 2005.

[5] H.-L. Lin, T. Chou, and C.-P. Chou, "Optimation of resistance spot welding process using Taguchi method and a neural network," *Experimental Techniques*, vol. 31, no. 5, pp. 30–36, 2007.

[6] K. D. Weiss, "Paint and coatings: a mature industry in transition," *Progress in Polymer Science*, vol. 22, no. 2, pp. 203–245, 1997.

[7] X. Deng, W. Chen, and G. Shi, "Three-dimensional finite element analysis of the mechanical behavior of spot welds," *Finite Elements in Analysis and Design*, vol. 35, no. 1, pp. 17–39, 2000.

[8] S.-H. Lin, J. Pan, T. Tyan, and P. Prasad, "A general failure criterion for spot welds under combined loading conditions," *International Journal of Solids and Structures*, vol. 40, no. 21, pp. 5539–5564, 2003.

[9] N. T. Williams and J. D. Parker, "Review of resistance spot welding of steel sheets: part 2—factors influencing electrode life," *International Materials Reviews*, vol. 49, no. 2, pp. 77–108, 2004.

[10] Y. Luo, J. Liu, H. Xu, C. Xiong, and L. Liu, "Regression modeling and process analysis of resistance spot welding on galvanized steel sheet," *Materials and Design*, vol. 30, no. 7, pp. 2547–2555, 2009.

[11] P. H. Thornton, A. R. Krause, and R. G. Davies, "Contact resistances in spot welding," *Welding Journal*, vol. 75, no. 12, pp. 402.s–412.s, 1996.

[12] A. G. Livshits, "Universal quality assurance method for resistance spot welding based on dynamic resistance," *Welding Journal*, vol. 76, no. 9, pp. 383–390, 1997.

[13] R. B. Hirsch, "Making resistance spot welding safer," *Welding Journal*, vol. 86, no. 2, pp. 32–37, 2007.

[14] J. Senkara, H. Zhang, and S. J. Hu, "Expulsion prediction in resistance spot welding," *Welding Journal*, vol. 83, no. 4, p. 123-S, 2004.

[15] S. Satonaka, K. Kaieda, and S. Okamoto, "Prediction of tensile-shear strength of spot welds based on fracture modes," *Welding in the World*, vol. 48, no. 5-6, pp. 39–45, 2004.

[16] S. Aslanlar, A. Ogur, U. Ozsarac, and E. Ilhan, "Welding time effect on mechanical properties of automotive sheets in electrical resistance spot welding," *Materials and Design*, vol. 29, no. 7, pp. 1427–1431, 2008.

[17] F. Hayat, "The effects of the welding current on heat input, nugget geometry, and the mechanical and fractural properties of resistance spot welding on Mg/Al dissimilar materials," *Materials and Design*, vol. 32, no. 4, pp. 2476–2484, 2011.

[18] T. Kim, H. Park, and S. Rhee, "Optimization of welding parameters for resistance spot welding of TRIP steel with response surface methodology," *International Journal of Production Research*, vol. 43, no. 21, pp. 4643–4657, 2005.

[19] R. S. Florea, D. J. Bammann, A. Yeldell, K. N. Solanki, and Y. Hammi, "Welding parameters influence on fatigue life and microstructure in resistance spot welding of 6061-T6 aluminum alloy," *Materials and Design*, vol. 45, pp. 456–465, 2013.

[20] C. Tsai and J. Papritan, "Modeling of resistance spot weld nugget growth," *Welding Journal*, vol. 71, no. 2, pp. 47–54, 1992.

[21] KAISER ALUMINUM, "Trent wood Works—WA 99215 CERTIFIED Serial Number 4315467," Article ID 4315467, TEST REPORT, December 2013.

5

Characterization and Optimization of Mechanical Properties of ABS Parts Manufactured by the Fused Deposition Modelling Process

Godfrey C. Onwubolu and Farzad Rayegani

School of Mechanical and Electrical Engineering & Technology, Faculty of Applied Science and Technology, Sheridan Institute of Technology and Advanced Learning, Brampton, ON, Canada L6Y 5H9

Correspondence should be addressed to Godfrey C. Onwubolu; godfrey.onwubolu@sheridancollege.ca

Academic Editor: Luigino Filice

While fused deposition modelling (FDM) is one of the most used additive manufacturing (AM) techniques today due to its ability to manufacture very complex geometries, the major research issues have been to balance ability to produce aesthetically appealing looking products with functionality. In this study, five important process parameters such as layer thickness, part orientation, raster angle, raster width, and air gap have been considered to study their effects on tensile strength of test specimen, using design of experiment (DOE). Using group method of data handling (GMDH), mathematical models relating the response with the process parameters have been developed. Using differential evolution (DE), optimal process parameters have been found to achieve good strength simultaneously for the response. The optimization of the mathematical model realized results in maximized tensile strength. Consequently, the additive manufacturing part produced is improved by optimizing the process parameters. The predicted models obtained show good correlation with the measured values and can be used to generalize prediction for process conditions outside the current study. Results obtained are very promising and hence the approach presented in this paper has practical applications for design and manufacture of parts using additive manufacturing technologies.

1. Introduction

Stratasys Inc. developed the fused deposition modeling (FDM) system which is one such layered manufacturing technology that produces parts with complex geometries by the layering of extruded materials, such as acrylonitrile butadiene styrene (ABS) thermoplastic [1, 2] as shown in Figure 1. In the FDM process, the build material is initially in the raw form of a flexible filament. The feedstock filament is then partially melted and extruded though a heated nozzle within a temperature controlled environment for building of the part. The material is extruded in a thin layer onto the previously built model layer on the build platform in the form of a prescribed two-dimensional (x-y) layer pattern. The deposited material cools, solidifies, and bonds with adjoining material. After an entire layer is deposited, the build platform moves downward along the z-axis by an increment equal to the filament height (layer thickness) and the next layer is deposited on top of it. The platen or table on which the build sheet is placed lies on the x-y plane.

The properties of built parts depend on settings of various process parameters fixed at the time of fabrication. Additive manufacturing (AM) is a new manufacturing technology, driven by computer-aided design (CAD) that makes it possible for companies to significantly cut design and manufacturing cycle times [1]. This new manufacturing paradigm is seamless because it has the capability of producing exactly any simple or complex part represented by a CAD system which is submitted to it. This paper first proposes a framework for additive manufacturing. In this framework, there are three main domains that drive AM: CAD-based, AM process planning and technologies, and materials/testing/inductive modeling. The designer chooses a CAD system for modeling from a suite of CATIA, SolidWorks, Inventor, and AutoCAD 3D. Then the design engineer utilizes design of experiment (DOE) for optimized process plan for different options of

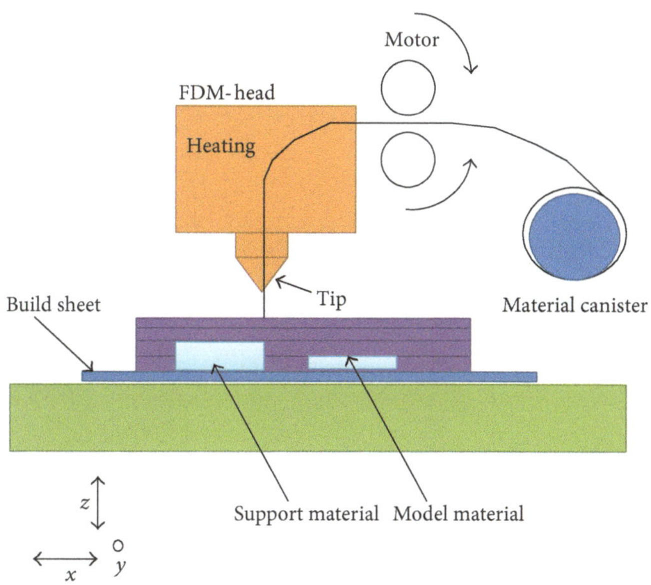

FIGURE 1: FDM process.

a part; intermediate software such as *Insight* is used for tool paths optimization. A decision has to be made regarding which type of machine to use depending on the material used for producing the part(s). For example, Fortus 900mc and 400mc are useful for producing parts made from thermoplastics. Spectrum Z510 are suited for powder materials, while Solidscape suits wax material. These are some of the decisions and choices to be made. The presentation of this paper follows this framework.

Fused deposition modelling (FDM) is a fast growing additive manufacturing (AM) technology due to its ability to build functional parts that have complex geometrical shapes in reasonable build time [1, 2]. Reduction of product development cycle time is a major concern in industries to remain competitive in the marketplace. Several sources have highlighted the advantages of additive manufacturing techniques such as additive manufacturing (AM) over traditional product development methodology [3–5]. Using AM speeds up the whole product development process especially when producing very complicated parts which may be challenging using traditional manufacturing methods. It is argued that since three-dimensional CAD is being used as the starting point and the transfer to AM is relatively seamless, there is less concern over interpretation of the design intent [1]. The seamlessness can also be seen in terms of reduction in process steps because, regardless of the complexity of parts to be built, building with an AM machine is generally performed in a single step unlike in most other manufacturing processes which normally require multiple and iterative stages to be carried out. In addition to these advantages, other advantages of using AM are significant reduction in the number of processes and resources required. Consequently, focus is fast shifting from traditional product development methodology to additive manufacturing technology [1]. References [1, 2] are excellent resources for the processes involved in and advantages of additive manufacturing.

Producing aesthetically appealing AM products that have complex shapes is not difficult, given available AM technologies. The challenge is to produce manufactured AM parts that are functionally reliable. Therefore, this paper reports the work that has been done to investigate the *functionality of manufactured AM parts*. The motivation for this emphasis is that no company will want to make commitment to a technology that may produce aesthetically appealing products that are not functionally reliable. A critical review of literature suggests that properties of AM parts are a function of various process related parameters and can be significantly improved with proper adjustment. Since mechanical properties are important for functional parts, it is absolutely essential to study the influence of various process parameters on mechanical properties so that improvement can be made through selection of best settings. The present study focuses on assessment of one of the mechanical properties, namely, tensile strength of fused deposition modelling- (FDM-) based fabricated parts, by first developing a model prediction and then optimizing the process parameter settings and responses.

2. Literature Review

Although AM is an efficient technology, full scale application has not gained much attention because of compatibility of presently available materials with AM technologies [6]. Despite several design and environmental advantages of AM technologies over other manufacturing processes, the adoption of AM as a means for fabricating end-use components has historically been setback by the technologies' narrow selection of available materials. The majority of materials used presently by modern AM techniques are proprietary polymers. There are also possibilities to use some metals (such as steel alloys and titanium) and ceramics. While there are different types of materials that can be used in AM, the material properties are typically not as strong as their conventionally manufactured counterparts due to the anisotropy caused by the layer-by-layer [7]. To overcome this limitation, one approach may be development of new materials having superior characteristics than conventional materials and its compatibility with technology. Another convenient approach may be suitably adjusting the process parameters during fabrication stage so that properties may improve [8, 9]. Several researchers have specifically considered the anisotropic characteristics of FDM parts in recent years. Rodríguez et al. [10] investigated the tensile strength and elastic modulus of FDM specimens with varying mesostructures in comparison with the properties of the ABS monofilament feedstock. The outcome of their investigation is that the tensile strength was the greatest for parts with fibers aligned with the axis of the tension force. The mechanical properties of FDM parts are not solely controlled by the build material of the original filament but are also significantly influenced by a directionally dependent production process that fabricates components with anisotropic characteristics associated with the inherent layering [11]. A material in which the elastic properties depend on the orientation of the sample is said to be anisotropic. Ahn et al. [11] designed a factorial experiment to quantify the effects of model temperature, bead width, raster orientation, air gap, and acrylonitrile butadiene styrene (ABS) color on both tensile and compressive strengths of FDM parts. They determined that both air gap and raster orientation had significant effects on the resulting tensile strength, while compressive strength was not affected by these factors. Their results include a set of recommended build rules for designing FDM parts. Sood et al. [12] carried out similar study, with varying factors of layer thickness, build orientation, raster angle, raster width, and air gap. They used response surface methodology (RSM) to analyze the functional relationship between process parameters and specimen strength and found that the tested factors influence the mesostructural configuration of the built part as well as the bonding and distortion within the part. Sood et al. [13] further examined the effect of the same five process parameters on the subsequent compressive strength of test specimens. Their results show the importance of fiber-to-fiber bond strength and the control of distortion during the build process. Lee et al. [14] concluded that layer thickness, raster angle, and air gap influence the elastic performance of compliant ABS prototypes manufactured by fused deposition. Lee et al. [15] focused on the compressive strength of layered parts as a function of build direction and determined that the compressive strength is greater for the axial FDM specimens than for the transverse. Panda et al. [16] performed experiments on the impact of process parameters on dimensional accuracy, surface roughness, and mechanical strengths, using response surface methodology (RSM) [17] for modelling and bacterial foraging for finding optimal process parameter settings and responses.

From the literature, it is found that good amount of work has been done in FDM strength modelling; however, little amount of work has been done to develop the strength model in terms of FDM process parameters for prediction purpose. Most published work on predictive model of FDM parts are based on response surface methodology (RSM) which is restrictive in modelling because it is not "data-driven"; that is, it is not inductive. Since the relation between a particular mechanical property and process parameters related to it is difficult to establish, attempt has been made in this paper to derive the empirical model between the processing parameters and mechanical properties using group method of data handling (GMDH) [18] which is one of the most robust existing inductive modelling. The present study reported in this paper uses the group method of data handling (GMDH) modelling approach to derive the required relationship among respective process parameters and tensile strength. The models derived serve as predictive models which can be used to anticipate the theoretical best parameter settings that would result in optimal response characteristic. The predictive models are therefore the objective functions, while the lower and upper bounds used from the design of experiment (DOE) [19] are the constraints, so that the problem becomes constrained optimization which can be solved using any of the existing optimization techniques. In our case, we used differential evolution (DE) [20] to solve the optimization problem. The solutions give the optimal tensile response and optimal process parameter settings.

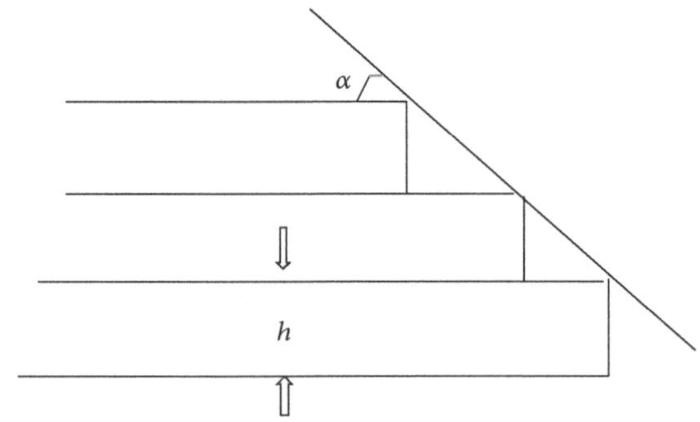

FIGURE 2: Height of slices or layout of layer thickness. Source: [26].

3. Design of Experiment (DOE) and Experimentation Setup

In this study, five important process parameters such as layer thickness (*A*), part orientation (*B*), raster angle (*C*), raster width (*D*), and air gap (*E*) have been considered to study their effects on tensile strength (TS). ABS material is used; that is, the material is constant. The temperature is also considered constant.

The definitions of FDM variable parameters in this study are as follows.

(*A*) The layer thickness which is recognized as the height of deposited slice from the FDM nozzle is shown in Figure 2. The layer thickness parameter is used to examine the influence of building thicker or thinner layers on the outcome quality.

(*B*) The orientation of part is defined as how the part should be positioned when produced as shown in Figure 3 (vertical position (a); inclined position (b); horizontal position (c)).

(*C*) Raster angle or orientation which is measured from the *x*-axis on the bottom part layer as shown in Figure 4. It also refers to the direction of the beads of material (roads) relative to the loading of the part. The deposited roads can be built at different angles to fill the interior part.

(*D*) The raster width or *road* width which refers to the width of the deposition path related to tip size. It also refers to the tool path width of the raster pattern used to fill interior regions of the part curves as shown in Figure 5. Narrow and wide filling pattern (roads) were considered to be examined.

(*E*) The air gap parameter which is defined as the space between the beads of deposited FDM material is shown in Figure 6. Hence, the influence of applying positive and negative gap between the deposited beads was investigated.

3.1. Experimental Procedure. The 3D models of specimen were modelled in SolidWorks and exported as STL file. The STL file is imported to FDM Insight software (Figure 7).

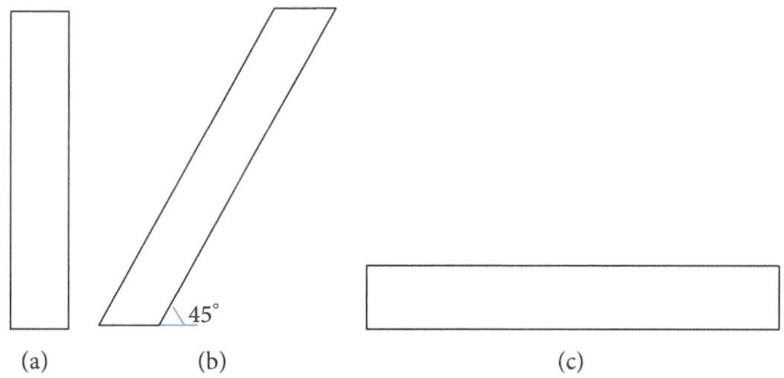

FIGURE 3: Orientation of part. Source: [11].

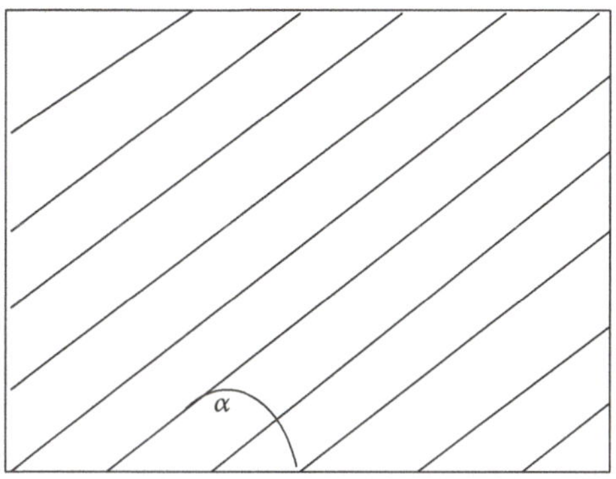

FIGURE 4: Raster angle parameter. Source: [11].

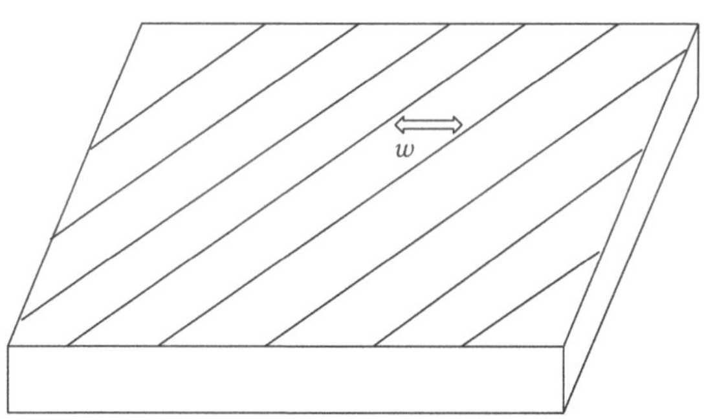

FIGURE 5: Raster width parameter. Source: [11].

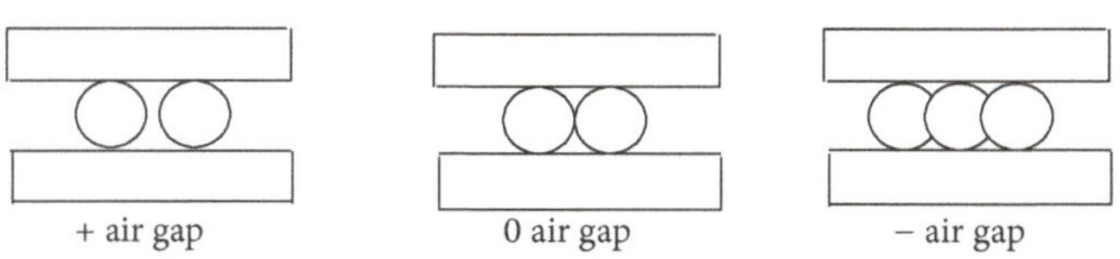

FIGURE 6: Air gap application. Source: [11].

Figure 8 reveals the internal structure of the model showing the raster width, raster angle, and air gap. This internal structure was obtained by slicing the model midway along its length for raster angle of 45°, raster width of 0.016″, and air gap of 0.02″ in order to clarify the concepts covered in the earlier part of this section. Understanding of these concepts is important because the design of experiments described in the later part of this section is based on the different settings

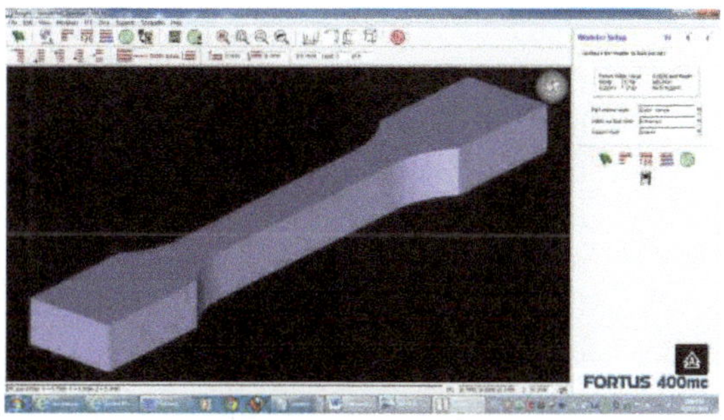

FIGURE 7: STL file of 3D model exported to FDM Insight software.

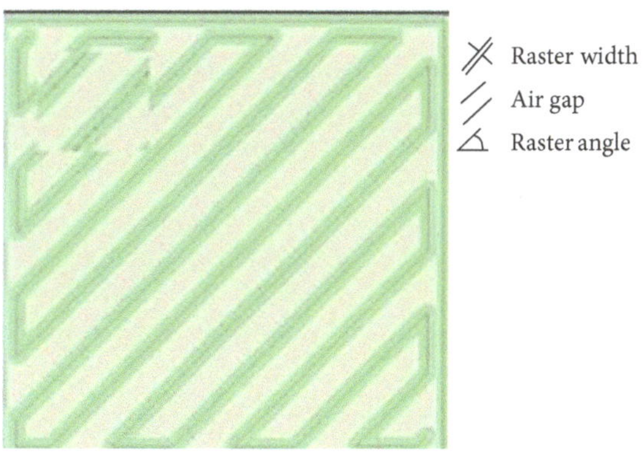

FIGURE 8: Internal structure of the model showing the raster width, raster angle, and air gap.

of the FDM parameters used for our experimentation, which are essentially raster angle, raster width, and air gap together with layer thickness and part orientation.

All testing specimens were constructed in a Stratasys FDM Fortus 400mc System (Figure 9) in the Advanced Manufacturing and Robotics laboratory at Sheridan Institute of Technology. The laboratory is equipped with both Fortus 400mc and 900mc Systems. The tensile test was performed using United Testing System (UTS), Model SSTM, Serial 1210555 with capacity of 20 kN (Figure 10) in accordance with ISO R527:1966 and ISO R178:1975, respectively.

3.2. Design of Experiment (DOE). In this work, factors as shown Table 1 are set as per experiment plan (Table 2)

Table 1: Variable process parameters and their selected low and high levels.

Variable parameter	Unit	Low level (−1)	High level (+)
Layer thickness (A)	in./mm	0.005″/(0.127)	0.013″/(0.3302)
Part orientation (B)	Degree	0°	90°
Raster angle (C)	Degree	0°	45°
Raster width (D)	in./mm	0.008″/(0.2032)	0.022″/(0.5588)
Air gap (E)	in./mm	−0.0001″/(−0.00254)	0.022″/(0.5588)

Table 2: Input process parameters that affect output responses.

Run	A [layer thickness]	B [part orientation]	C [raster angle]	D [raster width]	E [air gap]	Measured (MPa)
1	0.127	0	0	0.2032	−0.00254	**32.56**
2	0.127	0	0	0.2032	0.5588	**6.52**
3	0.127	0	0	0.5588	−0.00254	**24.81**
4	0.127	0	0	0.5588	0.5588	**10.44**
5	0.127	0	45	0.2032	−0.00254	**34.61**
6	0.127	0	45	0.2032	0.5588	**4.78**
7	0.127	0	45	0.5588	−0.00254	**30.15**
8	0.127	0	45	0.5588	0.5588	**8.08**
9	0.127	90	0	0.2032	−0.00254	**12.86**
10	0.127	90	0	0.2032	0.5588	**4.39**
11	0.127	90	0	0.5588	−0.00254	**23.55**
12	0.127	90	0	0.5588	0.5588	**9.56**
13	0.127	90	45	0.2032	−0.00254	**15.1**
14	0.127	90	45	0.2032	0.5588	**4.01**
15	0.127	90	45	0.5588	−0.00254	**24.3**
16	0.127	90	45	0.5588	0.5588	**8.09**
17	0.3302	0	0	0.4572	−0.00254	**31.46**
18	0.3302	0	0	0.4572	0.9652	**8.43**
19	0.3302	0	0	0.9652	−0.00254	**24.99**
20	0.3302	0	0	0.9652	0.9652	**9.39**
21	0.3302	0	45	0.4572	−0.00254	**31.42**
22	0.3302	0	45	0.4572	0.9652	**7.12**
23	0.3302	0	45	0.9652	−0.00254	**28.94**
24	0.3302	0	45	0.9652	0.9652	**11.07**
25	0.3302	90	0	0.4572	−0.00254	**20.55**
26	0.3302	90	0	0.4572	0.9652	**8.89**
27	0.3302	90	0	0.9652	−0.00254	**24.72**
28	0.3302	90	0	0.9652	0.9652	**11.28**
29	0.3302	90	45	0.4572	−0.00254	**24.96**
30	0.3302	90	45	0.4572	0.9652	**8.82**
31	0.3302	90	45	0.9652	−0.00254	**27.35**
32	0.3302	90	45	0.9652	0.9652	**11.63**

using design of experiment (DOE) methodology [19]. The engineering material used for test specimen fabrication is acrylonitrile butadiene styrene (ABS P400). The specimens are fabricated using FDM 400mc machine for respective strength measurement. The main FDM variable parameters are considered in this research in Table 1 to evaluate the correlation between these parameters and the proposed response characteristics. Tests on produced parts are carried out according to the American standard ASTM D638, at an ambient temperature of 23 ± 2°C and a relative humidity 50 ± 5%.

The rationale for considering the five variable process parameters for experimentation is given here. The layer thickness is known to affect the AM end-product because the smaller the layer thickness the stronger the finished AM part will be when subjected to axial load. Part orientation is important because when the part is built inclined, it will have the tendency to withstand greater loading in

Figure 9: FORTUS 400mc System used for the research.

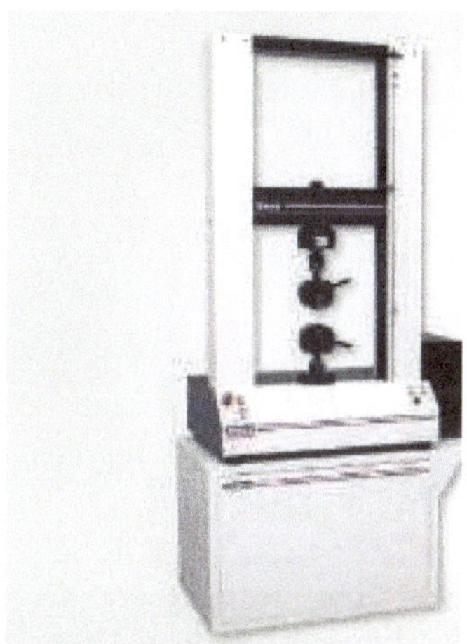

Figure 10: UTS System used for the research.

x-direction and y-direction. Raster angle will have the tendency to affect the internal structure of the finished product. Each neighbouring layer has a raster angle perpendicular to the immediate preceding layer. The raster width is known to affect the finished AM part in such a way that the larger the raster angle the greater the tendency for the finished part to withstand higher tensile stress. Negative air gap is known to produce finished AM part that withstands higher tensile stress. A number of researchers have considered negative air gap and reached these conclusions in their studies similar to our hypothesis. Bagsik and Schöppner [21] investigated the influence of the orientation and the air gap for the manufactured parts based on the mechanical data analysed. The outcome of the investigation was that best results were achieved for all directions by using a negative raster air gap, while with thick filaments better mechanical data was achieved for the x and z build direction, while a thinner filament improves the strength properties for y-specimen.

Using DOE approach the 32 full factorial conditions (2^5) were generated as shown in Table 2 for the experimental runs. For each run, three replications were made and averaged. Therefore, the reported measured values are averages of three readings per run. Each run in the design consists of combination of FDM parameters levels and each run result will contain response of tensile strength (UTS). As already

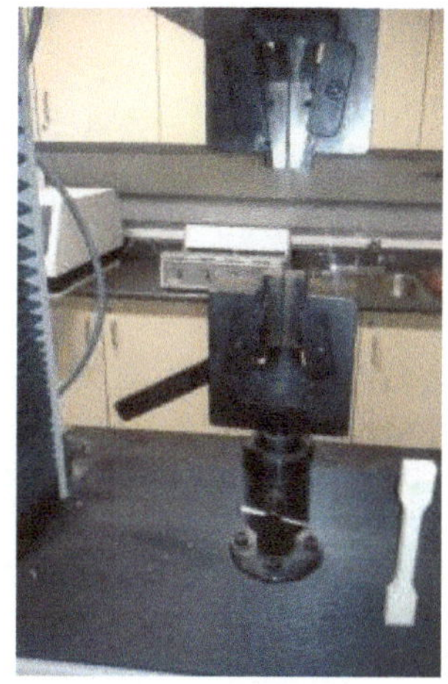

Figure 11: Tested tensile specimen (on the right side of UTS machine-bed).

discussed these parameters are clearly described in Figure 9. The parameters are varied to obtain the 32 different AM products. The values for layer thickness, raster width, and air gap given in Table 2 are in millimetre unit since the measured tensile strength (last column) is in MPa.

An image of tested tensile specimen on the table of the UTS machine is shown in Figure 11. The specimen shown is number 14 and this number is marked on the specimen to avoid being mixed up with other specimens. It is important that each specimen is carefully marked after being removed from the FDM machine so that they could be properly identified. Each specimen is 5″ (127 mm) long, cross-section of .3750″ × .3750″ (9.53 mm × 9.53 mm), while the middle having a width of .1875″ (4.76 mm). Fillet radius is .0900″ (2.2 mm).

4. Group Method for Data Handling (GMDH)

The framework for modeling chosen for this applied research is based on the Group Method for Data Handling (GMDH) introduced by Ivakhnenko (details are found in [18, 22, 23]) as a means of identifying nonlinear relations between input and output variables. The multilayered iteration (MIA) network is one of the variants of GMDH. The MIA relationship between the inputs and the output of a multiple inputs single output self-organizing network can be represented by an infinite Volterra-Kolmogorov-Gabor (VKG) polynomial of the form [18, 22, 23]

$$y_n = a_0 + \sum_{i=1}^{M} a_i x_i + \sum_{i=1}^{M}\sum_{j=1}^{M} a_{ij} x_i x_j + \sum_{i=1}^{M}\sum_{j=1}^{M}\sum_{k=1}^{M} a_{ijk} x_i x_j x_k \cdots, \tag{1}$$

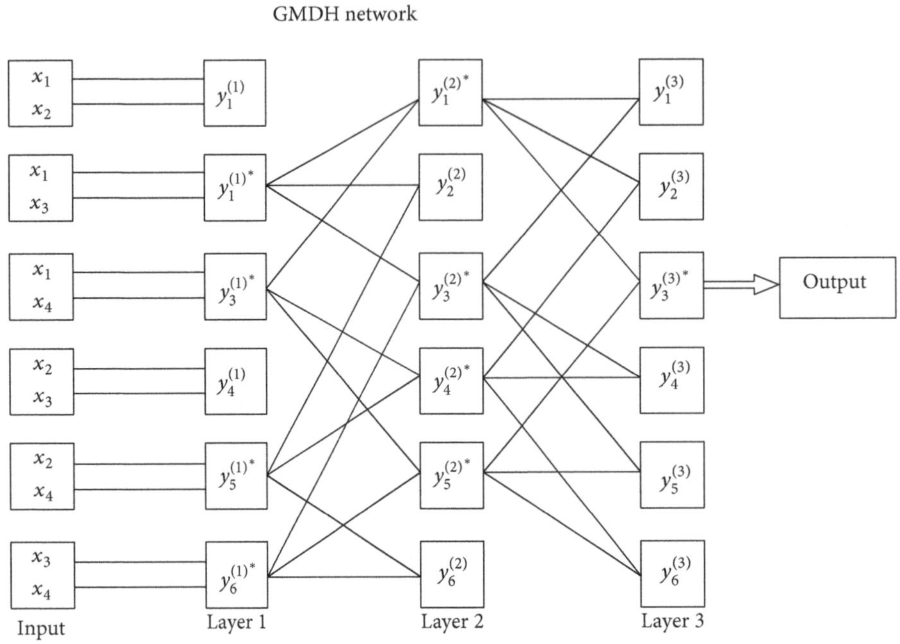

FIGURE 12: GMDH forward feed functional network.

where $X = (x_1, x_2, \ldots, x_M)$ is the vector of input variables and $A = (a_0, a_i, a_{ij}, a_{ijk}, \ldots)$ is the vector of coefficients or weights.

When the GMDH network is completed, there is a set of original inputs that filtered through the layers to the optimal output node. This is the computational network that is to be used in computing predictions (in our application, classifications are implied). The best nodes in the input layer (starred nodes in Figure 12) are retained and form the input to the next layer. The inputs for layer 1 are formed by taking all combinations of the surviving output approximations from the input layer nodes. It is seen that at each layer the order of the polynomial approximation is increased by two. Layer 2's best nodes for approximating the system output are retained and form layer 3 inputs. This process is repeated until the current layer's best approximation is inferior to the previous layer's best approximation.

4.1. Advantages of Basic GMDH Technique. The advantage of using pairs of input is that only six weights (coefficients) have to be computed for each neuron. The number of neurons in each layer increases approximately as the square of the number of inputs. During each training cycle, the synaptic weights of each neuron that minimize the error norm between predicted and measured values are computed and those branches that contribute least to the output of the neuron are discarded, the remaining branches being retained and their synaptic weights kept unchanged thereafter. A new layer is subsequently added and the procedure is repeated until the specified termination conditions are met.

There could be summarized that the GMDH-type polynomial networks influence the contemporary artificial neural network algorithms with several other advantages [23]:

(1) they offer adaptive network representations that can be tailored to the given task;

(2) they learn the weights rapidly in a single step by standard ordinary least square (OLS) fitting which eliminates the need to search for their values and which guarantees finding locally good weights due to the reliability of the fitting technique;

(3) these polynomial networks feature sparse connectivity which means that the best discovered networks can be trained fast.

4.2. Limitations of GMDH Technique. Although standard GMDH provides for a systematic procedure of system modeling and prediction, it has also a number of shortcomings. Anastasakis and Mort [24] have carried out a comprehensive study of the shortcomings of GMDH.

4.2.1. Selection of Input Arguments. One of the main features of GMDH is its ability to objectively select the most appropriate input arguments amongst a set of candidates. However, the identification of these candidate input arguments is not straightforward and may affect its performance.

4.2.2. Inaccuracies in Parameter Estimation. The method of least square estimates is the most popular method to calculate the coefficients of partial descriptions. If the data matrix is well defined its estimates will be accurate; however, in the majority of real world systems the data matrix is ill-defined and the least squares are biased.

4.2.3. Multicollinearity. Another problem found exclusively in multilayer algorithm, which affects the stability of coefficients, is that of multicollinearity.

4.2.4. Reduction of Complexity. Another shortcoming found GMDH approach is a tendency to generate quite complex polynomial (since the complexity of the network increases with each training and selection cycle through addition of new layers) for relatively simple systems (data input); also, an inclination to producing overly complex network (model) when dealing with highly nonlinear systems owing to its limited generic structure (quadratic two-variable polynomial).

4.2.5. Formulas of Partial Descriptions. Despite the wide range of partial descriptions majority of researchers follow the argument that Volterra series are capable of identifying any nonlinear system and therefore have adopted polynomial partial descriptions similar to Ivakhnenko polynomial. However, due to the complexity of the model and the requirement of including the theory behind the object, many modifications have been designed in order to adapt to system's properties.

4.2.6. Over Fitting. A consequence of complexity is the overfitting problem and poor generalization.

4.2.7. Partition of Data. The objectiveness of GMDH algorithm is based on the utilization of an external criterion to select the optimum model, which requires the partition of the data.

4.2.8. Low Accuracy in GMDH Method. In many cases and particularly in applications of long range prediction the GMDH has been observed to be inaccurate.

4.3. Hybrid-GMDH Network. Based on the shortcomings of the basic GMDH, hybrids of GMDH were proposed to significantly enhance the performance of GMDH [24]. For ease of reference and clarity to readers, the main design steps for the hybrid group method for data handling-genetic algorithm (GMDH-GA) based approach used for the work reported in this paper which applies to similar hybrid-types are summarized here (interested readers may refer to [25] for details on hybrid-GMDH).

Step 1 (configuration of input variable). The system input variables shown in Table 1 are the layer thickness (a), part orientation (b), raster angle (c), raster width (d), and air gap (e).

Step 2 (form training and testing data). Three-quarters of the dataset was used in designing the training data, while the remaining one-quarter was used in designing the testing data.

Step 3 (decision of initial information for constructing the hybrid GMDH-GA structure). The number of generations, population size, crossover rate, and mutation rate were chosen as 25, 50, 0.9, and 0.1, respectively.

Step 4 (determine polynomial neuron (PN) structure). The vector number of input variables of two, the polynomial order of Type 2 (quadratic), and the input variables were assigned to each node of the corresponding layer.

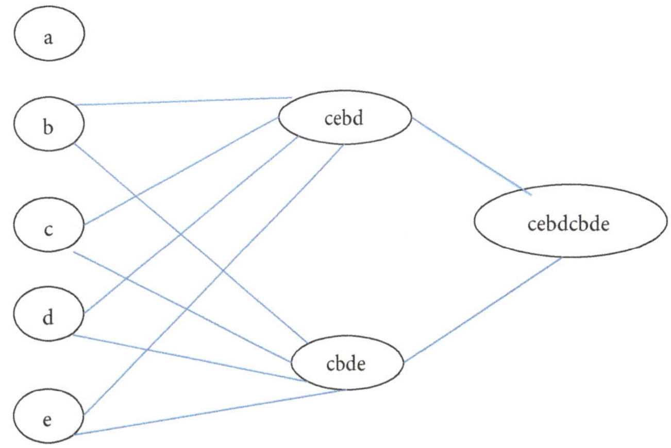

FIGURE 13: Hybrid GMDH-GA structural network for 5-input process parameters.

Step 5 (parametric optimization). Coefficient estimation of the polynomial corresponding to the selected node (PN): the vector of the coefficients of the partial descriptors (PDs) is determined using a standard mean square error for the training dataset subsets.

Step 6 (structural optimization). Select nodes (PNs) with the best predictive capability, and construct their corresponding layer: all nodes of the corresponding layer of the hybrid GMDH-GA architecture are constructed by optimization (see Figure 13).

Step 7 (termination criterion). After the iteration process, the final generation of population consists of highly fit solution population that provides optimum solutions.

5. GMDH-Based Predictive Model of Tensile Strength

The process parameters A, B, C, D, and E and tensile response of Table 2 were submitted to the MATLAB coded hybrid GMDH-GA software system to develop predictive model that relates the tensile strength to the process parameters. The last column of Table 3 shows the predicted values obtained from the GMDH modelling system, while the second to the last column shows the experimental values. Figure 14 shows the experimental (measured) and predicted values in a graphical form for the thirty two runs. As could be observed our GMDH modelling system predicts very well the behaviour of the tensile strength response with very little deviation.

The GMDH model for the part produced by the Stratasys FDM 400mc System based on Table 2 is

$$\begin{aligned} y_{\text{UTS}} = {} & 15.3577 + 54.622x_1 + 0x_2 + 0x_3 + 20.4923x_4 \\ & - 47.627x_5 - 0.1122x_1x_2 + 0.0774x_1x_3 \\ & - 37.5299x_1x_4 + 18.6233x_1x_5 - 0.0001x_2x_3 \\ & + 0.0996x_2x_4 + 0.0801x_3x_5 + 0.0066x_3x_4 \end{aligned}$$

TABLE 3: Input process parameters that affect output responses.

Run	A [layer thickness]	B [part orientation]	C [raster angle]	D [raster width]	E [air gap]	Measured (MPa)	Predicted (MPa)
1	0.127	0	0	0.2032	−0.00254	**32.56**	27.76
2	0.127	0	0	0.2032	0.5588	**6.52**	4.74
3	0.127	0	0	0.5588	−0.00254	**24.81**	28.14
4	0.127	0	0	0.5588	0.5588	**10.44**	8.58
5	0.127	0	45	0.2032	−0.00254	**34.61**	30.96
6	0.127	0	45	0.2032	0.5588	**4.78**	4.39
7	0.127	0	45	0.5588	−0.00254	**30.15**	31.96
8	0.127	0	45	0.5588	0.5588	**8.08**	8.84
9	0.127	90	0	0.2032	−0.00254	**12.86**	12.60
10	0.127	90	0	0.2032	0.5588	**4.39**	5.10
11	0.127	90	0	0.5588	−0.00254	**23.55**	23.89
12	0.127	90	0	0.5588	0.5588	**9.56**	8.58
13	0.127	90	45	0.2032	−0.00254	**15.1**	14.33
14	0.127	90	45	0.2032	0.5588	**4.01**	4.96
15	0.127	90	45	0.5588	−0.00254	**24.3**	26.67
16	0.127	90	45	0.5588	0.5588	**8.09**	8.78
17	0.3302	0	0	0.4572	−0.00254	**31.46**	28.32
18	0.3302	0	0	0.4572	0.9652	**8.43**	9.19
19	0.3302	0	0	0.9652	−0.00254	**24.99**	25.85
20	0.3302	0	0	0.9652	0.9652	**9.39**	11.35
21	0.3302	0	45	0.4572	−0.00254	**31.42**	32.05
22	0.3302	0	45	0.4572	0.9652	**7.12**	9.06
23	0.3302	0	45	0.9652	−0.00254	**28.94**	29.42
24	0.3302	0	45	0.9652	0.9652	**11.07**	10.96
25	0.3302	90	0	0.4572	−0.00254	**20.55**	22.30
26	0.3302	90	0	0.4572	0.9652	**8.89**	8.64
27	0.3302	90	0	0.9652	−0.00254	**24.72**	25.14
28	0.3302	90	0	0.9652	0.9652	**11.28**	10.53
29	0.3302	90	45	0.4572	−0.00254	**24.96**	24.92
30	0.3302	90	45	0.4572	0.9652	**8.82**	8.58
31	0.3302	90	45	0.9652	−0.00254	**27.35**	27.85
32	0.3302	90	45	0.9652	0.9652	**11.63**	10.34

$$- 0.0899x_3x_5 - 2.7021x_4x_5 - 60.0811x_1^2 - 0.0011x_2^2 + 0.0012x_3^2 - 6.874x_4^2 + 24.7908x_5^2, \quad (2)$$

where $x_1 = A$ (layer thickness); $x_2 = B$ (part orientation); $x_3 = C$ (raster angle); $x_4 = D$ (raster width); $x_5 = E$ (air gap).

In order to assess the quality of the model realized for the work reported in this paper, some statistical data are given. The sum of square residual, SS: residual: $J = 0.4525$; the sum of square deviation, SS: deviation from mean: $S = 9.6918e + 03$; hence the coefficient of determination given as $1 - J/S$ is given as r-squared value: $r^2 = 0.99995$. The regression sum of square, SS_regression = 0.00249; the error sum of square, SS_error = $6.1089e - 27$; the mean square regression, MSE_regression = $4.9999e - 04$; the mean square error, MSE_error = $2.1817e - 28$; the F-value = $2.2917e + 24$.

Figure 15 shows the surface plot of the process parameters for the runs 1–4 while Figure 16 shows the surface plot of the process parameters for the runs 1–32, confirming that the modelling search space is highly complex.

5.1. Examining the Breakage Modes of Specimens. Specimens 13 and 29 (left-most); 3 and 19 (middle); 5 and 21 (right-most) in Figures 15–17 were examined. It is noticed that each pack has three specimens used for experimentation.

The slice height of number 13 is 0.005″ (0.127 mm) while that of number 29 is 0.139″ (0.3302 mm), both printed with their longitudinal axis vertical; the raster angle for both is 45° while the raster widths are 0.008″ (0.2032 mm) and 0.018″ (0.4572 mm), respectively. The air gap is the same for both specimens (−0.001″ or −0.00254 mm). The broken areas confirm the internal structures represented in Table 3.

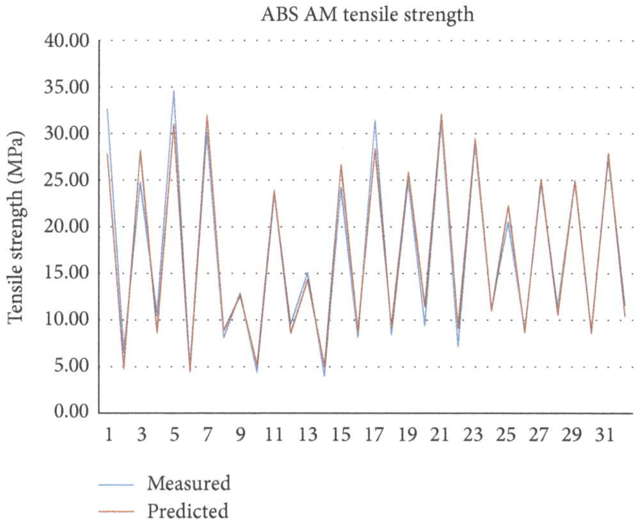

FIGURE 14: Measured and predicted tensile strength response.

FIGURE 15: Specimens 13 and 29 (left-most); 3 and 19 (middle); 5 and 21 (right-most).

Breakage took place virtually perpendicular to the direction of application of load.

The slice height of number 3 is 0.005″ (0.127 mm) while that of number 19 is 0.139″ (0.3302 mm), both printed with their longitudinal axis vertical; the raster angle for both is 0° while the raster widths are 0.022″ (0.5588 mm) and 0.038″ (0.965 mm), respectively. The air gap is the same for both specimens (−0.001″ or −0.00254 mm). The broken areas confirm the internal structures represented in Table 3. Breakage took place virtually perpendicular to the direction of application of load, in a very neat manner.

The parameters for numbers 5 and 21 are, respectively, the same as those for numbers 13 and 29, respectively, except that they are printed with the longitudinal axis horizontal. Breakage surface for number 5 is uneven (in a ridge form); this specimen has the largest tensile strength for all the specimens tested. Specimen number 21 shows slight uneven breakage surface.

6. Optimization of Process Parameters for Tensile Strength Prediction

6.1. Mathematical Formulation of the Tensile Strength Problem. The hybrid GMDH tensile strength model developed was utilized by the continuous DE for optimization in order to determine the optimal combinations of layer thickness (A), part orientation (B), raster angle (C), raster width (D), and air gap (E) that result in maximizing tensile strength for the FDM

FIGURE 16: Specimens as in Figure 15 but with more exposure of broken surfaces.

FIGURE 17: Specimens as in Figure 15 but with most exposure of broken surfaces.

part. The optimization problem is essentially that of using (2), as objective function with constraints taken from Table 2.

The tensile strength optimization problem can now be fully mathematically stated as follows:

$$\begin{aligned} y_{\mathrm{UTS}} = {} & 15.3577 + 54.622x_1 + 0x_2 + 0x_3 \\ & + 20.4923x_4 - 47.627x_5 - 0.1122x_1x_2 \\ & + 0.0774x_1x_3 - 37.5299x_1x_4 + 18.6233x_1x_5 \\ & - 0.0001x_2x_3 + 0.0996x_2x_4 + 0.0801x_3x_5 \\ & + 0.0066x_3x_4 - 0.0899x_3x_5 - 2.7021x_4x_5 \\ & - 60.0811x_1^2 - 0.0011x_2^2 + 0.0012x_3^2 \\ & - 6.874x_4^2 + 24.7908x_5^2 \end{aligned} \tag{3}$$

$$\begin{aligned} \text{s.t. } & 0.127 \le x_1 \le 0.3302 \\ & 0 \le x_2 \le 90 \\ & 0 \le x_3 \le 45 \\ & 02034 \le x_4 \le 0.9652 \\ & -0.00254 \le x_5 \le 0.9652. \end{aligned} \tag{4}$$

6.2. Differential Evolution Scheme. The differential evolution (DE) algorithm introduced by Storn and Price [20] is a novel parallel direct search method, which utilizes Np parameter vectors as a population for each generation G. DE is one of the extant evolutionary approaches used to solve complex real-life problems. It was primarily designed for continuous domain space formulation but was reformulated to solve permutative problems by Donald and Onwubolu [27]. The steps involved in the classical DE are summarized as follows:

Step 1: initialization,

Step 2: mutation,

Step 3: crossover,

Table 4: DE control parameters used for experimentation.

Population size, NP	50
Number of parameter	5
Mutation probability, F	0.20
Crossover probability, CR	0.60
Number of generations	500

Table 5: Optimal cutting parameters from DE.

Tensile strength	MPa	32.76
Layer thickness (A)	in./mm	0.01″/0.2778
Part orientation (B)	Degree	9.05°
Raster angle (C)	Degree	45°
Raster width (D)	in./mm	0.022″/(0.5588)
Air gap (E)	in./mm	0.0001″/(0.0025)

Step 4: selection,

Step 5: stopping criteria.

Using DE, which is one of the extant evolutionary approaches, the optimal parameter settings for the FDM tensile test were found. DE was used for optimizing the tensile strength optimization problem expressed in (3) and (4). The control conditions used for optimization are shown in Table 4. The approach of using the hybrid GMDH model (3), as objective functions with constraints given in (4) for optimizing the tensile strength problem, is more straightforward than when ANN is employed in modeling. This type of mathematical formulation makes the GMDH response models to be more useful to the end-user since the models for the problem being solved are transparent and could be used for future applications. Moreover, the mathematical models are easy to be used as the objective functions by most standard optimization techniques for determining optimal cutting and response conditions.

For the experimentation, the optimal process parameters and tensile strength optimization problem that the DE found are given in Table 5.

7. Results and Discussions

The results are discussed under specific and general cases. The specific discussion of the results is based on how tensile strength relates with the process parameters of layer thickness, part orientation, raster angle, raster width, and air gap. The general discussions relate to the conditions under which the weakest and strongest AM part is produced.

Effect of Layer Thickness on Tensile Strength. The highest tensile strength occurs for the minimum layer thickness of 0.005″/(0.127 mm) on run 5 (34.61 MPa). The next highest value also occurs for the minimum layer thickness of 0.005″/(0.127 mm) in run 1 (32.56 MPa). These values of tensile strength are greater than the highest tensile strength obtained for the maximum layer thickness of 0.013″/(0.3302). Consequently, we conclude that the maximum tensile strength occurs when the layer thickness is a minimum.

Effect of Part Orientation on Tensile Strength. Comparing the tensile strength values for run 1 (32.56 MPa) and run 9 (12.86 MPa), respectively, run 2 (6.52 MPa) and run 10 (4.39 MPa), respectively, and so forth, it is observed that part orientation significantly affects the tensile strength of AM parts. The zero angle part orientation is the strongest because the layers are parallel to the direction in which load is applied. Parts that have 90° orientation during production are the weakest.

Effect of Raster Angle on Tensile Strength. In this case, we compare the strength values for run 1 (32.56 MPa) and run 5 (34.61 MPa), respectively, run 2 (6.52 MPa) and run 6 (4.78 MPa), and so forth. This trend is repeated throughout the entire measured results. Therefore, it is concluded that tensile strength increases with increase in raster angle; howbeit, the increase is not of much magnitude.

Effect of Raster Width on Tensile Strength. The effect of raster width on tensile strength is studied by comparing cases in which only raster width changes. In this case we compare run 1 (32.56 MPa) and run 3 (24.81 MPa), respectively, run 5 (34.61 MPa) and run 5 (34.61 MPa), respectively,. . ., run 29 (24.96 MPa) and run 31 (27.35 MPa), respectively, and run 30 (8.82 MPa) and run 32 (11.63 MPa), respectively, and so forth. From the observations, we conclude that the tensile strength increases with minimum raster width.

Effect of Air Gap on Tensile Strength. To study the effect of air gap on tensile strength, two consecutive runs are compared: run 1 (32.56 MPa) and run 2 (6.52 MPa), respectively, run 3 (24.81 MPa) and run 4 (10.44 MPa), and so forth. It is observed that negative air gap results in the strongest AM part.

In general the following observations are made.

(1) The weakest AM parts are produced when air gap is maximum (runs 2, 6, 10, 14, and 16 for layer thickness of 0.005″/(0.127) and runs 18, 20, 22, 26, and 30 for layer thickness of 0.013″/(0.3302))

(2) The strongest AM parts are produced with negative air gap (NAG) (runs 1, 3, 5, and 7 for layer thickness of 0.005″/(0.127) and runs 17, 19, 21, 23, 27, 29, and 31 for layer thickness of 0.013″/(0.3302))

However, from our experimentation, for the maximum raster width of 0.022, an air gap of −0.001 gave very poor quality product, with “over-fill” in which the printed material spilled beyond the expected positions. The air gap was reduced to −0.0001 to give good quality results. To get good results, the percentage of the air gap to the maximum raster width should be 0.45% ((0.0001/0.022) ∗ 100).

8. Conclusions

In this work, functional relationship between process parameters and tensile strength for FDM process has been developed using group method for data modelling for prediction purpose. An initial test was carried out to determine whether part orientation and raster angle variations affect the tensile strength. It was found that both process parameters affect tensile strength response. For the more elaborate experimentation, the process parameters considered are layer thickness, orientation, raster angle, raster width, and air gap. The process parameters and the experimental results were submitted to a hybrid GMDH-GA system, resulting in predicted output, in which the predicted output values were found to correlate very closely with the measured values.

Since FDM process is a complex one, it is really challenging to determine good functional relationship between responses and process parameters. Using differential evolution (DE), which is one of the extant evolutionary approaches, optimal parameter settings were found. Our investigations have shown the following.

(1) Minimum layer thickness improves tensile strength, although is more costly due to more material usage for manufacturing parts.

(2) Negative air gap significantly improves the tensile strength.

(3) Minimum raster widths also improve tensile strength.

(4) Part orientation plays a major role as could be observed from the results. For zero part orientation (with the part orientation coinciding with the direction of tensile loading), maximum tensile strength is obtained.

(5) Increased raster angle also improves tensile strength, although not very significantly. The optimized solutions that DE found agree very reasonably with our observations from Table 2.

(6) Our experimental results in Table 2 show that maximum tensile strength is obtained in run 5 for which the part orientation is zero and raster angle and raster width are maximum, with negative air gap. DE optimal solutions match these observations, with part orientation being 9.05°.

Consequently, the conclusions reached in this research are reliable and can be applied to real life applications. Future work will include the investigation of other mechanical properties such as compressive strength and torsional strength. Microstructure investigation of broken parts during testing will be done using high powered microscope.

Nomenclature

x_1: A (layer thickness)
x_2: B (part orientation)
x_3: C (raster angle)
x_4: D (raster width)
x_5: E (air gap).

Conflict of Interests

The authors declare that there is no conflict of interests regarding the publication of this paper.

Acknowledgments

The two technologists Attila Nagy and Andrew Orton in the Sheridan Centre for Advanced Manufacturing & Design Technology (CAMDT) assisted the authors with the AM lab procedures, and they are gratefully appreciated.

References

[1] I. Gibson, D. W. Rosen, and B. Stucker, *Additive Manufacturing Technologies: Rapid Prototyping to Direct Digital Manufacturing*, Springer, Heidelberg, Germany, 2010.

[2] N. Hopkinson, R. J. M. Hagur, and P. H. Dickens, *Rapid Manufacturing: An Industrial Revolution for the Digital Age*, John Wiley & Sons, London, UK, 2006.

[3] B. Wiedemann and H.-A. Jantzen, "Strategies and applications for rapid product and process development in Daimler-Benz AG," *Computers in Industry*, vol. 39, no. 1, pp. 11–25, 1999.

[4] S. Upcraft and R. Fletcher, "The rapid prototyping technologies," *Assembly Automation*, vol. 23, no. 4, pp. 318–330, 2003.

[5] S. Mansour and R. Hague, "Impact of rapid manufacturing on design for manufacture for injection moulding," *Proceedings of the Institution of Mechanical Engineers, Part B: Journal of Engineering Manufacture*, vol. 217, no. 4, pp. 453–461, 2003.

[6] G. N. Levy, R. Schindel, and J. P. Kruth, "Rapid manufacturing and rapid tooling with layer manufacturing (LM) technologies, state of the art and future perspectives," *CIRP Annals: Manufacturing Technology*, vol. 52, no. 2, pp. 589–609, 2003.

[7] O. Ivanova, C. Williams, and T. Campbell, "Additive manufacturing (AM) and nanotechnology: promises and challenges," *Rapid Prototyping Journal*, vol. 19, no. 5, pp. 353–364, 2013.

[8] P. K. Jain, P. M. Pandey, and P. V. M. Rao, "Effect of delay time on part strength in selective laser sintering," *International Journal of Advanced Manufacturing Technology*, vol. 43, no. 1-2, pp. 117–126, 2009.

[9] K. Chockalingam, N. Jawahar, and U. Chandrasekhar, "Influence of layer thickness on mechanical properties in stereolithography," *Rapid Prototyping Journal*, vol. 12, no. 2, pp. 106–113, 2006.

[10] J. F. Rodríguez, J. P. Thomas, and J. E. Renaud, "Mechanical behavior of acrylonitrile butadiene styrene (ABS) fused deposition materials. Experimental investigation," *Rapid Prototyping Journal*, vol. 7, no. 3, pp. 148–158, 2001.

[11] S.-H. Ahn, M. Montero, D. Odell, S. Roundy, and P. K. Wright, "Anisotropic material properties of fused deposition modeling ABS," *Rapid Prototyping Journal*, vol. 8, no. 4, pp. 248–257, 2002.

[12] A. K. Sood, R. K. Ohdar, and S. S. Mahapatra, "Parametric appraisal of mechanical property of fused deposition modelling processed parts," *Materials and Design*, vol. 31, no. 1, pp. 287–295, 2010.

[13] A. K. Sood, R. K. Ohdar, and S. S. Mahapatra, "Experimental investigation and empirical modelling of FDM process for compressive strength improvement," *Journal of Advanced Research*, vol. 3, no. 1, pp. 81–90, 2012.

[14] B. H. Lee, J. Abdullah, and Z. A. Khan, "Optimization of rapid prototyping parameters for production of flexible ABS object,"

Journal of Materials Processing Technology, vol. 169, no. 1, pp. 54–61, 2005.

[15] C. S. Lee, S. G. Kim, H. J. Kim, and S. H. Ahn, "Measurement of anisotropic compressive strength of rapid prototyping parts," *Journal of Materials Processing Technology*, vol. 187-188, pp. 627–630, 2007.

[16] S. K. Panda, S. Padheee, A. K. Sood, and S. S. Mahapatra, "Optimization of fused deposition modelling (FDM) process parameters using bacterial foraging technique," *Intelligent Information Management*, vol. 1, pp. 89–97, 2009.

[17] K. Thrimurthulu, P. M. Pandey, and N. V. Reddy, "Optimum part deposition orientation in fused deposition modeling," *International Journal of Machine Tools and Manufacture*, vol. 44, no. 6, pp. 585–594, 2004.

[18] H. R. Madala and A. G. Ivakhnenko, *Inductive Learning Algorithms for Complex Systems Modelling*, CRC Press, Boca Raton, Fla, USA, 1994.

[19] D. C. Montgomery, *Design and Analysis of Experiments*, John Wiley & Sons, Singapore, 5th edition, 2003.

[20] R. Storn and K. Price, "Differential evolution—a simple and efficient heuristic for global optimization over continuous spaces," *Journal of Global Optimization*, vol. 11, no. 4, pp. 341–359, 1997.

[21] A. Bagsik and V. Schöppner, "Mechanical properties of fused deposition modeling parts manufactured with Ultem∗9085," in *Proceedings of the 69th Annual Technical Conference of the Society of Plastics Engineers (ANTEC '11)*, pp. 1294–1298, Boston, Mass, USA, May 2011.

[22] A. G. Ivakhnenko, "Polynomial theory of complex systems," *IEEE Transactions on Systems, Man and Cybernetics*, vol. SMC-1, no. 4, pp. 364–378, 1971.

[23] N. Y. Nikolaev and H. Iba, "Polynomial harmonic GMDH learning networks for time series modeling," *Neural Networks*, vol. 16, no. 10, pp. 1527–1540, 2003.

[24] L. Anastasakis and N. Mort, "The development of self-organization techniques in modelling: a review of the group method of data handling (GMDH)," Research Report 813, Department of Automatic Control & Systems Engineering, The University of Sheffield, Sheffield, UK, 2001.

[25] G. C. Onwubolu, Ed., *Onwubolu Hybrid Self-Organizing Modeling Systems*, Springer, 2009.

[26] G. C. Onwubolu, "Selection of drilling operations parameters for optimal tool loading using integrated response surface methodology: a Tribes approach," *International Journal of Production Research*, vol. 44, no. 5, pp. 959–980, 2006.

[27] G. C. Onwubolu and D. Davendra, Eds., *Differential Evolution: A Handbook for Global Permutation-Based Combinatorial Optimization*, Studies in Computational Intelligence, Springer, New York, NY, USA, 2009.

6

Implementation of Computer Aided Engineering for Francis Turbine Development in Nepal

Ravi Koirala, Sailesh Chitrakar, Amod Panthee, Hari Prasad Neopane, and Bhola Thapa

Turbine Testing Lab, Kathmandu University, Dhulikhel 45200, Nepal

Correspondence should be addressed to Ravi Koirala; ravikoirala@ku.edu.np

Academic Editor: Fu-Shiung Hsieh

The expansion of the existing industries involved in the production of components of hydropower to the Francis turbine manufacturer up to 5 MW unit size has been recognized as one of the most promising business models in Nepal. Given the current fact that the development of Francis turbines with the manufacturers of Nepal has not been done yet, due to lack of designing expertise and limitations in the available technology, this paper presents the use of different available manufacturing technologies, which is suitable in the Nepalese hydropower market. This is an experience based paper, in which the advanced manufacturing process implementing Computer Aided Simulation (CAS), Computer Aided Design (CAD), and Computer Aided Manufacturing (CAM) is introduced for turbine manufacturing. Moreover, CAD from Solidworks, 3D printing from Rapid Prototyping Machine (RPM), and manufacturing of three designs by three different methods, dye casting, lost wax casting, and forging in a local workshop, have been described. The outcome of this work is the identification of suitable Francis turbine development methodologies in context of Nepal, incorporating industrial revolution through research based products.

1. Introduction

Nepal is considered as a country having tremendous water resources. Despite the huge prospects of hydropower development, the past 100 years has shown a very poor track record in this country, with only about 2% of the total feasible potential harnessed so far. With the growing energy demand escalating worldwide, most of the hydropower development projects are shifting towards the unexplored regions of Asia and South America. Apart from the economic and political instabilities, the countries in these regions bear their own specific technical challenges with the prominent sediment erosion of turbine components.

Nepal has an inevitable problem of erosion wear of turbine parts due to abrasive sediments present in rivers, with high amount, particularly in Monson seasons. This has created several challenges in operation and maintenance of turbines causing financial losses [1]. Nepal also lacks competence to design and manufacture electromechanical components required for hydropower projects. In particular, Nepal does not have any experience of turbine manufacturing above 100 kW size. Both developers and manufacturers have identified need of local competence to design and fabricate turbines for future hydropower development in Nepal.

It has been forecasted that the demand of electrical energy in Nepal will reach 3,600 MW by 2027. Most of the turbines installed in Nepalese hydropower projects are imported from European manufacturers. Some of the power plants have also started using Korean and Chinese runners due to financial benefits, especially in small power plants. There are around 27 manufacturing companies developing micro hydro turbines (less than 100 kW size) in Nepal, but with competences only in the area of Pelton and Crossflow turbines. However, according to a survey [2], in Nepal, more than 50% of turbines under construction fall under the unit size of 1–5 MW, majority of which falls under low speed number (0.2–0.3) type of Francis turbines, which are compact and highly efficient types. In coming years, Nepal will also need such turbines with unit size up to 25 MW in significant number.

With the advancement in the manufacturing industries and technologies in the field of hydropower that has been observed, growing rapidly over the past few decades, Nepal has to catch up the pace by implementing technologies relevant to its economic condition. This paper has sought

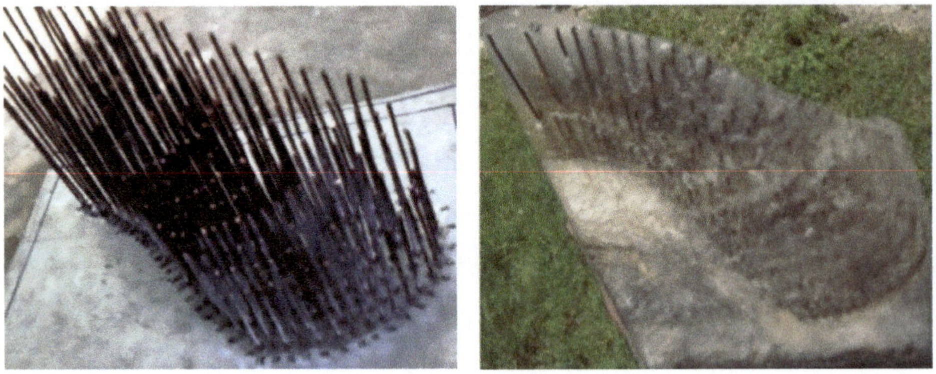

Figure 1: Wax pattern for visualization of Francis blade.

to discuss some recent endeavors carried out in design and manufacture of Francis turbines blending some state-of-the-art technologies with traditional approaches as a sensible path to move ahead.

2. Worldwide Evolution of Francis Turbine Design and Manufacturing

Experiments on the mechanics of reaction wheels were initially commenced by the Swiss mathematician Leonhard Euler and his son Albert in the 1750s. Later in 1826 Jean Victor Poncelet of Francis proposed the idea of an inward flowing radial turbine, the direct precursor of the modern water turbine. This machine has a vertical spindle and a runner with curved blades that was fully enclosed. A similar machine was patented in 1838 by Samuel B. Howd; this design was later improved by James B. Francis in 1849 by adding stationary guide vanes and shaping the blades so that water could enter at the correct angle. This design was named as Francis water turbine [3].

In the initial phase, Francis turbines were designed based upon specific requirements with no intense research related to the change in the blade shapes due to varying operating conditions. Gradually, the need of optimizing the efficiency with the working condition was felt, which resulted in implementing mathematical relations and numerical calculations in an iterative loop so as to obtain satisfactory values by changing the variables within the allowable range. However, during those times, the testing of the designs could not be done in virtual environment, nor could they be visualized three-dimensionally. Hence, the analytical methods could only be complemented through manufacturing. The turbine manufacturing trend has shifted from riveted plates and casting to fully automated manufacturing [4].

The manufacturing was performed in the form of single casting or part casting and assembly by unitary process. They were made of iron, steel/bronze. Most common methodology was to develop partly by sheet metal and partly by cast metal and unite them. C* Blade S.p.a. Forging and Manufacturing, an Italian company from 1963, implemented forging technology during its starting phase for power turbine manufacturing [5]. This forging technology is a component of the part casting where blades are forged whereas hub and shroud are casted and assembled together [6]. Figure 1 shows a wax model developed from radial and axial view coordinates of blade to create pattern for profile. Although this method is a type of traditional Prototyping Technology by using wire supports, it gave a useful understanding of how the relative velocity distribution in the blades results in the complex profiles along the stream-wise direction.

The traditional technologies had obvious limitations in terms of time consumption, accuracy, efficiency, and problems related to cavitation and vibrations. Hydraulic turbines and, especially, Francis turbines are considered to be tailor-made designs, where the profile of the runner blades, guide vanes, and stay vanes changes with the hydraulic design parameters. Hence, the design and manufacturing process was needed to be integrated with modern and suitable technologies for quality and time relevant designs.

Current design trend implements advance and more reliable process for turbine design and development technology. The modern trend used CAD/CAM/CAE integrated environment to increase the blade quality and decrease the time consumed for the design and the manufacturing processes. The use of CAD also enhances the ability of an organization to work in competitive environment to design site specific turbine [7]. There are various processes involved in blade manufacturing, including the development of solid model, testing, and optimization using numerical techniques and experimentations [8]. In order to carry out the manufacturing, a complex five-axis milling machine is used to develop the surface profile of the runner with high precision [9]. Figure 2 shows turbine manufactured by this process and Francis turbine of 700 MW installed in Three Gorges Power Plant (22.5 GW). These pictures show that the world has exhilarated rapidly in terms of manufacturing turbines by introducing new technologies within the past few decades.

3. Possibilities of Turbine Design and Manufacturing in Nepal

The feasibility study of turbine manufacturing in Nepal has shown that Nepalese manufacturer has an amiable future in the field of Francis turbine manufacturing [2]. At present,

Figure 2: Francis turbine manufacturing with new technologies [23] and the turbine in Three Gorges Power Plant of China [24].

(a) Turning turbine component

(b) Casting turbine blade

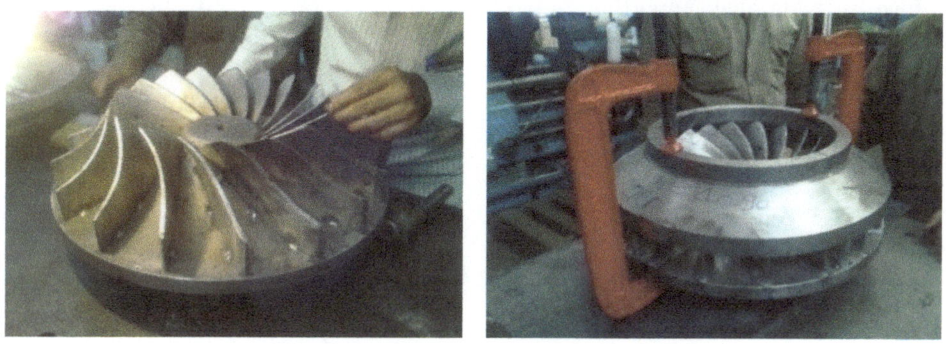

(c) Assembling runner hub, blade, and shroud together

Figure 3: 92 kW Francis turbines manufactured at NHE, Nepal.

the Francis turbines installed in the power plants of Nepal are being imported from manufacturers abroad. It is understandable that the turbines in the large hydropower plants need a certified company, equipped with design expertise and technological advancement. Although Nepal has around 27 manufacturing companies working in the field of hydropower sector and some of them having the capacity and motivation to manufacture turbines up to 5 MW, they are quite reluctant to bear the risk of Francis turbine manufacturing due to lack of competences.

At this moment, Nepal has developed sufficient background and technical competency to start up turbine manufacturing business. Kathmandu University has been putting its effort into development of hydro turbines in Nepalese context. Research for hydro turbines at KU had started in the late 90s with two miniature turbine laboratories named as Pico Turbine Laboratory and Waterpower Laboratory. Later in 2009, a new Turbine Testing Laboratory (TTL) was constructed with financial support from Norway. This lab possesses a long term goal of creating an excellence for turbine design, testing, and manufacturing Francis turbines in Nepal. It has performed several research based studies addressing the challenge faced by Nepalese hydropower plants, especially due to sediment erosion. Most of the computer aided technologies described in this paper was commenced from the research activities carried out in the lab, mostly in collaboration with Waterpower Laboratory at NTNU, which has experience of turbine testing for almost 100 years.

Some of the manufacturing experiences in Nepal have also shown that some manufacturing companies in Nepal have good competences in the field of Francis turbine manufacturing. Figure 3 shows processes involved in manufacturing a 92 kW Francis turbine at Nepal Hydro and Electric Company Pvt. Ltd. (NHE). The design of the turbine was carried out at TTL, with the motive of comparing the performance of conventional design with the optimized design, in terms of efficiency and erosion through testing. Although it was for the purpose of testing only, this turbine happens to be the first Francis turbine that was designed and manufactured in Nepal.

This initiation of turbine manufacturing has led to a need of endeavors for expanding the manufacturing arena of Nepal to another level. The processes used in manufacturing the turbine in the figure are discussed in latter sections of this paper.

4. Design and Computer Aided Simulation

4.1. Hydraulic Design. Turbine Testing Lab, since its earliest days of concept formulation, has good relationship with Water Power Laboratory, Norwegian University of Science and Technology regarding the technology and faculty exchange. Particularly, in the field of the design of Francis turbines, there have been close cooperation and activities between these partner institutions. With a combined effort of academic exchange program, the institutions were able to design a MATLAB based code called *Khoj* [10]. The prime design stage in Francis turbine is the iterative simulation for the suitable result. The head and flow are the two major design parameters which use additional supporting design variables like tangential component of velocities and their angles. Manual iteration for the suitable design outcome is complicated; hence it requires integration with state-of-the-art programming. In this case MATLAB based GUI system is developed for hydraulic design of Francis turbine. This step reduces the bulky iteration stage, since all the calculations will be performed with the selection of iteration driver.

4.2. Hydraulic Design Simulation. Once the design is developed from the design tool, it is also essential to interpret the performance of the design by simulations. The fluid domain involved in the turbine passage is simulated using Computational Fluid Dynamics (CFD). They are the numerical simulation on the discretized flow field for estimation of the performance. Although there is always certain level of deviance in the result compared to the experimental study and the actual condition, the result of CFD gives a comparative idea for estimation. Commercial tools or in-house codes can be used for this kind of fluid flow simulations. They use the basic governing equation from Navier Stokes equation with turbulence model along with the boundary conditions on the discretized geometry of the turbine system. The design given by the MATLAB code is imported to these solvers, where various parameters such as turbine efficiency, erosion, and cavitation can be observed in varying conditions [11]. Using this technology, a large range of the designs can be tested at low cost and based on the result; a design optimization can be carried out to proceed into the manufacturing process.

4.3. Mechanical Design. Along with the hydraulic performance, the rigidity of design is another significant prospect. Whereas the hydraulic simulation provides the pressure acting on the blade due to the water flow, the structural simulation shows whether the blade is able to withstand the pressure or not. This depends on the mechanical properties of the blade, such as its density, elasticity, geometry, and orientation. A separate computer simulation including Fluid Structural Interaction (FSI) is performed to relate the fluid flow with the structure [12]. Figure 4 shows some of the results of the Computer Aided Simulation done in the premises of ANSYS.

Once the result of iterative design and simulation is satisfactory in terms of performance, flow, and structural rigidity, the design is adopted and further processed for CAD development.

5. Computer Aided Design (CAD), Modeling, and Workshop Drawing

This is the stage in which the design performed is expressed in the form of 3D which latter is projected to 2D drafting. The main objective of this stage is expressing the design in the understandable form. 3D CAD visualizes the design and drafting creates comfort in the manufacturing.

5.1. 3D Drawing. The accepted hydraulic design with the allowable stresses on the thickness provided is primarily drawn in 3 dimensions in order to visualize the position and finalize the assembly in more realistic form. The features of the 3D CAD software allow us to predict the material use to preestimate the cost effectiveness of the design [13]. This operation is performed under the premises of Solidworks with easy and comfortable interface. Each component of the system is developed referencing the common origin in different files called part files. These files are later combined together in the assembly window of the workspace. The major features like sketch extrusion, revolution, reference, pattern, sweep, blend, and so forth are used for the 3D CAD.

5.2. Drafting. The next step in CAD is the design drafting. It is the process of the drawing that is understandable by the manufacturer. The standard guideline of either third angle or first angle is implemented for projection of the 3D drawing. The drawing window of the program produces the projection drawing in the relevant scale. The main advantage of adopting this kind of feature instead of traditional method of direct development of 2D is to prevent probable mistakes in the drawing since they are developed from 3D after observation and visualization of design. Figure 5 shows the difference in 3D modeling and drafting, in relation to the presentation of descriptive process in this method.

6. Computer Aided Manufacturing and Workshop Manufacturing

6.1. Computer Aided Manufacturing. This is the physical replication and product development stage involved in the manufacturing process. As discussed in the earlier sections, Francis turbine blades, guide vanes, and stay vanes are complex designs which need to be physically replicated as

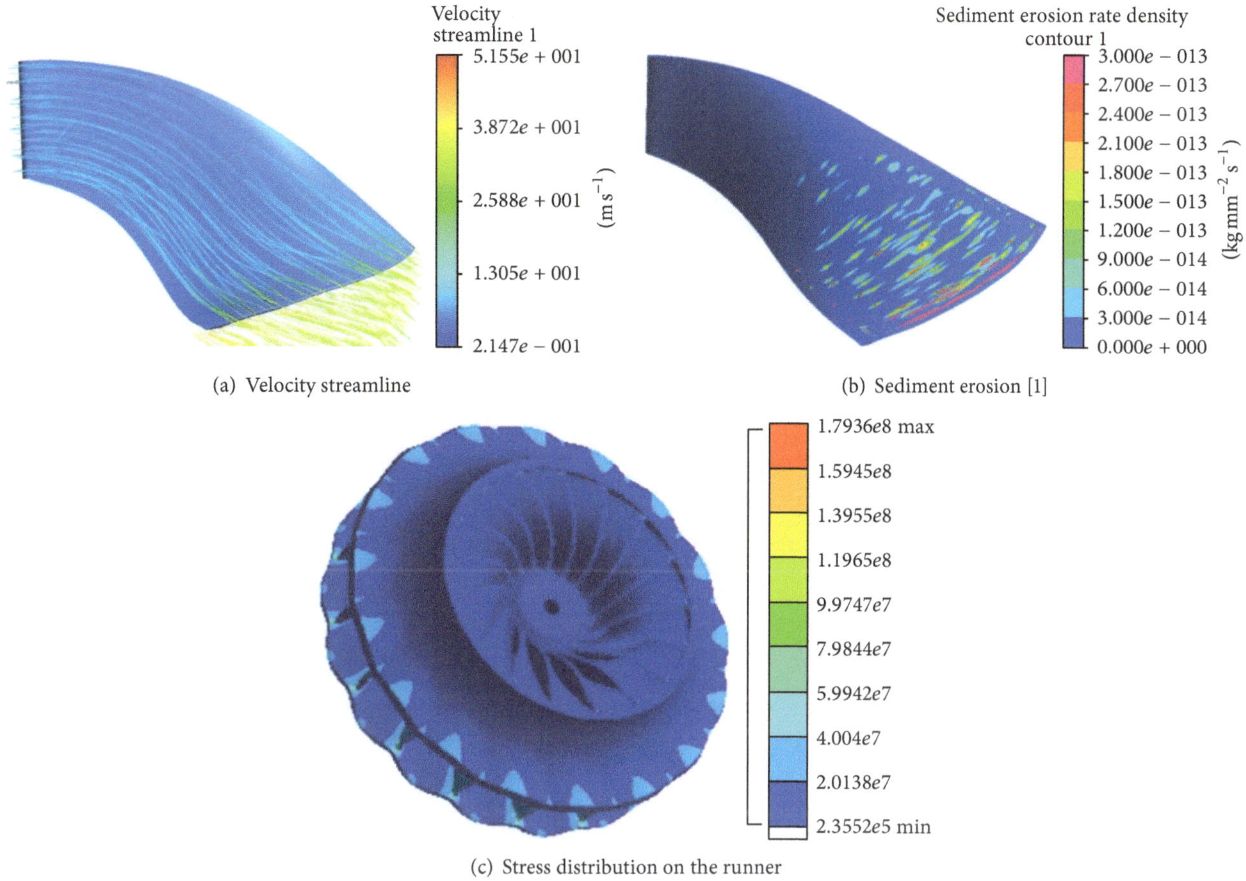

FIGURE 4: Computer Aided Simulations, [25].

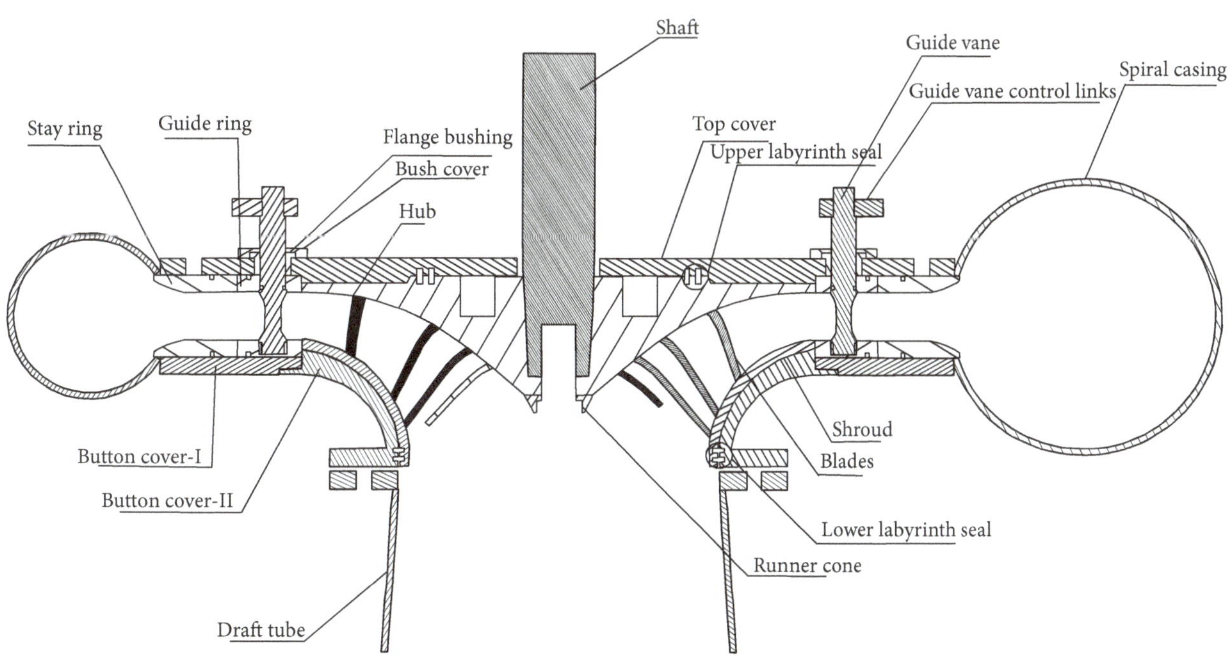

FIGURE 5: 3D modeling and drafting of a Francis turbine.

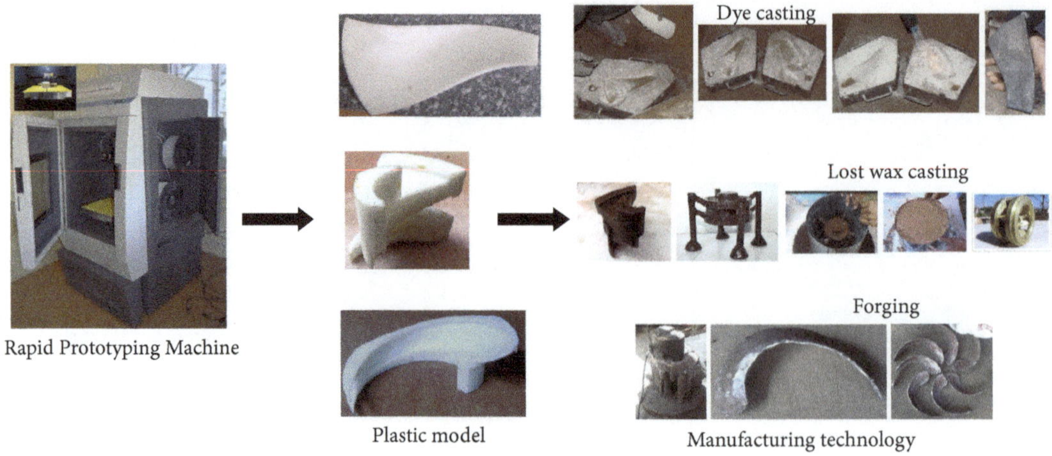

Figure 6: Computer Aided Manufacturing and Workshop Manufacturing.

accurately as possible for the expected performance. Hence, another advanced method called Computer Aided Manufacturing was needed, which performs Rapid Prototyping (RP) of a 3D model. RP is frontier technology which reduces time and increases accuracy for design verification by producing solid model from 3D CAD [14].

The complex geometry like runner blades or the entire runner is casted. To ascertain the design reliability accuracy of the designed CAD and its physical replication, dye plays the major role [15]. In such a scenario, Rapid Prototyping proves to be an efficient machine which helps to replicate the designed models in the form of plastic, which can be later converted into metal pieces at a nominal cost [16]. RP helps in minimizing the risks and maximizing the quality to overcome the gap of experience.

Currently, TTL acquires Inspire D290 3D printer which operates on Fused Decomposition Modeling (FDM). FDM is an additive process that involves heating a thermoplastic material's wire filament which is fed to a moving head and squeezing out of tube, which traces the area of the layer and deposits the filament on the surface. FDM is regarded as one of the effective RP techniques for fast, cost effective, and dimensionally correct product development. This method has been found effective in aerospace applications [17]. Once deposited, the material solidifies and adheres to the previous layer. The table moves through small distance in vertical direction (Z-direction) to create space for the formation of new layer while printing head is moving in X-, Y-direction. The material used in our case is Acrylonitrile Butadiene Styrene (ABS) polymer [18]. The product thus created will be very suitable for developing the mould in investment casting. So far the practices have showed that FDM plastic products have proven themselves to be effective casting molds [19].

The 3D design of the blade developed in Solidworks is now provided with required tolerance and file is converted into *.stl* file, which is readable to the printer. The printer has an integrated computer interface for printing arm initialization, height definition, fineness, temperature control, and so forth. Once the model is loaded, the major operations are performed to export the design to the SD card installed in the printer. Now, the printer displays the estimated time and materials for the printing permission. Once every requirement is defined, the printing starts. The printing device uses software packages: model wizard for 3D model processing, which is 3D slicing software. Table 1 shows the specification of the Rapid Prototyping Machine, which is at TTL.

Table 1: Specification of the Rapid Prototyping Machine.

Particular	Description
Build volume	320 * 290 * 255 mm (23.7 litres)
Layer thickness with 1-nozzle layer	100 microns/0.1 mm
Layer thickness with 2-nozzle layer	150 microns/0.15 mm
System size in mm	720 * 850 * 1650 ($W \times D \times H$) mm
Support material	ABS (Acrylonitrile Butadiene Styrene)
Operating speed	5–60 cm^3 adjustable
Power	220–240 V, 10 A
Software compatibility	Windows* XP/windows* 7/windows* 8

6.2. Workshop Manufacturing. The physically replicated guide vanes, stay vanes, and blades through the RP machine are then processed for manufacturing. The guide vanes and stay vanes are comparatively easier geometry; hence either of the casting technologies can result in a fine product with further possibility of postmachining. However, with the runner and the blade geometries, several constraints are to be considered, either single casting or part casting.

In this section, three of the major experiences of Turbine Testing Lab are presented.

A 92 kW runner designed under a NORAD funded project was manufactured by integrating two technologies.

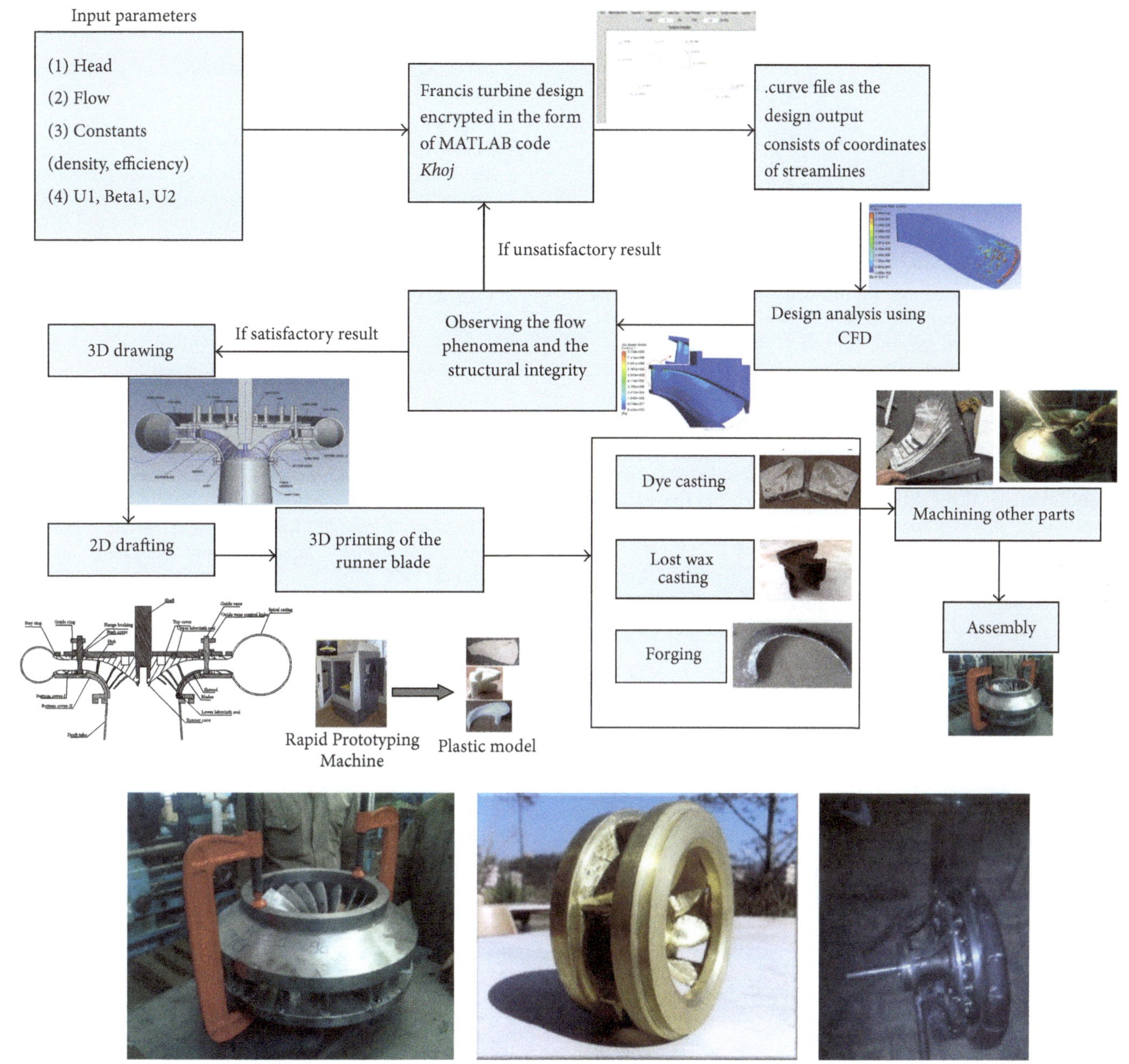

(a) (b) (c)

FIGURE 7: Flowchart of the methodology adopted in Francis turbine manufacturing and the turbine manufactured through (a) dye casting, (b) lost wax casting, and (c) forging.

The hub and shroud are machined and blades were dye casted. Later all these three sets of geometries were combined using welding technology [20].

In another activity for feasibility study of turbine manufacturing company in Nepal, a 2 kW Francis runner was casted using lost wax casting technology and finally buffed for surface finishing.

In one of the recent projects of 1.6 kW Reversible Pump Turbine the blades geometries were almost semicircular; hence it was very difficult to follow either of the above two technologies. Hence, a different but comparatively more traditional approach of forging was implemented, where the prototyped blade pattern was used as the quality control agent for forging. Figure 6 explains all three technologies implemented [21, 22]. All the other supporting components associated with Francis turbine are machined and finally assembled. Thus manufactured turbines will be assembled to the main turbine component comprising of spiral casing, guide vanes, stay vanes, and draft tube. Since the casting moulds and forged blades are created with allowance, postmachining allows for the assembly within the preestimated dimensions. The paper focuses on the manufacturing implementations; hence, performance-wise effectiveness of the process has not been considered.

7. Summary

This paper discussed possibilities of implementing Computer Aided Engineering in Francis turbine manufacturing in Nepal. It was discussed that, in the course of few years, most of the new hydropower plants in Nepal will need Francis turbines, based on the operating regime assessed. As the manufacturing companies presently available in Nepal have not manufactured Francis turbines, the future endeavor in making an attempt of up to 5 MW turbines can thrive the industries and hydropower market in Nepal. The design competences possessed by Kathmandu University can be blended with the available technologies discussed in this paper to manufacture turbines in much cheaper rates. Also, with the design optimization techniques, the turbines can address local problems of erosion, which would otherwise not be possible in the case of imported turbines. The different aspects of turbine manufacturing discussed in this paper are summarized in Figure 7. The experiences shared in this paper of manufacturing three different types of turbines from three methods urge the expansion of the manufacturing arena of Nepal to another level.

This work is more focused on uplifting the micro hydro status of Nepal by developing Francis turbine manfacturing facility in Nepal. The proposed methodologies can be implemented for turbine manufacturing by adopting the design and Rapid Prototyping facility available at Turbine Testing Lab. The manufacturing facility available with the local manufactueres is sufficient to manufacture it.

The future research should focus on the performance of the turbines manufactured with different processes. Additionally, economical and simple assembly design for promoting local manufacturing should be prioritized.

Conflict of Interests

The authors declare that there is no conflict of interests regarding the publication of this paper.

Acknowledgments

The author acknowledges the funding agency NORAD and UGC, Nepal, and all the project leaders, activity leaders, and researchers at Turbine Testing Lab for their contribution during the projects.

References

[1] B. S. Thapa, B. Thapa, and O. G. Dahlhaug, "Current research in hydraulic turbines for handling sediments," *Journal of Energy*, vol. 47, no. 1, pp. 62–69, 2012.

[2] Turbine Testing Lab, "Feasibility study of Turbine manufacturing company in Nepal," Dhulikhel Project Report, 2012.

[3] E. T. Layton, *From Rule of Thumb to Scientific Engineering: James B. Francis and the Invention of the Francis Turbine*, NLA Monograph Series, 1992.

[4] H. Brekke, "Design, performance and maintenance of Francis turbines," *Global Journal of Researches in Engineering A: Mechanical and Mechanics Engineering*, vol. 13, no. 5, pp. 29–40, 2013.

[5] C*Blade, C*Blade Forging and Manufacturing, http://www.cblade.it/history.html.

[6] G. A. Jessop and D. C. De Vine, "Manufacture of francis runners for hydraulic turbines," US Patent 1917037 A, 1933.

[7] F. Ali, B. V. Chowdary, and L. Gonzales, "An integrated design approach for rapid product development: a case study through application of reverse engineering, re-engineering and fast prototyping tools," *Journal of Engineering, Design and Technology*, vol. 11, no. 2, pp. 178–189, 2013.

[8] F. Ayancik, U. Aradag, E. Ozkaya, K. Celebioglu, O. Unver, and S. Aradag, "Hydroturbine runner design and manufacturing," *International Journal of Materials, Mechanics and Manufacturing*, vol. 1, no. 2, pp. 162–165, 2013.

[9] P. M. Pandey, N. V. Ready, and S. G. Dhande, "Slicing procedures in layered manufacturing: a review," *Rapid Prototyping Journal*, vol. 9, no. 5, pp. 274–288, 2003.

[10] G. Kristine, *Hydraulic design of Francis turbine exposed to sediment erosion [M.S. thesis]*, 2011.

[11] P. J. Gjostad, *Hydraulic design of francis turbine exposed to sediment erosion [M.S. thesis]*, Institutt for Energi og Prosessteknikk, 2011.

[12] H. P. Erichen, *Mechanical design of Francis turbine exposed to sediment erosion [M.S. thesis]*, 2011.

[13] N. Akgerman and T. Altan, "Application of CAD/CAM in forging turbine and compressor blades," *Journal of Engineering Gas Turbines Power*, vol. 98, no. 2, pp. 290–296, 2010.

[14] R. B. Gowda, C. S. Udayagiri, and D. D. Narendra, "Studies on the process parameters of rapid prototyping technique (Stereolithography) for the betterment of part quality," *International Journal of Manufacturing Engineering*, vol. 2014, Article ID 804705, 11 pages, 2014.

[15] I. Olama, C. B. Besant, and M. Ristic, "A CAD/CAM system for die design and manufacture," *The International Journal of Advanced Manufacturing Technology*, vol. 3, no. 2, pp. 21–29, 1988.

[16] C. K. Chua and K. F. Leong, *Rapid Prototyping Principles and Applications in Manufacturing*, World Scientific, 2000.

[17] D. Yagnik, "Fused deposition modeling—a rapid prototyping technique for product cycle time reduction cost effectively in aerospace applications," *IOSR Journal of Mechanical and Civil Engineering*, vol. 5, pp. 62–68, 2014.

[18] D-Inspire, *Manual for Inspire D 290 Rapid Prototyping Machine*, 2011.

[19] G. S. Bual and P. Kumar, "Methods to improve surface finish of parts produced by fused deposition modeling," *Manufactuirng Science and Technology*, vol. 2, no. 3, pp. 51–55, 2014.

[20] H. Prasad Neopane, O. Gunnar Dahlhaung, B. Singh Thapa, and A. Kayastha, "Development of hydraulic turbines with new design philosophy as foundation for turbine manufacturing in Nepal," KU Technical Report, 2013.

[21] S. Chitrakar, R. Koirala, N. Maharjan, and N. Gurung, "Capacity development for utilization of reversible pump turbines in Nepalese hydropower projects," Tech. Rep., Kathmandu University, Dhulikhel, Nepal, 2014.

[22] R. Koirala, S. Chitrakar, N. Maharjan, N. Gurung, and B. Prasad Aryal, "Design and development of test rig for reversible pump turbine," in *Proceedings of the Rentech Symposium Compendium (RENTECH '14)*, 2014.

[23] TOS Kurim Atlas Group, http://www.tos-kurim.eu/en/company/profile-of-company/.

[24] Inventor Atlas, Inventor Atlas, http://inventorspot.com/articles/final_turbine_chinas_three_gorges_dam_begins_testing.

[25] S. Chitrakar, M. Cervantes, and B. S. Thapa, "Fully coupled FSI analysis of Francis turbines exposed to sediment erosion," *International Journal of Fluid Machinery and Systems*, vol. 7, no. 3, pp. 101–109, 2014.

7

Failure Prediction in Bulk Metal Forming Process

Ameen Topa and Qasim H. Shah

Department of Mechanical Engineering, International Islamic University Malaysia, Jalan Gombak, 53100 Kuala Lumpur, Malaysia

Correspondence should be addressed to Ameen Topa; ameen.topa@live.iium.edu.my

Academic Editor: Hailiang Yu

An important concern in metal forming is whether the desired deformation can be accomplished without defects in the final product. Various ductile fracture criteria have been developed and experimentally verified for a limited number of cases of metal forming processes. These criteria are highly dependent on the geometry of the workpiece and cannot be utilized for complicated shapes without experimental verification. However, experimental work is a resource hungry process. This paper proposes the ability of finite element analysis (FEA) software such as LS-DYNA to pinpoint the crack-like flaws in bulk metal forming products. Two different approaches named as arbitrary Lagrangian-Eulerian (ALE) and smooth particle hydrodynamics (SPH) formulations were adopted. The results of the simulations agree well with the experimental work and a comparison between the two formulations has been carried out. Both approximation methods successfully predicted the flow of workpiece material (plastic deformation). However ALE method was able to pinpoint the location of the flaws.

1. Introduction

In metal forming processes, the term workability refers to the degree of deformation that can be achieved in the workpiece without the occurrence of a defect, that is, the appearance of surface or internal cracks. Failure usually occurs as ductile fracture in metalworking and rarely as brittle fracture. The propagation of the cracks is of little interest as the main objective is to avoid their initiation. These cracks usually appear within regions that are highly strained due to extensive plastic flow of the material during the metal forming process [1].

The initiation of ductile fracture is a multistep process consisting of microvoid nucleation, growth, and then their coalescence [2]. Microvoids usually nucleate due to the debonding of the matrix material followed by a stage of void growth in which the microvoids grow under the applied stress state. The final phase of the ductile fracture happens when neighbouring microvoids coalesce together into a crack [3].

The occurrence of ductile fracture in metal forming process is a major limiting factor [4]. The prediction of the initiation of the cracks is important as it allows prior modification of the forming process which can result in sound and reliable products [5]. Therefore, the prediction of ductile fracture in metal forming processes has attracted the attention of many researchers for more than five decades [6] and has been extensively studied in literature with contributions from both academic and industrial research institutions [7].

The basic feature of bulk metal forming processes is the high volume to surface area ratio. Forging is an important bulk metal forming process in which metals are shaped into useful complicated structures [8]. In this process, the workpiece is compressed between two dies and undergoes large plastic deformation. The material flow is confined in the cavity between the dies and the extra material flows out through the gap between them [9, 10]. In bulk metal forming, surface defects result from plastic instabilities associated with the localization phenomena while internal cracks result from the accumulation of damage during the deformation which depends on the stress and strain paths [4].

Due to the importance of ductile fracture in metal forming processes, tremendous efforts were made by different researchers to develop a valid criterion for surface and hidden ductile fractures [11]. Gouveia and coworkers studied four previously published ductile fracture criteria on ring, cylindrical, tapered, and flanged upset test samples [1]. Huang and Dong developed a new criterion for various deformation conditions based on the microvoid model and verified it in uniaxial tension, torsion, and upsetting conditions [12]. Reddy and coworkers investigated different ductile fracture criteria in axisymmetric drawing [4]. Kiran and Khandelwal proposed a new micromechanics ductile fracture criterion and calibrated it with ASTM A992 steels [11]. Goijaerts and coworkers focused on the ductile fracture initiation in metal blanking process [2]. Clift and coworkers experimentally evaluated various failure criteria in three types of metal forming processes: simple upsetting, axisymmetric extrusion, and strip compression and tension [5]. Ouali and coworkers studied the ductile fracture of aluminium alloys in bulk metal forming and attempted to predict the failure using multiscale and continuum damage mechanics models [13]. Rao and coworkers carried on a comparative evaluation of the theoretical failure criteria reported in the published literature. They concluded that no theoretical failure criterion can be described as truly geometry-independent for metal forming processes [6].

In the above mentioned work, it is clear that the available ductile failure criteria are not sufficient to predict the defects due to their dependence on various conditions and parameters such as the workpiece geometry, the temperature of the process, and the friction between workpiece and tool system. On the other hand, a combination of reduced experimental work and extensive numerical simulation may enable the user to obtain a reliable design. Therefore, in the suggested work, extensive finite element analysis would be carried out using arbitrary Lagrange Eulerian (ALE) and smooth particle hydrodynamics (SPH) formulations in LS-DYNA.

The main objective of conducting finite element analysis of metal forming processes is to provide necessary information for proper control and design of these processes [14]. The number of the required experimental trials is minimized. Thus, the development time and cost are reduced. Today, the development of inexpensive powerful computers technology and the application of finite element method (FEM) into user friendly programs have brought this technology forward. This evolution has more or less revolutionized the metal forming analysis [15, 16].

There are two methods to describe the deformation of a continuous medium. The first method is called the Lagrangian description in which the independent variables are the coordinate of a *particle* in the undeformed state. Lagrangian description is suitable for processes like forging, deep drawing, and sheet bending. The second method is the Eulerian description in which the independent variables are the coordinates of a *material* point in the deformed state. Eulerian description is convenient for processes like rolling, drawing, and extrusion. In the case of infinitesimal deformation, there is no need to make distinction between the two methods [17].

The Lagrangian approach is often not suitable in modelling fluid-like behaviour due the excessive deformation in the material. Severe mesh distortion can occur which may cause instability. However, in the arbitrary Lagrangian-Eulerian (ALE) formulation, the materials flow through a fixed or moving mesh. Thus, ALE is a much better choice for modelling fluid-like behaviour [18].

Meshfree or meshless methods are available in LS-DYNA software. In meshfree methods, the physical structure is discretized with particles instead of elements. The shape functions and the approximation solutions are solved at the particles. Therefore, very large deformation is permitted since there is no mesh. Smooth particle hydrodynamics (SPH) method was developed to avoid the limitations of mesh tangling during extreme deformation problem with the finite element method. SPH is proven to be a very useful tool in LS-DYNA that can simulate solid and fluid materials [19, 20].

2. Theoretical Background

2.1. Arbitrary Lagrangian Eulerian (ALE) Formulation. In large deformation problems, a reference system is required from which independent variables describing the kinematics and the statics of the material particles can be defined. In Lagrangian formulation, the reference system is connected to the body and evolves with it. It is denoted as Material Reference System (MRS). On the other hand, in Eulerian formulation, the reference system is fixed in space and every point of the material is connected to a set of coordinates in this invariant Spatial Reference System (SRS) [21].

The pros and cons of both formulations can be illustrated from Figure 1. The Lagrangian formulation has the advantage of less complex governing equations. However, the elements move with the deforming body. Thus, they may distort excessively which may result in numerical instabilities. Moreover, this makes it difficult to consider the non-material-associated boundary conditions. On the other hand, in Eulerian formulation, the mesh remains stationary and the material freely moves within. The elements are fixed in space thus the boundary of the deforming body does not coincide with the elements everywhere. As a result, taking the material-associated boundary conditions into account is complicated [21–23].

The above-motioned shortcomings are encountered in large deformation problems and can be overcome with the arbitrary Lagrangian Eulerian formulation. In ALE formulation, the movement of the mesh is represented by a set of self-governing variables and the merits of both the Lagrangian and Eulerian formulations are achieved [24]. The reference configuration consists of a set of grid points in arbitrary motion in space. The grid points can move with material or remain stationary in space or even move with any arbitrary velocity as desired [23]. As shown in Figure 2, the elements nodal point displacements ($d\vec{v}$) are not coupled with material displacements ($d\vec{u}$) and the body can flow through the elements however their shapes are being controlled to prevent unbounded distortions. ALE formulation has been used in fluid mechanics [25], solid-fluid interaction analysis [26, 27], and has been generalized to solve the flow problems with Navier-Stokes equations [28]. Recently, ALE has been successfully used in nonlinear analysis in solid mechanics [24, 29].

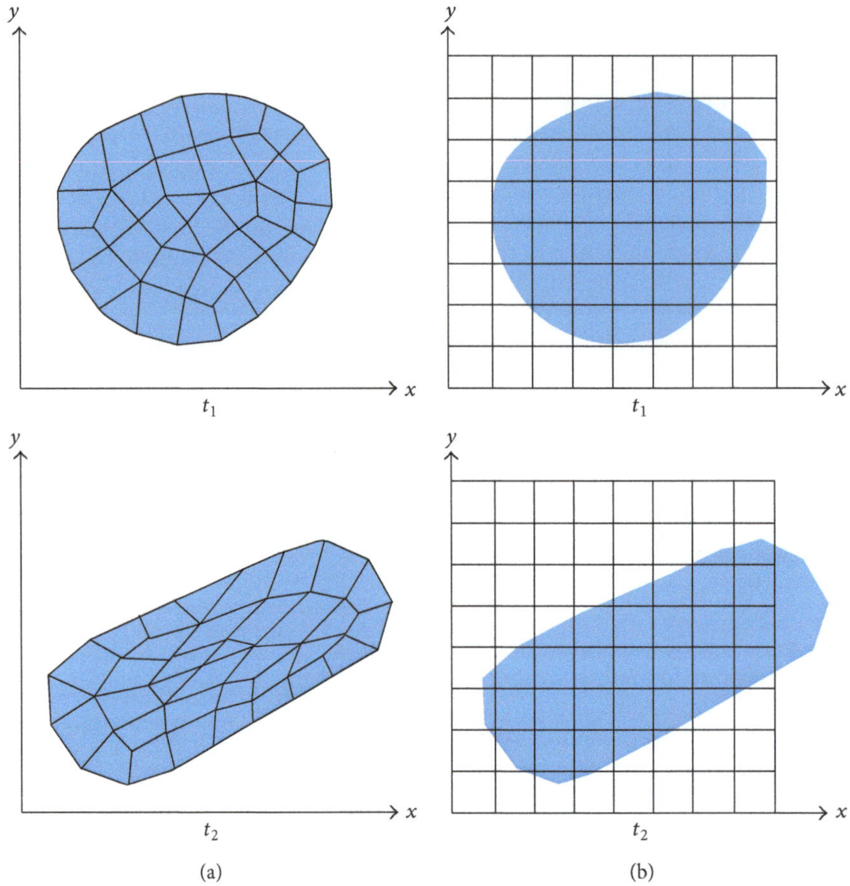

FIGURE 1: Lagrangian (a) and Eulerian (b) formulations.

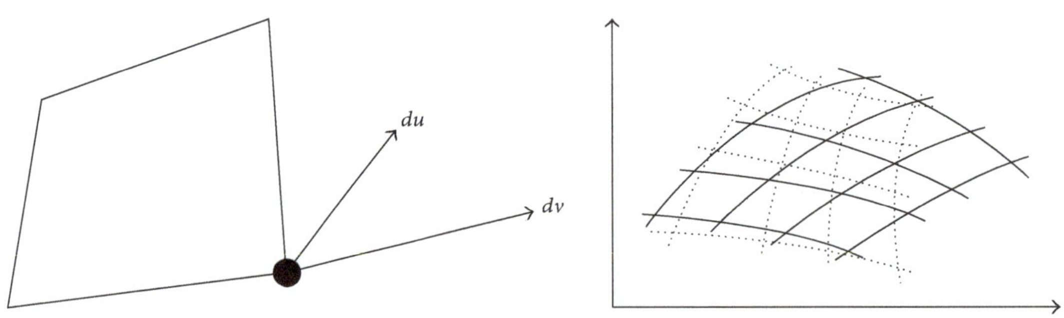

FIGURE 2: $d\vec{v}$ = mesh DOF, $d\vec{u}$ = material DOF, dotted line: material, and solid line: mesh [22].

2.2. Smooth Particle Hydrodynamics (SPH) Formulation. The Smooth particle hydrodynamics (SPH) formulation was first developed for astrophysical applications. It is a novel meshfree particle approach established on purely Lagrangian formulation [30]. However, it does not have a mesh as Lagrangian, Eulerian and arbitrary Lagrangian Eulerian methods. The body is defined by a number of mesh points or particles instead of a mesh and moves around in response to solid or fluid stresses produced by the interaction with neighbouring particles [31, 32] as illustrated in Figure 3. Since it is a particle based method and does not depend on fixed grids to track the material, it does not suffer from the inherent limitations of these methods. SPH is particularly well suited for flows with complex free surface behaviour and flows with complex physics such as the flow through industrial porous media [32].

In finite element methods, heavy remeshing might be required due to large plastic deformations in metal forming processes such as forging and extrusion. This increases the computational cost and may lead to inaccurate results. However, due to the meshfree nature of smooth particle hydrodynamics formulation, large deformations can be

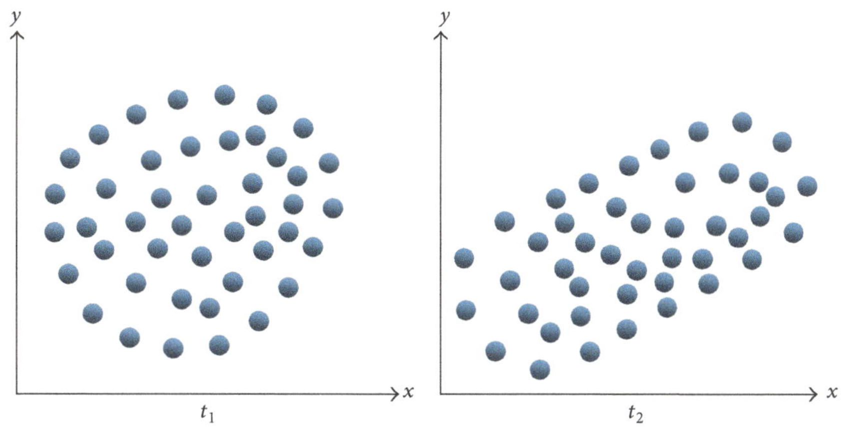

FIGURE 3: Smooth particle hydrodynamics formulation.

FIGURE 4: The hot forged part.

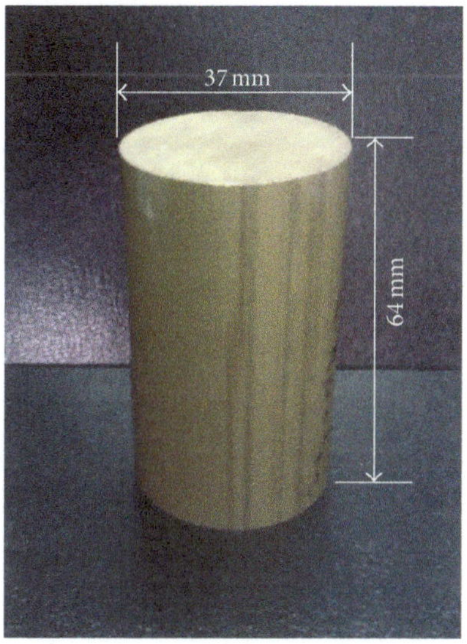

FIGURE 5: The workpiece geometry.

relatively easily handled. Moreover, it has the ability to follow the history dependent properties such as equivalent stress or plastic strain on a node-by-node basis, even during excessive deformation [33].

SPH formulation has been used widely in different applications. Droplet and thin film flow on smooth and rough fracture surfaces has been modelled with SPH method [34]. SPH formulation was adapted to simulate metal forming processes such as ring upsetting [35]. Crashworthiness optimization of foam filled tubes was performed using coupled SPH and finite element methods [36]. SPH method is used today in hypervelocity impact analysis [37], and its full potential is yet to be examined in detail.

3. Experimental Work

Hot forging process was carried out and the final forged part is shown in Figure 4. The workpiece is brass cylinder with a diameter of 37 mm and a height of 64 mm as illustrated in Figure 5. The density of the brass used in this study is 8490 kg/mm^3. The workpiece is preheated to 800°C and then placed between a stationary die and a moving punch as shown

FIGURE 6: Experimental setup of the forging process (cross section view).

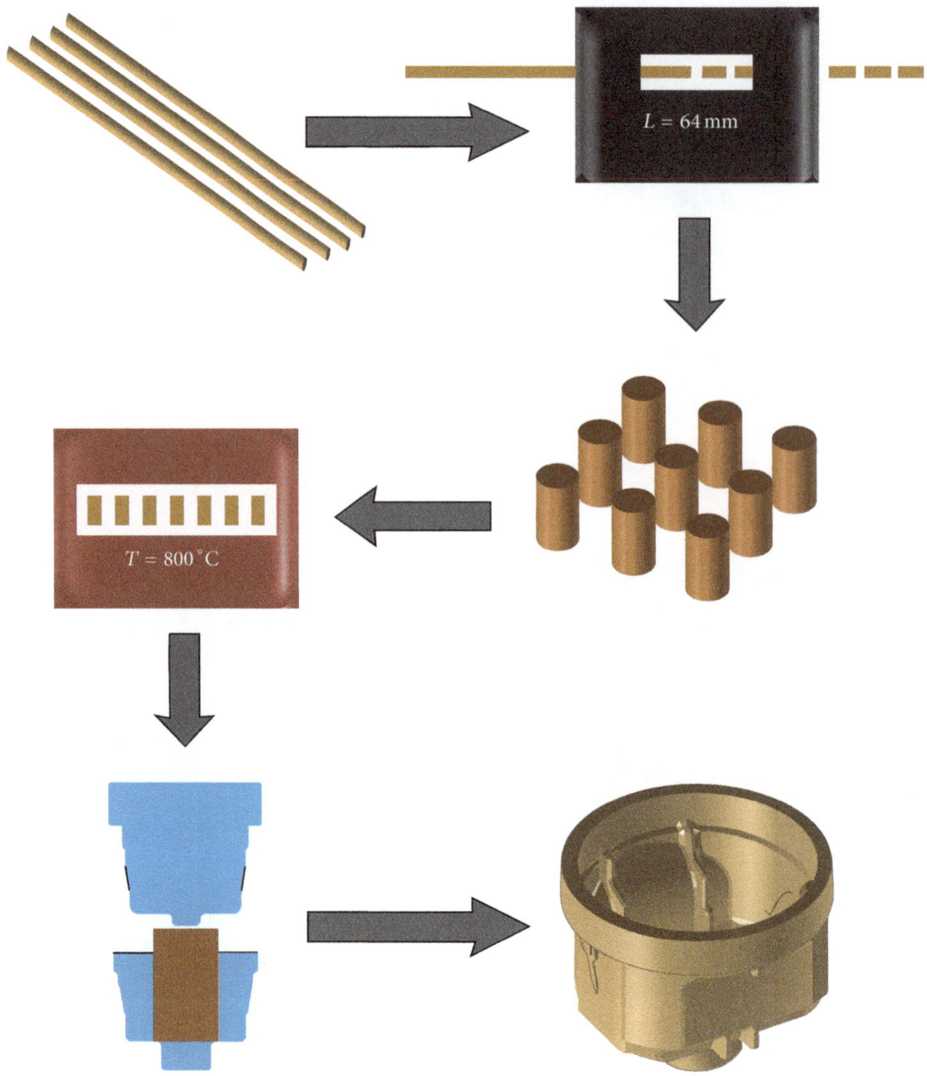

FIGURE 7: Schematic diagram of the experimental work.

in Figure 6. The punch compresses the workpiece at a velocity of approximately 10 m/s until it reaches a distance of 3.7 mm from the bottom. The experimental steps are depicted in the schematic diagram in Figure 7.

The approximate location of the defects is shown in Figure 8. Inspecting several defective products shows that the cracks usually appear within a certain range located between the two ridges. One of the defective products is cut down and the crack location is pinpointed in Figure 9. The crack-like flaws result in the formation of new surfaces. For example, the new surfaces are illustrated in the micrograph (×50) shown in Figure 10.

Upon thorough inspection, the wall thickness in the two ridges' side was found to be less than that in the rest of the product. This reduction in the thickness impedes the plastic flow of the material and is proposed to be the main source of the defects. The cross section of the product is shown in Figure 11, which clearly reveals the difference in the wall thickness of the product.

TABLE 1: Input values for the workpiece material model.

Parameter	Description	Value	Units
RO	Mass density	$8.490E-06$	Kg/mm^3
E	Young's modulus	101	GPa
PR	Poisson ratio	0.31	
SIGY	Yield strength	$0.65E-03$	GPa
ETAN	Tangent modulus	$0.225E-03$	GPa

4. Finite Element Modelling

The die and the punch were automatically meshed into shell rigid parts from the compatible CAD files (IGES) imported in the LS-DYNA preprocessor while the workpiece solid mesh was generated in the preprocessor as it is geometrically less complex. The material model used for the workpiece was MAT_PIECEWISE_LINEAR_PLASTICIY and the values used for the input parameters are shown in Table 1. The values

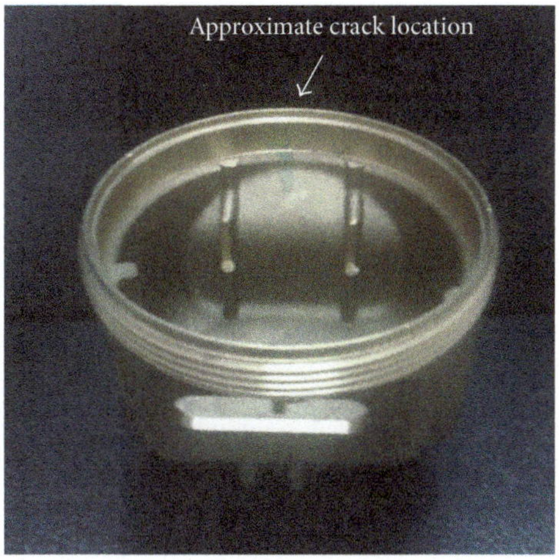

FIGURE 8: Approximate crack location.

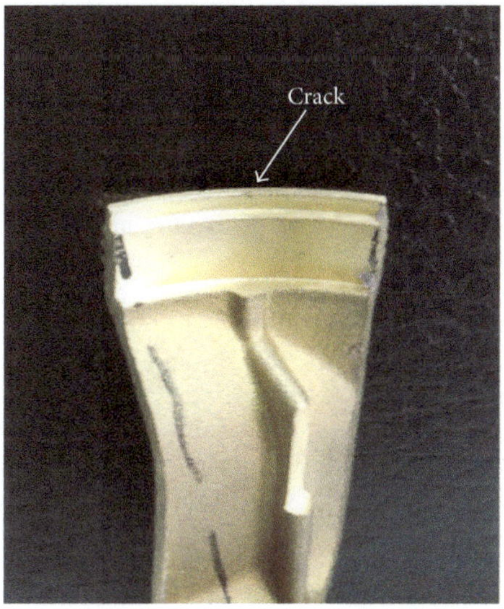

FIGURE 9: Identifying the location of the crack.

FIGURE 10: Free surface micrograph (×50) of the work piece edge in Figure 8.

FIGURE 11: The difference in the wall thickness of the product.

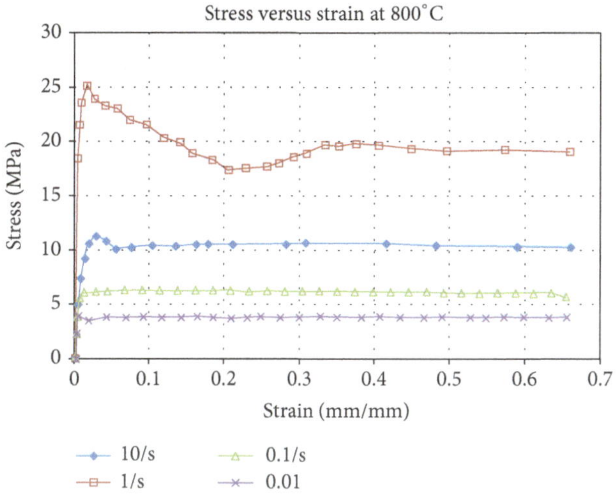

FIGURE 12: Stress-strain relationships for different strain rates at T = 800°C.

of Young's modulus and yield point at different strain rates were found in the literature [38]. The stress-strain curves for the material at different strain rates were obtained from previous work by Xiao and coworkers [39] and were plotted in Figure 12. Since the curves at different strain rates were used in the material model, the parameters C and P were ignored as they are used in calculating the strain rate effects [40].

4.1. Smooth Particle Hydrodynamics (SPH) Formulation. The cylindrical workpiece was discretized into particles while the die and tool were meshed using shell elements. In SPH, the bulk viscosity must be defined to get more realistic results. The bulk viscosity has two coefficients, the quadratic viscosity coefficient and the linear viscosity coefficient. The first coefficient is important to smear the shocks and prevent elements from collapsing while the second coefficient is used to damp oscillation [41, 42]. Both coefficients are increased by a factor of 10 for soft materials [43]. The default formulation of the particle approximation theory was used. Other approximations would cause calculation instabilities.

Table 2: Input values for CONTROL_ALE card.

Parameter	Description	Value
DCT	Default continuum treatment	2
NVAD	Number of cycles between advections	1
METH	Advection method	3
AFAC	ALE smoothing weight factor	−1

The contact was defined with CONTACT_AUTOMATIC_NODES_TO_SURAFCE. A node set which contains all the SPH particles was defined and used as the SLAVE (workpiece) in the contact card. Furthermore, for SPH particles representing a volume, the contact thickness STT is suggested to be 0.8–1.0 mm to avoid nodal release [44]. To prevent contact penetration, the penalty stiffness scale factors SFS and SFM must be increased to 10.

4.2. Arbitrary Lagrangian Eulerian (ALE) Formulation. ALE formulation is suitable for modelling large deformation problems and fluid-like behaviour as the material flows through a fixed or a moving mesh [18]. For simplicity, the effect of the air was ignored and it is assumed to be void in the simulation. Both the die and punch are modelled as rigid while the workpiece is modelled as solid ALE material which is permitted to deform and flow into the surrounding space (void). The mesh of the workpiece and the void should share the same nodes on their common boundaries [45].

Element formulation 12 was chosen for the workpiece. This formulation stands for one-point integration with single material and void. The nondefault input values of the CONTROL_ALE card are shown in Table 2. The selected values were used to obtain more reliable results [46, 47].

The interaction between the Lagrangian (die and tool) and ALE (workpiece and void) parts was defined under CONSTRAINED_LAGRANGE_IN_SOLID card. This card is very important and the values of its parameters should be chosen carefully. Otherwise, leakage problem and instabilities may occur. Table 3 shows the input values of this card. These values were adopted from previous works [48, 49].

Table 3: Input values for CONSTRAINED_LAGRANGE_IN_SOLID card.

Parameter	Description	Value
NQUAD	Number of quadrature points	3
CTYPE	Coupling type	4
DIREC	Coupling direction	2
PFAC	Penalty factor	1
ILEAK	Leakage control	0
FRCMIN	Minimum volume fraction to activate coupling	0.3
PLEAK	Leakage control penalty factor	0.1
PFACMM	Mass-based penalty stiffness factor	3

5. Results and Discussion

5.1. Smooth Particle Hydrodynamics (SPH) Formulation. The contours of the equivalent von Mises stress are shown in Figure 13. Any point of the material can be traced using the SPH method. Thus, the initial and final positions of any point in the material can be found. The rate of deformation varies with time. It can be observed that as the rate of deformation is higher near the end of the compression phase, the strain rate increases at this stage. However, since stress-strain relationship at different strain rates was defined in the material model, the strain rate effects are considered in the calculations.

The increase in the rate of deformation can be explained with the help of Figure 14. The flow velocity of the material depends on the volumetric flow rate and the cross sectional area perpendicular to the flow velocity [50]. The increase in the flow velocity of the material is due to the decrease of the flow area, which can be observed from Figure 14. As the compression reaches its end, material flow becomes more restricted due to reduced thickness through which it passes.

Analyzing Figure 13 shows that at time $t = 1$ ms, the maximum equivalent stress occurs almost at any point in workpiece model. Afterwards, the maximum equivalent stress occurs only in the internal points. This can be explained with the help of Figure 14. At the initial states, the process can be roughly considered as a standard one-dimensional compression test. However, as the compression continues, due to the convex shape of the punch and the concave shape of the die, the force distribution on the upper and lower surface of the workpiece will vary and shear stress will be generated.

5.2. Arbitrary Lagrangian Eulerian (ALE) Formulation. For the ALE formulation, the contours of the effective plastic strain are illustrated in Figure 15. ALE formulation gives clearer view of the shape of the material compared to SPH formulation. ALE method represents the volume (solid elements) of the material by surfaces while SPH method is represented by particles. Sharp edges are visible with ALE formulation in contrast to SPH formulation.

Towards the end of the compression, namely, at $t = 6$ ms, it can be observed that there is a separation in the workpiece material; that is, there is a brief discontinuity in the material flow. To have a better view of this phenomenon, the section view is presented in Figure 16. The start of discontinuity of the material flow can be seen at $t = 6$ ms. This can be interpreted as a result of the sudden increase in the flow velocity. As discussed earlier, the flow velocity increases as the compression comes to its end. This increase of flow velocity is more in the material on the free edges of the flow which causes the material on that portion to "splash" away. It is proposed that this phenomenon causes the cracks in the bulk metal forming. The cracks occur on the side of the ridges where the wall thickness is minimum as shown Figure 17. The wall thickness obtained from the simulations matches the experimental results.

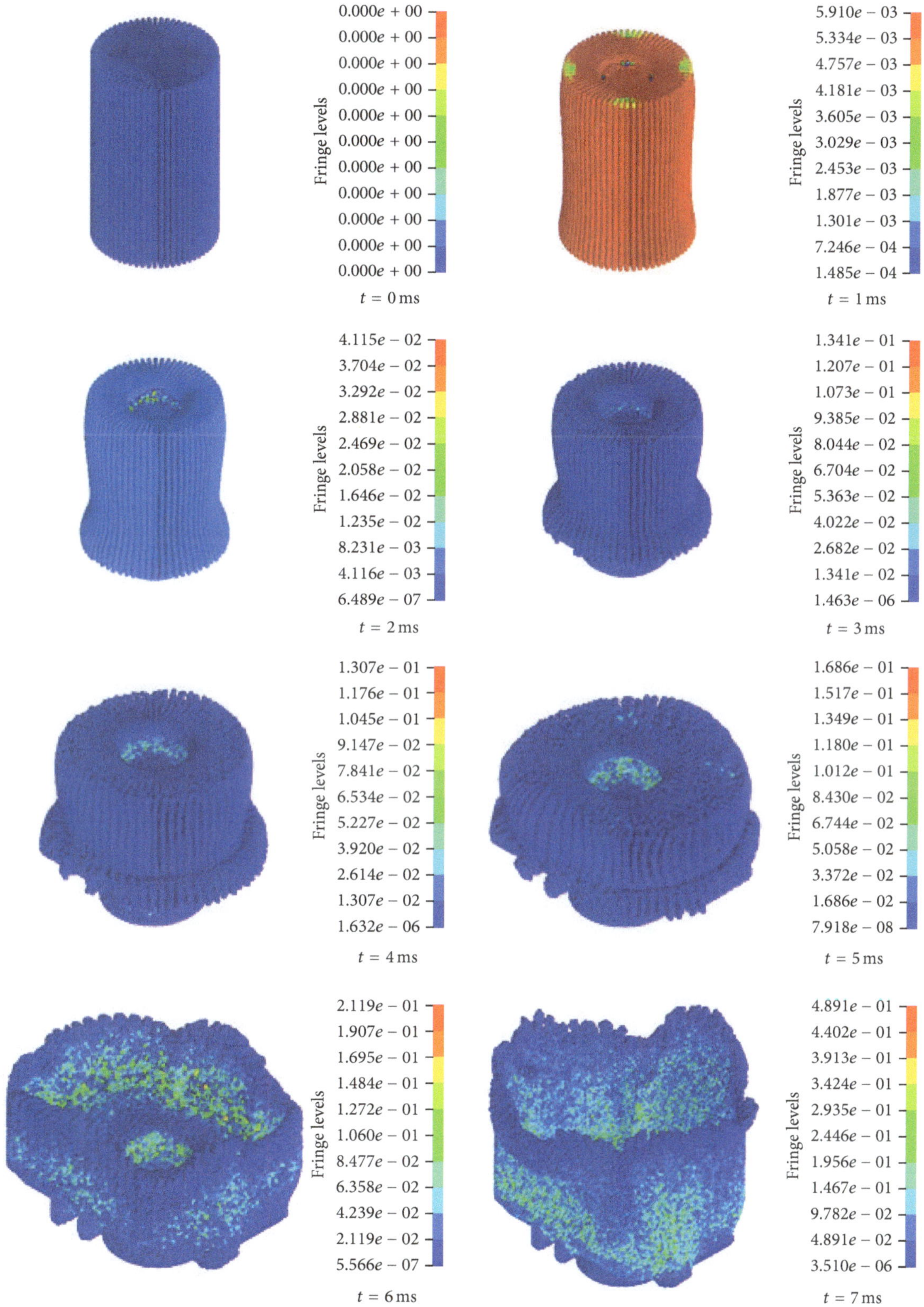

FIGURE 13: The deformation and the equivalent von Mises stress contours using SPH.

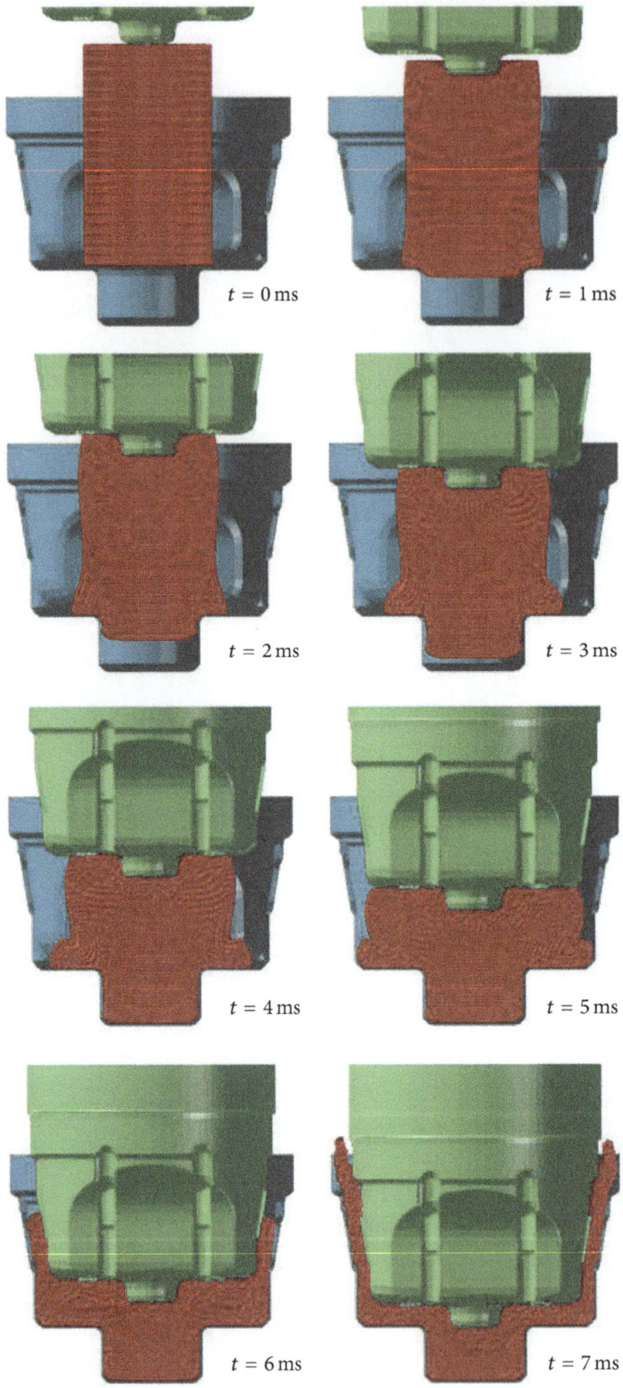

FIGURE 14: Section view normal the symmetry plane (SPH).

The velocity plot of four nodes initially located at the top surface of the workpiece is shown in Figure 18. It is observed that there is a sudden change in the velocity of three of these nodes towards the end of compression. These nodes travel from the top surface of the workpiece to the edges of the final product as shown in Figure 19. The sudden change in the velocity is not observed in node number 423615 which travels only from the top to the bottom of the workpiece. There is a tendency of the material at the edges to separate and form new surfaces. As the velocity of the workpiece material decreases at the end of the compression, the rejoining of the material occurs. However, some of the free surfaces which previously existed may not completely rejoin leaving flaws in the final product.

This "splash" effect occurs due to the fluid-like behaviour of the workpiece material. Thus, reducing the temperature of the process may prevent this effect. However, reducing the temperature means that more compression force is required.

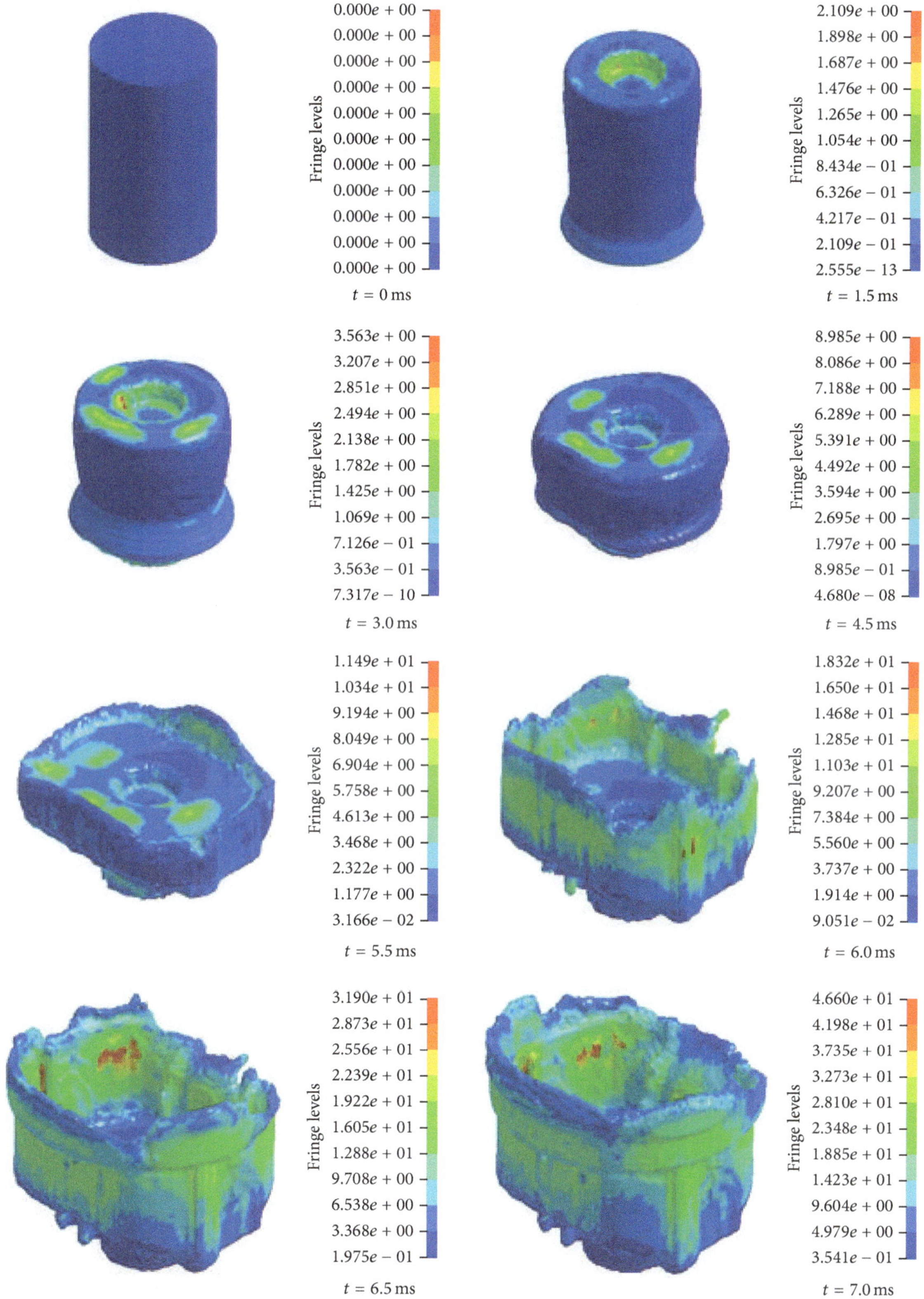

FIGURE 15: The deformation and the effective plastic strain using contours ALE.

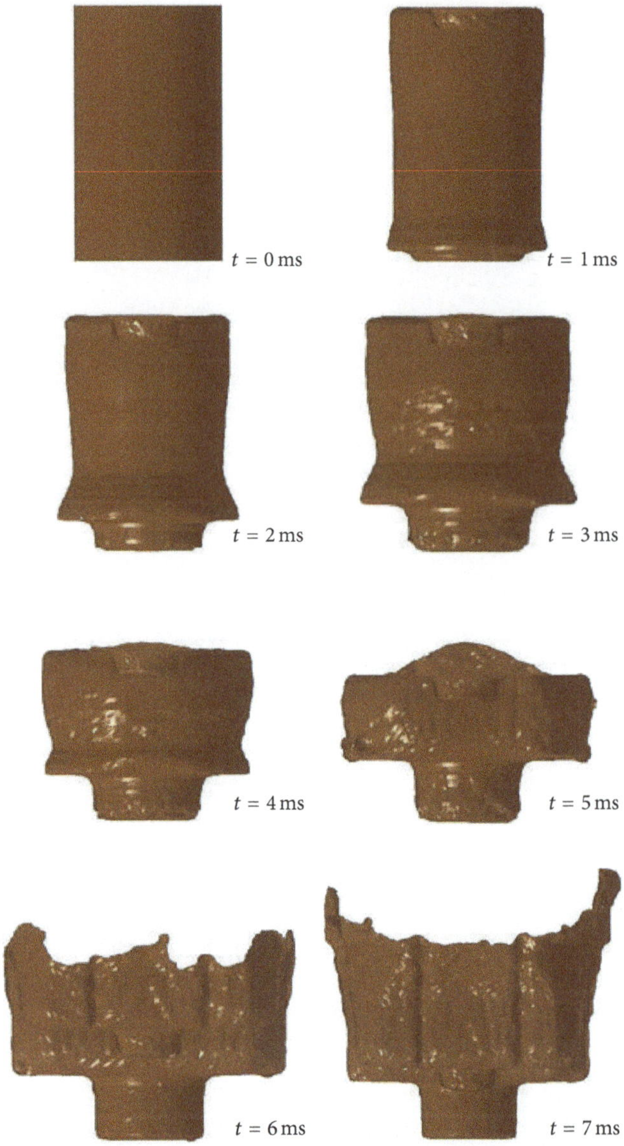

Figure 16: Section view normal the symmetry plane (ALE).

Figure 17: Cross section view showing the difference in the wall thickness near the ridges.

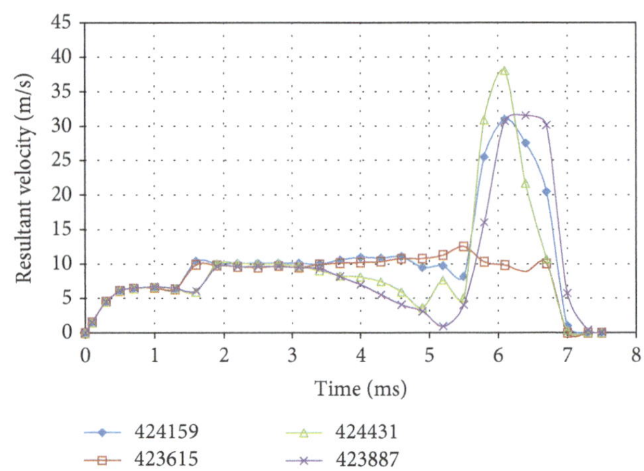

Figure 18: The resultant velocity of nodes initially located at top surface of the workpiece.

Therefore, it might not be a feasible solution. If the compression velocity were reduced, the magnitude of the sudden increase in the flow velocity would reduce. This might stop the formation of new surfaces. However, this would result in more production lead time. Another solution is to alter the design of the die and tool by increasing gap between them and thus increase the area through which the material flows. But this would require more material to produce the same product. A combination between the three previous solutions would give the optimum results.

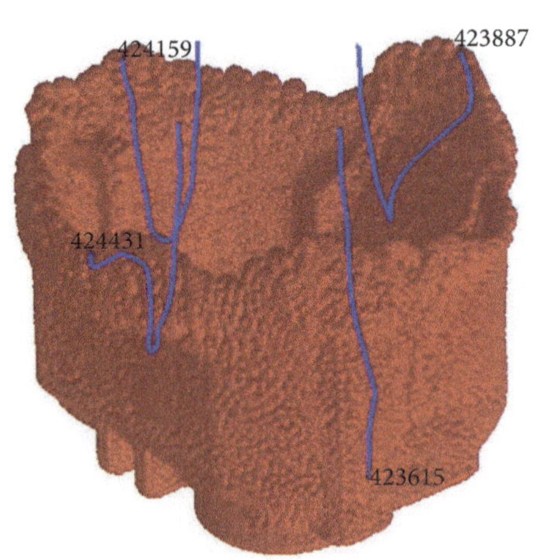

Figure 19: Tracing the displacement of four nodes originally located at the top surface of the workpiece.

5.3. *SPH versus ALE.* With the input parameters mentioned in Section 4, the simulation was solved successfully for both SPH and ALE methods. However, SPH formulation took considerably less computational time than ALE formulation and the time step was almost constant.

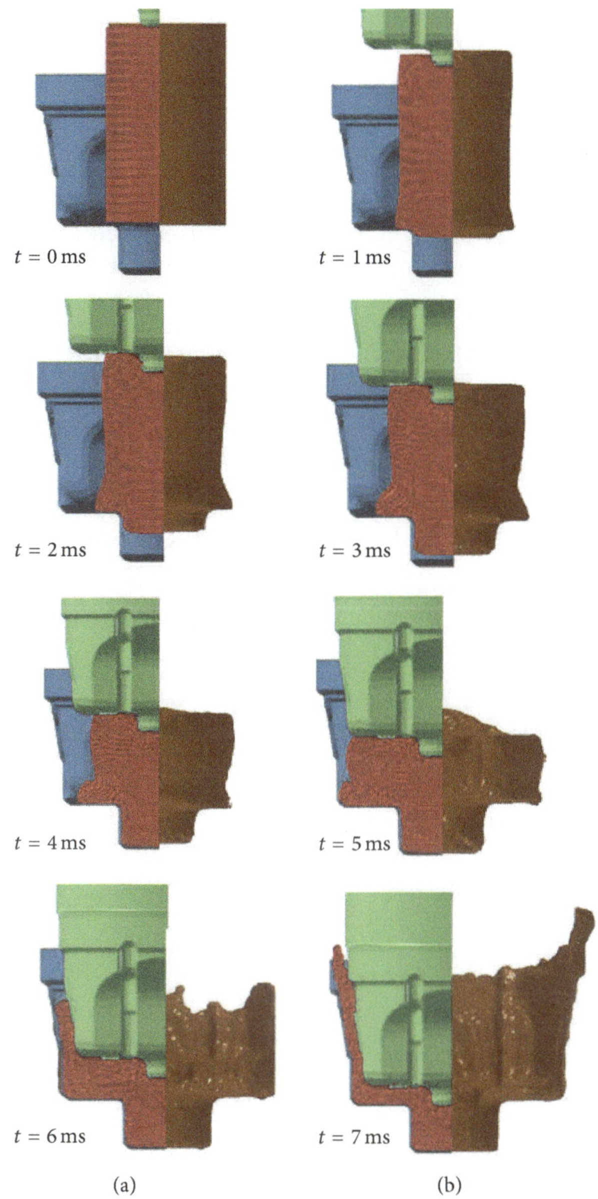

FIGURE 20: Comparison between the deformation with SPH (a) and ALE (b).

To compare the deformation in SPH and ALE, the section views were displayed side by side, as depicted in Figure 20. The material flow in SPH is more regular compared to the flow in ALE. The flow is roughly symmetric with SPH formulation as seen in Figure 14. However, in ALE formulation, the flow is unsymmetrical. With reference to Figure 20, it is observed that the initial and final states of the deformation are roughly the same. However, the deformation during the process is not similar.

One advantage of SPH method over ALE method is that the value of the contours at any point inside the material can be found by using a cutting plane. However, as mentioned before, ALE method represents the volume of the material with surfaces; thus, the state of the material under the surface cannot be found.

As stated in the introduction, bulk metal forming is a complex process yet a straightforward and symmetric deformation is achieved with SPH formulation. ALE formulation gives a more realistic deformation pattern and was able to predict the location of the possible flaws.

5.4. Ductile Fracture Model. A thorough study in the vast literature on ductile facture modelling reveals the limitation of the predictive capability of the ductile fracture models presently available. The drawback of these fracture models is their strong dependence on the application. However, Aretz and coworkers developed a simple ductile fracture model for problems of any stress triaxialities for sheet metal forming [51]. The failure in our study occurs when the thickness is greatly reduced; therefore the process can be considered as sheet metal forming in those regions. The developed criterion was established on Freudenthal ductile fracture model [5] in which the failure occurs when the plastic work density denoted by w^p reaches a critical value at fracture w_f^p as follows:

$$\int_0^{w_f^p} dw^p \leq w_f^p. \tag{1}$$

However, in engineering practice, the work-based Freudenthal model is not as convenient as the strain-based version which can be obtained by realizing the work-conjugated definition of equivalent plastic strain given by

$$dw^p = \boldsymbol{\sigma} : d\varepsilon^p = \overline{\sigma} \cdot \overline{\varepsilon}. \tag{2}$$

Therefore, the strain-based version of Freudenthal model is as follows:

$$\int_0^{\overline{\varepsilon}_f} d\overline{\varepsilon} \leq \overline{\varepsilon}_f, \tag{3}$$

where $\overline{\varepsilon}_f$ is the equivalent plastic strain at fracture. The back draw of this ductile facture model is the incapability of discriminating compressive and tensile stress states. However, ductile fracture in solids is considerably delayed in the existence of compressive stress states while it is accelerated in tensile ones. The modified Freudenthal ductile fracture model proposed by Aretz includes a transition function which depends on the loading mode parameter α. The transition function is given by

$$g(\alpha) = \left(\frac{1+\alpha}{2}\right)^A, \quad A \geq 0. \tag{4}$$

The parameter A is a material dependent parameter. The purpose of the transition function $g(\alpha)$ is to determine the damage accumulation by the stress states. In the case of pure compressive stress states ($\alpha = -1$), the transition function disappears. Thus, the contribution to damage accumulation becomes zero. The modified Freudenthal ductile fracture model becomes

$$\int_0^{\overline{\varepsilon}_f} g(\alpha)\, d\overline{\varepsilon} \leq \overline{\varepsilon}_f. \tag{5}$$

The present work mainly focuses on the prediction of failure through numerical simulations and highlights the important parameters in setting up the finite element model. Furthermore, it analyses the failure mechanism. The modified Freudenthal ductile fracture model will be used and coupled with the numerical simulations in future work. Moreover, the model will be utilized to eliminate the flaws in the product.

6. Conclusions

A complex shape product was produced from a simple cylindrical workpiece using hot forging process. Occasionally free surfaces and internal cracks were observed in the hot forged part. The forging process was simulated using SPH and ALE methods in LS-DYNA. The setup of the FEA modelling and the important parameters were discussed. The results of the numerical simulations show that both ALE and SPH methods are able to capture the large plastic deformation in the workpiece precisely. However, the ALE method played a prominent role in pinpointing the possible crack location.

The sudden increase of the material flow which causes the "splash" away effect is believed to be the reason for the occurrence of surface and internal cracks. Therefore, three solutions were proposed: to reduce the compression velocity, to reduce the temperature of the process, and to alter the design of the die and tool.

Conflict of Interests

The authors declare that there is no conflict of interests regarding the publication of this paper.

Acknowledgments

The authors gratefully acknowledge the support of Research and Management Centre (RMC) of International Islamic University Malaysia (IIUM). This work was supported by Research Grant (EDW B12-407-0885).

References

[1] B. P. P. A. Gouveia, J. M. C. Rodrigues, and P. A. F. Martins, "Ductile fracture in metalworking: experimental and theoretical research," *Journal of Materials Processing Technology*, vol. 101, no. 1, pp. 52–63, 2000.

[2] A. M. Goijaerts, L. E. Govaert, and F. P. T. Baaijens, "Prediction of ductile fracture in metal blanking," *Journal of Manufacturing Science and Engineering*, vol. 122, no. 3, pp. 476–483, 2000.

[3] K. Khandelwal and S. El-Tawil, "A finite strain continuum damage model for simulating ductile fracture in steels," *Engineering Fracture Mechanics*, vol. 116, pp. 172–189, 2014.

[4] N. V. Reddy, P. M. Dixit, and G. K. Lal, "Ductile fracture criteria and its prediction in axisymmetric drawing," *International Journal of Machine Tools and Manufacture*, vol. 40, no. 1, pp. 95–111, 2000.

[5] S. E. Clift, P. Hartley, C. E. N. Sturgess, and G. W. Rowe, "Fracture prediction in plastic deformation processes," *International Journal of Mechanical Sciences*, vol. 32, no. 1, pp. 1–17, 1990.

[6] A. V. Rao, N. Ramakrishnan, and R. K. Kumar, "A comparative evaluation of the theoretical failure criteria for workability in cold forging," *Journal of Materials Processing Technology*, vol. 142, no. 1, pp. 29–42, 2003.

[7] P. B. Woelke, M. D. Shields, N. N. Abboud, and J. W. Hutchinson, "Simulations of ductile fracture in an idealized ship grounding scenario using phenomenological damage and cohesive zone models," *Computational Materials Science*, vol. 80, pp. 79–95, 2013.

[8] A. S. Wifi, A. Abdel-Hamid, H. El-Monayri, and N. El-Abbasi, "Finite element determination of workability limits for disks and rings under different upsetting conditions," *Journal of Materials Processing Technology*, vol. 56, no. 1–4, pp. 918–932, 1996.

[9] S. Kobayashi, S. Oh, and T. Altan, *Metal Forming and the Finite-Element Method*, Oxford University Press, New York, NY, USA, 1989.

[10] R. H. Wagoner and J.-L. Chenot, *Fundamentals of Metal Forming*, John Wiley & Sons, New York, NY, USA, 1997.

[11] R. Kiran and K. Khandelwal, "A micromechanical model for ductile fracture prediction in ASTM A992 steels," *Engineering Fracture Mechanics*, vol. 102, pp. 101–117, 2013.

[12] J.-K. Huang and X.-H. Dong, "A new ductile fracture criterion for various deformation conditions based on microvoid model," *Journal of Iron and Steel Research International*, vol. 16, no. 5, pp. 92–97, 2009.

[13] M. O. Ouali, M. Almansba, and N. Hannachi, "Ductile fracture prediction in bulk metal forming using multiscale and continuum damage mechanics models," in *Proceedings of the International Conference on Computer Methods in Mechanics (CMM '11)*, Warsaw, Poland, May 2011.

[14] S. I. Oh, "Finite element analysis of metal forming processes with arbitrarily shaped dies," *International Journal of Mechanical Sciences*, vol. 24, no. 8, pp. 479–493, 1982.

[15] G. W. Rowe, C. E. N. Strugess, P. Hartley, and I. Pillinger, *Finite-Element Plasticity and Metalforming Analysis*, Cambridge University Press, Cambridge, UK, 1991.

[16] H. S. Valberg, *Applied Metal Forming Including FEM Analysis*, Cambridge University Press, Cambridge, UK, 2010.

[17] R. H. Wagoner and J.-L. Chenot, *Metal Forming Analysis*, Cambridge University Press, Cambridge, UK, 2001.

[18] LS-DYNA Aerospace Working Group, *Modeling Guidelines Document Version 13-1*, 2013.

[19] LS-DYNA, "Theory Manual," 2006.

[20] Y. Guo, "Meshless methods in LS-DYNA: an overview of EFG and SPH," in *Proceedings of the LS-DYNA Seminar*, 2010.

[21] P. J. G. Schreurs, F. E. Veldpaus, and W. A. M. Brekelmans, "Simulation of forming processes, using the arbitrary Eulerian-Lagrangian formulation," *Computer Methods in Applied Mechanics and Engineering*, vol. 58, no. 1, pp. 19–36, 1986.

[22] J.-P. Ponthot and M. Hogge, "The use of the Eulerian-Lagrangian FEM in metal forming applications including contact and aadaptive mesh," in *Proceedings of the ASME Symposium on Advances in Finite Deformation Problems in Material Processing and Structures*, vol. 125, pp. 49–63, 1991.

[23] S. Ghosh and N. Kikuchi, "An arbitrary Lagrangian-Eulerian finite element method for large deformation analysis of elastic-viscoplastic solids," *Computer Methods in Applied Mechanics and Engineering*, vol. 86, no. 2, pp. 127–188, 1991.

[24] W. K. Liu, T. Belytschko, and H. Chang, "An arbitrary lagrangian-eulerian finite element method for path-dependent materials," *Computer Methods in Applied Mechanics and Engineering*, vol. 58, no. 2, pp. 227–245, 1986.

[25] C. W. Hirt, A. A. Amsden, and J. L. Cook, "An arbitrary Lagrangian-Eulerian computing method for all flow speeds," *Journal of Computational Physics*, vol. 14, no. 3, pp. 227–253, 1974.

[26] J. Donea, P. Fasoli-Stella, and S. Giuliani, "Lagrangian and Eulerian finite element techniques for transient fluid-structure interaction problems," in *Transactions of the 4th International Conference on Structural Mechanics in Reactor Technology*, pp. 1–12, 1977.

[27] J. M. Kennedy and T. B. Belytschko, "Theory and application of a finite element method for arbitrary Lagrangian-Eulerian fluids and structures," *Nuclear Engineering and Design*, vol. 68, no. 2, pp. 129–146, 1982.

[28] J. Argyris, J. S. Doltsinis, H. Fischer, and H. Wüstenberg, "Ta panta rhei," *Computer Methods in Applied Mechanics and Engineering*, vol. 51, no. 1–3, pp. 289–362, 1985.

[29] D. J. Benson, "An efficient, accurate, simple ale method for nonlinear finite element programs," *Computer Methods in Applied Mechanics and Engineering*, vol. 72, no. 3, pp. 305–350, 1989.

[30] Y. Huang and Z. Dai, "Large deformation and failure simulations for geo-disasters using smoothed particle hydrodynamics method," *Engineering Geology*, vol. 168, pp. 86–97, 2014.

[31] M. F. Villumsen and T. G. Fauerholdt, *Simulation of Metal Cutting using Smooth Particle Hydrodynamics*, LS-DYNA Anwenderforum, Metallumformung III, Bamberg, Germany, 2008.

[32] M. Prakash, P. Cleary, and J. Grandfield, "Modelling of metal flow and oxidation during furnace emptying using smoothed particle hydrodynamics," *Journal of Materials Processing Technology*, vol. 209, no. 7, pp. 3396–3407, 2009.

[33] P. W. Cleary, M. Prakash, and J. Ha, "Novel applications of smoothed particle hydrodynamics (SPH) in metal forming," *Journal of Materials Processing Technology*, vol. 177, no. 1–3, pp. 41–48, 2006.

[34] J. Kordilla, A. M. Tartakovsky, and T. Geyer, "A smoothed particle hydrodynamics model for droplet and film flow on smooth and rough fracture surfaces," *Advances in Water Resources*, vol. 59, pp. 1–14, 2013.

[35] C. S. Li, X. H. Liu, and G. D. Wang, "Ring upsetting simulation by the meshless method of corrected smooth particle hydrodynamics," *Journal of Materials Processing Technology*, vol. 183, no. 2-3, pp. 425–431, 2007.

[36] H. Gedikli, "Crashworthiness optimization of foam-filled tailor-welded tube using coupled finite element and smooth particle hydrodynamics method," *Thin-Walled Structures*, vol. 67, pp. 34–48, 2013.

[37] Y. Michel, J.-M. Chevalier, C. Durin, C. Espinosa, F. Malaise, and J.-J. Barrau, "Hypervelocity impacts on thin brittle targets: experimental data and SPH simulations," *International Journal of Impact Engineering*, vol. 33, no. 1-12, pp. 441–451, 2006.

[38] Y. Wang, Y. Zhou, and Y. Xia, "A constitutive description of tensile behavior for brass over a wide range of strain rates," *Materials Science and Engineering A*, vol. 372, no. 1-2, pp. 186–190, 2004.

[39] Y.-H. Xiao, C. Guo, and X.-Y. Guo, "Constitutive modeling of hot deformation behavior of H62 brass," *Materials Science and Engineering A*, vol. 528, no. 21, pp. 6510–6518, 2011.

[40] Livermore Software Technology Corporation (LSTC), *LS-DYNA Keyword User's Manual, Version 971*, vol. 1, Livermore Software Technology Corporation (LSTC), 2007.

[41] S. Bala, "Viscosity for Shell and Solids in LS-DYNA," 2007, http://blog2.d3view.com/bulk-viscosity-for-shells-and-solids-in-ls-dyna/.

[42] C. Espinosa, J. L. Lacome, J. Limido, M. Salaun, C. Mabru, and R. Chieragatti, "Proceedings of the 10th International LS-DYNA Users Conference, Metal Forming," in *Proceedings of the 10th International LS-DYNA Users Conference*, vol. 12 of *Metal Forming (3)*, pp. 9–20, 2008.

[43] S. Bala and J. Day, *General Guidelines for Crash Analysis in LS-DYNA*, Livermore Software Technology Corporation, 2012.

[44] S. Bala, SPH Contact Definitions, 2008, http://blog2.d3view.com/sph-contact-definitions/.

[45] L. Olovsson, M. Souli, and I. Do, *LS-DYNA—ALE Capabilities, (Arbitrary-Lagrangian-Eulerian), Fluid-Structure Interaction Modeling*, 2003.

[46] D. Varas, R. Zaera, and J. López-Puente, "Numerical modelling of partially filled aircraft fuel tanks submitted to Hydrodynamic Ram," *Aerospace Science and Technology*, vol. 16, no. 1, pp. 19–28, 2012.

[47] J. Wang, *Simulation of Landmine Explosion Using LS-DYNA3D Software*, Aeronautical and Maritime Research Laboratory, 2001.

[48] M. A. Barsotti, "Modelling mind blast with SPH," in *Proceedings of the 12th International LS-DYNA Users Conference*, Blast/Impact (1), pp. 1–12, 2012.

[49] J. Day, *Guidelines for ALE Modeling in LS-DYNA*, 2009.

[50] B. Massey, *Mechanics of Fluids*, Taylor & Francis, Abingdon, UK, 2006.

[51] H. Aretz, S. Keller, R. Vogt, and O. Engler, "Modelling of ductile failure in aluminium sheet forming simulation," *International Journal of Material Forming*, vol. 4, no. 2, pp. 163–182, 2011.

8

Knowledge Management for Topological Optimization Integration in Additive Manufacturing

Nicolas Gardan

CAD/CAE Research Centre DINCCS of MICADO, 08 005 Charleville-Mézières, France

Correspondence should be addressed to Nicolas Gardan; nicolas.gardan@dinccs.com

Academic Editors: J.-Y. Hascoet and T. R. Kurfess

Engineering design optimization of mechanical structures is nowadays essential in the mechanical industry (automotive, aeronautics, etc.). To remain competitive in the globalized world, it is necessary to create and design structures that, in addition to complying specific mechanical performance, should be less expensive. Engineers must then design parts or assemblies that are a better compromise between mechanical and functional performance, weight, manufacturing costs, and so forth. In this context Additive Manufacturing (AM) process offers the possibility to avoid tools and manufacture directly the part. There are numerous technologies which are using different kind of material. For each of these, there are at least two materials: the production material and the support one. Support material is, in most cases, cleaned and becomes a manufacturing residue. Improving the material volume and the global mass of the product is an essential aim surrounding the integration of simulation in additive manufacturing process. Moreover, the layer-by-layer technology of additive manufacturing allows the design of innovative objects, and the use of topological optimization in this context can create a very interesting combination. The purpose of our paper is to present the knowledge management of an AM trade oriented tool which integrated the topological optimization of parts and internal patterns.

1. Introduction

The optimization of various processes involved in the product life cycle management becomes very crucial. This is the reason why firms have always faced the problem of continuous improvement of their structures in order to predict the optimal forms and structures in the early stages of design. In the research of this compromise, working methods have considerably changed in recent years. Hardware performance, associated with the new methods for solving optimization problems, leads to new methodologies in the design phase.

The optimization of mechanical structures and parts has for a long time been based on a "try/error" methodology. This approach leads to test prototypes whose design is based on the experience of the engineer.

This manual approach is unthinkable today because of the costs and numerous iterations it can cause. The mechanical engineers are required now to integrate the optimization of mechanical structures in their first design phases. This new approach integrates the design requirements and the specifications such as manufacturing constraints, constraints related to business processes, thermomechanical performance, weight requirements, and cost. The engineers can then automatically reproduce, through modeling and numerical optimization software, the work that a designer manually achieved, by adding many advantages [1]:

(i) opportunity to scan a wider design space;

(ii) possibility to implement design of experiments and thus to create approximation functions if needed;

(iii) ability to reach an optimum by using more efficient algorithms.

There are three main categories of shape optimization of mechanical structures [2]: parametric shape optimization, geometric shape optimization, and topological shape optimization. The last one provides, without any explicit or implicit restriction, the best shape possible even if topology changes. It is an appropriate method for the design phase of a new part, because it can explore new concepts and solutions in areas difficult to reach/work for engineers (see a basic example in Figure 1).

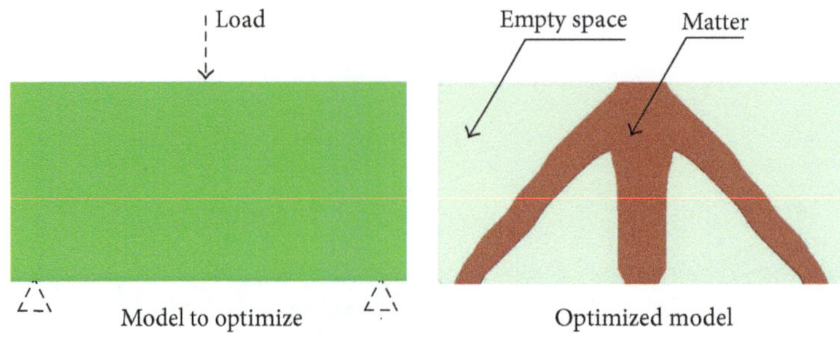

FIGURE 1: Simple example of a topological optimization.

The marriage between additive manufacturing, which can build almost any shape and topology optimization, seems obvious. Indeed, the topology optimization will provide innovative shapes but will require adaptation process from traditional manufacturing (typically a remodeling is required). The objective of this paper is to present the development of a methodology that will serve as a basis for developing a product that will be positioned upstream of the additive manufacturing machine. This software and the associated methodology are intended to be added on all types of additive manufacturing machines. The material and mass saving obtained through the digital optimization can apply for both plastics and metals. In the context of additive manufacturing centre NUM3D, we have access to a selective laser sintering (SLS) machine type, a photopolymerisation machine.... But the approach can be applied to another type of machine; the difference is in the knowledge capitalisation.

The aim of this paper is to present how we manage the knowledge for a good coupling between topological optimization and additive manufacturing. We consider the part and the material support as well (replacement of classic lattice) in our investigation.

2. Related Works

AM is nowadays widely used in industrial product development. The main advantage of the additive manufacturing concept used in AM is the ability to create almost any possible shape. This capacity is governed by the built up layer-by-layer process. There are several available technologies based on this additive machining concept [3]: stereolithography, photomasking, selective laser sintering, fused deposition modelling, 3D printing, and so on. Researchers work principally on the influence of part orientation, slicing strategy, and matching internal patterns to improve cost, product quality, built time, and so forth. Numerical topological optimization is a technical break which allows the modelling of really innovative shapes, based on trade knowledge. The union of the two technologies, AM and numerical topological optimization, seems to be very promising and more particularly for steel machining (for a real ROI on mass, e.g.).

The use of optimization in additive manufacturing [4] is generally in the context of the optimization of the build direction [5], parameter optimization trades, optimization construction layers algorithm, and so on. The optimization of the quantity of material used is an important goal. This optimization can match both the product material and the support material. Figure 2 shows the case of using a topology optimization on both the part and the support used (2 optimizations are performed separately. Optimization in the "design" zone is the area that can be optimized and the "nondesign" zone that cannot be changed).

AM machines generally offer the possibility of reducing the mass by using honeycomb shapes, lattices, and so forth. These algorithms model the simplified form without taking into account the specifications of mechanical strength. They are on the majority applied for internal gain of matter. Actually, there are many researches on the influence of cellular structures. Reference [6] has studied the influence of circular and rectangular shapes on the polyamide through compression tests. The study shows the influence of two types of geometric shape based on their use. The circular structure is able to absorb 43.5% more energy than a rectangular structure that behaves better still in high deformation rates (useful for dynamic fast as crash, explosions, etc.). Reference [7] investigated the use of lattice structures including rapid prototyping to lighten sandwich panels while maintaining their mechanical strength. The study enabled to determine the anisotropy directions of the lattice. The lattice modeling can be adjusted according to the specifications of mechanical strength. Other studies develop specific structures as curved [4], honeycomb [8], cell shape "tetrachirales," [9] or "hexachirales" [10]. However, these studies did not use the opportunity to integrate notion of mechanical strength allowing one to optimize the best shape. The topological optimization through numerical simulation can solve this problem. Reference [11] shows the interest to integrate the topological optimization but also highlights some difficulties such as the following:

(i) the difficulty to manage the drainage system of the part support;

(ii) the size of the CAD file and the difficulty of implementation.

The majority of research mainly focuses on the implementation of new forms (honeycomb lattice) but uses little power topology optimization in numerical simulation which allows one to model the shape automatically. Knowledge management is however necessary to obtain innovative form in the trade context. It is important to note the necessity to

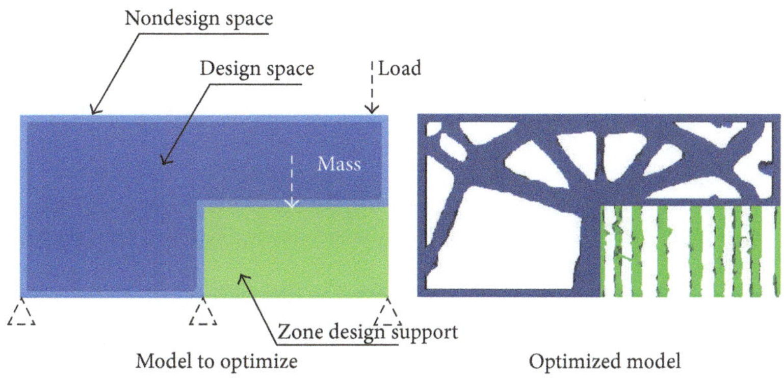

FIGURE 2: Simple example of part and support optimization.

manage the drainage system of the support part. Recently [12] developed a methodology which allows the production of topological optimized part by a low cost FDM. This methodology use the classic optimization process top optimize the mass of the part (included the skin of the part). There is no dedicated optimization tool for AM. Our aim is to develop such a tool. First we have to model knowledge of AM in the topological optimization context.

3. Methodology and Knowledge Management

Knowledge management becomes a true stake for the companies. The loss of consciousness due to retirement or turnover and the intrinsic dispersion of knowledge involve high notorious costs. Computer science and in particular artificial intelligence (AI) attempted to provide solutions to help the management of knowledge. AI, while trying in its beginnings to model human intelligence, opened research orientations to set up a decision-making help. The researchers introduce for this purpose the concept of knowledge bases which allow the separation between declarative and procedural statements. The first concrete results appeared with the rules based expert systems, for example, with the software of decision-making medical help MYCIN in 1974 [13]. To reduce production time and production cost, industries are interested in different technologies which depend on research on AI. Knowledge based system (KBS) is born due to AI researches. A KBS is a system which models the knowledge to use it in an application. For example, system expert is a special KBS. There are many researches in using a KBS to CAD modelling. Reference [14] developed an approach for generating assembly drawings from 3D computer assembly models of mechanical products. This approach uses a specific KBS which saves the strategies of an expert to reuse it. Reference [15] used KBS which includes an expert system to model machine tool. The implemented system uses specific functional features, parametric modeling, and an expert system to modify assembly of a machine tool by applying trade rules. Reference [16] developed a KBS for selection of cutting tools and conditions of turning operations for process planning. The KBS is composed of a base of rules, a base of constraints, a base of facts, and a module called "way of thinking" which represents the logical procedure of an expert for decreasing the solution space. A state of the art on KBS research is done in [17].

3.1. The Knowledge Based System. The knowledge based system (KBS) is dedicated to the AM process and topological optimization. The KBS is using production rules and constraints to represent the declarative knowledge. It is the most usual representation to model heuristic know-how. We are particularly interested in the scenario model [18], that is, the decomposition of a problem in a series of tasks [19]. We base ourselves on the fact that each time an expert resolves a problem, he runs in an intuitive way a scenario. Schank and Abelson [20] set up this structure by affirming that there are thousands of scenarios in the human memory. The different scenarios have been collected by tracking the work of experts.

Our structure is made up of a knowledge base including the following.

(i) A base of scenarios: as we said before experts translate their step of topological optimization and AM process into diagrams of sequences of tasks. They consequently use a scenario of optimization and AM process.

(ii) A base of rules: it contains the production rules and uses the base of constraints for this purpose. These rules are associative, which means that each formalized rules must contain at the same time the context of application of the rule and the condition of the processing of the rule. The rules are rules of production like If (Conditions) Then (Conclusions). The part "Conditions" is the process of the rules and the part "Conclusions" describes the actions to be started in the event of release.

(iii) A base of constraints: it contains all the mechanical laws and the message interaction. The base of constraints contains the set of the constraints having a relationship with the considered field. For example, a constraint can be like Wall_Thickness > 1 mm to point out the limit of manufacturing to avoid collapsing.

The KBS allows the user to describe the boundary conditions, loads, materials, and so forth. CAD system and CAE system (mesh generator and analysis manager) are fully knowledge driven. They are usable by either CAD expert or mechanical engineer and ensure a good usage and the results quality.

Parametric modeling put constraints on geometric model to design modification or to engineering analysis [21]. Constraints can be geometrical (one entity is tangent to another) or knowledge based (e.g., trade rules and constraints). Functional design has to be represented by the vocabulary of the domain. This vocabulary is then used in the definition of trade rules, constraints, and scenarios. Thus, we define specific expression called graphonumerical expression (GNE) and specific parameters called graphonumerical parameters (GNP). The particularity of GNE/GNP is to link the vocabulary to knowledge and to geometry. For example, a sentence like "We have to create at the bottom draft of 20 degrees" is transformed in GNE expression to "Create a the bottom draft of 20°." The GNE expression is saved in a dedicated module in the CAD software (here Rhinoceros3D) which automatically creates functional parameter (GNP). For this example, GNP will be "create (draft)(20°)(#GeomRef)" (#GeomRef is a keyword to interact with the CAD user to select geometrical entities).

A GNE contains three classes of different terms. These terms are different according to the context.

Here are the three classes with some examples of terms (several environments included):

(i) action class (create, destroy, modify, consult, remove, apply, to radiate, etc.);

(ii) object class (part, clearance, hole, rib, face, line, edge, point, etc.);

(iii) constraint class (at a tangent to, parallel to, symmetric to, diameter, mass, length, etc.).

A GNE can contain one or several GNP. A GNP is constructed automatically by the graphonumerical engine.

Capitalization has focused around the entire process from design to topological optimization then to AM. Each trade rule affecting the geometry was modelized as EGN. The central idea is to rely the AM parameters to get all information necessary for preparation of the topological optimization and by the way he CAD model. Thus, the knowledge necessary to set data, such as loads, boundary conditions, materials, was modeled to integrate the behavior of the part to the CAD model (Figure 3).

3.2. Knowledge Management in AM Process Machining. The aim of the AM knowledge management is to characterize functional parameters for an integration into topological optimization. It is necessary to identify the manufacturing limits to pack the topological optimization solver: it can be the minimum allowed thickness, the possible height without including reinforcement, the dimension of cleaning channels, and so on. We focus, for instance, on the knowledge capitalization on two AM machining processes: SLS and photopolymerization.

Many researches [22–26] present a process of AM machine qualification by manufacturing a test part to quantify defects and grounds. However, these approaches doing not allow pushing to the technological limits of the machine.

This work aims to quantify the defects inherent in each process by the parallel between possible measures in metrology and process-related settings. Our approach is different (and complementary) since we determinate influential parameters and their critical values according to using context, based on [27] research.

The experimental process to recover AM knowledge is based on two types of specimens:

(i) iso campus norm ones manufactured by AM process: their aims are to recover material behaviour (typically elastoplastic law) to use it in simulation (topological optimization). We use specific tools for the measure, for instance, 3D deformation Aramis camera;

(ii) dedicated shapes ones also manufactured by AM process: their aims are to capitalize knowledge management like determining influences parameters: thickness/height limit, pocket depth allowed for powder evacuation, and so on.

We can see on Figure 4 different manufacturing direction and shape of the test parts. Those configurations allow the determination of risk factors.

Our approach involves the study of three very important factors for the topological optimization:

(i) the minimum thickness printable and cleanable without part deterioration: we seek to maximize the minimum thickness of the wire cloth (final material) without loss of geometric and morphological qualities of the part;

(ii) the minimum diameter printable and cleanable without mechanical cleaning: the objective is to size the best channels dimensions for cleaning the internal structure of the piece (allow the powder evacuation);

(iii) the maximum height, in fact the ratio between the projected length and height of the part which may cause a falling down of the matter.

We develop design of experiment (DOE) for different tests: test of the laser temperature impact (for SLS machine), test thickness and height allowed (with cleaning process), test of the manufacturing orientation, test of the plate placement, and so on.

For example, in the case of SLS machine, the tested material is a rapid steel which is metallic powder coat with a polymer. The SLS process involves much knowledge capitalisation because the part is involved in the laser phase, then debinding phase, then infiltration phase, and so on. Before the infiltration phase the part is very breakable and requires specific dimension management. The DOE is based on features analysis. Many features have been studied: holes, slots, ribs, pads, pockets, and so on. They are used in different contexts. For example, the rib is studied in a free way and in an embedded way. We define fourteen tests parts which include this feature (as seen, e.g., in Figure 5) and 3 DOE to drive the experiment. For example, the first eight parts are dedicated to three specific features: vertical holes, slots, free and embedded ribs. The aim is to verify their manufacturability and ability

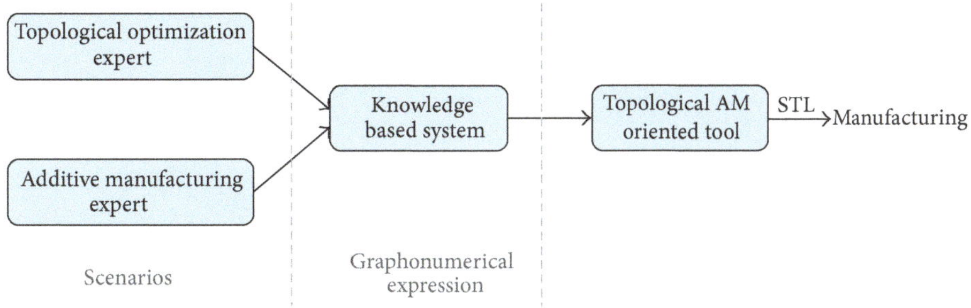

FIGURE 3: Global process.

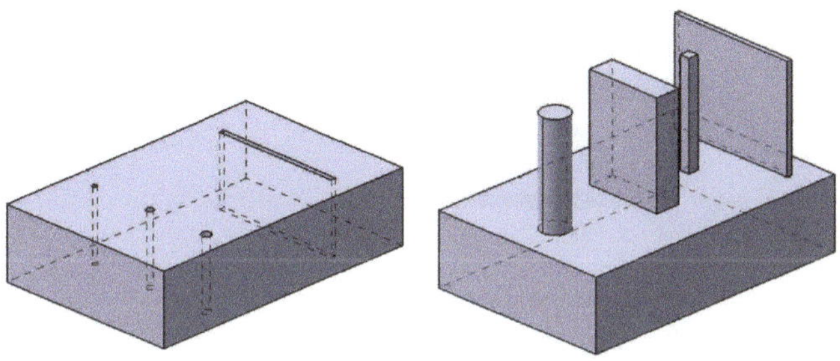

FIGURE 4: View of the different method of manufacturing.

to be cleaned up. The next algorithm presents a view of the different parameters used in one of the DOE.

[DOE plan 1 - Parts 1–8]

Hole_Diameter | Slot_Width = 3 mm | 5 mm

Hole_ Height | Slot_Height = 7 mm | 12 mm

Part_Height = 15 mm | 30 mm

PartSide_Distance = 5 mm | 30 mm

Laser_Power = 33 W | 35 W

Powder_Temperature = 80°C | 85°C

ScanningLaser_Direction = X | Y

Table 1 shows the result for free ribs which are used in optimization parameters.

To avoid too many manufacturing tests, we developed an associated numerical simulation test based on the DOE experiment. We used iso campus specimens in tensile testing machine to determine elastoplastic material behaviour.

3.3. Knowledge Management in Topological Optimization. In topological optimization we try to solve a specific structural problem (as seen in (1)):

$$\min \sum_{i=1}^{n} w_i \left(\frac{f_i(x) - T_i}{T_i} \right)^2 g_j(x) \le 0 \quad j = 1, m. \tag{1}$$

In this equation, $f_i(x)$ are structural responses obtained from a classic finite element analysis (FEA) and $f_i(x)$ is the constraints function. During the calculation, if $g = 0$ then constraints are active, if $g < 0$ then constraints are considered as inactive, and if $g > 0$ constraints are defined as violated, the calculation cannot reach the objective as well, defined by the target values T_i. The values of the weighting factors, w_i are chosen by experiment, in order to obtain a suitable rapid and stable convergence of the scheme [28]. In topology optimization design variables are element densities and are managed in our work by SIMP method. As seen before, specific techniques need to be introduced to penalize intermediate densities and to force final design to be represented by densities of 0 or 1 for each element. This particular adjustment is really important for the knowledge management. The penalization technique used is the power law representation of elasticity properties which can be expressed as shown in (2), where $\underline{K}$ is the penalized stiffness matrix, K the real stiffness matrix, ρ the density, and P the penalization factor which has to be superior to 1:

$$\underline{K}(\rho) = \rho^P K. \tag{2}$$

This penalized factor will have a very relevant impact on our problem. Indeed, in the case of AM, we have to manage the minimum length which represents the thickness of the different walls. Belonging to the machine process, this value is determined by experiment data. A simple example is shown in Figure 6: on a simple C-CLIP optimization, the penalization factor management allows a bad or a good management of knowledge.

A topology optimization problem relative to an AM process can be defined by the following.

(i) Design spaces: a design space corresponds to the interior of the objects and a nondesign space corresponds to the skin of the object (or any other area that should not be modified such as the apertures for cleaning). These areas are identified in CAD model.

(ii) Design variables: they are the set of parameters of the design space related to the AM process to define

TABLE 1: Results for rib DOE analysis.

Parameter	Operator	Value	Result
Thickness	<	1 mm	Nonfeasible (matter collapsing)
Thickness	=	2 mm	Deformation for height > 10 mm
Thickness	=	3 mm	Deformation for height > 40 mm
Width	>	15 mm	Cleaning constraint
To help the global strength, pins can be brought back			

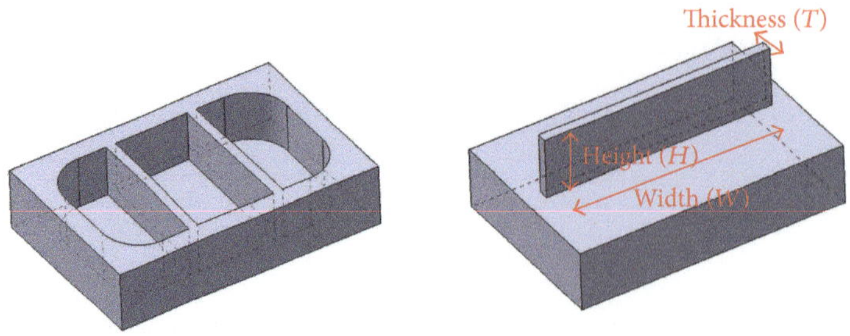

FIGURE 5: Two kinds of tested rib features.

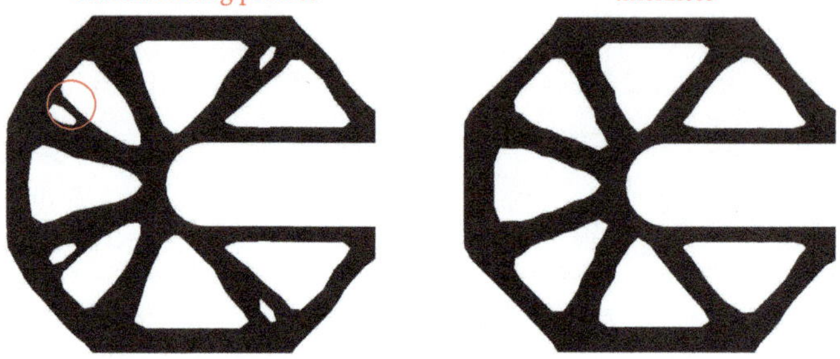

FIGURE 6: Penalization factor management on a C-CLIP example.

the initialization problem of topological optimization. We find here the penalization factor, the pattern repetition, and so on.

(iii) Responses: responses correspond to structural responses, calculated in a finite element analysis, or combinations of these responses to be used as objective and constraint functions in a structural optimization. Available responses could be for example, static displacement, mass, volume, temperature, natural frequency, and so on.

(iv) Constraints: constraints are based on responses by marking them with specific values.

(v) Objective: the objective function is, as we have seen before, the minimisation (or the maximisation) of the problem, here specific responses (e.g., the aim is to manage one response by objective function).

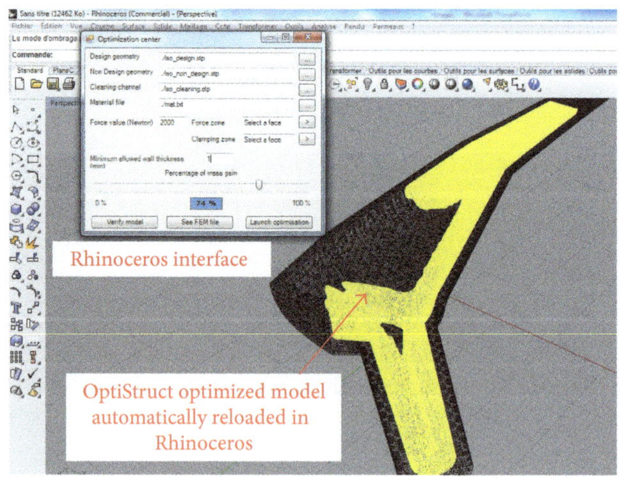

FIGURE 7: View of Rhinoceros program and half of the hip model.

4. Applications

To validate our methodology and prepare the software integration, we first verified our assertion with commercial software. We developed in Rhinoceros3D the KBS (in Python) and a basic interface which helps the designer to prepare the CAD model and launch in background Optistruct solver (see Figure 7).

For instance, the topological optimization scenario is limited to one objective: the weight gain.

The first step is to define design variables like the penalization factor as we explained before. This penalization factor is defined according to the minimal thickness obtain by test. We define then two specific responses:

(i) compliance response: the compliance is the strain energy of the structure and can be considered a reciprocal measure for the stiffness of the structure.

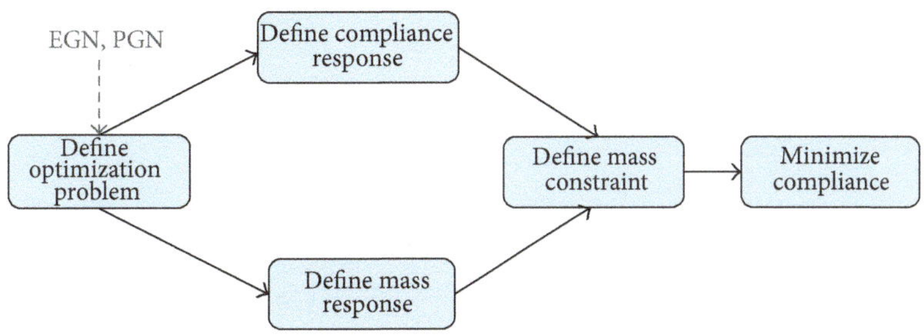

FIGURE 8: Mass scenario for topological optimization in AM.

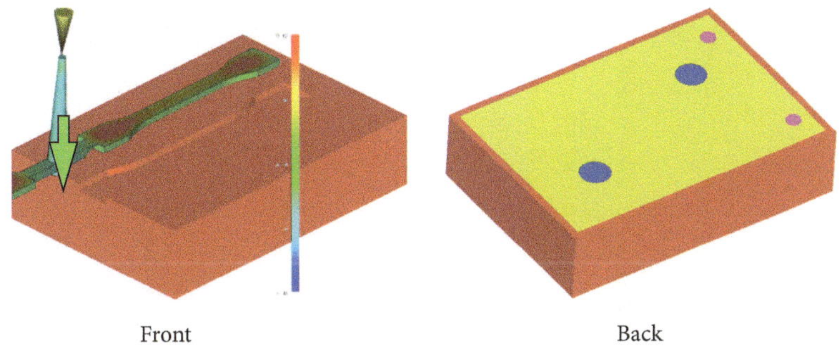

FIGURE 9: View of the insert and the different design spaces.

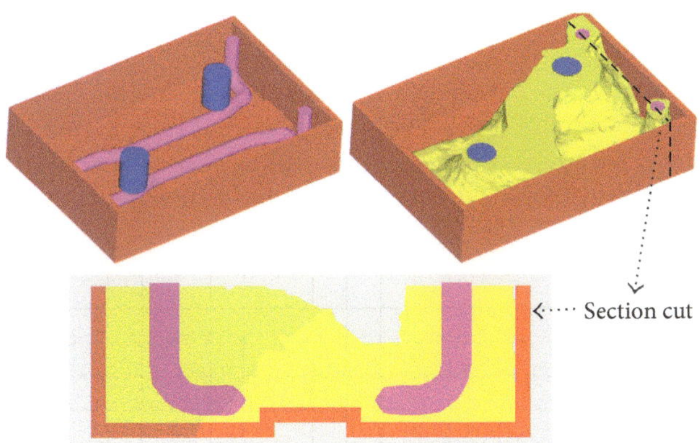

FIGURE 10: Set of the cleaning channels in the numerical model.

A global measure of the displacements is the compliance of the structure under the prescribed boundary conditions.

(ii) fraction of mass response: the fraction mass response is the material fraction of the designable material mass. It corresponds to a global response with values between 0 and 1. This allows the user to specify intuitive question like "I want to gain 30% of mass."

The next step is to minimize the compliance. Generally, in optimization, the compliance is used to evaluate the stiffness. Minimizing the compliance means to have a stiffer structure. The lower the compliance, the higher the stiffness of the structure. So, the problem statement involves the objective functional of the strain energy which has to be minimized (see the scenario in Figure 8).

A various number of application has been done to test the different scenarios. We limited our application for instance into to AM process available in the NUM3D platform: a SLS machine and a photopolymerisation machine (resp. DTM and EDEN 260V).

One of our examples is to obtain a significant weight gain on a mold insert for manufacture test specimen. The manufacturing is done on an EDEN 260 V. The objective of weight gain is fixed to 50% and the insert has to resist to temperature, pressure, and so on of the injection process. The knowledge relative to the EDEN 260 V and the constraints due to the using of the insert is modeled in the KBS. We can see on Figure 9 the nondesign spaces and the design space (in yellow).

One of the difficulties of this object is to correctly integrate the cooling channel especially for the support cleaning. We see on Figure 10 on the left the cooling channel and on the right the topological result.

Figure 11 shows the manufacturing process of the obtained shape with almost 47% of weight gain. The difference with the theoretical 50% target is explained by numerical adaptation like mesh size, material behavior and so on. It is important to note that there is no transformation between the optimized shape and the manufacturing process. We keep the AM principle based on the "one click" manufacturing based on the STL format.

5. Conclusion

We present in this paper the knowledge management developed to link topological optimization to additive manufacturing. This coupling is very interesting for many reasons: topological optimization provides very innovative shape

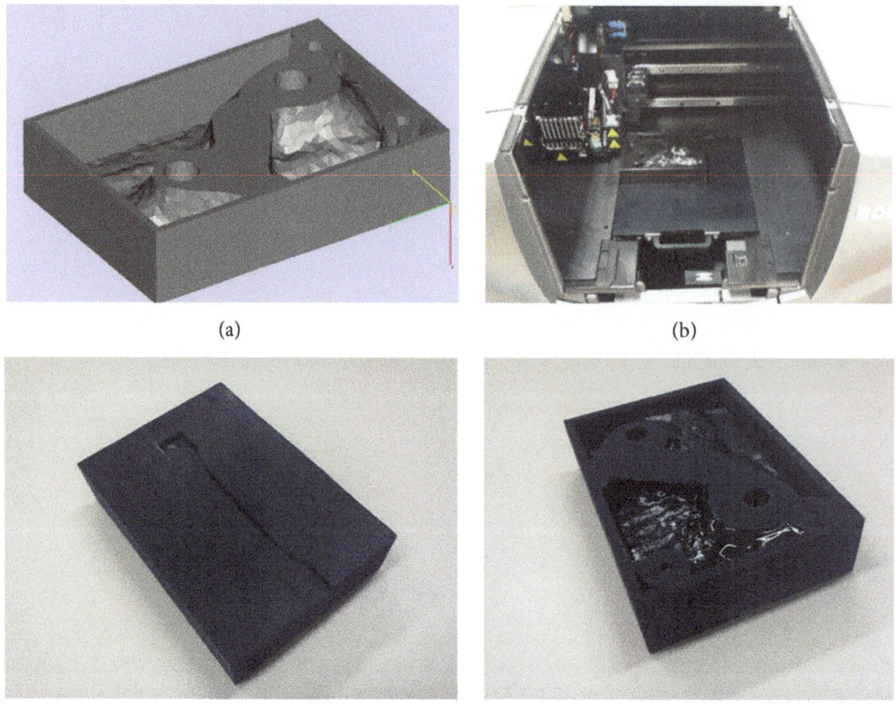

(a) (b) (c) (d)

Figure 11: View of manufacturing: (a) numerical model obtained with topological optimization dedicated to AM, (b) view of the manufacturing process, and (c) and (d) view of the manufactured insert.

but sometimes is difficult to manufacture with traditional manufacturing process and additive manufacturing, with the layer-by-layer concept, allows the production of almost any kind of shape.

Our approach is based on the definition of scenarios and the integration of AM knowledge with the help of graphonumerical expression. The demonstrator, developed for instance, in commercial software, allows the optimization in AM context of the part (traditional topological optimization but in AM context) and the interior of the part as well (to replace classic honeycomb lattice). This particularity allows keeping external skin and has a really weight gain by removing matter with full respect to mechanical strength. For instance, we test a dozen of parts which give very good results like a prosthetic implant used in a hip replacement surgical procedure. One of the improvements is to manage automatically the drainage system which has to be included manually, for instance.

Actually the project is currently growing up in a huge project. Many industrial partners are interested in the development of dedicated AM topological optimization tool.

Conflict of Interests

The author declares that there is no conflict of interests regarding the publication of this paper.

Acknowledgments

This project is partly based on the platform NUM3D (partners: UTT, ENSAM, and CReSTIC for URCA). The author also thanks the Champagne-Ardenne region and the General Council. Advice on AM material from Julien Gardan of Aztech society and help from Alexandre Schneider from the University of Reims deserve the authors' heartfelt thanks.

References

[1] G. Allaire, C. Dapogny, and P. Frey, "Topology and geometry optimization of elastic structures by exact deformation of simplicial mesh," *Comptes Rendus Mathematique*, vol. 349, no. 17-18, pp. 999–1003, 2011.

[2] A. Thakur, A. G. Banerjee, and S. K. Gupta, "A survey of CAD model simplification techniques for physics-based simulation applications," *Computer-Aided Design*, vol. 41, no. 2, pp. 65–80, 2009.

[3] T. Wohlers, "Additive manufacturing and 3D printing—state of the industry," Wohlers Report: Annual Worldwide Progress Report, 2012.

[4] L. M. Galantucci, F. Lavecchia, and G. Percoco, "Study of compression properties of topologically optimized FDM made structured parts," *CIRP Annals*, vol. 57, no. 1, pp. 243–246, 2008.

[5] A. M. Phatak and S. S. Pande, "Optimum part orientation in rapid Prototyping using genetic algorithm," *Journal of Manufacturing Systems*, vol. 31, pp. 395–402, 2012.

[6] M. Vesenjak, L. Krstulović-Opara, Z. Ren, and Ž. Domazet, "Cell shape effect evaluation of polyamide cellular structures," *Polymer Testing*, vol. 29, no. 8, pp. 991–994, 2010.

[7] Y. Sugimura, "Mechanical response of single-layer tetrahedral trusses under shear loading," *Mechanics of Materials*, vol. 36, no. 8, pp. 715–721, 2004.

[8] H. Abramovitch, M. Burgard, L. Edery-Azulay et al., "Smart tetrachiral and hexachiral honeycomb: sensing and impact

detection," *Composites Science and Technology*, vol. 70, no. 7, pp. 1072–1079, 2010.

[9] W. Miller, C. W. Smith, F. Scarpa, and K. E. Evans, "Flatwise buckling optimization of hexachiral and tetrachiral honeycombs," *Composites Science and Technology*, vol. 70, no. 7, pp. 1049–1056, 2010.

[10] D. Prall and R. S. Lakes, "Properties of a chiral honeycomb with a Poisson's ratio of—1," *International Journal of Mechanical Sciences*, vol. 39, no. 3, pp. 305–314, 1997.

[11] P. Rochus, J.-Y. Plesseria, M. van Elsen, J.-P. Kruth, R. Carrus, and T. Dormal, "New applications of Rapid Prototyping and Rapid Manufacturing (RP/RM) technologies for space instrumentation," *Acta Astronautica*, vol. 61, no. 1-6, pp. 352–359, 2007.

[12] R. Rezaie, M. Badrossamay, A. Ghaie, and H. Moosavi, "Topology optimization for fused deposition modeling process," *Procedia CIRP*, vol. 6, pp. 521–526, 2013.

[13] E. H. Shortliffe and B. G. Buchanan, *Rule-Based Expert Systems: MYCIN Experiments of the Stanford Heuristic Programming Project*, Addison-Wesley, Reading, Mass, USA, 1984.

[14] K.-Z. Chen, X.-A. Feng, and L. Ding, "Intelligent approaches for generating assembly drawings from 3-D computer models of mechanical products," *Computer-Aided Design*, vol. 34, no. 5, pp. 347–355, 2002.

[15] S. Myung and S. Han, "Knowledge-based parametric design of mechanical products based on configuration design method," *Expert Systems with Applications*, vol. 21, no. 2, pp. 99–107, 2001.

[16] S. C. Park, "Knowledge capturing methodology in process planning," *Computer-Aided Design*, vol. 35, no. 12, pp. 1109–1117, 2003.

[17] S.-H. Liao, "Expert system methodologies and applications—a decade review from 1995 to 2004," *Expert Systems with Applications*, vol. 28, no. 1, pp. 93–103, 2005.

[18] N. Gardan and Y. Gardan, "An application of knowledge based modelling using scripts," *Expert Systems with Applications*, vol. 25, no. 4, pp. 555–568, 2003.

[19] A. Aamodt and E. Plaza, "Case-based reasoning: foundational issues, methodological variations, and system approaches," *AI Communications*, vol. 7, no. 1, pp. 39–59, 1994.

[20] R. C. Schank and R. P. Abelson, *Scripts, Plans, Goals and Understanding: An Inquiry Into Human Knowledge Structures*, Lawrence Erlbaum Associates, Hillsdale, NJ, USA, 1977.

[21] F. Dancsi, N. Gardan, Y. Gardan, and M. Reimeringer, "P4LM: a methodology for product lifecycle management," *Computers in Industry*, vol. 59, no. 2-3, pp. 304–317, 2008.

[22] A. Bernard and A. Fischer, "New trends in rapid product development," *CIRP Annals*, vol. 51, no. 2, pp. 635–652, 2002.

[23] D. Kochan, C. C. Kai, and D. Zhaohui, "Rapid prototyping issues in the 21st century," *Computers in Industry*, vol. 39, no. 1, pp. 3–10, 1999.

[24] J.-P. Kruth, M. C. Leu, and T. Nakagawa, "Progress in additive manufacturing and rapid prototyping," *CIRP Annals*, vol. 47, no. 2, pp. 525–540, 1998.

[25] D. Scaravetti, P. Dubois, and R. Duchamp, "Qualification of rapid prototyping tools: proposition of a procedure and a test part," *The International Journal of Advanced Manufacturing Technology*, vol. 38, no. 7-8, pp. 683–690, 2008.

[26] T. H. C. Childs and N. P. Juster, "Linear and geometric accuracies from layer manufacturing," *CIRP Annals*, vol. 43, no. 1, pp. 163–166, 1994.

[27] B. H. Lee, J. Abdullah, and Z. A. Khan, "Optimization of rapid prototyping parameters for production of flexible ABS object," *Journal of Materials Processing Technology*, vol. 169, no. 1, pp. 54–61, 2005.

[28] M. P. Bendsøe and N. Kikuchi, "Generating optimal topologies in structural design using a homogenization method," *Computer Methods in Applied Mechanics and Engineering*, vol. 71, no. 2, pp. 197–224, 1988.

9

An Investigation of Abrasive Water Jet Machining on Graphite/Glass/Epoxy Composite

Deepak Doreswamy,[1] Basavanna Shivamurthy,[2] Devineni Anjaiah,[1] and N. Yagnesh Sharma[1]

[1]*Department of Mechanical and Manufacturing Engineering, Manipal Institute of Technology, Manipal University, Manipal 576104, India*
[2]*School of Engineering & IT, Department of Mechanical Engineering, Manipal University Dubai Campus, Dubai International Academic City, P.O. Box 345050, Dubai, UAE*

Correspondence should be addressed to N. Yagnesh Sharma; yagnesh.sharma@manipal.edu

Academic Editor: SeinLeung Soo

In the present research work, the effect of abrasive water jet (AWJ) machining parameters such as jet operating pressure, feed rate, standoff distance (SOD), and concentration of abrasive on kerf width produced on graphite filled glass fiber reinforced epoxy composite is investigated. Experiments were conducted based on Taguchi's L_{27} orthogonal arrays and the process parameters were optimized to obtain small kerf. The main as well as interaction effects of the process parameters were analyzed using the analysis of variance (ANOVA) and regression models were developed to predict kerf width. The results show that the operating pressure, the SOD, and the feed rate are found to be significantly affecting the top kerf width and their contribution to kerf width is 24.72%, 12.38%, and 52.16%, respectively. Further, morphological study is made using scanning electron microscope (SEM) on the samples that were machined at optimized process parameters. It was observed that AWJ machined surfaces were free from delamination at optimized process parameters.

1. Introduction

Fiber reinforced polymer composite is used in product manufacturing due to its distinct advantages such as lower weight, higher strength and stiffness, ability to mold into complex shapes, better corrosion resistance, and damping properties. In recent days, nanofillers such as graphite particles are impregnated with glass fiber reinforced polymer (GFRP) to enhance specific properties. Shivamurthy et al. [1], found that mechanical properties of glass/epoxy composite, namely, Young's modulus, tensile strength, flexural strength, impact strength, and wear resistance, show improvement with addition of graphite flakes. Such composites are highly suitable for manufacturing of bearing liners, gears, seals, cams, wheels, brakes, rollers, clutches, bushings, and so forth. Studies on electrical properties of graphite filled composites by Goyal and Kadam [2] and Bhattacharya et al. [3] revealed that these composites are also suitable to shield electromagnetic interference in electronic devices.

Traditional machining of GFRP composites for secondary operations (trimming and drilling) is more difficult than machining of metals. A review on traditional machining of composites by Teti [4] and Abrão et al. [5] highlights the problems such as exposure of fibers to environmental attack, material degradation due to localized heating, poor dimensional accuracy, shorter tool life, delamination, and fiber pullout due to anisotropic, nonhomogeneous, and abrasive nature of such composite materials. To address these challenges, advanced machining methods have been explored. Study of machining of carbon fiber reinforced polymer (FRP) composite by electrical discharge machining [6] and laser beam machining [7] reported the formation of recast layer on cut surface due to relatively higher localized temperature. AWJ machining is a relatively new manufacturing tool which has been realized to address limitations mentioned earlier. In AWJ machining, the machined surface does not suffer from thermal damage not only due to low heat that is generated during machining but also due to water that acts as coolant. The impacting abrasives exert smaller cutting force on the localized spots of the work piece. Machining requires simple fixtures to support the workpiece, and hence the process does not produce mechanical distortion on the cut surface. One of the major problems encountered while machining of FRP composite by traditional methods is formation of the fine dust which leads to air pollution that causes serious lung related health problems to the machine operators. In AWJ machining, the dust generated is carried away by the water jet and also no hazardous chemical is used in the process. Hence, AWJ is considered as environmentally friendly machining process. This machine tool can be effectively used for machining of FRP composites [8]. In AWJ machining, the mixture of abrasive and water is directed on the target material. The nozzle is attached to CNC control to produce required profiles on workpiece. Material removal rate (MRR) in AWJ machining depends on operating parameters and properties of target material. In addition, other parameters such as hardness, shape, size, and quantity of abrasives laced in the water jet play a vital role in MRR [9]. Huttunen-Saarivirta et al. [10] studied the effect of particle shape on AWJ erosion process using silica sand abrasive. It was observed that spherical shaped or blunt edged abrasives tend to create ductile fracture with low MRR and angular shaped abrasive with sharp edges resulting in brittle fracture contributing to higher MRR. The effect of particle size on erosion of titanium workpiece was studied by Yerramareddy and Bahadur [11] and ElTobgy et al. [12], using experimental and finite element analysis, respectively. The erosion rate was found to increase considerably for particle size up to 200 μm and remain constant for further increase in the particle size. Machining performance of various abrasives on glass workpiece was investigated by Khan and Haque [13]. Taper of the machined surface was found to increase with decrease in hardness of abrasive. Abrasive particles with higher hardness have better machining capability, but they have limitations like accelerating the wear of machine components and abrasive embedment on cut surface. Hence, AWJ machining industry uses garnet abrasive due to its specific advantages like low nozzle wear rate, good machinability, and economical availability [14–17].

In addition to abrasive properties, machining performance is also influenced by operating parameters such as jet angle, SOD, feed rate, number of cutting passes, jet pressure, abrasive flow rate, and nozzle geometry. The effect of abrasive impact angle on machining of ceramic material was investigated by Srinivasu et al. [18], using silicon carbide as abrasive. The effect of particle impact angle was also studied by Junkar et al. [19], using finite element analysis. Authors found that maximum material removal occurs at jet impact angle of 90°. Studies on the effect of jet pressure on cotton FRP composite by Wang and Guo [20] reported that the delamination occurs due to incapability of jet penetration into composite at lower operating pressures. Study of kerf taper angle produced on glass and graphite reinforced epoxy composite was made by Shanmugam et al. [21] and also developed a model to predict delamination length. Authors observed that, at higher operating pressures, considerable surface taper and delamination are found on the cut surface due to higher feed rates, flow turbulence, and brittle nature of composite material. Investigations on the effect of process parameters by Azmir and Ahsan [22] on glass fiber reinforced epoxy composites infer that abrasive hardness, operating pressure, SOD, and jet traverse rate were significant control factors which affect surface roughness (R_a) and a mathematical model was developed by authors to predict R_a. Further analysis of machined surface by Azmir and Ahsan [23] shows that at a jet angle of 90° glass fibers were found to be perfectly chopped. Alberdi et al. [24] studied the suitability of machinability model developed for metals to use it in composite materials. The machinability index was found to vary with thickness and composition of the composite. A comparative analysis of AWJ machining of metals in air and in submerged conditions is made by Haghbin et al. [25]. The study shows that machining under submerged conditions produced narrower kerf than the free jet machining.

In AWJ machining, the kerf profile produced depends on jet energy, jet exposure time on the workpiece, jet orientation, and material properties. Axinte et al. [26] developed a geometrical model to predict jet footprint (kerf) in maskless controlled milling applications. AWJ milling experiments were conducted on silicon carbide ceramic material at 90° jet impingement angle at various jet feed rates to validate the model. Kong et al. [27] developed a mathematical model to predict the jet foot prints for arbitrarily moving jets in single straight paths. Vundavilli et al. [28] used fuzzy logic based expert system such as simulated annealing and genetic algorithm to optimize process parameters and develop a mathematical model to predict depth of cut. Zain et al. [29] also used soft computing techniques to optimize the process parameters that produced low surface roughness. Billingham et al. [30] developed model that predicts jet foot prints of the overlapped single and multiple straight paths. Narayanan et al. [31] developed model to predict the jet energy distribution by considering parameters such as abrasive particle size distribution and the effect of particle fracture. Nouraei et al. [32] developed surface evolution model that can predict the shape features in micromachining of brittle material such as borosilicate glass. These models were found to be powerful tools to develop advanced jet path strategies on complex geometries using CAD/CAM by considering various process parameters including jet exposure time and orientation. From the literature review, it is noticed that generally addition of graphite improves mechanical, tribological, and electrical properties of GFRP composite material due to which the composite finds a wide range of industrial applications. AWJ machining is the suitable method for processing of such composite materials. The literature review establishes the need to investigate the effect of operating parameters on the machinability characteristics of graphite laced GFRP. Hence, this paper attempts to explore the effect of process parameters, namely, jet operating pressure, SOD, feed rate, and abrasive concentration on kerf width, while machining of graphite filled GFRP composite and their optimization using Taguchi method. In addition, morphological study is carried out using SEM on the cut surface machined at optimum and few selected process settings.

2. Materials and Methods

2.1. Materials Used. Properties of polymer composites are enhanced by impregnating the additives. In the present research, graphite of 200 μm size in the particulate form was used as filler in glass fiber reinforced epoxy laminate. Bisphenol-A based epoxy resin was used as matrix; and bidirectional glass fabric of 200 g·m^{-2} was used as reinforcement to prepare the composite. Addition of graphite to the glass epoxy laminate improves wear resistance with low coefficient of friction as well as toughness. Since graphite also acts as solid lubricant, the machinability of the laminate improves. Machining is carried out by using garnet abrasive of the size 80 mesh.

2.2. Fabrication of Composite Specimen. Graphite filled GFRP laminates were prepared using hand lay-up process as shown in Figure 1. The material compositions for composite were graphite 3% by weight, glass fibers 50% by weight, and epoxy 47% by weight. Bidirectional glass fabric was cut into square shape of 250 mm × 250 mm. Fabric was impregnated with the graphite filled epoxy and layered one above the other. The layered green laminate was transferred to hot compression molding press and compressed with uniform pressure. Curing was done for 2 hours in compression mode at 140°C. Postcuring was done at 180°C for 8 hours in free hanging condition in the furnace. The final thickness of laminate was found to be 3 ± 0.1 mm.

2.3. Characterization of Composite and Garnet Particles. The fabricated composite was cut into specimen size as per ASTM D638 standard. Tensile properties were estimated using a universal testing machine (Instron 3366). The hardness of the composite was measured using computerized microhardness tester (Matsuzawa-MMTX7A). The morphology of garnet abrasive and AWJ machined surfaces were analyzed using scanning electron microscope (Zeiss EVO 18, Germany). The particle size analysis of garnet abrasive was carried out as per ASTM D422 standard. The chemical composition of garnet abrasive was estimated by elemental mapping method using SEM. The kerf widths of machined surfaces were measured by tool room microscope (Mitutoyo) and the surface roughness was measured by Surtronic 3+ (Taylor Hobson).

2.4. Plan of Experiments. Design of experiment (DOE) is a powerful statistical tool to optimize the machining process parameters economically. Many researchers have used DOE to optimize the AWJ process parameters for machining various materials [23, 33]. In the present work, experiments were designed using Taguchi's fractional factorial orthogonal array. The AWJ machining parameters and levels of the experiments were chosen based on the literature review followed by trial experiments. Table 1 shows the machining parameters that were chosen to study the performance of AWJ machining of graphite filled GFRP composite.

The main effect of operating pressure (A), feed rate (B), abrasive concentration (C), and SOD (D) as well as interaction effect of AB, AC, and BC on the response parameter (kerf width) is considered for designing the experiment. The degree of freedom for the present experimental condition is 21 (which means the minimum number of experiments that must be conducted to estimate the effect is 21). The nearest suitable fractional orthogonal array for four factors of each three levels is L_{27} (4^3). This design is used for experiments conducted. The maximum material removal from the workpiece occurs at jet impact angle of 90° [18] and the garnet is the best suitable abrasive for AWJ machining polymer composite [13]. In the present investigation, irregular and angular shaped garnet abrasive particles were uniformly dispersed with water in the mixing chamber and jet was expelled through the nozzle at 90° for machining. Combinations of machining parameters used in the present study are shown in Table 4.

Table 1: Machining parameters.

Machining parameter/levels	1	2	3
Operating pressure (MPa)	90	120	150
Feed rate (mm/min)	75	100	125
Abrasive concentration (wt.%)	6	10	14
Standoff distance (mm)	1	3	5

Table 2: Mechanical properties of the target material.

Tensile strength	Young's modulus	% elongation	Hardness
255 MPa	8.3 GPa	3.2%	VHN 30

2.5. Experimental Setup. A three-axis CNC AWJ machine (supplied by Dardi Water Cutter Co. Ltd., make: DWJ1525-FA, maximum operating pressure: 400 MPa, repeatability accuracy: ±0.025 mm) was used for experimentation. The experimental setup used for machining is shown in Figure 2. The garnet abrasive water jet was directed at 90° on to the surface of graphite glass epoxy composite through a carbide nozzle of 0.8 mm focus tube diameter. A length of 20 mm cut was made on workpiece in each trial.

3. Results and Discussion

3.1. Mechanical Properties of Target Material. The tensile properties of graphite filled GFRP laminates are determined using universal testing machine (model: Instron 3366). The tensile tests are carried out on three test specimens (size as per ASTM D-638 of standard Type 5) and their average results are tabulated in Table 2. The surface hardness of the composite is determined using Vickers hardness test as per the standard loading procedure. A load of 300 grams is applied on the test specimen using 136° diamond indenter for a duration of 20 seconds and the Vickers hardness number is measured using computerized microhardness tester (make: Matsuzawa-MMTX7A). This test is conducted on 10 different locations on the specimen and the mean value of the Vickers hardness number is found. The mechanical properties of the workpiece are as shown in Table 2.

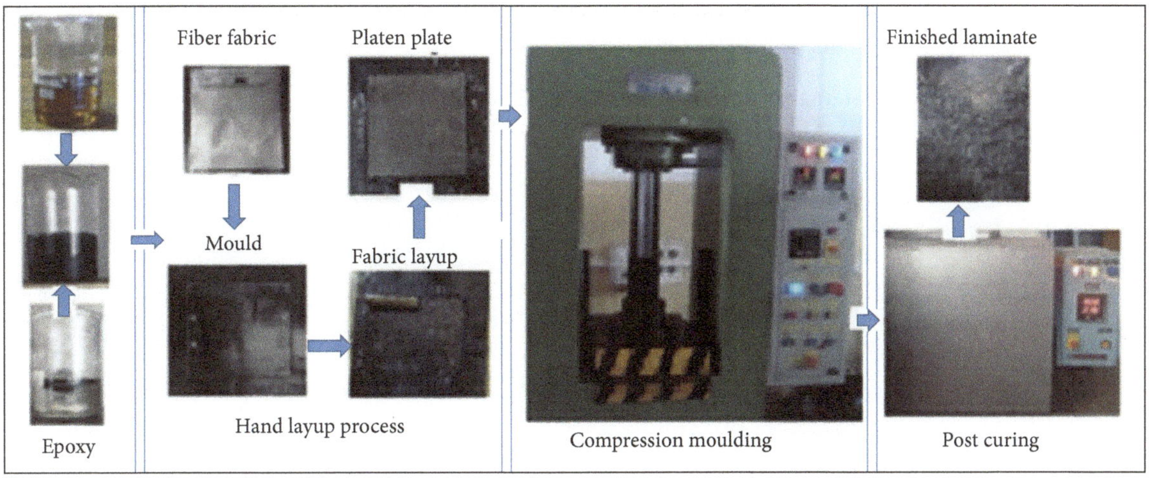

FIGURE 1: Fabrication process of graphite filled glass epoxy composite laminate.

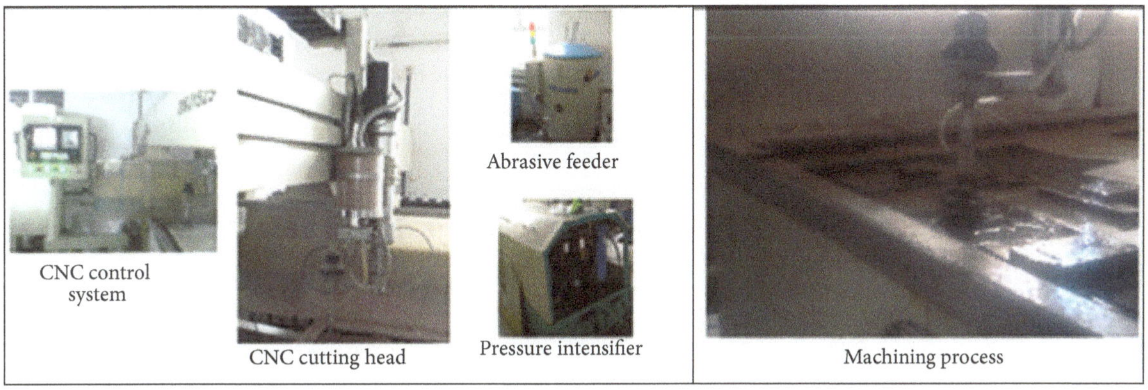

FIGURE 2: Experimental setup.

TABLE 3: Mineral composition of garnet abrasive.

Element	C	O	Mg	Al	Si	Ca	Fe
Quantity (wt.%)	6.62	42.84	1.93	5.82	9.13	1.23	32.45

3.2. Characterization of Garnet Abrasive Particles. The morphology, particle size analysis, and chemical composition of the garnet abrasive were investigated using SEM (Zeiss EVO 18 special edition) before mixing into water jet. The garnet abrasives particles were coated with thin layer of conducting material (gold) by sputtering method to prevent charging of a specimen with an electron beam in SEM. Figure 3 shows the SEM image of garnet abrasive. It is observed that the abrasives have irregular sharp edges. The particle size of the garnet abrasive was investigated as per ASTM D 422. It was found that the range of particle size was 80–240 μm and average particle size was around 174 μm. The mineral composition of garnet abrasive as per elemental mapping method by weight percent is shown in Table 3.

3.3. Effect of AWJ Process Parameters on Kerf Width. In the present work, the effect of AWJ process parameters, namely, jet pressure, feed rate, abrasive concentration, and SOD, on the top and bottom kerf width is investigated using L_{27} orthogonal array. The number of experiments conducted was 27 with two replications. Levels of process parameters in each run were set as mentioned in Table 4. In each run, a cut of 20 mm length was made on the test sample using a 3-axis CNC AWJ machine. The width of cut at upper and lower surface is called top kerf width (TKW) and bottom kerf width (BKW), respectively. Kerf widths (responses) were measured by using tool room microscope with an accuracy of

FIGURE 3: SEM image of garnet abrasive.

TABLE 4: Plan of experiment and measured output parameters.

Trial	Process parameters				Average response values	
	Pressure (MPa) (1)	Feed rate (mm/min) (2)	Abrasive concentration (wt.%) (5)	SOD (mm) (9)	TKW (mm)	BKW (mm)
1	90	75	6	1	0.845	0.803
2	90	75	10	3	0.905	0.892
3	90	75	14	5	0.964	0.936
4	90	100	6	3	0.896	0.853
5	90	100	10	5	0.950	0.912
6	90	100	14	1	0.847	0.827
7	90	125	6	5	0.870	0.835
8	90	125	10	1	0.823	0.816
9	90	125	14	3	0.860	0.850
10	120	75	6	3	0.960	0.913
11	120	75	10	5	1.176	1.150
12	120	75	14	1	0.890	0.868
13	120	100	6	5	1.104	1.055
14	120	100	10	1	0.875	0.845
15	120	100	14	3	0.981	0.977
16	120	125	6	1	0.855	0.803
17	120	125	10	3	0.917	0.852
18	120	125	14	5	0.987	0.935
19	150	75	6	5	1.203	1.155
20	150	75	10	1	0.938	0.918
21	150	75	14	3	1.052	1.005
22	150	100	6	1	0.903	0.872
23	150	100	10	3	0.965	0.887
24	150	100	14	5	1.150	1.015
25	150	125	6	3	0.892	0.880
26	150	125	10	5	1.065	0.992
27	150	125	14	1	0.882	0.871

0.005 mm. In each sample, the kerf widths were measured at 5 equidistant points along the length of the cut and average values were tabulated in Table 4.

3.3.1. Effect of Jet Pressure on Kerf Width. Figure 4(a) shows the effect of pressure on top and bottom kerf widths. The percentage contribution of jet pressure on TKW is 24.72% and on BKW is 19.68%. Both top and bottom kerf widths increase significantly with increase in operating pressure. This is due to the influence of increased kinetic energy and flow turbulence of the water jet at a higher pressure. The flow turbulence promotes interabrasive particle collision and jet expansion leading to increased kerf widths. Further, as the jet proceeds through the graphite filled glass epoxy composite, it loses part of its kinetic energy. Also, the abrasive particles impacting on the surface lose their sharpness and a few abrasive particles penetrate into the surface while a few other particles retard towards the top surface. Because of these reasons, the bottom kerf width is smaller than the top kerf width.

3.3.2. Effect of Feed Rate on Kerf Width. Figure 4(b) shows a significant effect of jet feed rate on kerf widths. At lower feed rate, a larger number of abrasive particles participate in the cutting action and hence produce wider kerfs. As the feed rate is increased, it is found that a reduced number of abrasive particles participate in cutting action and also tend to reduce the possibility of interaction of rebounding of particles with machined surface. Because of these reasons, the kerf widths reduce with increase in feed rate. The percentage contribution of feed rate on top and bottom kerf width is 12.38% and 15.31%, respectively.

3.3.3. Effect of Concentration of Abrasive on Kerf Width. Marginal increase in kerf width is observed with increased concentration of abrasives as shown in Figure 4(c). Increase in abrasive concentration in the water increases the mass density and marginally reduces the jet velocity. At higher concentration, the dense packing of abrasives in the jet increases the possibility of interparticular collision that results in jet expansion by disintegration of the flow. This

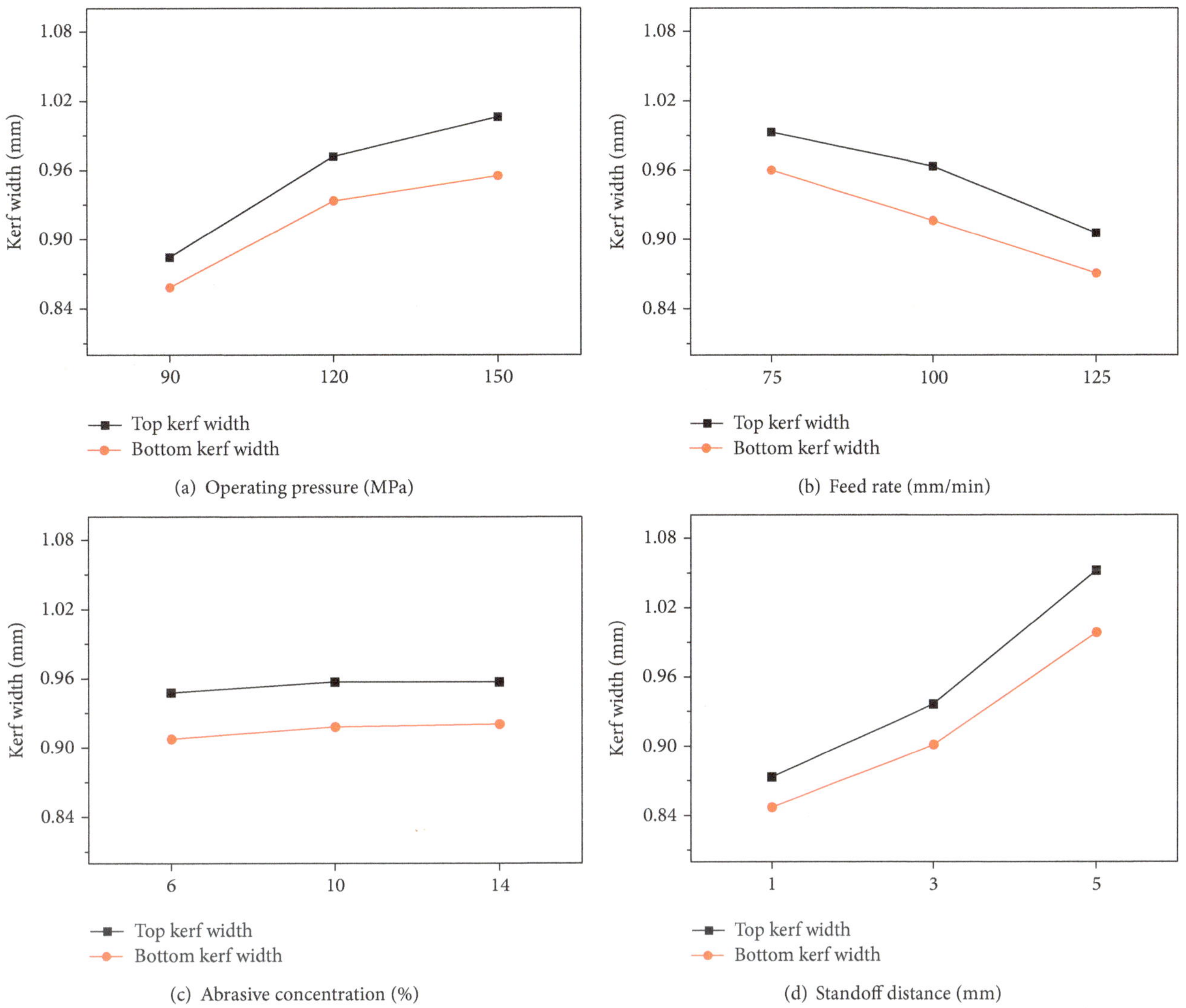

FIGURE 4: Main effect plots of top and bottom kerf width.

causes more abrasives particulates to impinge on the cutting surface during machining, thus contributing to increase in kerf width. However, the percentage contribution of abrasive concentration on top and bottom kerf width is 0.19% and 0.35%, respectively.

3.3.4. Effect of SOD on Kerf Width. Figure 4(d) represents the effect of SOD on top and bottom kerf widths. Increase in SOD results in significant increase in kerf widths. As the jet is discharged through the nozzle, expansion and disintegration of the jet takes place due to outside boundary interaction. This leads to increase in top kerf width. As the jet kinetic energy reduces because of disintegration and expansion of the jet, the bottom kerf width is found to decrease. The percentage contribution of SOD on top and bottom kerf width is 52.16% and 44.92%, respectively. The jet expansion increases with increase in the distance of jet traverse in air after ejecting from nozzle. Higher jet expansion leads to increase in top kerf width and also higher loss of jet kinetic energy. Hence, at higher SOD, the machining capability of the jet also decreases.

Further, it is found from the main effects plots shown in Figure 4 that the mean value of the top kerf width (0.954 mm) is higher than the bottom kerf width (0.913 mm). The difference in the mean kerf widths is observed to be 0.041 mm.

3.4. Analysis of Variance (ANOVA) Results. The kerf widths were analyzed using ANOVA method to find the effect of process parameters at 95% confidence level using F (variance ratio) statistics as shown in Table 5. It is observed that $F_{\text{calculated}}$ is greater than $F_{(\alpha=.005,6)}$ for the machining parameters such as SOD, jet pressure, and feed rate. A small variation of these parameters contributes to a considerable variation on the kerf width. Hence, it is clear from ANOVA that jet pressure, SOD, and feed rate are significant process parameters. However, for abrasive concentration, $F_{\text{calculated}}$ is found to be less than $F_{(\alpha=.005,6)}$. The effect of variation in

Table 5: ANOVA for top and bottom kerf width.

Source	DF	Top kerf width (TKW)				Bottom kerf width (BKW)			
		SS	MS	F	% SS	SS	MS	F	% SS
Pressure (A)	2	0.07027	0.03514	14.08	24.72	0.04636	0.02318	9.14	19.68
Feed rate (B)	2	0.03521	0.01761	7.05	12.38	0.03609	0.01805	7.12	15.31
Abrasive concentration (C)	2	0.00054	0.00027	0.11	0.19	0.00084	0.00042	0.17	0.35
SOD (D)	2	0.14829	0.07415	29.70	52.16	0.10588	0.05294	20.87	44.92
$A * B$	4	0.00368	0.00092	0.37	1.29	0.01212	0.01215	1.19	5.14
$A * C$	4	0.00481	0.00537	0.48	1.68	0.00618	0.00618	0.61	2.62
$B * C$	4	0.02149	0.00121	1.40	7.55	0.01299	0.01299	1.28	5.15
Error	6	0.01497	0.00249	—	5.13	0.01521	0.00253	—	6.45

DF: degree of freedom, MS: mean square, SS: sum of square, and % SS: percentage contribution.

Table 6: Mean value of top kerf widths (mm).

Control parameter/levels	1	2	3
Pressure (MPa)	**0.885**	0.972	1.01
Feed rate (mm/min)	0.995	0.964	**0.906**
Abrasive concentration (%)	**0.948**	0.958	0.957
SOD (mm)	**0.874**	0.937	1.053

Table 7: Results of confirmation experiments.

Trials	1	2	3	4	5
Top kerf width	0.780	0.802	0.792	0.781	0.778
Bottom kerf width	0.761	0.792	0.758	0.749	0.749

Table 8: Test results of regression model.

Trial	Experimental		Predicted	
	TKW	BKW	TKW	BKW
1	0.785	0.761	0.845	0.794
2	0.776	0.748	0.864	0.827
3	0.849	0.833	0.934	0.895
4	0.821	0.802	0.856	0.831
5	0.910	0.902	0.990	0.934

abrasive concentration on kerf width is not significant. Based on the F values, the significance of machining parameters can be ranked as follows: SOD (1), pressure (2), feed rate (3), and abrasive concentration (4). Further, it is observed from the interaction plot shown in Figure 5 and ANOVA values that there is no significant interaction effect between jet pressure with feed rate and jet pressure with abrasive concentration as well as between feed rate and abrasive concentration.

3.5. Optimization and Regression Modeling of AWJM Process Parameters. Process parameters were optimized to produce smallest kerf width on the workpiece. Using the experimental results shown in Table 4, the mean value of the top kerf width is calculated at each level of process parameters as shown in Table 6. It is seen that smallest kerf width is obtained at an operating pressure of 90 MPa, feed rate of 125 mm/min, SOD of 1 mm, and 6% abrasive concentration. At these optimum settings of process parameters, the predicted top kerf width is 0.756 ± 0.068 mm and bottom kerf width is 0.731 ± 0.068 mm. Further, the confirmation experiments were conducted and test results are shown in the Table 7. It is seen that kerf width obtained by experiments agree well with the expected optimized kerf width.

A mathematical model is developed to establish relationship between the process parameters (x_1: pressure, x_2: feed rate, x_3: abrasive concentration, and x_4: SOD) and response parameters, that is, TKW: y_{TKW} and BKW: y_{BKW}. The main effect, squared terms, and interaction effect of process parameters are considered in the modeling. The coefficients of determination (R^2) for the fitted regression equations are 96.86% and 91.52% for TKW and BKW, respectively. The predicting accuracy of regression model is tested by conducting confirmation experiments within the range of operating parameters, that is, $90 \geq x_1 \leq 150$; $75 \geq x_2 \leq 125$; $6 \geq x_3 \leq 14$; $1 \geq x_4 \leq 5$. The kerf widths predicted by regression model and the kerf widths obtained by experiments are shown in Table 8. The developed regression models are given below:

$$\begin{aligned} y_{\text{TKW}} = {} & 0.086850 + 0.008726x_1 + 0.005988x_2 \\ & - 0.010031x_3 + 0.000050x_4 - 0.000020x_1x_2 \\ & + 0.000076x_1x_3 + 0.000552x_1x_4 + 0.000096x_2x_3 \\ & - 0.000559x_2x_4 - 0.000486x_3x_4 - 0.000030x_1^2 \\ & - 0.000023x_2^2 - 0.000302x_3^2 + 0.006542x_4^2, \end{aligned} \quad (1)$$

$$\begin{aligned} y_{\text{BKW}} = {} & 0.107786 + 0.009745x_1 + 0.001111x_2 \\ & + 0.005242x_3 + 0.050461x_4 - 0.000017x_1x_2 \\ & - 0.000032x_1x_3 + 0.000317x_1x_4 + 0.000119x_2x_3 \\ & - 0.000614x_2x_4 - 0.002175x_3x_4 - 0.000029x_1^2 \\ & - 0.000001x_2^2 - 0.000260x_3^2 + 0.005417x_4^2. \end{aligned} \quad (2)$$

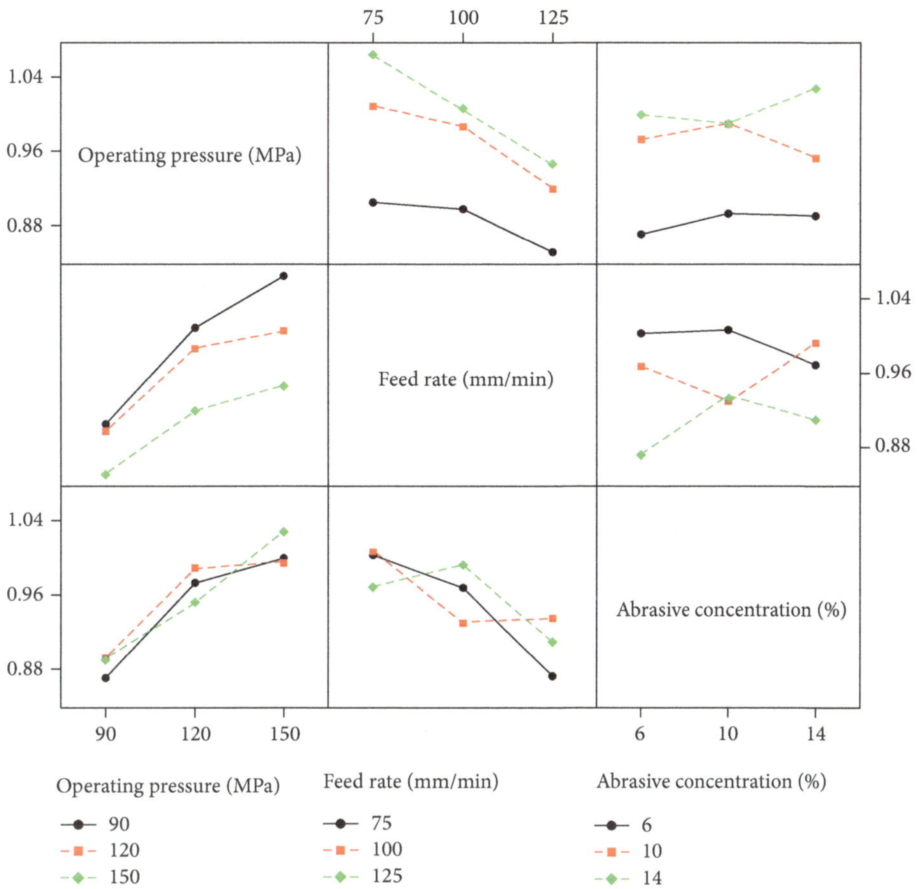

FIGURE 5: Interaction plot of process parameters.

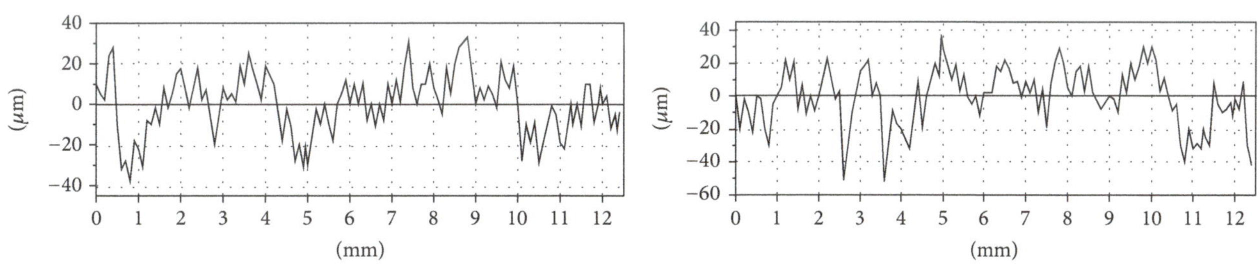

(a) R_a values at the upper cut region ($R_a = 3.73\,\mu$m)

(b) R_a values at the lower cut region ($R_a = 4.82\,\mu$m)

FIGURE 6: Plots of surface roughness on cut surface.

3.6. Surface Roughness and Morphology

3.6.1. Surface Roughness. The roughness (R_a) of the surface generated at optimum settings of machining parameters was measured at a distance of 1 mm from top and bottom edge of the machined surface. The surface roughness plots obtained by Taylor Hobson Surtronic 3^+ are shown in Figures 6(a) and 6(b). The plots reveal that the cut surface at jet entry region is smoother compared to jet exit. This is due to synergetic effect of loss of jet energy and addition of residual chips combined with abrasives when AWJ is penetrating through the workpiece.

3.6.2. Morphology. Morphological study has been carried out using SEM on the specimens that were machined at the optimum process settings which generate small kerf width. It is seen from Figure 7(a) that the top and bottom edges of the kerf are free from delamination. It is also observed from Figure 7(b) that there are inconsistent and nonuniform pits at few locations on top kerf due to collision between abrasive particles which richochiates from the edges. Further, SEM analysis is carried out to understand the microlevel of destruction in the composite. Figures 7(c) and 7(d) show the morphology of cut surface when fibers are oriented at 90° and are parallel to jet traverse, respectivly. It is seen that

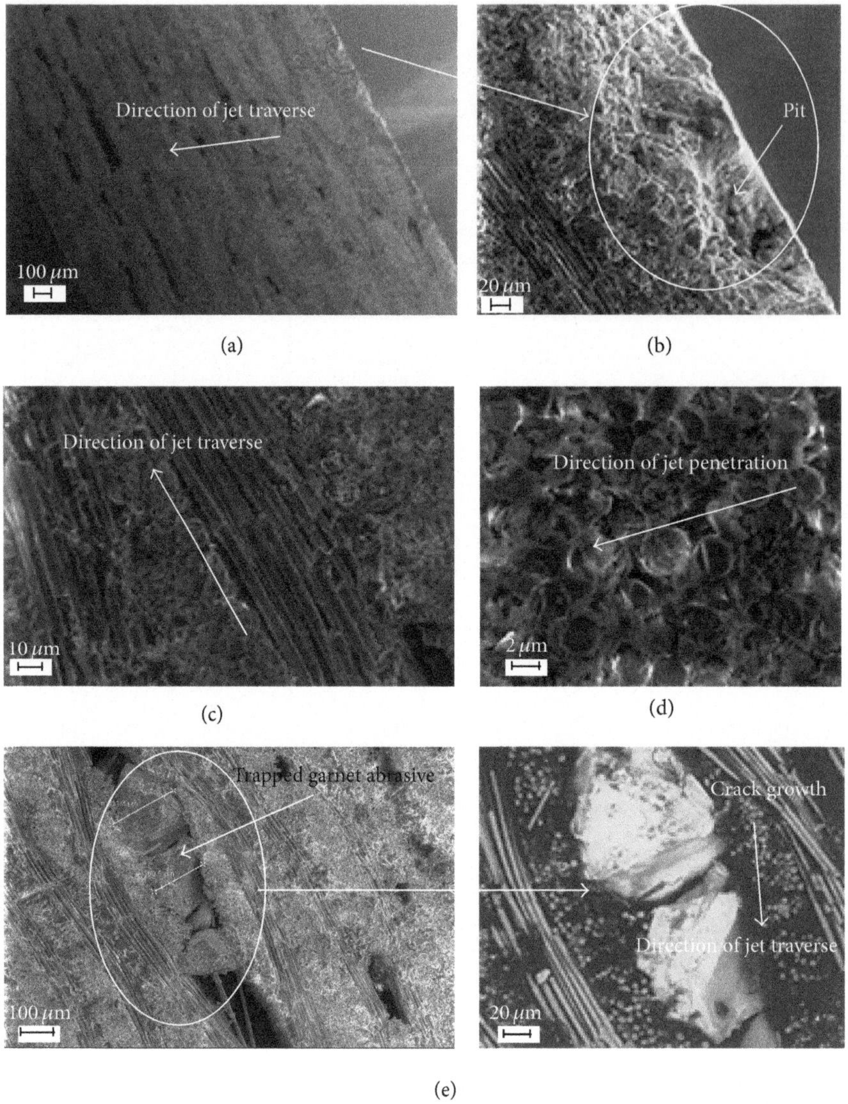

FIGURE 7: SEM images of (a) machined surface at the optimum process settings, (b) magnified view of top kerf damage, (c) fibers in plane of jet traverse, (d) fibers at normal to direction of jet traverse, and (e) Abrasive embedment on the surface machined at nonoptimum settings.

the fibers are cut accurately and no abrasive embeddment is found accoss the machined surface.

In addition, morphological analysis is also carried out on the cut surfaces that were machined at selected nonoptimum process settings (jet operating pressure: 150 MPa, feed rate: 75 mm/min, and SOD: 5 mm) to compare the surface morphology. The cut surface exhibited the abrasive embeddment as shown in Figure 7(e) due to the increase in flow turbulance of the jet at higher operating pressure and SOD. A crack (length of 6 to 8 times the diameter of the abrasive particle) is found along the direction of jet traverse due to abrasive embedment. Further, it is seen from Figure 7(e) that the crack has widened due to the penetration and wedging action of abrasives into the surrounding fibers and the matrix leading to delamination. Also, a few chopped fibers are found in the crack and a few fibers were carried away by the jet. From the morphological analysis, it is concluded that the surface machined at optimum process settings produced the cut surface which is free from delamination.

4. Conclusions

The following conclusions have been drawn from the present research work.

(i) The kerf width increased with an increase in operating pressure (contribution: TKW, 24.72% and BKW, 19.68%) and standoff distance (contribution: TKW, 52.16% and BKW, 44.92%), but it decreased with increase in feed rate (contribution: TKW, 12.38% and BKW, 15.31%). Abrasive concentration has marginal effect on the kerf width.

(ii) The interaction effects of the process parameters are found to be insignificant. Hence, process parameters can be independently varied to get required kerf width.

(iii) Process parameters are optimized to get small kerf width and regression models are developed to predict the top and bottom kerf width with R^2 values of 96.86% and 91.52% for TKW and BKW, respectively.

(iv) The morphological study revealed that there is no delamination of the surface machined at optimum process settings chosen to obtain small kerf as compared to that of the surface machined at nonoptimum process parameters.

Conflict of Interests

The authors declare that there is no conflict of interests regarding the publication of this paper.

Acknowledgments

The authors would like to acknowledge the support rendered by Industrial Mineral Co., Tuticorin, India, and Manipal University for their support in this research work.

References

[1] B. Shivamurthy, K. Udaya Bhat, and S. Anandhan, "Mechanical and sliding wear properties of multi-layered laminates from glass fabric/graphite/epoxy composites," *Materials and Design*, vol. 44, pp. 136–143, 2013.

[2] R. K. Goyal and A. Kadam, "Polyphenylene sulphide/graphite composites for emi shielding applications," *Advanced Materials Letters*, vol. 1, no. 2, pp. 143–147, 2010.

[3] S. Bhattacharya, R. P. Tandon, and V. K. Sachdev, "Electrical conduction of graphite filled high density polyethylene composites: experiment and theory," *Journal of Materials Science*, vol. 44, no. 9, pp. 2430–2433, 2009.

[4] R. Teti, "Machining of composite materials," *CIRP Annals—Manufacturing Technology*, vol. 51, no. 2, pp. 611–634, 2002.

[5] A. M. Abrão, P. E. Faria, J. C. C. Rubio, P. Reis, and J. P. Davim, "Drilling of fiber reinforced plastics: a review," *Journal of Materials Processing Technology*, vol. 186, no. 1–3, pp. 1–7, 2007.

[6] Y. H. Guu, H. Hocheng, N. H. Tai, and S. Y. Liu, "Effect of electrical discharge machining on the characteristics of carbon fiber reinforced carbon composites," *Journal of Materials Science*, vol. 36, no. 8, pp. 2037–2043, 2001.

[7] Z. L. Li, H. Y. Zheng, G. C. Lim, P. L. Chu, and L. Li, "Study on UV laser machining quality of carbon fibre reinforced composites," *Composites Part A: Applied Science and Manufacturing*, vol. 41, no. 10, pp. 1403–1408, 2010.

[8] J. Wang, "Abrasive waterjet machining of polymer matrix composites cutting performance, erosive process and predictive models," *International Journal of Advanced Manufacturing Technology*, vol. 15, no. 10, pp. 757–768, 1999.

[9] W. M. Andreas and R. Kovacevic, *Principles of Abrasive Water Jet Machining*, Spinger, London, UK, 1998.

[10] E. Huttunen-Saarivirta, F. H. Stott, V. Rohr, and M. Schütze, "Particle angularity effects on the elevated-temperature erosion-oxidation behaviour of aluminium diffusion coatings on 9% Cr steel," *Wear*, vol. 261, no. 7-8, pp. 746–759, 2006.

[11] S. Yerramareddy and S. Bahadur, "Effect of operational variables, microstructure and mechanical properties on the erosion of Ti-6Al-4V," *Wear*, vol. 142, no. 2, pp. 253–263, 1991.

[12] M. S. ElTobgy, E. Ng, and M. A. Elbestawi, "Finite element modeling of erosive wear," *International Journal of Machine Tools and Manufacture*, vol. 45, no. 11, pp. 1337–1346, 2005.

[13] A. A. Khan and M. M. Haque, "Performance of different abrasive materials during abrasive water jet machining of glass," *Journal of Materials Processing Technology*, vol. 191, no. 1–3, pp. 404–407, 2007.

[14] M. K. Babu and O. V. K. Chetty, "A study on recycling of abrasives in abrasive water jet machining," *Wear*, vol. 254, no. 7-8, pp. 763–773, 2003.

[15] G. Holmqvist and U. Honsberg, "Sensitivity analysis of abrasive waterjet cutting economy," in *Proceedings of the 19th International Conference on Water Jetting*, pp. 273–287, Nottingham University, Nottingham, UK, October 2008.

[16] J. Folkes, "Waterjet—an innovative tool for manufacturing," *Journal of Materials Processing Technology*, vol. 209, no. 20, pp. 6181–6189, 2009.

[17] F. Boud, C. Carpenter, J. Folkes, and P. H. Shipway, "Abrasive waterjet cutting of a titanium alloy: the influence of abrasive morphology," *Journal of Materials Processing Technology*, vol. 210, no. 15, pp. 2197–2205, 2010.

[18] D. S. Srinivasu, D. A. Axinte, P. H. Shipway, and J. Folkes, "Influence of kinematic operating parameters on kerf geometry in abrasive waterjet machining of silicon carbide ceramics," *International Journal of Machine Tools and Manufacture*, vol. 49, no. 14, pp. 1077–1088, 2009.

[19] M. Junkar, B. Jurisevic, M. Fajdiga, and M. Grah, "Finite element analysis of single-particle impact in abrasive water jet machining," *International Journal of Impact Engineering*, vol. 32, no. 7, pp. 1095–1112, 2006.

[20] J. Wang and D. M. Guo, "A predictive depth of penetration model for abrasive waterjet cutting of polymer matrix composites," *Journal of Materials Processing Technology*, vol. 121, no. 2-3, pp. 390–394, 2002.

[21] D. K. Shanmugam, T. Nguyen, and J. Wang, "A study of delamination on graphite/epoxy composites in abrasive waterjet machining," *Composites Part A*, vol. 39, no. 6, pp. 923–929, 2008.

[22] M. A. Azmir and A. K. Ahsan, "Investigation on glass/epoxy composite surfaces machined by abrasive water jet machining," *Journal of Materials Processing Technology*, vol. 198, no. 1–3, pp. 122–128, 2008.

[23] M. A. Azmir and A. K. Ahsan, "A study of abrasive water jet machining process on glass/epoxy composite laminate," *Journal of Materials Processing Technology*, vol. 209, no. 20, pp. 6168–6173, 2009.

[24] A. Alberdi, A. Suárez, T. Artaza, G. A. Escobar-Palafox, and K. Ridgway, "Composite cutting with abrasive water jet," *Procedia Engineering*, vol. 63, pp. 421–429, 2013.

[25] N. Haghbin, J. K. Spelt, and M. Papini, "Abrasive waterjet micro-machining of channels in metals: comparison between machining in air and submerged in water," *International Journal of Machine Tools and Manufacture*, vol. 88, pp. 108–117, 2015.

[26] D. A. Axinte, D. S. Srinivasu, J. Billingham, and M. Cooper, "Geometrical modelling of abrasive waterjet footprints: a study for 90° jet impact angle," *CIRP Annals—Manufacturing Technology*, vol. 59, no. 1, pp. 341–346, 2010.

[27] M. C. Kong, S. Anwar, J. Billingham, and D. A. Axinte, "Mathematical modelling of abrasive waterjet footprints for arbitrarily moving jets: part I—single straight paths," *International Journal of Machine Tools and Manufacture*, vol. 53, no. 1, pp. 58–68, 2012.

[28] P. R. Vundavilli, M. B. Parappagoudar, S. P. Kodali, and S. Benguluri, "Fuzzy logic-based expert system for prediction of depth of cut in abrasive water jet machining process," *Knowledge-Based Systems*, vol. 27, pp. 456–464, 2012.

[29] A. M. Zain, H. Haron, and S. Sharif, "Estimation of the minimum machining performance in the abrasive waterjet machining using integrated ANN-SA," *Expert Systems with Applications*, vol. 38, no. 7, pp. 8316–8326, 2011.

[30] J. Billingham, C. B. Miron, D. A. Axinte, and M. C. Kong, "Mathematical modelling of abrasive waterjet footprints for arbitrarily moving jets: part II—overlapped single and multiple straight paths," *International Journal of Machine Tools and Manufacture*, vol. 68, pp. 30–39, 2013.

[31] C. Narayanan, R. Balz, D. A. Weiss, and K. C. Heiniger, "Modelling of abrasive particle energy in water jet machining," *Journal of Materials Processing Technology*, vol. 213, no. 12, pp. 2201–2210, 2013.

[32] H. Nouraei, K. Kowsari, J. K. Spelt, and M. Papini, "Surface evolution models for abrasive slurry jet micro-machining of channels and holes in glass," *Wear*, vol. 309, no. 1-2, pp. 65–73, 2014.

[33] M. K. Babu and O. V. K. Chetty, "A study on the use of single mesh size abrasives in abrasive waterjet machining," *The International Journal of Advanced Manufacturing Technology*, vol. 29, no. 5-6, pp. 532–540, 2006.

10

Enhancing Multistage Deep-Drawing and Ironing Manufacturing Processes of Axisymmetric Components: Analysis and Experimentation

F. Javier Ramírez,[1,2] Rosario Domingo,[3] Michael S. Packianather,[4] and Miguel A. Sebastian[3]

[1] *EXPAL Systems, Avenida Partenón 16, 28042 Madrid, Spain*
[2] *School of Industrial Engineering, University of Castilla-La Mancha, Albacete, Spain*
[3] *School of Industrial Engineering, UNED, 28042 Madrid, Spain*
[4] *School of Engineering, Cardiff University, Cardiff, UK*

Correspondence should be addressed to F. Javier Ramírez; franciscoj.ramirez@uclm.es

Academic Editor: Jean-Yves Hascoet

An optimization technique for combined processes of deep-drawing and ironing has been created in order to improve the total process time and cost in manufacturing procedures of axisymmetric components. The initial solution is optimized by means of an algorithm that minimizes the total time of the global process, based on relationship between lengths, diameters, and velocities of each stage of a multistage process and subject to constraints related to the drawing ratio. The enhanced solution offers a significant reduction in time and cost of the global process. The final results, applied to three cases, are compared with experimental results, showing the accuracy of the complete solution.

1. Introduction

The industry of metallic components manufacturing requires developments to be more efficient, in particular in the deep-drawing procedures, where it is important to decrease the process times and costs as in mass production. Thus, it is necessary to devise specific algorithms that will satisfy these demands. These algorithms should be based on technological and scientific basis that will provide solutions that are ready for transfering to the industries. The deep-drawing process has been analysed with this objective in mind due to its convenient nature as a global model which includes all stages of the process, namely, drawing, redrawing, and ironing.

The majority of literature contributions are focused on the study of properties of process, in particular, the prediction of the limiting drawing ratio (LDR) [1–3], the blank design using different methodologies, such as parametric NURBS surfaces [4], upper bound method [5], or artificial neural network [6, 7], the effect of die radius on the blank holder force and drawing ratio [8], the predicted thickness distribution of the deep drawn circular cup of stainless steel [9], the improvement of drawability by means of technological parameters [10, 11] or the formability with different thickness [12]. However, some efforts have been realised about the parts design [13] or generation of algorithms, mainly related to the process planning; Ramana and Rao [14] developed a framework based on knowledge related to design-process planning integration for sheet metal components, although there is no evidence of its application. Also, Vosniakos et al. [15] devised an intelligent system to process design of sheet parts. As can be seen, the researches of deep-drawing processes are not focused on the reduction of time, despite frequently being used on mass production due to the characteristics of the parts.

This paper presents a model that provides a comprehensive analysis of those phenomena occurring in the multistage processes of axisymmetric geometry and applied to the manufacturing of this type of components. The scientific development stems from the work done by Leu [1] and Sonis et al. [2] which provides LDR solutions based on normal anisotropy value, strain hardening exponent, and others, applied to the drawing and redrawing stages. The model [16, 17] has a scientific foundation based on the literature covering the plastic deformation processes and, in particular, drawing [18], redrawing [19], and ironing. This study focuses on analyzing geometries of axisymmetric components, manufactured by a multistage deep-drawing process.

The drawing processes from a blank [6, 8] have been researched and their governing equations are well known [3, 5]. However, the redrawing and ironing phenomena have received relatively less attention [2, 20]. This work contributes to the definition of a global process of drawing, redrawing, and ironing combined process, together with a common focus on the drawing process in order to obtain a global solution.

The model is used to carry out a quantitative and integrated analysis in a multistage deep-drawing process based on a scientific solution and real measurements. Also, the model permits the modification for some process variables to predict their influence on the process [9]. In addition, it is based on the definition of limiting conditions to guarantee the stability of the process. The simultaneous accomplishment of limiting situations of each process (drawing, redrawing, and ironing) allows for fixing a boundary for the values of each stage which gives the initial solution. The model permits the optimisation of the initial solution from several points of view: total process time, manufacturing cost, among others, such as punch and die wear. This optimisation is based on the resolution of an algorithm by means of recursive functions, which explores all the possibilities of the process and selects the most adequate one. The algorithm is supported by a software tool that provides the optimized solution [21]. The optimized solution is compared with the experimental results.

The remainder of the paper is organised as follows. In Section 2, the methodology used is formulated and applied to some industrial cases. In Section 3, the initial solution found from the scientific model is presented. Section 4 defines the optimization process. In Section 5, the optimization algorithm for solving the multistage process is proposed. In Sections 6 and 7, the simulated and experimental results are compared in order to show the reliability of the complete solution found by the proposed algorithm.

2. Methodology

The methodology used in the definition and resolution of the algorithm is based on a model that departs from an initial solution, defined by the technological constraints that characterize the multistage deep-drawing processes (see Section 3). The use of these processes in mass production will require the minimization of the total process time and the quantity of raw material, thus acting as constraints.

The algorithm resolution has been carried out by the software called deep-drawing tool (DDT) developed by the author. This tool allows the user to perform the total multistage process; it allows the evaluation of the most important variables of the process and the implementation of the optimization algorithm which is the objective of this paper.

Table 1: Brass UNS C26000 properties.

Material	UNS C26000
Density, ρ	8.53 Kg/dm^3
Material rigid-plastic constant, C	895.0 MPa
Strain hardening exponent, n	0.485
Tensile strength, yield, S_y	435.0 MPa
Tensile strength, ultimate, S_u	525.0 MPa
Normal anisotropy value, R	0.83

Table 2: Dimensions of parts (in mm).

	Case A	Case B	Case C
External diameter, d_n	98.5	134.4	102.7
Length, l_n	620	650	1,480
Bottom thickness, s_n	13.5	13.8	15
Wall thickness, e_n	2	1.45	0.95

The algorithm results are contrasted experimentally by means of the resolution of several industrial cases. The applied product type is an axisymmetric component produced in brass by multistage deep-drawing and ironing processes. The use of this material is justified due to its good properties suitable for these multistage processes. Table 1 shows the characteristics of brass UNS C26000 used on the resolution of the industrial cases presented in Section 5.

The solutions provided by this algorithm have been tested against three industrial cases and geometries as shown in Table 2.

The resolution of these types of cases requires the combination of different processes of drawing, redrawing, and ironing. Figure 1 shows the drawing and redrawing/ironing press used on the manufacturing processes of the industrial cases. Figure 2 shows some of the manufacturing stages for Case A.

3. Model Approach

Figure 3 shows the flow diagram of the model providing the initial solution: model phases, input variables required to define the "new project," and the flow information between different stages. The model develops the required solution from the input data. These input data correspond to the dimensions and material of the final piece. The required input data are the external diameter (d_n), the length (l_n), bottom thickness (s_n), wall thickness (e_n), and the material type. The model permits the selection of the material type and its mechanical characteristics needed to form the part: density, ρ (kg/m^3); ultimate tensile strength, S_u (MPa); yield tensile strength, S_y (MPa); material rigid-plastic constant, C (MPa); strain hardening exponent, n; and normal anisotropy value, R. From the input data, the dimensions of the initial part or blank are calculated based on the incompressibility condition of the plastic deformation. This condition considers that the volume of the piece is unchanged throughout the deformation process [18]. A detailed formulation about the previous model that determines the initial solution can be found in

FIGURE 1: Redrawing and ironing press.

FIGURE 2: Case A: manufacturing stages.

Ramirez et al. [22]. The dimensions of this blank are the starting point to develop the drawing, redrawing, and ironing processes until the required final dimensions are achieved. In this way, data from the blank becomes the input for the next step: drawing.

3.1. Drawing. The model presents a calculation procedure for each process: drawing, redrawing, and ironing. The determination of the initial solution is treated independently for each of the threads that may occur. It is important to note that, depending on the type of piece, in some cases, its geometry is not necessary to consider redrawing or ironing stages. For determining the initial solution of the drawing, a hypothesis that would later be amended in the adjustment of the technological process is used, and it is the size of the diameter of the punch corresponding to that stage. The LDR is a measure of deep draw ability which defines the largest blank that can be drawn without tearing. It is the ratio between the maximum blank diameter and punch diameter. The design of this punch will depend on the LDR and the final dimensions of the piece. It is also possible that this diameter is changed later in the adjustment of the technological process in order to allow a clearance between the punch and the workpiece for subsequent stages of redrawing and ironing [20]. The solution for the initial stage of drawing is calculated by considering two limiting conditions as described below. The model selects the drawing diameter or the die, d, by using the largest diameter ($d_{1,1}$ and $d_{1,2}$) obtained from the two drawing conditions. Once the diameter is known, the model determines all the dimensions needed to define the drawing stage. Verification of the drawing stage dimension is conducted using two experimental conditions based on empirical studies and collected field data. The corresponding flow diagram is shown in Figure 4.

The first limiting drawing condition is related to the maximum force ($F_{e,\max}$) executed by the punch on the workpiece during the drawing process, which must be less than the cracking load of the material (F_{cr}), according to (1). These forces establish the range in which the force can be obtained using from the efficiency coefficient (η_{def}), ultimate tensile strength, blank thickness (s_0), and the initial (d_0) and final (d_1) diameters (note that the mean wall diameter is $d_{m,1} = d_1 + s_0$), and they are based on the experimental expression from Siebel and Beisswänger [23]. This condition requires the following:

$$F_{e,\max} = \pi \cdot d_{m,1} \cdot s_0 \left[1.1 \cdot \frac{1.3 S_u}{\eta_{\text{def}}} \left(\ln \frac{d_o}{d_1} - 0.25 \right) \right], \qquad (1)$$

$$F_{cr} = \pi \cdot d_{m,1} \cdot s_0 \cdot S_u.$$

The second drawing limiting condition is calculated by using the expression developed by Leu [1], which showed good agreement between theoretical and experimental results [2, 3]. Considering the condition of constant volume throughout the process of plastic deformation, LDR can be defined as follows:

$$\text{LDR} = \sqrt{e^{(2fe^{-n\sqrt{(1+R)/2}})} + e^{(2n\sqrt{(1+R)/2})} - 1}. \qquad (2)$$

This expression estimates the LDR as a function of normal anisotropy, the strain hardening exponent, and efficiency (f).

In this manner, upper and lower bounds are established, considering the materials proprieties and the drawing capacity. The two limiting conditions use new data from the blank geometry and process efficiency (η_{def} and f).

3.2. Redrawing. Once the dimensions of the drawing stage are obtained, the model provides these data as input values for the next phase of the process: redrawing. The goal of this step is to obtain the final dimensions of the piece needed to perform the next step: ironing. The solution for the initial stage of redrawing is calculated from the consideration of three limiting conditions as described below. The model selects the diameter of the die from the largest of the solutions obtained in the three redrawing constraints ($d_{1\text{-}n,1}$, $d_{1\text{-}n,2}$, and $d_{1\text{-}n,3}$). Once the diameter is known, the model determines all dimensions needed to define this stage. Figure 5 presents the flow diagram of this redrawing algorithm.

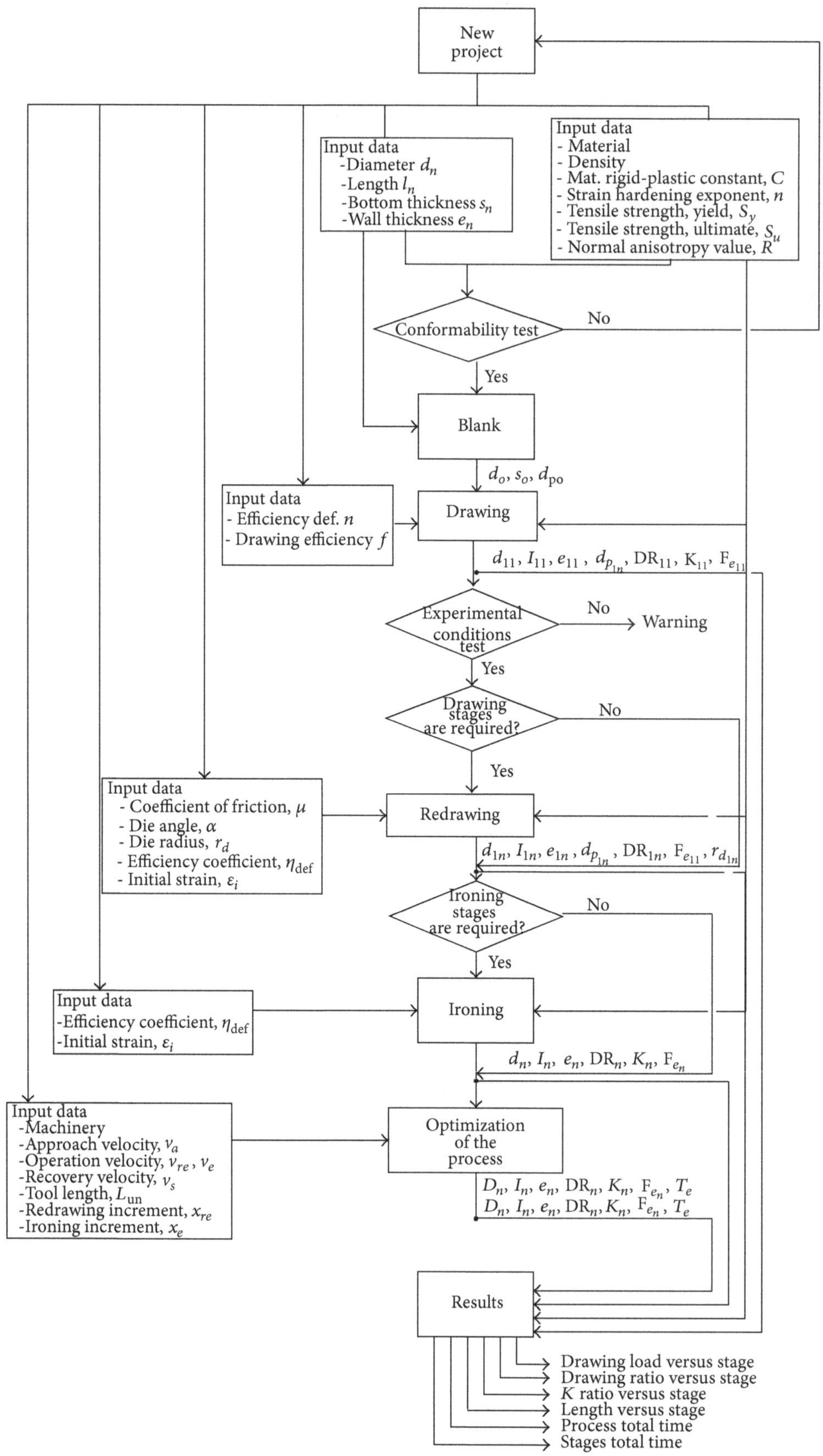

FIGURE 3: Flow diagram of the model.

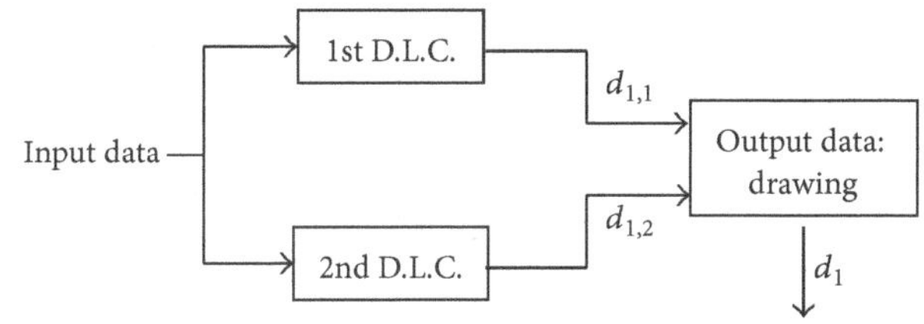

(D.L.C.: drawing limiting condition)

FIGURE 4: Flow diagram of the drawing algorithm.

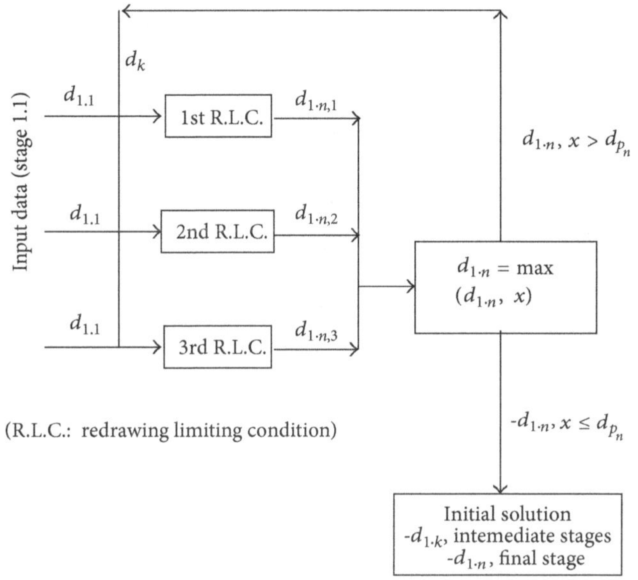

(R.L.C.: redrawing limiting condition)

FIGURE 5: Flow diagram of the redrawing algorithm.

The first redrawing limiting condition is related to the redrawing load of the punch during the process of redrawing ($F_{\mathrm{re,max}}$), which must be less than the cracking load of the material (F_{cr}), according to (3) and (4), respectively. Consider

$$F_{\mathrm{re,max}} = \pi \cdot d_{m,1\cdot n} \cdot s_{1\cdot n} \cdot 1.3 S_u \left[2e^{\mu\pi/2} \frac{d_{p,1\cdot n} - d_{m,1\cdot n}}{d_{p,1\cdot n} + d_{m,1\cdot n}} 1.1 + \frac{\mu}{\tan\alpha} + \frac{s_{1\cdot n}}{2r_D} \right], \quad (3)$$

$$F_{\mathrm{cr}} = \pi \cdot d_{m,1\cdot n} \cdot s_{1\cdot n} \cdot S_u. \quad (4)$$

The first condition is based on Siebel and Beisswänger [23], where μ is the friction coefficient, α is the die half-angle, d_p is the punch diameter, and r_D is the die radius.

The second redrawing limiting condition assumes the *rigid-plastic condition of the material.* Assuming that the material behaves according to a parabolic law that approximates the potential behaviour of metallic materials cold-manufactured, then it is possible to determine an expression relating the external diameter of the part of a generic step of the process of redrawing, $d_{1\cdot n}$, to the diameter of the previous stage. Assuming the material is annealed ($\varepsilon_i = 0$) and considering the total deformation between the initial and final states, the condition takes the following form:

$$d_{1\cdot n} = d_{1\cdot n-1} e^{-\varepsilon_f/2}. \quad (5)$$

The third redrawing limiting condition is referred to as the *limiting drawing ratio.* In applying the third limiting condition of drawing, the model applies the formulation defined by Sonis et al. [2] about the LDR study in the operations of redrawing. The model considers the effects of the normal anisotropy of the material, friction coefficient, coefficient of strain hardening, and the radius of the input die (r_d). The LDR is used in this model as a variable to determine the required number of redrawing steps and size of the stages. It is assumed that the material is rigid-plastic [2]. Moreover, assuming that the material is rotationally symmetric, the same properties are based on the existence of normal anisotropy and planar isotropy. The Sonis model [2] is based on the tension that is created in the area of the radius of the die redrawing causing instability in the plastic wall of the cup, which is equal to the radial tension in the drawing area of the flange, due to the continuity of tension throughout the piece. Based on the Sonis model, the expression for LDR_i can be written as

$$\begin{aligned} f(\mathrm{LDR}_i) &= -\frac{C_1 \cdot r_{c_{(i-1)}}}{r_{c_{(i-1)}} + \mathrm{LDR}_{(i-1)} \cdot r_d} + \frac{C_3 \cdot r_{c_{(i-1)}}}{r_d + r_{c_{(i-1)}}/\mathrm{LDR}_{(i-1)}} \\ &\quad + C_4 \ln\left(\frac{r_{c_{(i-1)}}}{r_d + r_{c_{(i-1)}}/\mathrm{LDR}_{(i-1)}}\right) \\ &\quad + C_5 \ln\left(\frac{r_{c_{(i-1)}}}{r_d + r_{c_{(i-1)}}/\mathrm{LDR}_{(i-1)}}\right) \\ &\quad - \frac{1}{\left(r_d + r_{c_{(i-1)}}/\mathrm{LDR}_{(i-1)}\right)^2 \left(\left(R_{(i+1)}/r_{cd}\right)^2 - 1\right)} \\ &\quad \times \left[r_{c_{(i-1)}} \times \sqrt{\left(r_d + \frac{r_{c_{(i-1)}}}{\mathrm{LDR}_{(i-1)}}\right)^2 \left(\left(\frac{R_{(i+1)}}{r_{cd}}\right)^2 - 1\right) + r^2_{c_{(i-1)}}} \right. \\ &\quad - \left(r_d + \frac{r_{c_{(i-1)}}}{\mathrm{LDR}_i}\right)^2 \left(\frac{R_{(i+1)}}{r_{cd}}\right) \\ &\quad \left. - \left(r^2_{c_{(i-1)}} - \left(r_d + \frac{r_{c_{(i-1)}}}{\mathrm{LDR}_i}\right)\right)^2 \right] \\ &\quad + C_5 \ln\left[r_{c_{(i-1)}} + \sqrt{\left(r_d + \frac{r_{c_{(i-1)}}}{\mathrm{LDR}_{(i-1)}}\right)^2 \left(\left(\frac{R_{(i+1)}}{r_{cd}}\right)^2 - 1\right) + r^2_{c_{(i-1)}}} \right] \\ &\quad - C_5 \ln\left(\left(\frac{r_d + r_{c(i-1)}}{\mathrm{LDR}_i}\right)\left(1 + \frac{R_{(i+1)}}{r_{cd}}\right)\right) = 0, \end{aligned} \quad (6)$$

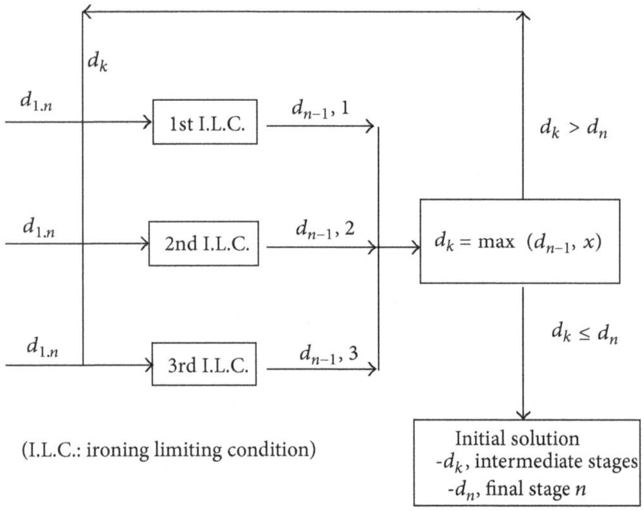

FIGURE 6: Flow diagram of the ironing algorithm.

where C_1, C_3, C_4, and C_5 are constants, r_c is the corner radius, and r_{cd} is the die opening radius [2].

Based on the previous expressions in the function under the LDR, an expression depending on the LDR is obtained, and the nonlinear equations are solved by the Newton-Raphson method. In this way, it is possible to determine the values of LDR for each stage of redrawing, starting from an initial die radius and by the consideration that the die radius of each redrawing step will be reduced until 80% of the corresponding value of the previous phase. Once the LDR for each redrawing phase is found, the model determines the diameter of each stage.

3.3. Ironing. The number of stages of ironing depends on the size of the final part. The model is based on the performance of three limiting conditions to be drawn from stage 2 to stage n. The calculation of the process variable is the external diameter of the piece, according to the block diagram, illustrated in Figure 6.

By considering separately the merits of the three limiting conditions for ironing, the solution can be determined for phases 2 to n.

The First Ironing Limiting Condition. The maximum ironing load ($\sigma_{e,\max}$) must be less than the cracking load (σ_r) of the material. This condition is expressed as the following:

$$F_{e,n-1} < \frac{\pi}{4} \cdot \left(d_{n-2}^2 - d_{n-1}^2\right) \cdot n \cdot S_u. \quad (7)$$

Second Ironing Limiting Condition. This condition assumes that the maximum load in the ironing process (σ_e) must be less than the yield tensile strength ($Y(\varepsilon_f)$). This condition provides good results in parts of drawing [24] and it requires

$$\frac{C}{n+1}\left(\varepsilon_f^{\ n+1} - \varepsilon_i^{\ n+1}\right) < Y\left(\varepsilon_f\right) \equiv C \cdot \varepsilon_f^{\ n}. \quad (8)$$

Third Ironing Limiting Condition. This refers to the value of the thickness reduction ratio limit (K). This ratio is widely used in the calculation and design of the processes of drawing [5]. The model performs the calculation for each K stage, which mainly depends on the stage of drawing and the type of material used.

4. Optimization Process

The model is based on the minimization of the total process time, which is the time to approach the punch from the initial stage, plus the time of the operation, plus the time to recover the punch to its initial position, according to the following objective function:

$$F_o = \min t_e = \min \sum_{i=1}^{n} t_i = \min \sum_{i=1}^{n} \left(t_{a,i} + t_{o,i} + t_{s,i}\right), \quad (9)$$

where F_o is the objective function, t_e is the total process time, t_i is the total process time at stage i, $t_{a,i}$ is the approach time at stage i, $t_{o,i}$ is the operation time at stage i, and $t_{s,i}$ is the recovery time at stage i.

The process time at stage i is given by the expression

$$t_i = \frac{1.1 \cdot l_{i-1}}{v_{a,i}} + \frac{L_{ui} + l_i}{v_{e,i}} + \frac{1.1 \cdot l_{i-1} + L_{ui} + l_i}{v_{s,i}}, \quad (10)$$

where l_{i-1} is the part length in stage $i-1$, L_{ui} is the punch length in stage i, l_i is part length in stage i, $v_{a,i}$ is the approach velocity of the press in the stage i, $v_{e,i}$ is the ironing velocity in the stage i, and $v_{s,i}$ is the recovery velocity of the press in the stage i.

As can be seen, the time is defined by means of the velocities involved in each operation and the punch or part length of each stage. Moreover, the part length in stage i is related to the bottom thickness, the blank diameter, the punch diameter, and the optimized diameter. Therefore, the model presents an optimization problem with an objective function which is subject to some constraints as follows:

$$F_o = \min t_e = \min \sum_{i=1}^{n} t_i$$
$$= \min \sum_{i=1}^{n} \left(\frac{1.1 \cdot l_{i-1}}{v_{a,i}} + \frac{L_{ui} + l_i}{v_{e,i}} + \frac{1.1 \cdot l_{i-1} + L_{ui} + l_i}{v_{s,i}}\right) \quad (11)$$

subject to

$$d_n \leq D_n \leq d_n\left(1 + \Delta DR_e\right), \quad (12)$$

where

$$l_i = \frac{s_i\left(d_o^2 - d_{p,i}^2\right)}{\left(D_i^2 - d_{p,i}^2\right)}, \quad (13)$$

and D_i is the optimized diameter in stage i, s_i is the bottom thickness in stage i, $d_{p,i}$ is the punch diameter in stage i, and ΔDR_e is the drawing surplus ratio between the stages n and m, defined by the expression

$$\Delta DR_e = DR_m - DR_n. \quad (14)$$

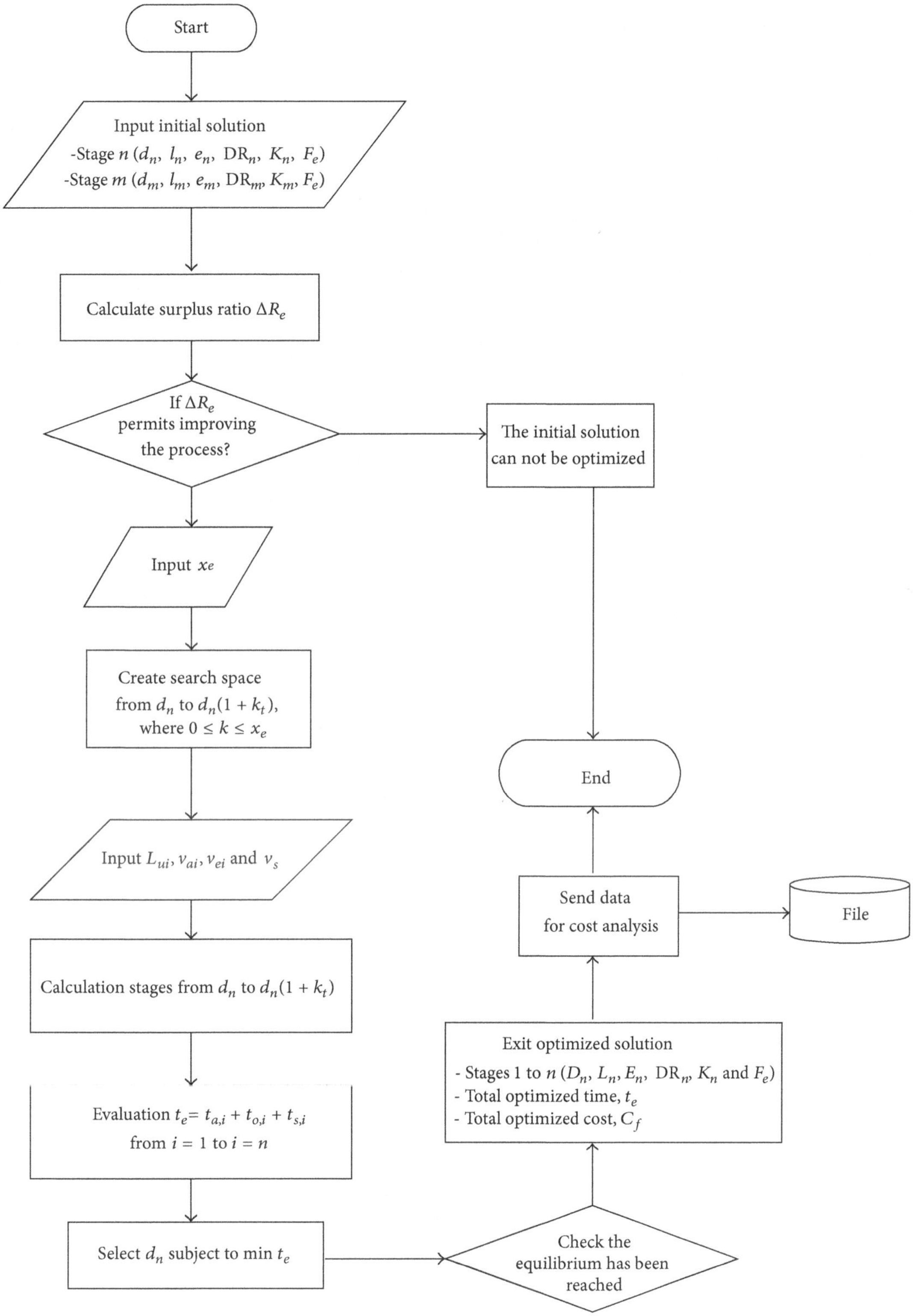

FIGURE 7: Flowchart of the algorithm.

The optimization requires a distribution between the stages 1 to n of the drawing surplus ratio, ΔDR_e. This distribution is conditioned to the minimization of the total process time as the sum of each stage time. Thus, an improvement is achieved in each drawing stage. In this manner, the optimized drawing conditions are separate from the maximum limits fixed by the initial drawing conditions.

The resolution of the recursive function is carried out according to the values of all the possible diameters between d_n and $d_n(1 + \Delta DR_e)$. The algorithm identifies which combination is the most suitable, and it selects the optimized diameter D_n in each stage, such that the condition of the objective function is satisfied; that is, the combination of diameters must give the minimum total process time.

The algorithm also permits the modification of the velocity parameters in each press required in the multistage process (approach velocity, operation velocity, and recovery velocity of the punch). Accordingly, it is possible to realise a more adequate distribution of available presses in the facility.

5. Algorithm Resolution Process

The algorithm proposed in Figure 7 performs the resolution in the following steps.

Step 1. Definition of incremental factor t, given by the following expression:

$$t = \frac{\Delta DR_e}{x_e}, \quad (15)$$

where ΔDR_e is the surplus ratio of drawing and x_e is the number of times that the drawing surplus ratios are fractioned.

Step 2. Progressive increment of diameter of each process stage, from d_n to $d_n(1+kt)$, where k is a parameter that varies from 0 to factor x_e.

Step 3. Resolution of the recursive function is based on obtaining all possible substages, by means of the modification of each stage diameters, from d_n to $d_n(1 + \Delta DR_e)$.

Step 4. Once all the possible diameters of each stage are defined, the algorithm searches the arrangement that allows for the minimizing of the total process time.

Step 5. The search stops and the best solution up to the current iteration is given as the output.

6. Experimental Results

The deep-drawing tool has allowed for the verification of this algorithm's integrity. Computational and experimental tests have been carried out in brass, in particular in UNS C26000 alloy, applied to three parts. The dimensions of the parts are shown in Table 2.

For the experimental results of the three cases presented in Table 2, its analytical resolution has been carried out by the software DDT according to the flowchart presented in Figure 3. The final result of using the software tool is an improved solution that optimizes the total process time and reduces the manufacturing cost of the presented industrial cases. These costs use the information from the results provided by the previous steps and are based on the total time of the deformation process and the labour cost per hour.

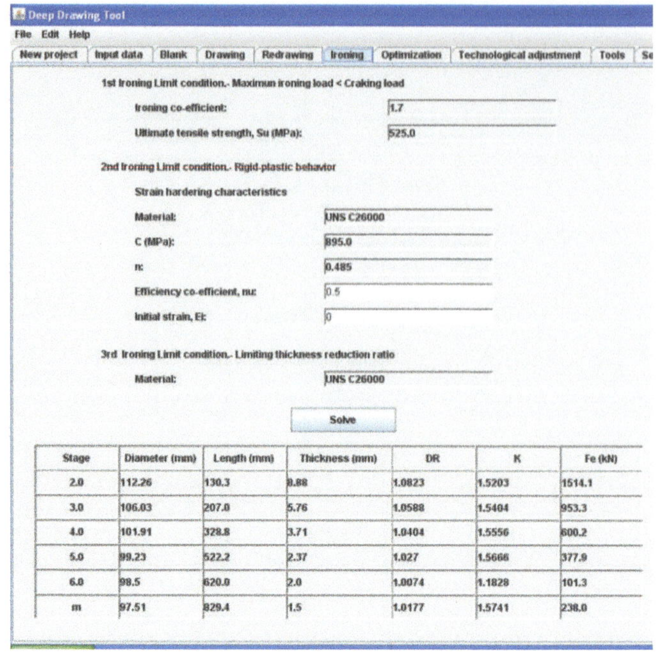

FIGURE 8: Case A: initial solution results.

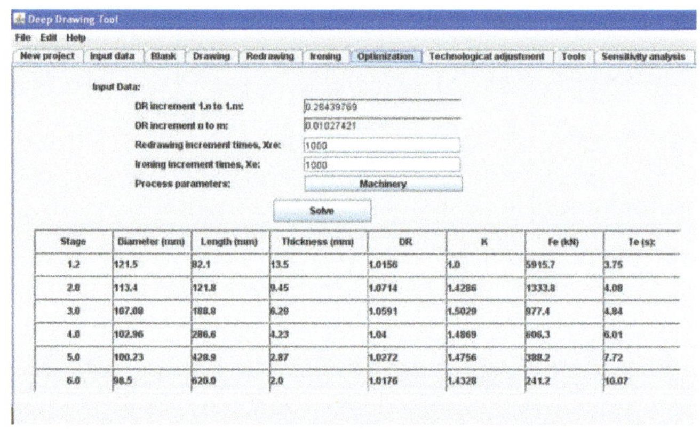

FIGURE 9: Case A: optimization process results.

Figure 8 shows the results of the tool for the initial solution corresponding to Case A.

Figure 9 presents the calculation of the optimized process by means of the DDT using the proposed algorithm. Values for x_{re} and x_e in operations of redrawing and ironing are given.

The resolution of the optimization process using the proposed algorithm requires the introduction of the machinery operation parameters involved in the process.

The computer tool requires the input of these data (Figure 10), according to the flowchart of the algorithm presented in Figure 3.

One of the most important variables in the analysis of such processes is the variable "ratio of wall thickness reduction." The optimization process allows the user to obtain a greater stability of the process maintaining the balance of the different phases. Table 3 shows the evolution of the drawing load for the industrial Cases A, B, and C under study.

TABLE 3: Evolution of drawing load.

Stages	Case A		Case B		Case C	
	Initial solution (IS)	DDT algorithm solution	Initial solution (IS)	DDT algorithm solution	Initial solution (IS)	DDT algorithm solution
Drawing						
Stage 1.1	3188.5	3188.5	3175.0	3175.0	3692.1	3692.1
Redrawing						
Stage 1.2	5915.7	5915.7	—	—	—	—
Ironing						
Stage 2.0	1514.1	1346.4	2082.3	1859.8	1803.8	1710.1
Stage 3.0	953.3	962.6	1311.1	1320.5	1135.7	1141.1
Stage 4.0	600.2	626.8	825.5	859.3	715.1	718.6
Stage 5.0	377.9	381.4	519.8	523.2	450.2	452.6
Stage 6.0	101.3	229.8	327.3	329.5	283.5	285.0
Stage 7.0			15.8	189.3	178.5	179.4
Stage 8.0					32.4	112.4

TABLE 4: Evolution of wall thickness.

Stages	Case A		Case B		Case C	
	Initial solution (IS)	DDT algorithm solution	Initial solution (IS)	DDT algorithm solution	Initial solution (IS)	DDT algorithm solution
Drawing						
Stage 1.1	1	1	1	1	1	1
Redrawing						
Stage 1.2	1	1	—	—	—	—
Ironing						
Stage 2.0	1.520	1.434	1.535	1.450	1.518	1.477
Stage 3.0	1.540	1.494	1.552	1.505	1.538	1.516
Stage 4.0	1.555	1.511	1.564	1.519	1.554	1.520
Stage 5.0	1.566	1.474	1.572	1.479	1.565	1.512
Stage 6.0	1.182	1.412	1.578	1.442	1.573	1.492
Stage 7.0			1.028	1.345	1.578	1.459
Stage 8.0					1.11	1.409

Table 4 shows the evolution of the wall thickness. The results show that the optimized process achieves a better balance so that it is more stable. This stability is transmitted in a better distribution of the capabilities of drawing, redrawing, and ironing between the different stages, subsequently resulting in improved processing time and reduced manufacturing cost.

Once the evolution of different variables that influence the process has been analyzed, the results for "process time" and "manufacturing cost" are presented in order to study the improvements that could be achieved by the implementation of the proposed algorithm.

Tables 5 and 6 show the evolution of these improvements on the variables "total process time" (t_e) and "manufacturing cost" (C_m) in terms of number of iterations performed by the algorithm.

Similarly, the evolution of "total process time" and the "manufacturing cost" for Cases B and C has been analyzed. Figures 11 and 12 show the improvements evolution obtained for the three Cases A, B, and C. It can be observed that the greater improvements are achieved in the first iterations (first five according to Case A), but the results improve further as the iteration increases.

7. Analysis and Discussion of the Results

The resolution of the industrial Cases A, B, and C by means of the DDT Algorithm presents important advantages. As shown in Tables 7, 8, and 9, the DDT algorithm produces improvements between 5.17% and 8.18% compared with the initial solution for the process time variable and between 4.40% and 7.78% for the manufacturing cost. In the same manner, the experimental results obtained an improvement between 6.55% and 9.34% for the process time and from 6.60% to 11.55% for the manufacturing cost. These improvements are more relevant in mass production. The most

TABLE 5: Process time evolution as a function of the algorithm iteration number.

x_e	5	25	50	100	250	500	1000
$t_{e,1}$	42.054	42.054	42.054	42.054	42.054	42.054	42.054
$t_{e,n}$	40.634	40.109	39.998	39.952	39.889	39.813	39.797
% reduction	3.67%	5.02%	5.31%	5.43%	5.59%	5.79%	5.83%

TABLE 6: Manufacturing cost evolution as a function of the algorithm iteration number.

x_e	5	25	50	100	250	500	1000
Cm_1	34.974	34.974	34.974	34.974	34.974	34.974	34.974
Cm_n	33.554	33.029	32.918	32.872	32.809	32.733	32.717
% reduction	4.06%	5.56%	5.88%	6.01%	6.19%	6.41%	6.45%

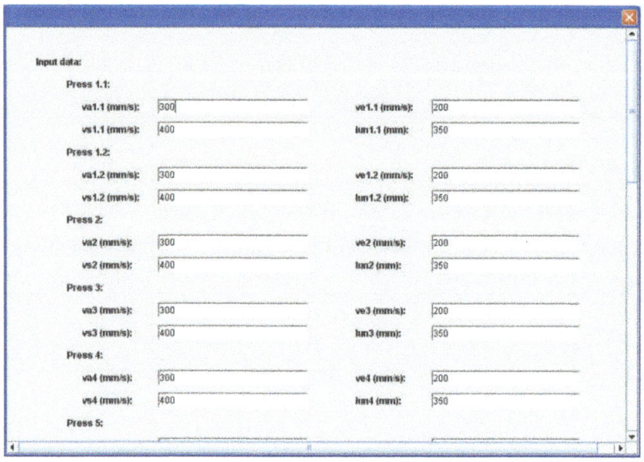

FIGURE 10: Machinery parameters.

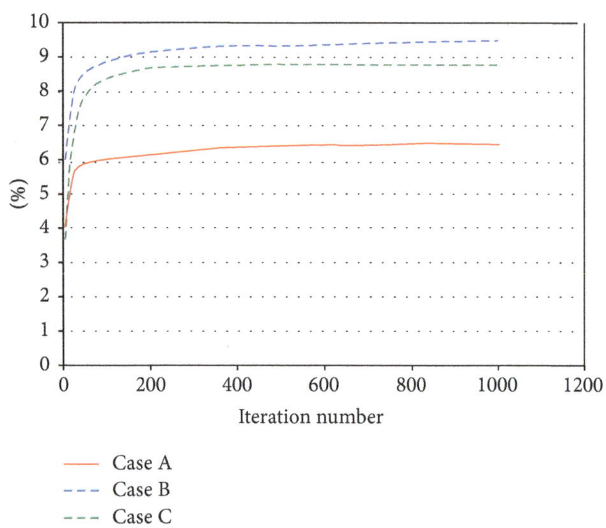

FIGURE 11: Cases A, B, and C: process time improvements as function of the algorithm iteration number.

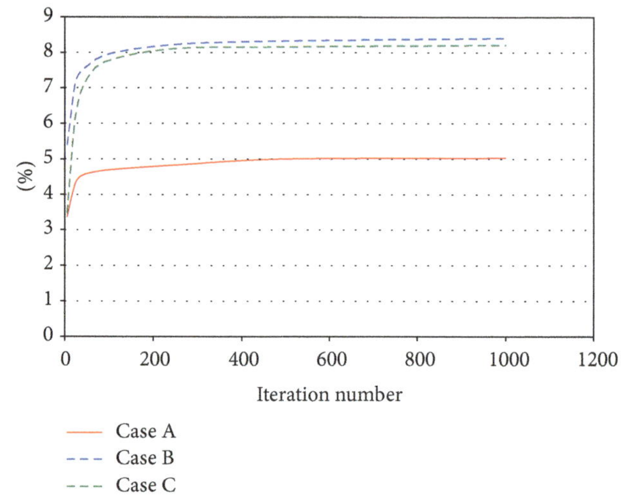

FIGURE 12: Cases A, B, and C: manufacturing cost improvements as function of the algorithm iteration number.

TABLE 7: Case A: analysis of the results.

Comparative analysis	Process time	Manufacturing cost per unit
Initial solution (IS)	34.83 s	0.318€
DDT algorithm solution	33.03 s	0.304€
Experimental solution (EXS)	32.55 s	0.297€
DDT versus IS improvement	5.17%	4.40%
EXS versus IS improvement	6.55%	6.60%
EXS versus DDT accuracy	98.55%	97.7%

significant advances have been achieved in Case B, which represent the highest progress according to the algorithm iteration number (see Figures 11 and 12). Also there is similarity between this algorithm iteration number and Cases A and C where the trend is the same in all cases. The final geometry of the part does not seem to have a particular effect on the results. This is due to a good definition of initial solution that allows processing the blank.

Comparing the algorithm results with the experimental results shows that the accuracy of the process is very high, obtaining values from 98.41% to 98.74% for the process time variable and from 95.92% to 98.13% for the manufacturing cost. Although there are some disparities between the cases analyzed, these differences are not significant because the accuracy is always above 95% and normally very close to 99%. Thus, the mathematical minimization of the total process

TABLE 8: Case B: analysis of the results.

Comparative analysis	Process time	Manufacturing cost per unit
Initial solution (IS)	42.4 s	0.398€
DDT algorithm solution	38.93 s	0.367€
Experimental solution (EXS)	38.44 s	0.352€
DDT versus IS improvement	8.18%	7.78%
EXS versus IS improvement	9.34%	11.55%
EXS versus DDT accuracy	98.74%	95.92%

TABLE 9: Case C: analysis of the results.

Comparative analysis	Process time	Manufacturing cost per unit
Initial solution (IS)	76.58 s	0.688€
DDT algorithm solution	70.95 s	0.639€
Experimental solution (EXS)	69.82 s	0.627€
DDT versus IS improvement	7.35%	7.12%
EXS versus IS improvement	8.83%	8.88%
EXS versus DDT accuracy	98.41%	98.13%

time through the technological variables provides a solution which is very close to the experimental outcomes.

8. Conclusions

In this paper, an algorithm that allows the reduction of the total process times and cost in the manufacturing of axisymmetric components has been presented. The algorithm is based on the minimization of total process time, defined by means of the part dimensions and process velocities, and costs by considering the reduction of material usage, through constraints related to drawing surplus ratio. The algorithm has been enhanced with the use of technological parameters. The simulation results of the algorithm achieved a good agreement with the experimental results obtained for the three cases and caused significant improvements in manufacturing times and costs in the deep-drawing process of axisymmetric parts. The comparison of analytical results obtained with the experimental results has demonstrated the high accuracy of the algorithm, which is of interest for real industrial applications.

Future works will aim to get a more efficient process, from a perspective of sustainable energy, thus achieving an integral solution, in terms of scientific and technological basis.

Conflict of Interests

The authors declare that there is no conflict of interests regarding the publication of this paper.

Acknowledgments

The authors gratefully acknowledge the company EXPAL Systems, S.A., as well as its factory in Navalmoral de la Mata (Cáceres, Spain) for supporting the experimental results presented in this paper.

References

[1] D. K. Leu, "The limiting drawing ratio for plastic instability of the cup-drawing process," *Journal of Materials Processing Technology*, vol. 86, no. 1–3, pp. 168–176, 1998.

[2] P. Sonis, N. V. Reddy, and G. K. Lal, "On multistage deep drawing of axisymmetric components," *Journal of Manufacturing Science and Engineering*, vol. 125, no. 2, pp. 352–362, 2003.

[3] R. K. Verma and S. Chandra, "An improved model for predicting limiting drawing ratio," *Journal of Materials Processing Technology*, vol. 172, no. 2, pp. 218–224, 2006.

[4] R. Padmanabhan, M. C. Oliveira, A. J. Baptista, J. L. Alves, and L. F. Menezes, "Blank design for deep drawn parts using parametric NURBS surfaces," *Journal of Materials Processing Technology*, vol. 209, no. 5, pp. 2402–2411, 2009.

[5] A. Agrawal, N. V. Reddy, and P. M. Dixit, "Determination of optimum process parameters for wrinkle free products in deep drawing process," *Journal of Materials Processing Technology*, vol. 191, no. 1-3, pp. 51–54, 2007.

[6] M. Haddadzadeh, M. R. Razfar, and M. R. M. Mamaghani, "Novel approach to initial blank design in deep drawing using artificial neural network," *Proceedings of the Institution of Mechanical Engineers B: Journal of Engineering Manufacture*, vol. 223, no. 10, pp. 1323–1330, 2009.

[7] A. Chamekh, S. Ben Rhaiem, H. Khaterchi, H. Bel Hadj Salah, and R. Hambli, "An optimization strategy based on a metamodel applied for the prediction of the initial blank shape in a deep drawing process," *International Journal of Advanced Manufacturing Technology*, vol. 50, no. 1–4, pp. 93–100, 2010.

[8] S. Sezek, V. Savas, and B. Aksakal, "Effect of die radius on blank holder force and drawing ratio: A model and experimental investigation," *Materials and Manufacturing Processes*, vol. 25, no. 7, pp. 557–564, 2010.

[9] R. Padmanabhan, M. C. Oliveira, J. L. Alves, and L. F. Menezes, "Influence of process parameters on the deep drawing of stainless steel," *Finite Elements in Analysis and Design*, vol. 43, no. 14, pp. 1062–1067, 2007.

[10] K. Mori and H. Tsuji, "Cold deep drawing of commercial magnesium alloy sheets," *CIRP Annals-Manufacturing Technology*, vol. 56, no. 1, pp. 285–288, 2007.

[11] L. M. A. Hezam, M. A. Hassan, I. M. Hassab-Allah, and M. G. El-Sebaie, "Development of a new process for producing deep square cups through conical dies," *International Journal of Machine Tools and Manufacture*, vol. 49, no. 10, pp. 773–780, 2009.

[12] H. C. Tseng, C. Hung, and C. C. Huang, "An analysis of the formability of aluminum/copper clad metals with different thicknesses by the finite element method and experiment," *International Journal of Advanced Manufacturing Technology*, vol. 49, no. 9–12, pp. 1029–1036, 2010.

[13] B. C. Hwang, S. M. Han, W. B. Bae, and C. Kim, "Development of an automated progressive design system with multiple processes (piercing, bending, and deep drawing) for manufacturing products," *International Journal of Advanced Manufacturing Technology*, vol. 43, no. 7-8, pp. 644–653, 2009.

[14] K. V. Ramana and P. V. M. Rao, "Data and knowledge modeling for design-process planning integration of sheet metal components," *Journal of Intelligent Manufacturing*, vol. 15, no. 5, pp. 607–623, 2004.

[15] G. C. Vosniakos, I. Segredou, and T. Giannakakis, "Logic programming for process planning in the domain of sheet metal forming with progressive dies," *Journal of Intelligent Manufacturing*, vol. 16, no. 4-5, pp. 479–497, 2005.

[16] F. Javier Ramírez and R. Domingo, "Application of an aided system to multi-step deep drawing process in the brass pieces manufacturing," in *Proceedings of the 3rd Manufacturing Engineering Society International Conference, MESIC 2009*, pp. 370–379, Alcoy, Spain, June 2009.

[17] F. J. Ramirez, R. Domingo, and M. A. Sebastian, "Design of an aided system to optimise times and costs in deep drawing process," in *Proceedings of the 2nd IPROMS International Researchers Symposium*, vol. 1, pp. 191–196, Ischia, Italy, 2009.

[18] K. Lange, *Handbook of Metal Forming*, McGraw-Hill, New York, NY, USA, 1985.

[19] S. Y. Chung and S. H. Swift, "An experimental investigation into the re-drawing of cylindrical shells," *Proceedings of the Institution of Mechanical Engineers B: Journal of Engineering Manufacture*, vol. 1, pp. 437–447, 1975.

[20] M. A. Sebastián and A. M. Sanchez-Perez, " Diseño asistido por ordenador de los útiles para la embutición profunda de piezas cilíndricas huecas," Internal Report, ETSII, UPM, Madrid, Spain, 1980.

[21] F. J. Ramirez, R. Domingo, M. A. Sebastian, and M. S. Packianather, "The development of competencies in manufacturing engineering by means of a deep-drawing tool," *Journal of Intelligent Manufacturing*, vol. 24, no. 3, pp. 457–472, 2011.

[22] F. J. Ramirez, R. Domingo, and M. A. Sebastian, "A technological model applied to multi-stage deep drawing process of axisymmetric components," in *Proceedings of the 21st International Computer-Aided Production Engineering Conference (CAPE '10)*, Edinburgh, UK, 2010.

[23] E. Siebel and H. Beisswänger, *Deep Drawing*, Carl Hanser, Munich, Germany, 1995.

[24] E. M. Rubio, M. Marín, R. Domingo, and M. A. Sebastián, "Analysis of plate drawing processes by the upper bound method using theoretical work-hardening materials," *International Journal of Advanced Manufacturing Technology*, vol. 40, no. 3-4, pp. 261–269, 2009.

11

Development of HPDC Advanced Dies by Casting with Reinforced Tool Steels

I. Vicario,[1] J. K. Idoiaga,[2] E. Arratibel,[3] I. Erauskin,[1] L. M. Plaza,[1] I. Crespo,[1] and P. Caballero[1]

[1] *Tecnalia Research & Innovation, 20009 San Sebastián, Spain*
[2] *Industrias Lebario, 48213 Izurza, Spain*
[3] *Fundiciones Inyectadas Alavesas S.A., 01230 Nanclares de la Oca, Spain*

Correspondence should be addressed to I. Vicario; iban.vicario@tecnalia.com

Academic Editor: Jean-Yves Hascoet

High pressure die casting (HPDC) dies are nowadays manufactured with high quality forged steels. Cavities are made by electrical discharge machining (EDM) or by high speed milling. The average life of an aluminium HPDC die is about 125.000 injections. Refrigeration circuits have simple configurations, because they are produced by drilling the die with straight holes. They are limitations in the distances and diameters of holes. Sensors are placed where the geometry of the die permits an easy machining. In order to obtain complex figures, several rapid prototyping methods have been developed. However, there is a limitation in the life of the dies produced by this technique, from several parts to thousands. A new method to obtain semifinished high pressure die casting dies in a steel of higher mechanical properties and with the refrigeration circuits and sensors embedded into it is described in this paper. The method consists in producing a molten steel alloy with micro-nano-special ceramic particles inserted in it and casting the composite material in sand moulds of the desired geometry. The resultant solidified near-net shape die with the cooling tubes and sensors embedded into it. A use-life and a productivity about 50% and 10% higher are obtained.

1. Introduction

In the last years HPDC cast parts have developed a quick development due to the high automation, the new die materials, and sensoring and construction methods and also by the advanced control and foundry systems.

HPDC are generally made in H11 or H13 steel alloys by high speed machining combined with electrical discharge machining (EDM). Dies are expensive, due to the expensive tooling and the high energy and man power demanded in these operations.

Rapid prototyping processes are coming more and more employed in HPDC moulds, but they are employed normally for first series, because the properties obtained with those parts produced by this technology cannot obtain the same life time than with standard fabricated dies. According to bibliography [1–3], there are different direct metal part production methods to made dies by rapid prototyping process. Between them, the direct metal laser sintering (DMLS) and selective laser sintering (SLS) offers the possibility to adapt the cooling channels to the contour of a cavity without limitations in the form, optimizing die temperature control and increasing the die life. Cycle time reductions can be also obtained by optimizing the cooling circuits, and quality is increased by the nonquality parts reject reduction due to a better thermal stability of the die. However, there are several constraints to employ some alloys as the H11, and the metallurgical quality of produced parts is not as good as forged pieces used for dies.

There are also several indirect metal part production methods [4]. These processes have been employed for insert fabrication, but their applications are very limited.

The steel sand casting technique has been also appointed as a possible way to construct cavities and dies [5]. There is a reduction in the fabrication time, is necessary less raw material and machining operations are reduced. However, the application of those products is limited for first series and for

Table 1: Temper temperatures and materials.

	460°C	580°C	600°C
Base alloy	●		●
Reinforced alloy	●	●	●

the supporting parts of the dies, due to the limited lifetime of the dies and the necessity of machining the refrigerating circuits.

One of the main mechanisms that define the total life of a die is the thermal fatigue cracking [6, 7]. In order to increase die life, new hot work tool steels with improved properties for HPDC applications are being continuously developed [8, 9].

Between them, the metal matrix composites have the potential to increase the thermal fatigue property [10, 11].

This work deals mainly with the development of a foundry process to produce HPDC dies with customised internal refrigeration and high resistance new steel. The objective is to obtain an economical process to produce advanced HPDC dies.

2. Design of Experiments

The base material for the tests has been chosen from the most employed steels used by HPDC die makers. The alloy 1.2344 (H13) was finally elected.

Based on the previous experience, in order to reinforce the base material, a Fe(TiMo)C master-alloy produced by self-propagating high temperature synthesis (SHS) has been chosen to be added to liquid steel.

To compare the new developed material properties in function of different tempering temperatures, some test bars were casted, heat-treated, and machined. An industrial induction furnace was used to prepare 300 kg of molten steel from selected scraps. When the scraps were melted at 1650°C, the steel was dedrossed and a 7.5% in weight of Fe(TiMo)C master alloy was added until a good dissolution was obtained. Test bars were poured in sand moulds at 1.550°C. After the solidification, parts were demoulded and separated from runners and overflows. Test wars were quenched at 890°C for 2 hours and tempered at temperatures between 460 and 600°C for 2 hours. After sand blasting, parts were machined in a high speed milling machine.

For the mechanical characterization, standard EN-10002.1, EN-10045.1, and EN-10003.1 bars were machined to test tensile, resilience, and hardness properties, respectively. Wearing properties were also checked by pin on disk test. Thermal test bars have a dimension of 150 × 50 × 50 mm with internal channels of 20 mm of width, 5 mm of depth, and 110 mm length (see Figure 1(a)). The test bars were essayed in a test machine during 100.000 cycles and with a variation of temperature from 600 to 250°C every 30 s, by induction heating and followed then by water and air spraying.

The two cast materials were quenched at 890°C and tempered at different temperatures (see Table 1) in order to determinate the differences between materials and the optimum temper temperature for HPDC die applications.

Table 2: Tube sizes, fillers, and positions.

Ref.	Outside diameter (mm)	Inside diameter (mm)	Filler	Tube position
1	10	7	Gas	Vertical
2	18	9	Sand	Vertical
3	17	11	Sand	Vertical
4	8	5.4	—	Horizontal
5	21.5	16	—	Horizontal

In order to define a method to obtain directly from casting the cooling circuits a 125 mm radius and 300 mm length cylinder cavity to be produced into a chemically bounded sand mould was designed with placements for tubes. In this sand mould tubes with different diameters and materials were tested, with and without internal sand as protective material for the tubes, and also a continuous gas flow was introduced to cool the tube during casting solidification. Definitively, standard AISI 316 stainless steel tubes were chosen for the internal cooling circuits, due to the low mismatch in composition between base alloy and AISI 316. On the other hand, zirconia sand was selected as inside filler for the cooling tubes due to its high refractoriness.

We can observe in Table 2 the different tubes, fillers, and placement in respect to the parting line of the sand mould employed for casting trials as we can observe in Figure 2.

After pouring and subsequent solidification of the liquid composite material, casting part was cut and inspected to determinate if the tubes resisted the molten metal.

Optical microscopy has been used to determinate the structure of different alloys after heat treatments and also to determinate the coherency between the alloy and the tube material, in order to understand the obtained properties.

In order to test the materials in real production, a design and a cooling optimisation of real parts were made (see Figure 3(a)) and the patterns for sand mould were prepared, having in mind the support system for cores, as we can observe in Figure 3(b).

Internal tubes were cold deformed to obtain the desired forms and also in few cases arc welding technique was employed to obtain complex shapes.

3. Experimental

In order to determinate a robust process, different cooling tubes in size and position and with and without internal insulating material (zirconia) defined in Table 2 were tested by using an induction furnace to cast steel over sand moulds with test tubes places into the mould cavity.

We can observe in Figure 4 how the tubes of Table 2 were placed in the sand test mould.

H13 base alloy was poured at 1560°C (see Figure 5(a)), and after solidification, the test cylinder was cut (Figure 5(b)).

We can observe in Figure 5(a) how the metal solidifies into the sand die and in Figure 5(b) how only tubes filled with

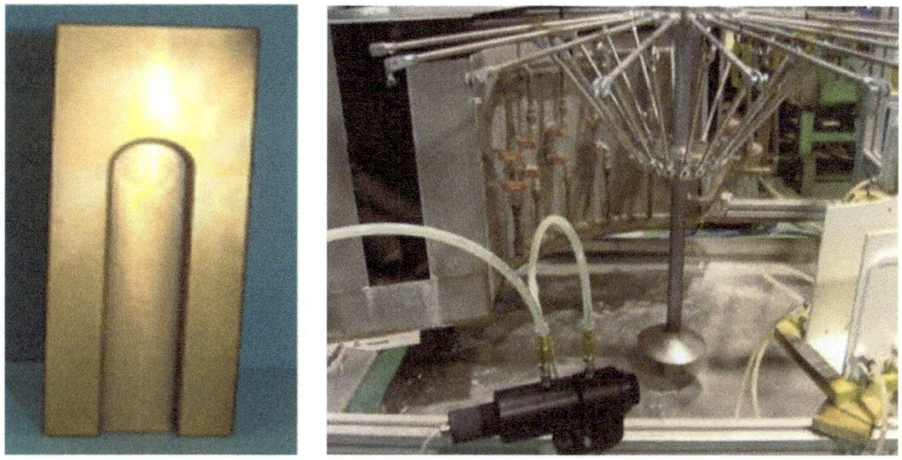

(a) (b)

FIGURE 1: (a) Thermal fatigue test bar, (b) thermal fatigue test machine.

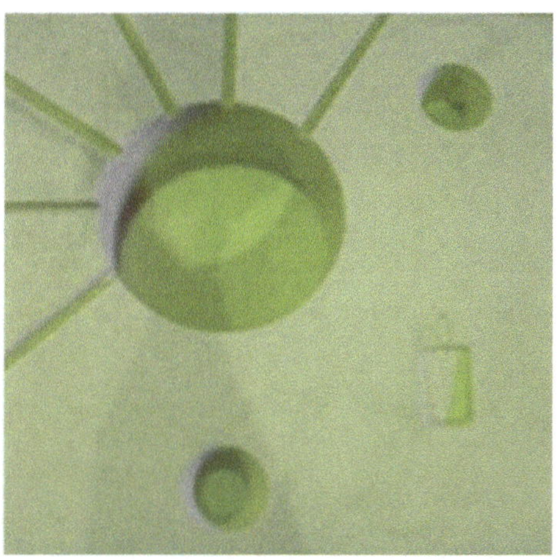

FIGURE 2: Die sand mould with placements for tubes.

TABLE 3: Resistance to metal penetration in tubes.

Ref.	Tube filling	Tube position	Resistant to metal penetration
1	Gas	Vertical	No
2	Sand	Vertical	Yes
3	Sand	Vertical	Yes
4	—	Horizontal	No
5	—	Horizontal	No

zirconia sand were capable of resisting the attack of molten metal.

The obtained results are resumed in Table 3.

With these results the external diameter of cooling tubes was set to 10 mm, the internal diameter to 7 mm, and zirconia sand as tube filler.

In order to define the real behaviour of the developed process, dies and cores for industrial validation were defined. Die

TABLE 4: Base alloy composition (% in weight).

Element	C	Mn	Si	Cr	Ni	Mo
%	0.22	0.80	1.00	3.00	0.70	0.35

sand mould patterns were constructed and internal cooling tubes conformed. The mould patterns are 5 mm oversized in comparison with finished die, due to the necessity of having enough material to machine the exterior of the casting to obtain the desired quality and dimensions.

Demonstrators were tested with a H13 base steel having the composition shown in Table 4.

The Fe(TiMo)C master-alloy was produced by powder metallurgy and promoting a SHS reaction. The powders of Ti (48%, 100 μm size), C (12%, 1–3 μm), Mo (10%, 10 μm), and Fe (30%, 30 μm) were mixed and they were then pressed. The briquette was introduced into a reactor and was directly heated by an electrical current. Once reaction started, electrical current was stopped and there was an autopropagation of the reaction without any external energy contribution. We can observe in Figure 6 a master-alloy sample.

Using an insulation furnace, 250 Kg of H13 base alloy was melted at 1650°C. The master-alloy Fe(TiMo)C was added to molten steel in a 7.5% concentration in weight. When the master-alloy was dissolved in the base alloy, the composite material (steel + ceramic particles) was cleaned and poured at 1660–1665°C in the sand moulds.

Once the castings are solidified, cooled, and demoulded as we can see in Figure 7, in order to make easier, the machining is necessary to decrease the hardness by an annealing process. Annealing temperature was established in 980°C with 4 hours of residence time. After that, the parts were shot-blasted in order to eliminate the skin of the metal that had been in contact with the sand mould and also to detect possible superficial shrinkage cavities and other foundry defects, the risers of the feeding and filling channels were cut. We can observe in Figure 8(a) the annealing process

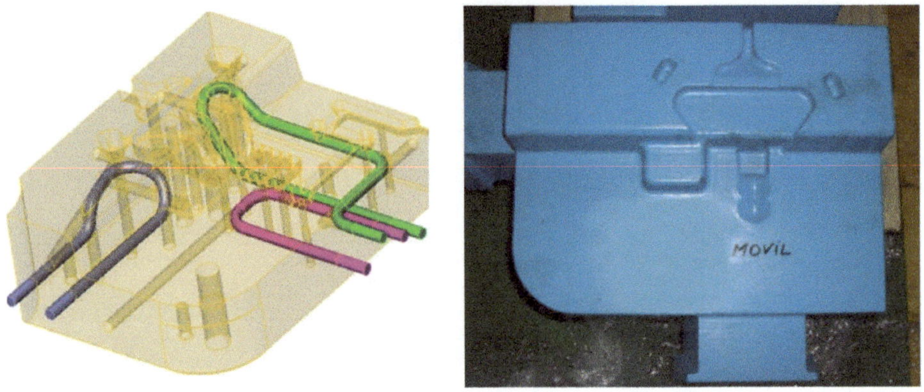

(a) (b)

FIGURE 3: (a) Cooling optimisation, (b) sand mould pattern.

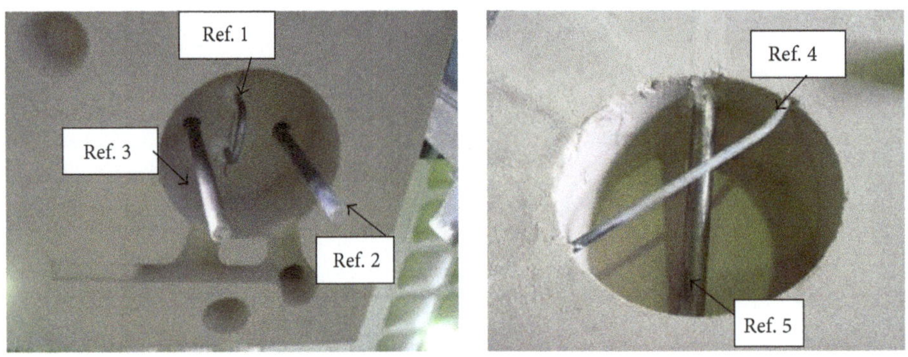

FIGURE 4: Placement of cooling tubes in the test sand mould.

in a heat treatment four and in Figure 8(b) one part after shot blasting, without risers, channels, and so forth.

When the part it's machined is possible to observe the external skin that has been in contact with sand mould, which it's very hard because it contains very hard oxides (see Figure 9(a)). After machining, the part is quenched in air at 890°C (4 hours) and tempered at 580–600°C (4 hours). The tempering heat treatment reduces the hardness of the part to 45–48 HRC. A very fine machining is often required in order to get final dimensions, due to the distortions caused by heat treatments on the part. In Figure 9(b) we can observe the final HPDC core sample.

4. Results and Discussion

The thermal fatigue bars were used to determinate the process parameters and properties of the obtained parts. To determinate the most suitable temper temperature, the conventional base alloy and the reinforced alloy were tempered at different temperatures. In Figure 20 we can observe how the hardness decreases when we increase the temper temperature in base alloy (B.Alloy) and reinforced alloy (R.Alloy).

In Figure 21 we can observe how in both cases the resilience value is increased by using a higher temper temperature.

In order to determinate the optimum temper temperature, we can observe in Figure 22 how at 580°C and 600°C balance properties of hardness and resilience are quite similar.

Indeed, pin on disk tests have been performed with the alloys, showing that when we increase the tempering temperature, the wearing speed is increased. Also, the reinforced materials have a much better behaviour to wearing than base alloys, as shown in Figure 23.

With the previous results, the reinforced material with a temper temperature of 580°C has been elected to be tested in tensile test and to be employed in validation test. There is not a big difference between a temperature of 580°C or 600°C, because there is a balance between the obtained properties.

Table 5 shows the ultimate tensile strength and yield strength expressed in MPa and the elongation in % for the base alloy and for the reinforced alloy with a temper temperature of 580°C. We can observe a slight increase in the ultimate tensile strength (UTS), as well as an important increase of yield strength (YS). However the elongation drops from 19% to 11%.

As much as the main properties that influence the life of a die are the thermal shock resistance and the hardness, the two alloys were tested during 100.000 cycles in the thermal fatigue test machine. Test bars were analysed under magnetic fields in order to determinate the severity of the cracks generated by thermal shocks. The reinforced alloy did not show any crack

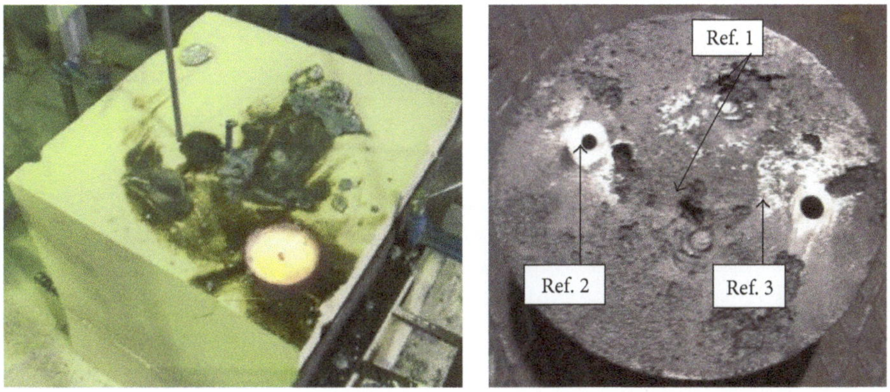

(a) (b)

FIGURE 5: (a) Sand mould with poured metal, (b) study of metal penetration.

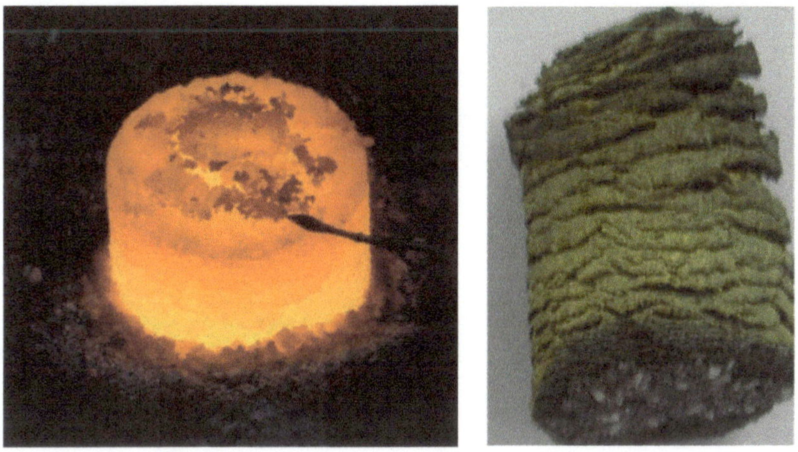

FIGURE 6: SHS produced Fe (TiMo) C master-alloy.

TABLE 5: Ultimate and tensile strength for base and reinforced alloy.

	UTS (MPa)	YS (MPa)	Elongation (%)
Base alloy (tempering temp. 580°C)	1.050	775	19
Reinforced alloy (tempering temp. 580°C)	1.070	980	11

after 100.000 cycles. We can observe in Figure 10 the base alloy test bar with cracks after sand blasting (Figure 10(a)) and under magnetic fields (Figure 10(b)).

4.1. Microstructure. The steel microstructure is based on a tempered martensitic matrix with inserted and dispersed ceramic carbide particles, as it can be seen in Figure 11(a).

In Figure 12 we can notice the difference between base steel and the reinforced one. The reinforced material has well distributed primary ceramic carbides.

4.2. Validation Tests. The best treated material was used to obtain real die casting pieces with a sampling die. Production and quality rates and total die life was compared with standard materials and cooling circuits. For that, validation tests were made at Fiasa with a comparable die with 4 identical cavities. Three cores were made with base alloy and another was made with the reinforced alloy (Figure 19). A very aggressive alloy for the die, an Al Si10 Mg with a Fe concentration minor than 0.1%, was selected in order to work in the worst conditions.

The minimum distance between the surface of the mould and the refrigeration tube was established in only 10 mm, where at least 25 mm is required in standard dies (see Figure 13).

A thermal camera was employed in order to determinate the different core external temperature, maintaining the same

FIGURE 7: Example of cast HPDC parts with internal cooling circuits.

(a) (b)

FIGURE 8: (a) Annealing furnace, (b) shot blasted part with superficial defects.

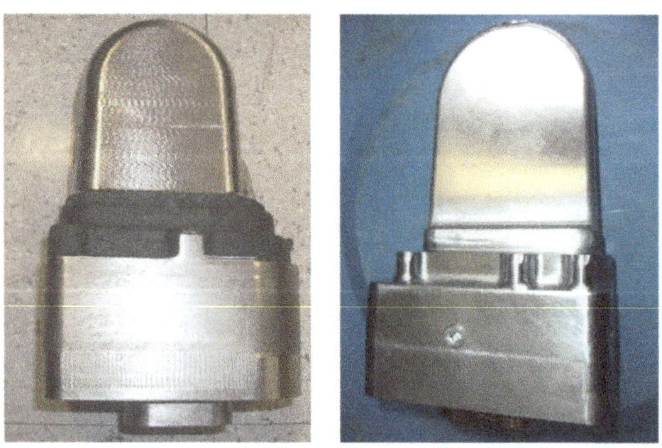

(a) (b)

FIGURE 9: (a) Partial machined part with contact skin, (b) finished HPDC core.

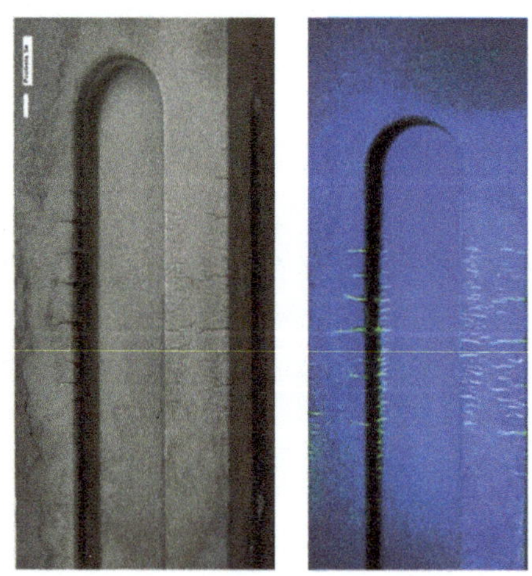

(a) (b)

FIGURE 10: Thermal stress produced cracks observed (a) under magnetic particles, (b) after sand blasting.

water flow in all the cores. We can see in Table 6 the average values in the same testing area.

After only 3.505 injections, the cores were extracted and compared, in order to determine the performance of the new alloy. In Figure 13 we can observe how the erosion has been much higher in the base alloy (Figure 14(a)) than in the reinforced alloy (Figure 14(b)).

Some heat cracks due to high thermal stresses were detected in the core area where the distance between the cooling circuit was nearer to the surface of the core (see Figure 15). The cracks copy the internal curve shape of the cooling circuit, and what it shows is that the minimum

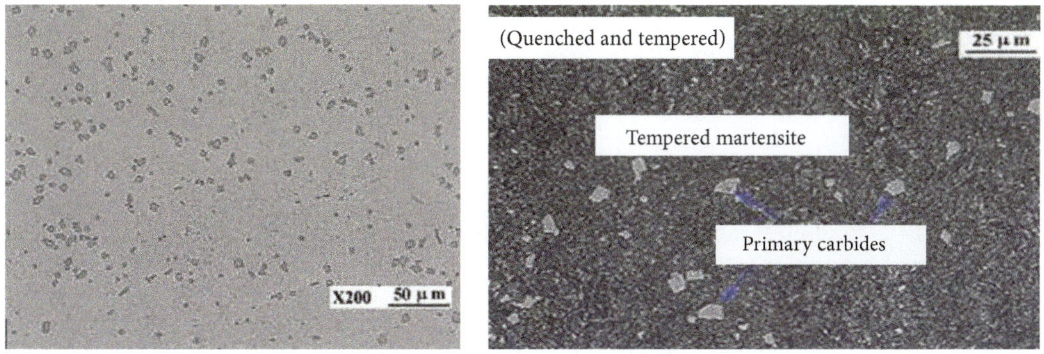

(a) (b)

FIGURE 11: (a) Micrograph without attack, (b) attacked micrograph.

(a) (b)

FIGURE 12: (a) Micrograph without attack, (b) attacked micrograph.

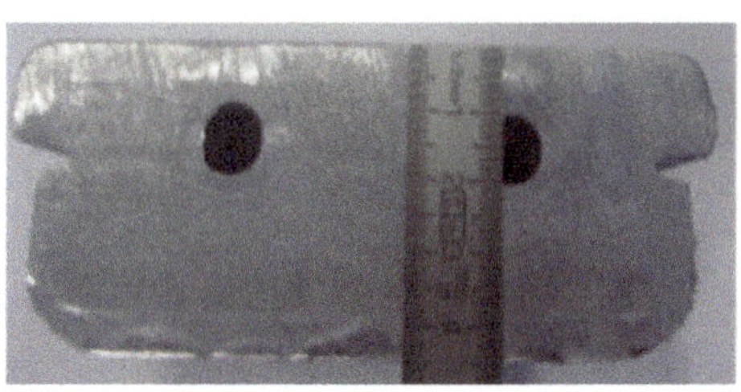

FIGURE 13: Reinforced alloy core with the internal cooling place at 10 mm and 20 mm from surface.

TABLE 6: External temperature of cores.

	Reinf. alloy core 1	Base alloy core 2	Base alloy core 3	Base alloy core 4
Temperature (°C)	140	178	180	180

distance from the cooling channel should be at least of 20 mm.

In order to study the cracks and the interphase between the internal cooling tubes and the reinforced alloy, the part was divided in 6 different areas (as shown in Figure 16).

TABLE 7: Micro-hardness of reinforced alloy core.

Ref. 1	Ref. 2	Ref. 3	Ref. 4	Ref. 5	Ref. 6
320	308	317	319	318	317

In Figure 17 we can observe how the cracks advance from the external surface of the core to the cooling circuit perpendicular to the surface.

Also we can see there is a metallurgical continuity between the tube and the reinforced alloy material, as shown in Figure 18. No porosity or brittle interphases are detected.

Micro-hardness (HV_{10}) values were measured in order to determinate a possible temper action in the core by repetitive injections. We can see in Table 7 the average values for the six different areas. Micro-hardness was determined by 3 measurements in 6 different points for every sample, from the external surface to the tube. There are not sensible differences between the values in the surface or in the inside area.

The average micro-hardness before testing the core in the foundry was 329 HV10. So there are not big differences between the samples before and after working.

The injected parts quality was compared also after 3.505 injections. A clear difference was determined in the injected part. The part obtained with the unreinforced material show a poorer surface quality, with cracks in the surface and thicker

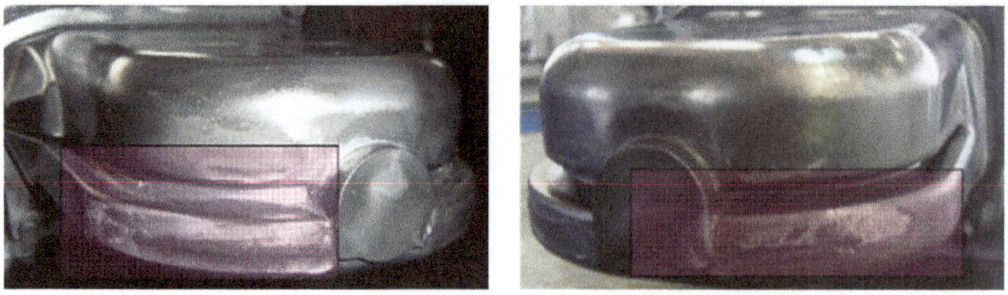
(a) (b)

FIGURE 14: (a) Base alloy core, (b) reinforced alloy core.

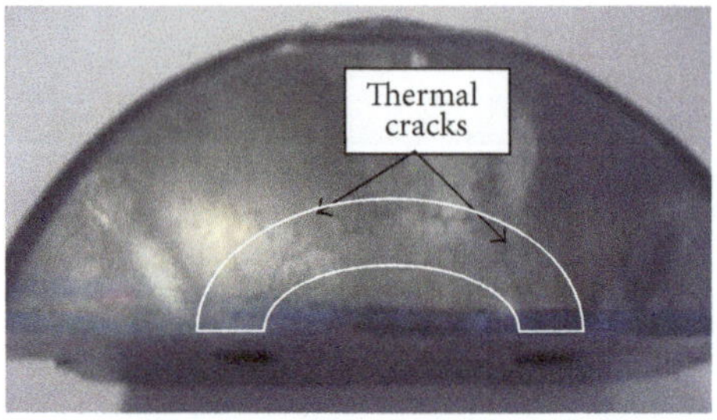

FIGURE 15: Heat cracks with the internal cooling circuit shape.

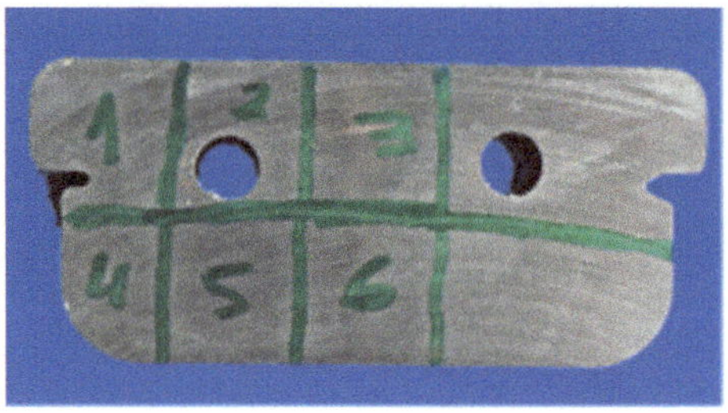

FIGURE 16: Heat cracks with the internal cooling circuit shape.

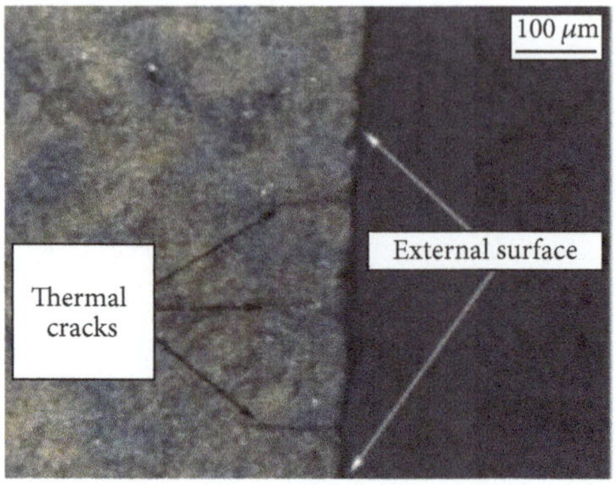

FIGURE 17: Heat cracks with the internal cooling circuit shape.

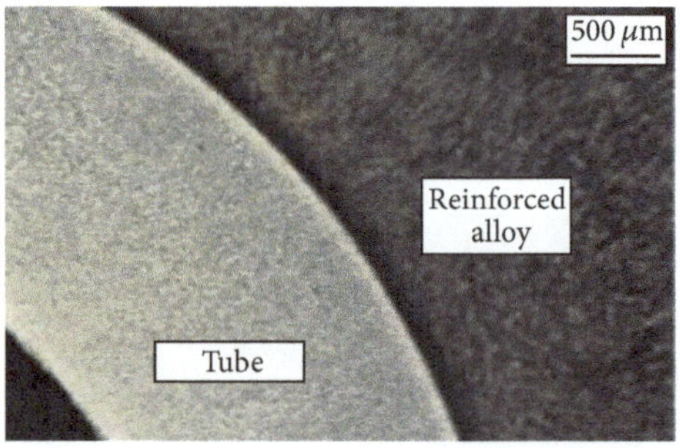

FIGURE 18: Interphase between tube and reinforced material.

over material in the eroded area, in coincident with the eroded and cracks areas showed in Figures 14 and 15.

5. Conclusion

There are several methods to construct dies, but nowadays it is necessary large machining operations in order to obtain the desired shapes and internal cooling circuits with several constraints in the design. Rapid prototyping process cannot afford by the moment the same results than machining dies from 3D forged hot work steels.

The SHS process allows developing a ceramic Fe(TiMo)C master alloy that is dispersed in the molten steel and after solidification gives a martensitic matrix reinforced with particles of carbides (TiMo)C well distributed.

This composite material allows increasing the UTS, YS, hardness, and wearing, with a decrease in elongation and to impact. Thermal fatigue is also increased with the new reinforced alloy. These properties are obtained producing the near-net shape dies by pouring the liquid material into sand moulds, what permits reducing drastically the machining operations and costs, reducing delivery times in about 2 weeks.

The use of stainless steel tubes filled with zirconia sand allows developing customized cooling circuits, reducing the necessities of machining operations to construct the cooling circuits, given a better thermal control. Also a good continuity between the tube and the reinforced alloy is observed.

Verification tests of new reinforced alloy on casting pieces show an increase on the life time, a decrease in core temperature in working conditions, but also a limit in the distance from the cooling circuits to the surface of the part, due to heat cracking.

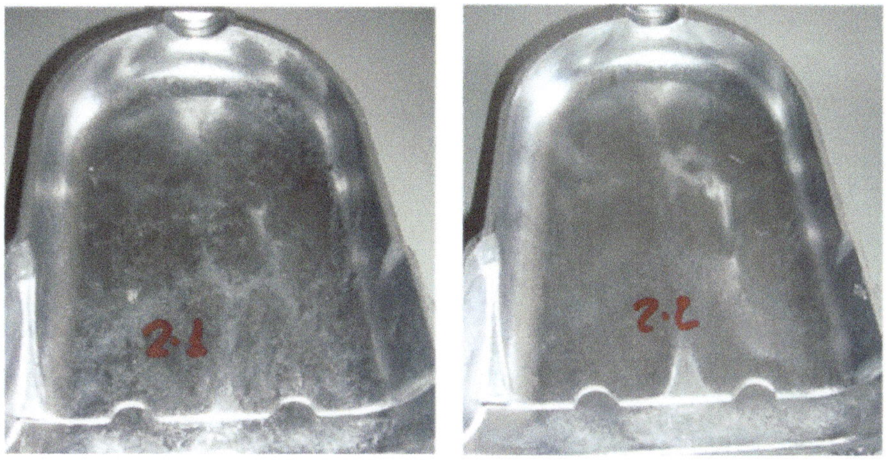

(a) (b)

FIGURE 19: (a) Base alloy cast part, (b) reinforced alloy cast part.

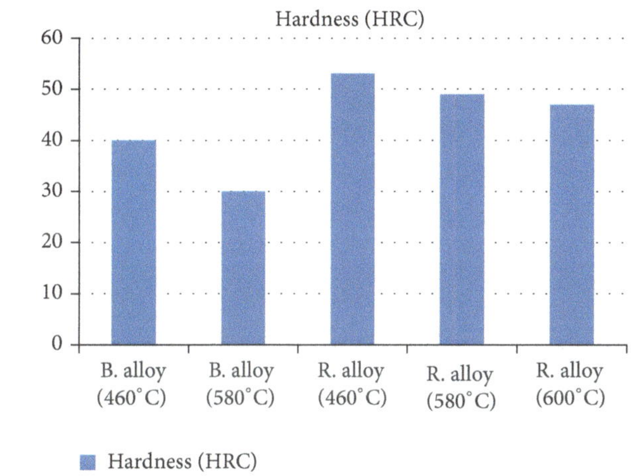

FIGURE 20: Hardness variation in function of temper temperature.

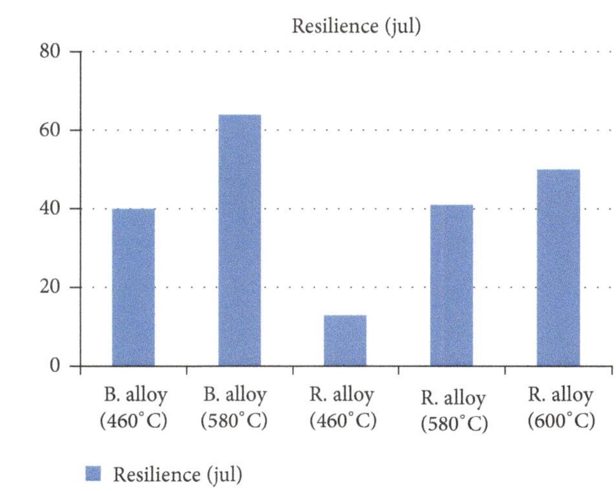

FIGURE 21: Hardness variations in function of temper temperature.

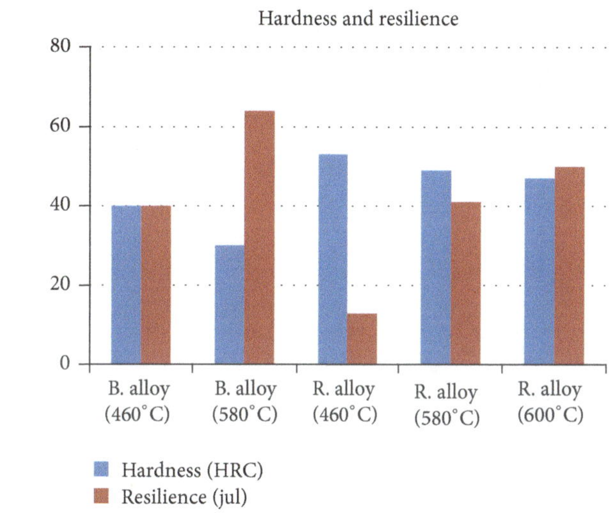

FIGURE 22: Hardness and resilience variations in function of temper temperature.

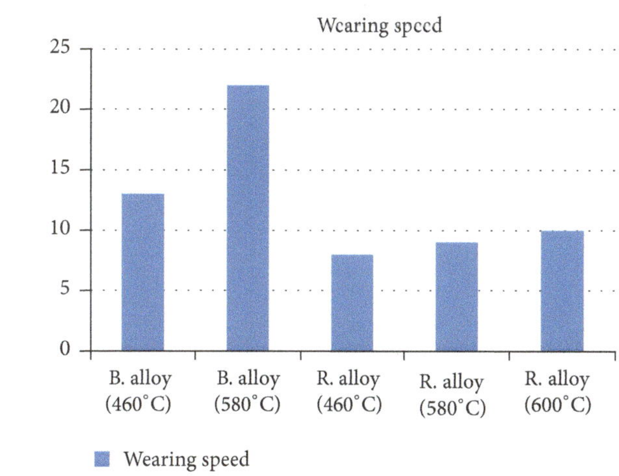

FIGURE 23: Pin on disk wearing speed comparison.

Conflict of Interests

The authors declare that there is no conflict of interests regarding the publication of this paper.

Acknowledgments

This work has been partially funded by the Basque government through the Project nos. IG-2010/0000118 and IG-2011/0000678 and CDTI through the Project CDTI IDI-20100385. The authors are also much grateful to the personnel of Industrias Lebario and Fundiciones Inyectadas Alavesas, S.A., companies that have collaborated in aspects related to the design, test, and analysis of results.

References

[1] J. Meckley and R. Edwards, "A study on the design and effectiveness of conformal cooling channels in rapid tooling inserts," *The Technology Interface Journal*, vol. 10, no. 1, pp. 1–28, 2009.

[2] B. R. Srikrishna and D. Chandrababu, "Utilization of dies and molds using intelligent methods of management," *International Journal of Mechanical Engineering and Computer Applications*, vol. 1, no. 3, pp. 55–60, 2013.

[3] T. Altan, B. Lilly, and Y. C. Yen, "Manufacturing of dies and molds," *CIRP Annals: Manufacturing Technology*, vol. 50, no. 2, pp. 404–422, 2001.

[4] W. Shambley, "From RP to AM: a review of prototyping methods," *Moder Casting*, vol. 100, no. 7, pp. 30–33, 2010.

[5] C. Mus, "Aluminum die casting rapid tooling: state of the art and innovative approaches," in *Proceedings of the 4th International Conference of High Tech Die Casting*, Montichiari, Italy, April 2008.

[6] M. Muhič, J. Tušek, F. Kosel, D. Klobčar, and M. Pleterski, "Thermal fatigue cracking of Die-casting dies," *Metalurgija*, vol. 49, no. 1, pp. 9–12, 2010.

[7] B. Kosek, M. Sokovic, and G. Kosek, "Failure analysis of dies f or aluminium alloys die-casting," in *Proceedings of the 13th International Scientific Conference on Achievements in Mechanical and Materiala Engineering*, Glibice-Wisla, Poland, May 2005.

[8] E. Haberling, K. Rasche, and I. Schruff, *New Materials and Processes for Tooling*, edited by H. Berns, H. Nordberg and H. J. Fleischer, Schürmann & Klagges, Bochum, Germany, 1989.

[9] K.-D. Fuchs, "Hot-work tool steels with improved properties for die casting applications," in *Proceedings of the 6th International Tooling Conference*, pp. 15–22, Karlstad, Sweden, 2002.

[10] A. Mortensen and T. W. Clyne, "Metal-matrix composites: challenges and opportunities," in *European White Book of Fundamental Research in Materials Science*, chapter 6.5, pp. 210–213, 2002.

[11] J. I. Erausquin, *Aceros reforzados con partículas de carburos primarios elaboradas por SHS. Obtención mediante metalurgia líquida, conformado por moldeo y caracterización [Ph.D. thesis]*, ESII de Bilbao, Bilbao, Spain, 2009.

12

Optimization of Fusion Zone Grain Size, Hardness, and Ultimate Tensile Strength of Pulsed Current Microplasma Arc Welded AISI 304L Sheets Using Genetic Algorithm

Siva Prasad Kondapalli,[1] Srinivasa Rao Chalamalasetti,[2] and Nageswara Rao Damera[3]

[1] *Department of Mechanical Engineering, Anil Neerukonda Institute of Technology & Sciences, Visakhapatnam 530 003, India*
[2] *Department of Mechanical Engineering, AU College of Engineering, Andhra University, Visakhapatnam 531 162, India*
[3] *Centurion University of Technology & Management, Odisha 761 211, India*

Correspondence should be addressed to Siva Prasad Kondapalli; kspanits@gmail.com

Academic Editor: Shia-Chung Chen

Austenitic stainless steel sheets have gathered wide acceptance in the fabrication of components, which require high temperature resistance and corrosion resistance, such as metal bellows used in expansion joints in aircraft, aerospace, and petroleum industry. In case of single pass welding of thinner sections of this alloy, Pulsed Current Microplasma Arc Welding (PCMPAW) was found beneficial due to its advantages over the conventional continuous current process. The quality of welded joint depends on the grain size, hardness, and ultimate tensile strength, which have to be properly controlled and optimized to ensure better economy and desirable mechanical characteristics of the weld. This paper highlights the development of empirical mathematical equations using multiple regression analysis, correlating various process parameters to grain size, and ultimate tensile strength in PCMPAW of AISI 304L sheets. The experiments were conducted based on a five-factor, five-level central composite rotatable design matrix. A genetic algorithm (GA) was developed to optimize the process parameters for achieving the desired grain size, hardness, and ultimate tensile strength.

1. Introduction

AISI 304L is an austenitic stainless steel with excellent strength and good ductility at high temperature. Typical applications include aeroengine hot section components, miscellaneous hardware, tooling, and liquid rocket components involving cryogenic temperature. AISI 304L can be joined using a variety of welding methods, including Gas Tungsten Arc Welding (GTAW), Plasma Arc Welding (PAW), Laser Beam Welding (LBW), and Electron Beam Welding (EBW). Of these methods, low current PAW (Micro-PAW) has attracted particular attention and has been used extensively for the fabrication of metal bellows, diaphragms which require high strength and toughness. PAW is conveniently carried out using one of two different current modes, namely, a continuous current (CC) mode or a pulsed current (PC) mode.

Pulsed current MPAW involves cycling the welding current at selected regular frequency. The maximum current is selected to give adequate penetration and bead contour, while the minimum is set at a level sufficient to maintain a stable arc [1, 2]. This permits arc energy to be used effectively to fuse a spot of controlled dimensions in a short time producing the weld as a series of overlapping nuggets. By contrast, in constant to current welding, the heat required to melt the base material is supplied only during the peak current pulses allowing the heat to dissipate into the base material leading to narrower heat affected zone (HAZ). Advantages include improved bead contours, greater tolerance to heat sink variations, lower heat input requirements, reduced residual stresses and distortion, and refinement of fusion zone microstructure and reduced width of HAZ. There are four independent parameters that influence the process and they are peak current, back current, pulse, and pulse width.

Table 1: Chemical composition of AISI 304L (weight%).

C	Si	Mn	P	S	Cr	Ni	Mo	Ti	N
0.021	0.35	1.27	0.030	0.001	18.10	8.02	—	—	0.053

In this investigation, experiments conducted using the design of experiments concept were used for developing mathematical models to predict such variables. Many works were reported in the past for predicting bead geometry, heat affected zone, bead volume, and so forth, using mathematical models for various welding processes [3–5]. Usually, the desired welding process parameters are determined based on the experience of skilled workers or from the data available in the handbook. This does not ensure the formation of optimal or near optimal weld pool geometry [6]. It has been proven by several researchers that efficient use of statistical design of experiment techniques and other optimization tools can impart scientific approach in welding procedure [7, 8]. These techniques can be used to achieve optimal or near optimal bead geometry from the selected process parameters.

Kim et al. reviewed that optimization using regression modeling, neural network, and Taguchi methods could be effective only when the welding process was set near the optimal conditions or at a stable operating range [9], but near-optimal conditions cannot be easily determined through full-factorial experiments when the number of experiments and levels of variables are increased. Also, the method of steepest ascent based upon derivatives can lead to an incorrect direction of search due to the nonlinear characteristics of the welding process. Genetic algorithm, being a global algorithm, can overcome the above problems associated with full-factorial experiments and the objective function to be optimized using GA need not be differentiable [10].

In the present study, a sequential genetic algorithm has been used to optimize the process parameters and achieve minimum fusion zone grain size, maximum hardness, and ultimate tensile strength of PCMPAW AISI 304L sheets.

2. Experimental Procedure

Austenitic stainless steel (AISI 304L) sheets of 100 × 150 × 0.25 mm are welded autogenously with square butt joint without edge preparation. The chemical composition of SS304L stainless steel sheet is given in Table 1. High purity argon gas (99.99%) is used as a shielding gas and a trailing gas right after welding to prevent absorption of oxygen and nitrogen from the atmosphere. The welding has been carried out under the welding conditions presented in Table 2. From the literature four important factors of pulsed current MPAW as presented in Table 3 are chosen. A large number of trail experiments were carried out using 0.25 mm thick AISI 304L sheets to find out the feasible working limits of pulsed current MPAW process parameters. Due to a wide range of factors, it has been decided to use four-factor, five-level rotatable central composite design matrix to perform the number of experiments for investigation. Table 4 indicates the 31 set of coded conditions used to form the design matrix. The first sixteen experimental conditions (rows) have been formed for main effects. The next eight experimental conditions are called corner points and the last seven experimental conditions are known as center points. The method of designing such matrix is dealt elsewhere [11, 12]. For the convenience of recording and processing the experimental data, the upper and lower levels of the factors are coded as +2 and −2, respectively, and the coded values of any intermediate levels can be calculated by using the expression [13]

Table 2: Welding conditions.

Power source	Secheron Microplasma Arc Machine (Model: PLASMAFIX 50E)
Polarity	DCEN
Mode of operation	Pulse mode
Electrode	2% thoriated tungsten electrode
Electrode diameter	1 mm
Plasma gas	Argon and hydrogen
Plasma gas flow rate	6 Lpm
Shielding gas	Argon
Shielding gas flow rate	0.4 Lpm
Purging gas	Argon
Purging gas flow rate	0.4 Lpm
Copper nozzle diameter	1 mm
Nozzle to plate distance	1 mm
Welding speed	260 mm/min
Torch position	Vertical
Operation type	Automatic

$$X_i = \frac{2\left[2X - \left(X_{\max} + X_{\min}\right)\right]}{\left(X_{\max} - X_{\min}\right)}, \quad (1)$$

where X_i is the required coded value of a parameter X. The X is any value of the parameter from $X_{\min}$ to $X_{\max}$, where $X_{\min}$ is the lower limit of the parameter and $X_{\max}$ is the upper limit of the parameter.

3. Recording the Responses

3.1. Measurement of Grain Size. Three metallurgical samples are cut from each joint, with the first sample being located at 25 mm behind the trailing edge of the crater at the end of the weld and mounted using Bakelite. Sample preparation and mounting are done as per ASTM E 3-1 standard. The samples are surface grounded using 120 grit size belt with the help of belt grinder and polished using grade 1/0 (245 mesh size), grade 2/0 (425 mesh size), and grade 3/0 (515 mesh size) sand paper. The specimens are further polished by using aluminum oxide initially and by utilizing diamond paste and velvet cloth

TABLE 3: Important factors and their levels.

Serial number	Input factor	Units	Levels -2	-1	0	+1	+2
1	Peak current	Amperes	6	6.5	7	7.5	8
2	Back current	Amperes	3	3.5	4	4.5	5
3	Pulse rate	Pulses/second	20	30	40	50	60
4	Pulse width	%	30	40	50	60	70

TABLE 4: Design matrix and experimental results.

Serial number	Peak current (Amperes)	Back current (Amperes)	Pulse rate (pulses/second)	Pulse width (%)	Grain size (Microns)	Hardness (VHN)	Ultimate tensile strength (UTS) (Mpa)
1	−1	−1	−1	−1	20.812	198	713
2	1	−1	−1	−1	30.226	190	705
3	−1	1	−1	−1	21.508	199	718
4	1	1	−1	−1	27.536	193	706
5	−1	−1	1	−1	27.323	193	706
6	1	−1	1	−1	25.206	195	710
7	−1	1	1	−1	25.994	195	705
8	1	1	1	−1	23.491	197	706
9	−1	−1	−1	1	26.290	194	705
10	1	−1	−1	1	29.835	190	700
11	−1	1	−1	1	20.605	200	715
12	1	1	−1	1	27.764	193	708
13	−1	−1	1	1	30.095	190	698
14	1	−1	1	1	26.109	194	706
15	−1	1	1	1	27.385	193	704
16	1	1	1	1	25.013	195	710
17	−2	0	0	0	20.788	196	710
18	2	0	0	0	25.830	195	706
19	0	−2	0	0	31.663	188	701
20	0	2	0	0	27.263	193	708
21	0	0	−2	0	25.270	195	712
22	0	0	2	0	26.030	194	705
23	0	0	0	−2	24.626	195	711
24	0	0	0	2	26.626	194	705
25	0	0	0	0	24.845	196	710
26	0	0	0	0	24.845	196	710
27	0	0	0	0	20.145	200	720
28	0	0	0	0	24.845	195	710
29	0	0	0	0	20.045	201	718
30	0	0	0	0	24.845	195	710
31	0	0	0	0	20.445	198	712

in a polishing machine. The polished specimens are etched by using 10% Oxalic acid solution to reveal the microstructure as per ASTM E407. Micrographs are taken using metallurgical microscope (Make: Carl Zeiss, Model: Axiovert 40MAT) at 100x magnification. The micrographs of parent metal zone and weld fusion zone are shown in Figures 1 and 2.

Grain size of parent metal and weld joint is measured by using Scanning Electron Microscope (Make: INCA Penta FETx3, Model: 7573). Figures 3 and 4 indicate the measurement of grain size for parent metal zone and weld fusion zone. Average values of grain size are presented in Table 4. From Figure 2, it is understood that the weldments consist

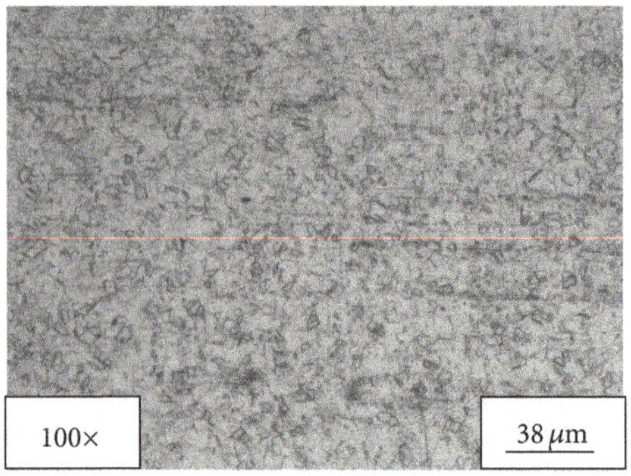

Figure 1: Microstructure of parent metal zone.

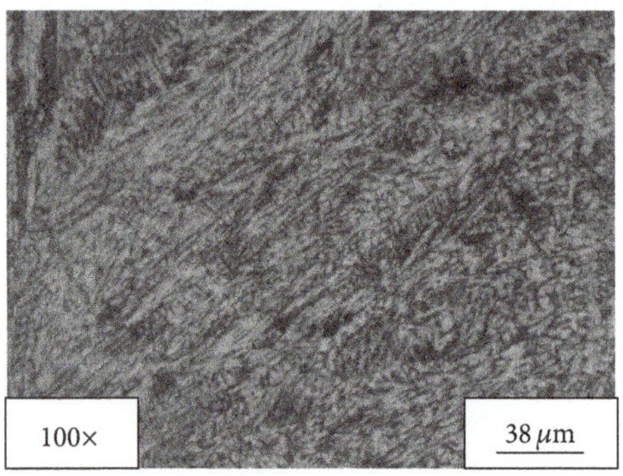

Figure 2: Microstructure of weld fusion zone.

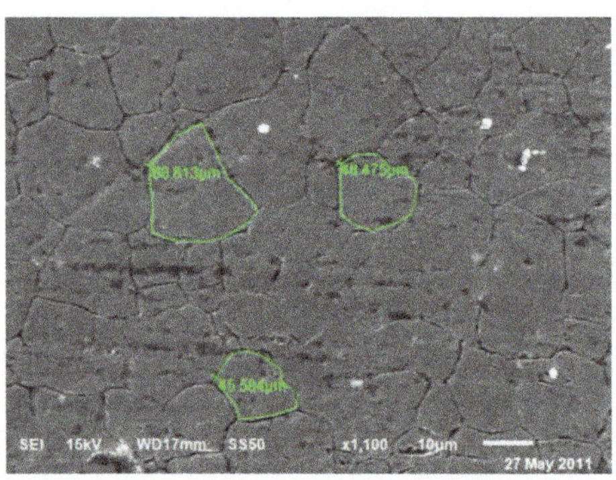

Figure 3: Grain size of parent metal.

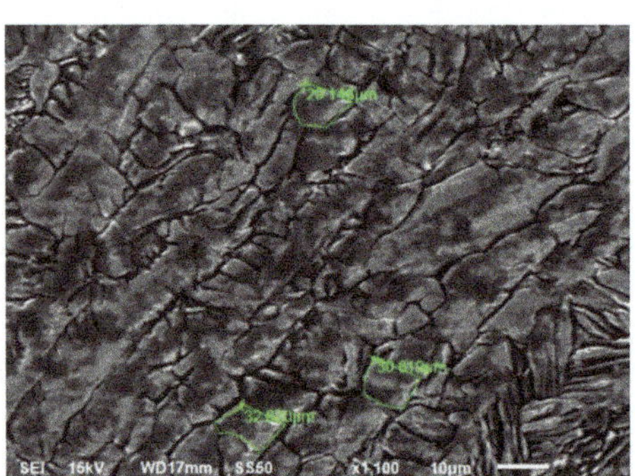

Figure 4: Grain size of weld fusion zone.

of a columnar dendritic structure. A pronounced acicular δ-ferrite formation (dark etching) is observed around the cellular dendritic austenitic grain.

3.2. Measurement of Hardness. Vickers's microhardness testing machine (Make: METSUZAWA CO., LTD, JAPAN, Model: MMT-X7) was used to measure the hardness at the weld fusion zone by applying a load of 0.5 Kg as per ASTM E384. Average values of three samples of each test case are presented in Table 4.

3.3. Measurement of Ultimate Tensile Strength. Three transverse tensile specimens are prepared as per ASTM E8 M-04 guidelines and the specimens after wire cut Electro Discharge Machining are shown in Figure 2. Tensile tests are carried out in 100 KN computer controlled Universal Testing Machine (ZENON, Model number: WDW-100) as shown in Figure 3. The specimen is loaded at a rate of 1.5 KN/min as per ASTM specifications, so that the tensile specimens undergo deformation. From the stress strain curve, the ultimate tensile strength of the weld joints is evaluated and the average of three results is presented in Table 4.

4. Developing Mathematical Models

The output response of the weld joint (Y) is a function of peak current (A), back current (B), pulse (C), and pulse width (D). It can be expressed as [14–16]

$$Y = f(A, B, C, D). \tag{2}$$

The second order polynomial equation used to represent the response surface "Y" is given by [11]

$$Y = b_o + \sum b_i x_i + \sum \beta_{ii} x_i^2 + \sum \sum b_{ij} x_i x_j + \varepsilon. \tag{3}$$

Using MINITAB 14 statistical software package, the significant coefficients were determined and final model is developed using significant coefficients to estimate grain size, hardness, and ultimate tensile strength values of weld joint.

The final mathematical models are given by grain size (G)

$$\begin{aligned} G = {} & 22.859 + 1.052X_1 - 1.058X_2 \\ & + 0.315X_3 + 0.625X_4 + 1.640X_2^2 - 2.320X_1X_3, \end{aligned} \tag{4}$$

hardness (H)

$$\begin{aligned} H = {} & 197.286 - 0.708X_1 + 1.292X_2 \\ & - 0.292X_3 - 0.542X_4 - 1.603X_2^2 + 2.188X_1X_3, \end{aligned} \tag{5}$$

and ultimate tensile strength (T)

$$\begin{aligned} T = {} & 713.143 - 0.833X_1 + 1.833X_2 \\ & - 1.583X_3 - 1.417X_4 - 1.369X_1^2 \\ & - 2.119X_2^2 + 3.250X_1X_3 + 1.750X_2X_4, \end{aligned} \tag{6}$$

where X_1, X_2, X_3, and X_4 are the coded values of peak current, back current, pulse rate, and pulse width.

TABLE 5: ANOVA table.

Source	DF	Seq SS	Adj SS	Adj MS	*F*	*P*
			Grain size			
Regression	14	249.023	249.023	17.7873	6.10	0.000
Linear	4	65.207	65.207	16.3018	5.59	0.005
Square	4	91.443	91.443	22.8608	7.84	0.001
Interaction	6	92.372	92.372	15.3954	5.28	0.004
Residual error	16	46.639	46.639	2.9149		
Lack of fit	10	9.750	9.750	0.9750	0.16	0.994
Pure error	6	36.889	36.889	6.1481		
Total	**30**	**295.661**				
			Hardness			
Regression	14	228.18	228.18	16.299	5.67	0.001
Linear	4	61.17	61.17	15.292	5.32	0.006
Square	4	83.64	83.64	20.910	7.27	0.002
Interaction	6	83.38	83.38	13.896	4.83	0.005
Residual error	16	46.01	46.01	2.876		
Lack of fit	10	10.58	10.58	1.058	0.18	0.991
Pure Error	6	35.43	35.43	5.905		
Total	**30**	**274.19**				
			Ultimate tensile strength			
Regression	14	676.218	676.218	48.3013	6.87	0.000
Linear	4	205.667	205.667	51.4167	7.31	0.002
Square	4	210.551	210.551	52.6379	7.48	0.001
Interaction	6	260.000	260.000	43.3333	6.16	0.002
Residual error	16	112.524	112.524	7.0327		
Lack of fit	10	1.667	1.667	0.1667	0.01	1.000
Pure error	6	110.857	110.857	18.4762		
Total	**30**	**788.742**				

DF: degrees of freedom, SS: sum of squares, MS: mean squares, and *F*: Fishers ratio.

5. Checking the Adequacy of the Developed Models

The adequacy of the developed models was tested using the analysis of variance technique (ANOVA). As per this technique, if the calculated value of the F_{ratio} of the developed model is less than the standard F_{ratio} (from F table) value at a desired level of confidence (say 99%), then the model is said to be adequate within the confidence limit. ANOVA test results are presented in Table 5 for all the models. From the table it is understood that the developed mathematical models are found to be adequate at 99% confidence level. The value of coefficient of determination "R^2" for the above developed models is found to be about 0.85.

From Table 5 it is understood that "F" value for all the three output parameters are less than 7.87 (specified value in Fishers table for 99% confidence level). Hence we confirm the model is adequate at 99% confidence level.

6. Optimization Procedure

The purpose of optimization is to minimize grain size and maximize hardness and ultimate tensile strength. The tool used for optimization is GA. GA is an adaptive search and optimization algorithm that mimics principles of natural genetics. Due to their simplicity, ease of operation, minimum requirements, and global perspective they have been successfully used in a wide variety of problem domains [17].

6.1. Introduction to GA. GA simulates the survival of the fittest among individuals over consecutive generations for solving a problem. Each generation consists of a population of characters. Each individual represents a point in a search space and a possible solution. The individuals in the population are then made to go through a process of evolution. The basic concept of GA is to encode a potential solution to a problem as a series of parameters. A single set of parameter value is termed as the genome of an individual solution. An initial population of individuals is generated randomly. In every generation the individuals in the current population are decoded according to a fitness function. The chromosomes with the highest population fitness are selected for mating. The genes of the parameters are allowed to exchange to produce new ones. These new ones then replace the earlier ones in the next generation. Thus the old population is discarded and the new population becomes the current population. The

Table 6: Parameters used in GA.

		Grain size	Hardness	Ultimate tensile strength
1	Sample size	30	40	40
2	Crossover probability	0.7	0.96	0.2
3	Mutation probability	0.1	0.7	0.01
4	Number of generations	100	100	100
5	Type of crossover	Single	Single	Single

current population is checked for acceptability or solution. The iteration is stopped after the completion of maximum number of generations or on the attainment of the best results [17–20].

The GA algorithm for the optimization problem is given below.

Step 1: choose a coding to represent problem parameters, a selection operator, a crossover operator, and a mutation operator.

Step 2: choose population size (n), crossover probability (Pc), and mutation probability (Pm).

Step 3: initialize a random population of strings of size L. Choose a maximum allowable generation number Gmax. Set $G = 0$. Evaluate each string in population.

Step 4: if $G >$ Gmax or other termination criteria are satisfied, terminate or perform reproduction on the population.

Step 5: perform crossover on random pair of strings.

Step 6: perform mutation on every string.

Step 7: evaluate strings in the new population; set $G = G + 1$ and go to step 3.

After applying the GA operators, a new set of population is created. Then they are decoded and objective function values are calculated. This completes one generation of GA. The number of generations is continued till the termination criterion is achieved. The parameters used in GA are presented in Table 6.

6.2. Objective Function. The optimization of grain size, hardness, and ultimate tensile strength was carried with the help of its mathematical equation. The mathematical equation is considered as objective function. The source code was developed using Turbo C. It is desirable to maximize hardness and ultimate tensile strength and minimize grain size. The objective function for minimizing grain size, as in (4), and maximizing hardness and ultimate tensile strength, as in (5) and (6), was taken as their fitness functions.

6.3. Results of GA

6.3.1. Grain Size. Figure 5 shows the results obtained by running the C program. The initial variation in the curve is due to the search for optimum solution. It is evident that the minimum grain size of 22.13219 Microns is observed at the 2nd iteration. It then converges to the same value up to 100 generations.

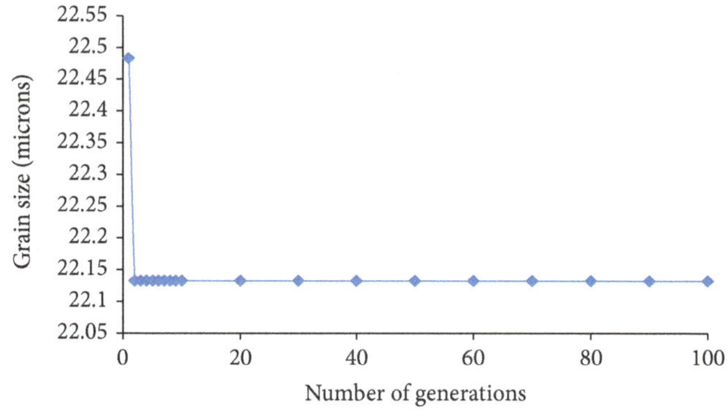

Figure 5: Variation of grain size with number of generations.

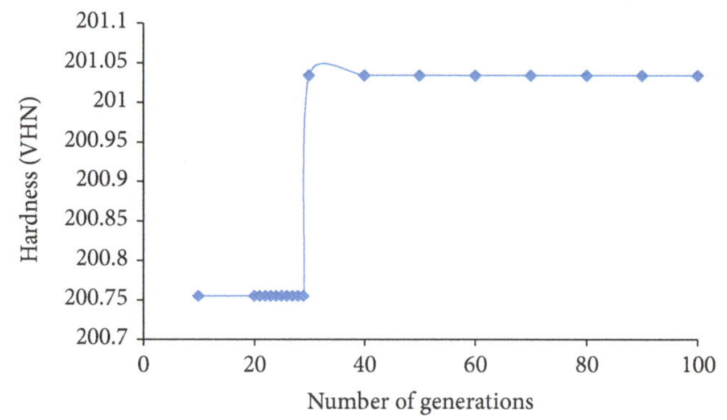

Figure 6: Variation of hardness with number of generations.

6.3.2. Hardness. Figure 6 shows the results obtained by running the C program. The initial variation in the curve is due to the search for optimum solution. It is evident that the maximum hardness of 201.033981 VHN is observed at the 30th iteration. It then converges to the same value up to 100 generations.

6.3.3. Ultimate Tensile Strength. Figure 7 shows the results obtained by running the C program. The initial variation in the curve is due to the search for optimum solution. It is evident that the maximum ultimate tensile strength of 718.7376 MPa is observed at the 7th iteration. It then converges to the same value up to 100 generations.

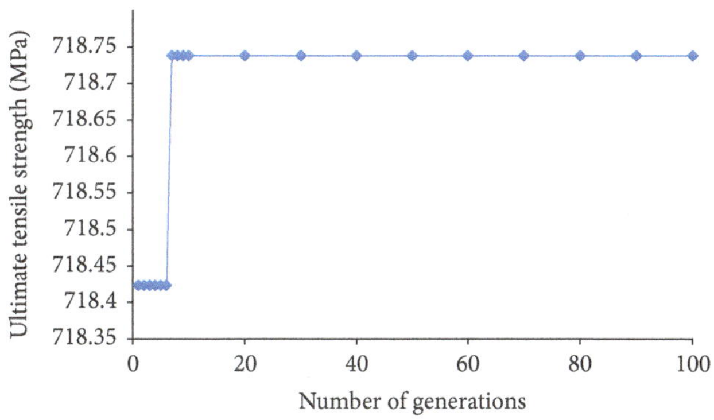

FIGURE 7: Variation of ultimate tensile strength with number of generations.

TABLE 7: Comparison of GA and experimental values.

Welding parameters	Experimental values	GA values
Grain size		
Peak current (Amperes)	7	6.634
Back current (Amperes)	4	3.851
Pulse rate (pulses/second)	40	36.386
Pulse width (%)	50	55.011
Grain size (Microns)	20.045	22.13219
Hardness		
Peak current (Amperes)	7	6.517
Back current (Amperes)	4	4.117
Pulse rate (pulses/second)	40	30.276
Pulse width (%)	50	40.4843
Hardness (VHN)	201	201.033981
Ultimate tensile strength		
Peak current (Amperes)	7	6.514
Back current (Amperes)	4	3.912
Pulse rate (pulses/second)	40	30.098
Pulse width (%)	50	40.004
Ultimate tensile strength (Mpa)	720	718.7376

7. Results and Discussions

Based on the results obtained in GA, the optimum values of PCMPAW input parameters and output responses like grain size, hardness, and ultimate tensile strength are computed and presented in Table 7. The optimum grain size of 20.045 Microns is obtained for the welding input parameter combination of peak current 6.634 Amperes, back current 3.851, pulse rate 36.383 pulses/second, and pulse width 55.011%. The optimum hardness of 201.033981 VHN is obtained for the welding input parameter combination of peak current 6.517 Amperes, back current 4.117 Amperes, pulse rate 30.276 pulses/second, and pulse width 40.4842. The optimum ultimate tensile strength of 718.7376 MPa is obtained for the welding input parameter combination of peak current 6.514 Amperes, back current 3.912 Amperes, pulse rate 30.098 pulses/second, and pulse width 40.004%. The optimum values obtained using genetic algorithm coincide with the experimental values

8. Conclusions

Empirical models are developed for grain size, hardness, and ultimate tensile strength for PCMPAW AISI 304L austenitic steel sheets using RSM CCD design matrix. The weld quality parameters like grain size, hardness, and ultimate tensile strength are optimized using GA. The objective is to minimize grain size and maximize hardness and ultimate tensile strength. The optimum values obtained using GA are 22.13219 Microns grain size, 201.033981 VHN hardness, and 718.7376 MPa ultimate tensile strength. The optimum values obtained using GA are in good agreement with experimental values

Conflict of Interests

The authors declare that there is no conflict of interests regarding the publishing of this paper.

Acknowledgments

The authors would like to thank Shri. R. Gopla Krishnan, Director, M/s Metallic Bellows (I) Pvt Ltd, Chennai, India, and Mr. P. V. Vinay, Associate Professor, GVP College for PG courses (Technical Campus), Visakhapatnam, for their support to prepare this paper.

References

[1] M. Balasubramanian, V. Jayabalan, and V. Balasubramanian, "Effect of process parameters of pulsed current tungsten inert gas welding on weld pool geometry of titanium welds," *Acta Metallurgica Sinica*, vol. 23, no. 4, pp. 312–320, 2010.

[2] M. Balasubramanian, V. Jayabalan, and V. Balasubramanian, "Optimizing the pulsed current gas tungsten arc welding parameters," *Journal of Materials Science and Technology*, vol. 22, no. 6, pp. 821–825, 2006.

[3] K. Marimuthu and N. Murugan, "Prediction and optimisation of weld bead geometry of plasma transferred arc hardfaced valve seat rings," *Surface Engineering*, vol. 19, no. 2, pp. 143–149, 2003.

[4] V. Gunaraj and N. Murugan, "Prediction and comparison of the area of the heat-affected zone for the bead-on-plate and bead-on-joint in submerged arc welding of pipes," *Journal of Materials Processing Technology*, vol. 95, no. 1–3, pp. 246–261, 1999.

[5] V. Gunaraj and N. Murugan, "Application of response surface methodology for predicting weld bead quality in submerged arc welding of pipes," *Journal of Materials Processing Technology*, vol. 88, no. 1, pp. 266–275, 1999.

[6] S. C. Juang and Y. S. Tarng, "Process parameter selection for optimizing the weld pool geometry in the tungsten inert gas welding of stainless steel," *Journal of Materials Processing Technology*, vol. 122, no. 1, pp. 33–37, 2002.

[7] T. T. Allen, R. W. Richardson, D. P. Tagliabue, and G. P. Maul, "Statistical process design for robotic GMA welding of sheet metal," *Welding Journal*, vol. 81, no. 5, pp. 69s–172s, 2002.

[8] I.-S. Kim, J.-S. San, and Y.-J. Jeung, "Control and optimization of bead width for mutipass-welding in robotic arc welding processes," *Australian Welding Journal*, vol. 46, pp. 43–46, 2001.

[9] D. Kim, M. Kang, and S. Rhee, "Determination of optimal welding conditions with a controlled random search procedure," *Welding Journal*, vol. 84, no. 8, pp. 125s–130s, 2005.

[10] D. Kim and S. Rhee, "Optimization of arc welding process parameters using a genetic algorithm," *Welding Journal*, vol. 80, no. 7, pp. 184s–189s, 2001.

[11] D. C. Montgomery, *Design and Analysis of Experiments*, John Wiley & Sons, New York, NY, USA, 3rd edition, 1991.

[12] E. P. Boxg, W. H. Hunter, and J. S. Hunter, *Statistics for Experiments*, John Wiley & Sons, New York, NY, USA, 1978.

[13] S. Babu, T. S. Kumar, and V. Balasubramanian, "Optimizing pulsed current gas tungsten arc welding parameters of AA6061 aluminium alloy using Hooke and Jeeves algorithm," *Transactions of Nonferrous Metals Society of China*, vol. 18, no. 5, pp. 1028–1036, 2008.

[14] W. G. Cochran and G. M. Cox, *Experimental Designs*, John Wiley & Sons, London, UK, 1957.

[15] T. B. Barker, *Quality by Experimental Design*, Marcel Dekker, ASQ Quality Press, 1985.

[16] W. P. Gardiner and G. Gettinby, *Experimental Design Techniques in Statistical Practice*, Horwood, Chichester, UK, 1998.

[17] P. Palanisamy, I. Rajendran, and S. Shanmugasundaram, "Optimization of machining parameters using genetic algorithm and experimental validation for end-milling operations," *International Journal of Advanced Manufacturing Technology*, vol. 32, no. 7-8, pp. 644–655, 2007.

[18] N. Baskar, P. Asokan, G. Prabhaharan, and R. Saravanan, "Optimization of machining parameters for milling operations using non-conventional methods," *The International Journal of Advanced Manufacturing Technology*, vol. 25, no. 11-12, pp. 1078–1088, 2005.

[19] K. S. Prasad, C. S. Rao, and D. N. Rao, "Multi objective optimization of weld bead geometry parameters of pulsed current micro plasma arc welded inconel 625 sheets using enhanced non-dominated sorting genetic algorithm," *Journal of Materials & Metallurgical Engineering*, vol. 3, no. 1, pp. 231–3818, 2013.

[20] K. S. Prasad, C. S. Rao, and D. N. Rao, "Optimization of pulsed current parameters to minimize pitting corrosion in pulsed current micro plasma arc welded AISI 304L sheets using genetic algorithm," *International Journal of Lean Thinking*, vol. 4, no. 1, pp. 9–19, 2013.

13

Studies on the Process Parameters of Rapid Prototyping Technique (Stereolithography) for the Betterment of Part Quality

Raju Bangalore Singe Gowda,[1] Chandra Sekhar Udayagiri,[2] and Drakshayani Doulat Narendra[3]

[1] *Department of Mechanical Engineering, REVA ITM, Bangalore, Karnataka 560037, India*
[2] *Institution of Engineers (India), Gachibowli, Hyderabad 500032, India*
[3] *Department of Mechanical Engineering, Sir MVIT, Bangalore, Karnataka 562157, India*

Correspondence should be addressed to Raju Bangalore Singe Gowda; rajubsgowda@gmail.com

Academic Editor: Konstantinos Salonitis

Rapid prototyping (RP) has evolved as frontier technology in the recent times, which allows direct transformation of CAD files into functional prototypes where it tremendously reduces the lead time to produce physical prototypes necessary for design verification, fit, and functional analysis by generating the prototypes directly from the CAD data. Part quality in the rapid prototyping process is a function of build parameters such as hatch cure depth, layer thickness, orientation, and hatch spacing. Thus an attempt was made to identify, study, and optimize the process parameters governing the system which are related to part characteristics using Taguchi experimental design techniques quality. The part characteristics can be divided into physical part and mechanical part characteristics. The physical characteristics are surface finish, dimensional accuracy, distortion, layer thickness, hatch cure, and hatch file, whereas mechanical characteristics are flexural strength, ultimate tensile strength, and impact strength. Thus, this paper proposes to characterize the influence of the physical build parameters over the part quality. An L_9 orthogonal array was designed with the minimum number of experimental runs with desired parameter settings and also by analysis tools such as ANOVA (analysis of variance). Establishment of experimentally verified correlations between the physical part characteristics and mechanical part characteristics to obtain an optimal process parameter level for betterment of part quality is obtained. The process model obtained by the empirical relation can be used to determine the strength of the prototype for the given set of parameters that shows the dependency of strength, which are essential for designers and RP machine users.

1. Introduction

Due to the advances in electronics and computers, there has been a significant growth in communication, information technology, and worldwide networking, which leads to globalization and opening of markets [1, 2]. Thus in product development, rapid prototyping (RP) and rapid product development have turned out to be the key instruments to save time and money with respect to the development of innovative products [2, 3]. Stereolithography (SLA) is one of the RP techniques, which involve fabrication of intricate shape of a plastic monomer directly from computer aided design (CAD) data by depositing material layer by layer by photopolymerization process [4]. The SLA process involves the following steps: conversion of the CAD model to the standard triangulation language (STL) file format; slicing the STL file into thin cross sectional layers; constructing the model by one layer on top of another layer; cleaning and finishing off the model. SLA prototypes have wide applications in aerospace, automobile, and manufacturing sectors especially in rapid tooling. Strength plays a very important role in rapid tooling [5] where the components have to withstand high pressure during the test of fitment and also when used as a die in injection moulding, where the dies prepared through SLA

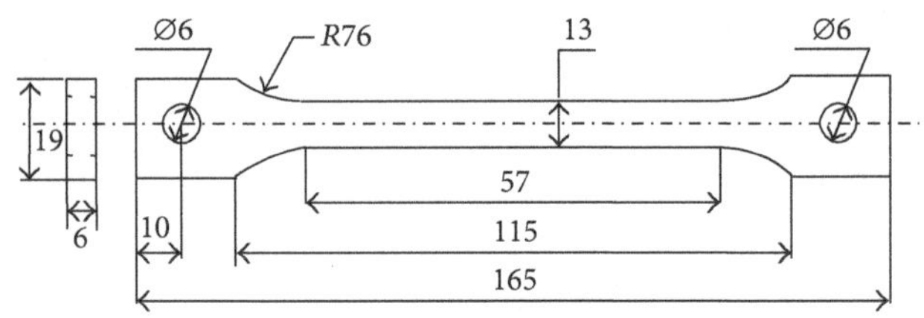

FIGURE 1: Tensile test specimen as per ASTM D638-01.

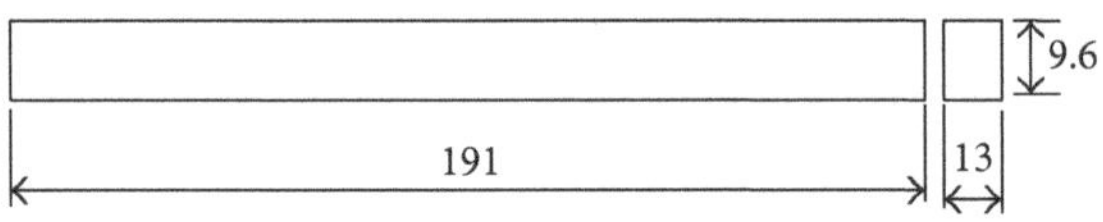

FIGURE 2: Flexural test specimen as per ASTM D790-03.

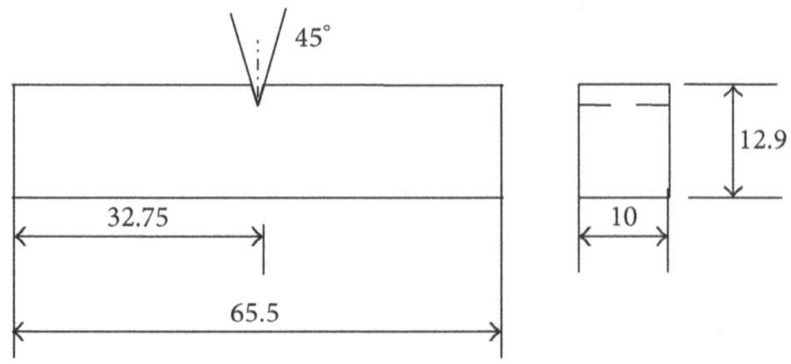

FIGURE 3: Impact test specimen as per ASTM D256-04.

process will be subjected to high tension due to high injection pressure. Taguchi technique is a powerful tool for the design of high quality systems [6, 7]; it provides a simple efficient and systematic approach to optimize design for performance, quality, and cost. The methodology is valuable when design parameters are qualitative and discrete. Taguchi parameter design can optimize the performance characteristic through the setting of design parameters and reduce the sensitivity of the system performance to source of variation [8]. Dingal et al. used Taguchi method to find out the significant factors influencing density, porosity, and hardness on selective laser sintering of iron powder [9]. Guharaja et al. made an attempt to obtain optimal settings of green sand casting parameters using Taguchi method [10]. Rao and Padmanabhan used Taguchi method and ANOVA in optimization of process parameters for material removal rate in electrochemical machining of Al/5% SiC composites [11]. Nataraj et al. used risk analysis Taguchi method to find optimum conditions of design parameters [12]. Barua et al. used the Taguchi method to optimize the mechanical properties of vacuum casting process [13]. Hence, an attempt is made in order to achieve high strength of the prototypes with the specified process parameters which gives prior information of the part strength before fabricating the actual SLA prototypes. Hence, parameter optimization of SLA process is investigated and evaluated through a standard test specimen [14, 15].

2. The Experimental Methods and Methodology

The experimental building material adopted is CIBATOOL 5530 epoxy resin. The experimental building models are categorized into three specimens, namely, the tensile test (Figure 1), flexural test (Figure 2), and impact test (Figure 3), and are characterized using ASTM D638-01 [16], ASTM D790-03 [17], and ASTM D256-04 [18] specifications, respectively. The STL format is generated by CATIA V5 R16 and sent to the 3D system SLA 5000 rapid prototyping machine. The various conditions in preprocessing steps such as STL verification, deposition layer thickness, orientation, building interior structure form, supporting method, and building deposition direction are incorporated by means of 3D light year software [19] provided by 3D system of Valencia, USA, followed by the layer slicing process to generate the building path with ACES build style. Building quality characteristics or attributes include the-larger-the-better (LB) for the strength of the SLA prototypes.

2.1. Experimental Apparatus. The major experimental apparatus adopted includes 3D system SLA5000 rapid prototyping machine produced by the Valencia, USA, where it uses CIBATOOL 5530 epoxy resin to build geometrical shape of the work piece by photopolymerization process. Similarly, the tensile and flexural tests were conducted using Instron Universal Testing Machine, UK, Model 5582. The impact test was conducted using impact tester, Aditya Instruments, Bangalore, Model IT-30.

2.2. Taguchi Quality Engineering. The orthogonal array is employed for the Taguchi method as the experimental analysis basis. The experimental factors and their corresponding levels are identified. Then the experimental results are manipulated and validated by analysis of variance (ANOVA), in order to determine each factor effect versus the response variable strength of the SLA prototypes. The experimental procedures are given as follows: identification of SLA process parameter that influences the response variable and determining the various levels of the factors. Based on the factors and their levels, the degree of freedom is calculated and the suitable orthogonal array is selected and the experiment proceeds according to the variable factor layout of the orthogonal array. The experimental results are obtained and the signal to noise ratio (S/N ratio), the ANOVA, and the corresponding contribution are computed, obtaining the empirical relationship for the response variable under different parameter settings.

In the Taguchi method, parameter design converts the objective value to S/N ratio, which is known as quality characteristic evaluation index [20–22], with the S/N ratio where the least variation and the optimal quality design can be obtained. The S/N ratio is beneficial in increasing factor weighting effect, decreasing mutual action, simultaneously processing the average and variation, and improving the quality. The higher the S/N ratio is, the more stable the quality

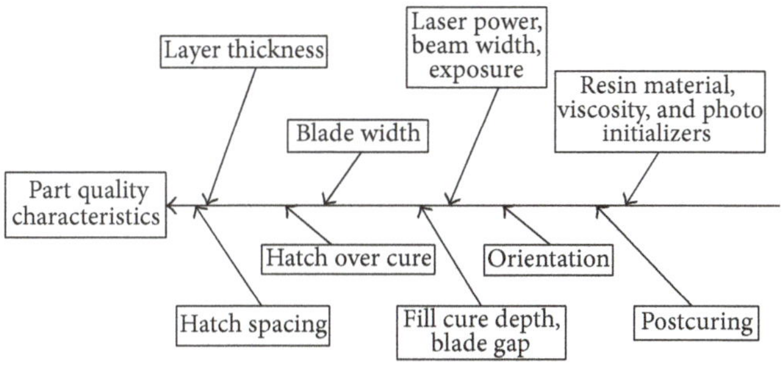

FIGURE 4: Cause and effect diagram of SLA process parameters.

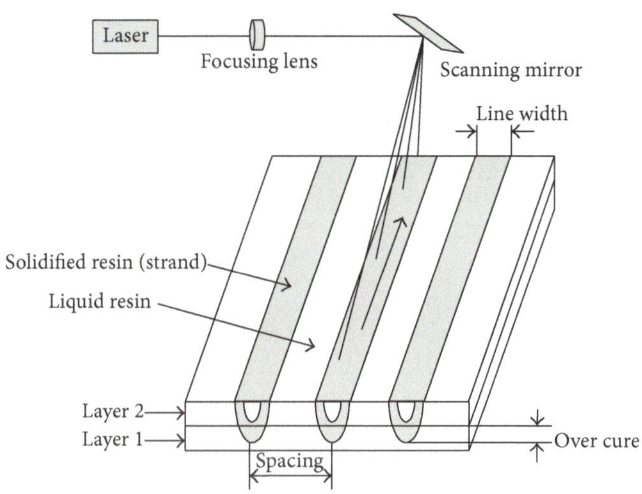

FIGURE 5: Process parameter in SLA process.

can be obtained. According to the response variable, the-larger-the-better (LB) is used. In the LB, the objective optimal value is larger better for the strength of the SLA prototypes:

$$\eta = -10 \log \left[\frac{1}{n} \sum_{i=1}^{n} \frac{1}{y_i^2} \right], \quad (1)$$

where η is S/N ratio; y_i is the ith result of the experiment; n is the number of replications.

Figure 4 shows the probable parameters (causes) that influence the part quality characteristics (effects) in the SLA process. Figure 5 represents the various process parameters of SLA process, and among these the layer thickness (L_t) is the thickness where the model is sliced in the Z direction, orientation (O) is the position in which the prototype is built, and hatch spacing (H_s) is the narrow region solidified by the laser scanning. If the strand is located at the top or bottom surface of part, spacing is called fill spacing otherwise hatch spacing, which are the parameters that influence the strength of the SLA parts [23]. Table 1 provides the three levels of the process parameters for the experimentation.

The total number of experiments in full factorial design for "m" parameters in each set of "L" levels is L^m and it increases exponentially with L & m. Taguchi suggested the use of orthogonal array which will be used for conducting the fractional factorial experiments [24]. The Taguchi orthogonal

FIGURE 6: SLA prototypes of tensile test specimen.

TABLE 1: Description of experimental control parameters.

Symbol	Control parameters	Level 1	Level 2	Level 3
A	Layer thickness (L_t)	0.075	0.1	0.125
B	Orientation (O)	0° (H_x)	45° (VH_{xy})	90° (V_y)
C	Hatch spacing (H_s)	0.01	0.015	0.02

TABLE 2: L_9 orthogonal array.

Experimental run (j)	Levels		
	A Layer thickness (L_t)—mm	B Orientation (O)—°	C Hatch spacing (H_s)—mm
1	1	1	1
2	1	2	2
3	1	3	3
4	2	1	2
5	2	2	3
6	2	3	1
7	3	1	3
8	3	2	1
9	3	3	2

array adopted in the research experiment is L_9 for three-factor three-level settings as shown in Table 2.

3. Experimentation

3.1. Experimental Analysis for Tensile Strength (Ts). The nine tensile specimens as per the ASTM standards (ASTM D638-01) were built for L_9 orthogonal array setting using epoxy resin CIBATOOL SL5530 in SL5000 machine of three replications each. The dimensional details of the test specimen are given in Figure 1 [16] and the SLA prototypes are as shown in Figure 6. The tensile strength is calculated using the ratio of ultimate load to cross sectional area. The experimental results are given in Table 3.

3.1.1. Prediction of Optimal Levels of Process Parameters. S/N ratio is an evaluation measure for the process parameters at each of their process levels where the signal represents the desirable target (LB of tensile strength) and noise indicates the undesirable value which is defined in (1). The average S/N ratio for each process parameter (control variable) at each level is an average of n_j at a defined level. Table 4 provides the average S/N ratios for the process parameters (L_t, O, and H_s)

TABLE 3: Ultimate tensile strength for OA settings.

j	1	2	3	4	5	6	7	8	9
Ts_j (N/mm^2)	55.46	54.57	55.07	58.46	54.51	58.59	58.62	55.34	61.73

TABLE 4: S/N ratio for different levels for tensile strength.

Parameters	Level	Exp^m run	Ts_j (N/mm^2)	n_j	n_{avg}
		1	55.46	34.89	
	1 **(0.075)**	2	54.57	34.74	**34.82**
		3	55.07	34.82	
		4	58.46	35.34	
Layer thickness	2 **(0.1)**	5	54.51	34.73	**35.14**
		6	58.59	35.36	
		7	58.62	35.36	
	3 **(0.125)**	8	55.34	34.86	***35.34***
		9	61.73	35.81	
		1	55.56	34.89	
	1 **(0°)**	4	58.46	35.34	**35.2**
		7	58.62	35.36	
		2	54.57	34.74	
Orientation	2 **(45°)**	5	54.51	34.73	**34.77**
		8	55.34	34.86	
		3	55.07	34.82	
	3 **(90°)**	6	58.59	35.36	***35.33***
		9	61.73	35.81	
		1	55.56	34.89	
	1 **(0.01)**	6	58.59	35.36	**35.04**
		8	55.34	34.86	
		2	54.57	34.74	
Hatch spacing	2 **(0.015)**	4	58.46	35.34	***35.29***
		9	61.73	35.81	
		3	55.07	34.82	
	3 **(0.02)**	5	54.51	34.73	**34.97**
		7	58.62	35.36	

at the three levels. The graph represented in Figure 7 shows the variation of average *S/N* ratio with respect to the various levels. The main objective is to maximize the tensile strength of the parts produced by SLA process; in order to achieve this *S/N* ratio should be more. Hence, the level having higher *S/N* ratio is selected as the optimum level, which is contributing higher strength to the part. Therefore, the optimum levels contributing to the higher strength of the part are as follows:

layer thickness: 0.125 mm (level 3, *S/N* ratio: 35.34),

orientation: 90° V_y (level 3, *S/N* ratio: 35.33),

hatch spacing: 0.015 (level 2, *S/N* ratio: 35.29).

3.1.2. Identification of Significance and Percentage of Contribution of Process Parameters. The process parameters (L_t, O, and H_s) which have more influence on the response variable are identified through the percentage of contribution of each parameter. The parameter which has more percentage of contribution is the significant parameter to the response variable [25–28]. They have mentioned that ANOVA is widely used for determining the significance of the independent variables in influencing the dependent variables and also in determination of percentage of contribution of these dependent variables to the response variable.

Table 5 shows the percentage of contribution of the parameters to the response variable with the ANOVA parameters. Hence, from the ANOVA table the significance of each parameter is identified.

3.2. Experimental Analysis for Flexural Strength (Fs). The analysis carried out for the flexural strength is identical to the one in Section 3.1. The nine flexural test specimens as per the ASTM standards (ASTM D790-03) were built for L_9 orthogonal array setting using epoxy resin CIBATOOL

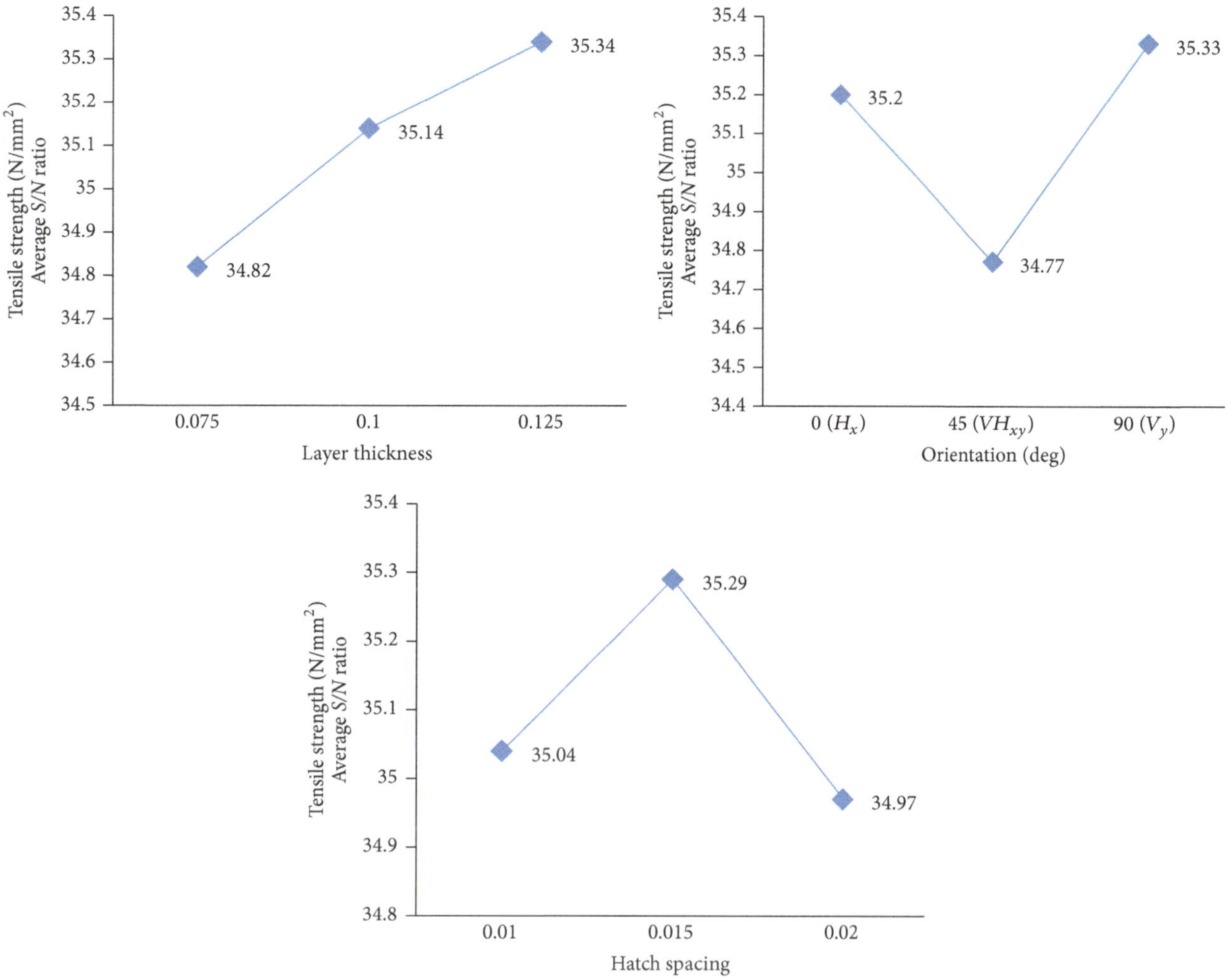

FIGURE 7: *S/N* ratio graph for tensile strength analysis.

TABLE 5: The % of contribution of the parameters to the tensile strength along with the estimated ANOVA parameter.

Parameter "*i*"	Sum of squares	Degree of freedom	Mean sum of squares	*F* statistics	*F* tabulated $F_{(0.1,2,2)}$	% of contribution
L_t	18.57	2	9.285	9.622*		36.93
O	21.72	2	10.86	11.25*	9	43.19
H_s	8.06	2	4.03	4.176		16.03
Error	1.93	2	0.965			3.83
Total	50.28	8	25.14			

*Significance at 90% confidence level (*F* statistics > *F* tabulated).

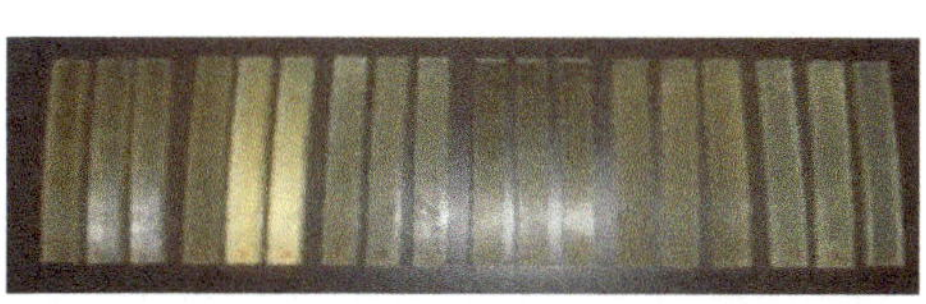

FIGURE 8: SLA prototypes of flexural test specimen.

SL5530 in SL5000 machine of three replications each. The dimensional details of the test specimen are given in Figure 2 [17] and the SLA prototypes are as shown in Figure 8. The flexural specimens are subjected to point load at the midpoint between the supports, which are placed at 190 mm apart with a particular load where the fatigue occurs. The flexural strength at fracture (δ) is determined by

$$\delta = \frac{3WL}{2bd^2}, \quad (2)$$

where W is load in Newton, L is the length of the test specimen in mm, b is the breadth of the test specimen in mm, and d is the thickness of the test specimen in mm. The

TABLE 6: Flexural strength for OA settings.

j exp., run	1	2	3	4	5	6	7	8	9
FS_j (N/mm^2)	116.67	113.92	114.08	115.7	110.0	115.6	115.8	114.8	118.7

TABLE 7: S/N ratio for different levels for flexural strength.

Parameters	Level	Exp^n run	Fs_j (N/mm^2)	n_j	n_{avg}
		1	116.67	41.33	
	1 **(0.075)**	2	113.92	41.13	**41.2**
		3	114.08	41.14	
		4	115.7	41.26	
Layer thickness	2 **(0.1)**	5	110.0	40.82	**41.11**
		6	115.6	41.25	
		7	115.8	41.27	
	3 **(0.125)**	8	114.8	41.20	***41.32***
		9	118.7	41.49	
		1	116.67	41.33	
	1 **(0°)**	4	115.7	41.26	**41.28**
		7	115.8	41.27	
		2	113.92	41.13	
Orientation	2 **(45°)**	5	110.0	40.82	**41.05**
		8	114.8	41.20	
		3	114.08	41.14	
	3 **(90°)**	6	115.6	41.25	***41.30***
		9	118.7	41.49	
		1	116.67	41.33	
	1 **(0.01)**	6	115.6	41.25	**41.26**
		8	114.8	41.20	
		2	113.92	41.13	
Hatch spacing	2 **(0.015)**	4	115.7	41.26	***41.29***
		9	118.7	41.49	
		3	114.08	41.14	
	3 **(0.02)**	5	110.0	40.82	**41.08**
		7	115.8	41.27	

experimental results are tabulated in Table 6 which shows the flexural strength. Table 7 shows the S/N ratio calculations for each of the levels in each factor which predicts the optimum level among the selected 3 levels and Figure 9 shows the variation of S/N ratio for all the controllable factors.

The level which has higher S/N ratio is selected as the optimum level contributing higher flexural strength to the part. Hence the optimal parameters are as follows:

layer thickness: 0.125 mm (level 3, S/N ratio: 41.32),

orientation: 90° V_y (level 3, S/N ratio: 41.3),

hatch spacing: 0.015 (level 2, S/N ratio: 41.29).

Table 8 represents the percentage of contribution of each factor for flexural strength along with the estimated ANOVA parameters. From the ANOVA table the significance of each parameter is identified.

3.3. Experimental Analysis for Impact Strength. The analysis carried out for the impact strength is identical to the one in Section 3.1. The nine impact test specimens as per the ASTM standards (ASTM D256-04) are built for L_9 orthogonal array setting using epoxy resin CIBATOOL SL5530 in SL5000 machine of three replications each. The dimensional details of the test specimen are given in Figure 3 [18] and the SLA prototypes are shown in Figure 10. In Izod test method, the specimen is placed vertically and is broken by a single swing of the pendulum weight with a contact point at a fixed distance from the centerline of the notch. The impact strength is obtained through the ratio of energy absorbed by the specimen during the break and the width of the specimen using Izod impact tests. The experimental results of impact strength are tabulated in Table 9. Table 10 shows the S/N ratio for each of the levels in each factor and Figure 11 shows the variation of S/N ratio for all the controllable factors.

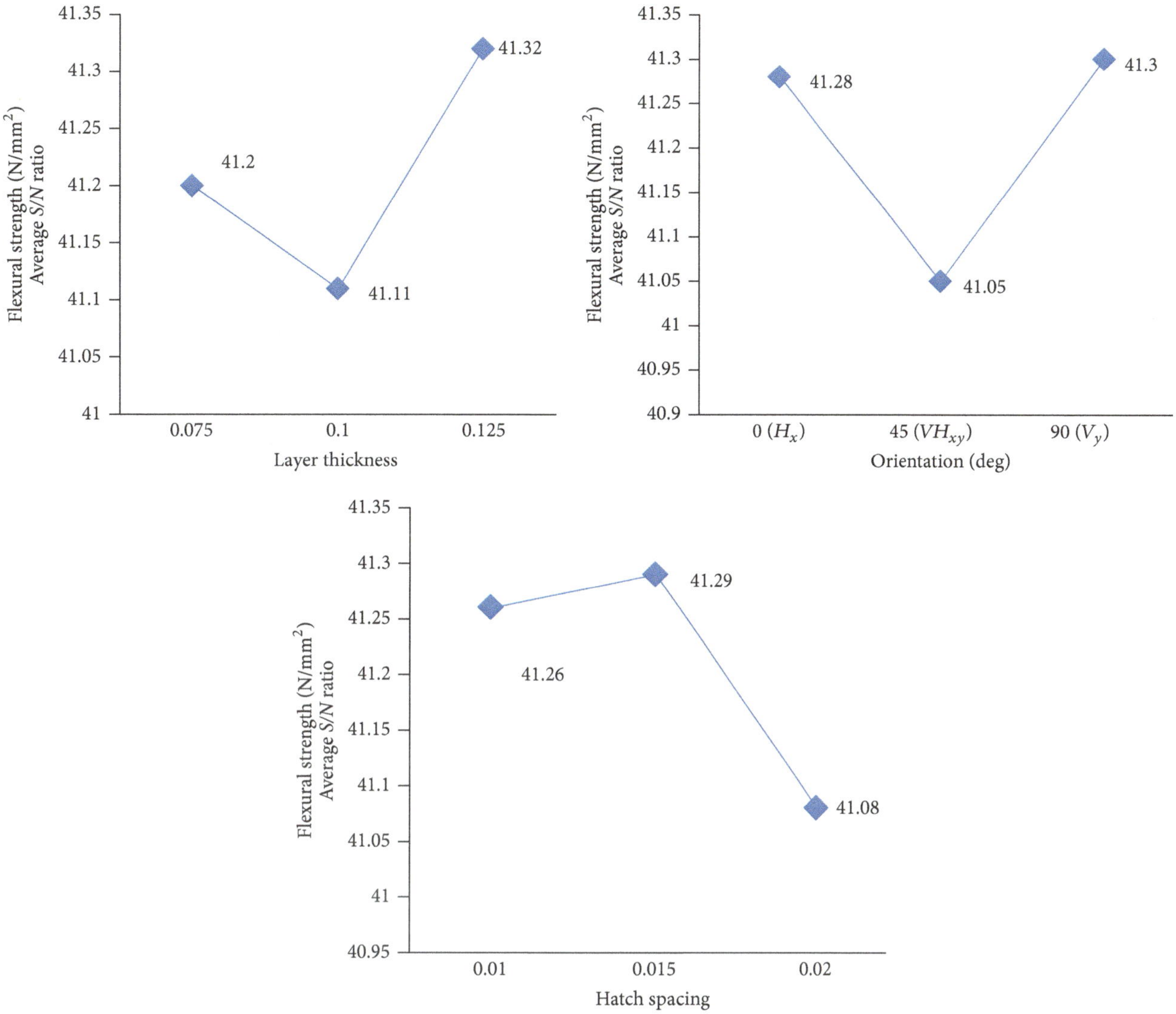

FIGURE 9: *S/N* ratio graph for flexural strength analysis.

TABLE 8: The % of contribution of the parameters to the flexural strength along with the estimated ANOVA parameter.

Parameter "*i*"	Sum of squares	Degree of freedom	Mean sum of squares	*F* statistics	*F* tabulated $F_{(0.01,2,2)}$	% of contribution
L_t	10.75	2	5.375	82.69		23.88
O	20.3	2	10.15	156.15*	99	45.09
H_s	13.84	2	6.92	106.45*		30.75
Error	0.13	2	0.065			0.28
Total	45.02	8	22.507			

*Significance at 99% confidence level (*F* statistics > *F* tabulated).

TABLE 9: Impact strength for OA setting.

j exp., run	1	2	3	4	5	6	7	8	9
IS_j (J/m)	20.8	22	21.1	20.3	17.9	21.3	21.4	19.9	23.6

TABLE 10: S/N ratio for different levels for impact strength.

Parameters	Level	Experimental run	Is_j (J/m)	n_j	n_{avg}
		1	20.8	26.36	
	1 (**0.075**)	2	22	26.85	**26.56**
		3	21.1	26.48	
		4	20.3	26.15	
Layer thickness	2 (**0.1**)	5	17.9	25.06	**25.93**
		6	21.3	26.57	
		7	21.4	26.61	
	3 (**0.125**)	8	19.9	25.97	***26.68***
		9	23.6	27.46	
		1	20.8	26.36	
	1 (**0°**)	4	20.3	26.15	**26.37**
		7	21.4	26.61	
		2	22	26.85	
Orientation	2 (**45°**)	5	17.9	25.06	**25.96**
		8	19.9	25.97	
		3	21.1	26.48	
	3 (**90°**)	6	21.3	26.57	***26.84***
		9	23.6	27.46	
		1	20.8	26.36	
	1 (**0.01**)	6	21.3	26.57	**26.3**
		8	19.9	25.97	
		2	22	26.85	
Hatch spacing	2 (**0.015**)	4	20.3	26.15	***26.82***
		9	23.6	27.46	
		3	21.1	26.48	
	3 (**0.02**)	5	17.9	25.06	**26.05**
		7	21.4	26.61	

TABLE 11: The % of contribution of the parameters to the impact strength along with the estimated ANOVA parameter.

Parameter "i"	Sum of squares	Degree of freedom	Mean sum of squares	F statistics	F tabulated $F_{(0.01,2,2)}$	% of contribution
L_t	5.5	2	2.75	2.69		28.47
O	6.44	2	3.22	3.16*	3	33.33
H_s	5.34	2	2.67	2.617		27.64
Error	2.04	2	1.02			10.55
Total	19.32	8	9.66			

*Significance at 75% confidence level (F statistics > F tabulated).

FIGURE 10: SLA prototypes of impact test specimen.

The level which has higher S/N ratio is selected as the optimum level contributing higher impact strength to the part. Hence the optimal parameters are as follows:

layer thickness: 0.125 mm (level 3, S/N ratio: 26.68),

orientation: 90° V_y (level 3, S/N ratio: 26.84),

hatch spacing: 0.015 (level 2, S/N ratio: 26.82).

Table 11 represents the percentage of contribution of each factor for impact strength along with the estimated ANOVA parameters. From the ANOVA table the significance of each parameter is identified.

4. Establishment of Process Model (Regression Equation)

ANOVA reveals that the layer thickness, orientation, and hatch spacing are contributing significantly to mechanical

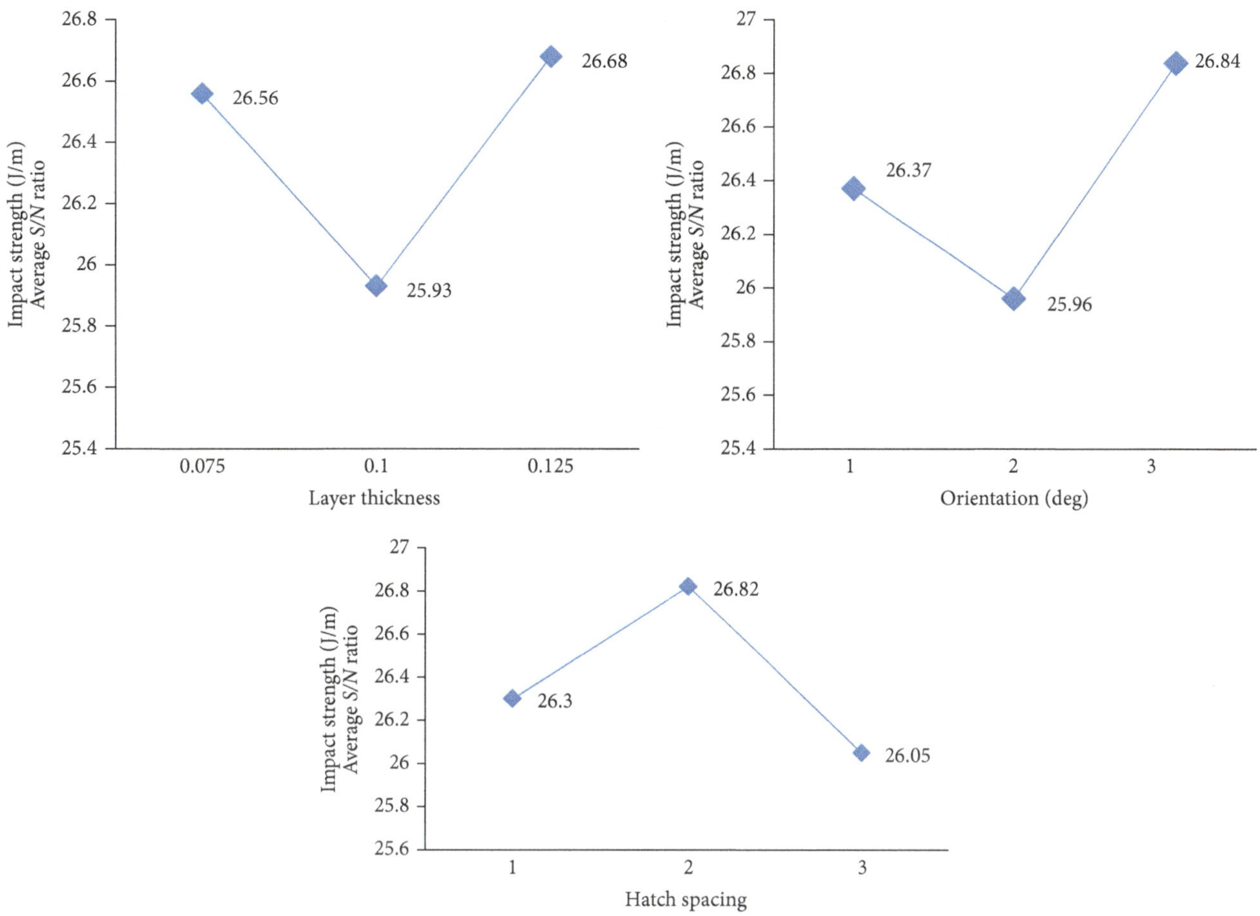

FIGURE 11: S/N ratio graph for impact strength analysis.

properties. Hence, establishment of a process model (empirical relationship/regression model) for mechanical properties (tensile/flexural/impact strength) as a function of process parameters (layer thickness, orientation, and hatch spacing) is used to predict the strength for the given set of process parameters, which provides the prior information of the strength before fabricating the SLA prototype and is useful for rapid designers as well as RP machine users. Montgomery [24] suggests the orthogonal polynomial which is very useful for developing the process model with the L_9 orthogonal data. A quadratic polynomial model is proposed to establish the process model between the response variable and process parameter as shown in

$$\mathrm{RV} = \beta_0 + \sum_{i=1}^{9} \left[\beta_{1i} P_{1(i)} + \beta_{2i} P_{2(i)} \right] + \epsilon, \quad (3)$$

where RV is response variable, i is process parameter identifier, β_0 is constant coefficient, β_{1i} is linear coefficient for the ith parameter, β_{2i} is nonlinear coefficient for the ith parameter, ϵ is error component, $P_{1(i)}$ is 1st order orthogonal polynomial of parameter, and $P_{2(i)}$ is 2nd order orthogonal polynomial of the parameter.

4.1. Empirical Relation for Tensile Strength versus Process Parameters. Layer thickness, orientation, and hatch spacing influence the tensile strength and the response variable. The regression equation for the tensile strength is as given by

$$\begin{aligned}
\mathrm{Ts}\,(\mathrm{RV}) = {} & \beta_0 + \beta_{1L_t} * \lambda_1 \left[\frac{L_t - \overline{m}_{L_t}}{d_{L_t}} \right] + \beta_{2L_t} \\
& * \lambda_2 \left\{ \left[\frac{\left(L_t - \overline{m}_{L_t}\right)^2}{d_{L_t}^2} \right] - \left[\frac{L_{L_t}^2 - 1}{12} \right] \right\} + \beta_{1O} \\
& * \lambda_1 \left[\frac{O - \overline{m}_O}{d_O} \right] + \beta_{2O} \\
& * \lambda_2 \left\{ \left[\frac{\left(LO - \overline{m}_O\right)^2}{d_O^2} \right] - \left[\frac{O^2 - 1}{12} \right] \right\} + \beta_{1H_s} \\
& * \lambda_1 \left[\frac{H_s - \overline{m}_{H_s}}{d_{H_s}} \right] + \beta_{2H_s} \\
& * \lambda_2 \left\{ \left[\frac{\left(H_s - \overline{m}_{H_s}\right)^2}{d_{H_s}^2} \right] - \left[\frac{L_{H_s}^2 - 1}{12} \right] \right\}.
\end{aligned} \quad (4)$$

TABLE 12: Orthogonal contrast coefficients for linear and nonlinear terms at different levels.

Levels	Linear	Nonlinear
Lower	−1	1
Medium	0	−2
Higher	1	1

Among the above process parameters, layer thickness and hatch spacing are based on quantitative measures with equal spacing and orientation is based on qualitative measures. Hence the coded value of the orthogonal array is used with two extremes and the center value, that is, lower, higher, and middle values which are coded as −1, 0, and 1, respectively [29, 30]. Table 12 provides the values of orthogonal contrast coefficient for linear and nonlinear terms. Hence, the mean value of the levels of process parameters ($\overline{m_i}$) becomes zero and the spacing between the levels of process parameters (d_i) becomes one. Therefore, the study with three parameters will have $\lambda_1 = 1$ and $\lambda_2 = 3$ [16]. Substituting the value of ($\overline{m_i}$), d_i, λ_1, and λ_2, the equation becomes as shown in

$$\begin{aligned}\text{Ts (RV)} = {} & \beta_0 + \beta_{1L_t} * [L_t] + \beta_{2L_t} * 3\left\{L_t^2 - \left[\frac{L_{L_t}^2 - 1}{12}\right]\right\} \\ & + \beta_{1O} * [O] + \beta_{2O} * 3\left\{O^2 - \left[\frac{LO^2 - 1}{12}\right]\right\} \\ & + \beta_{1H_s} * H_s + \beta_{2H_s} * 3\left\{H_s^2 - \left[\frac{L_{H_s}^2 - 1}{12}\right]\right\}.\end{aligned} \tag{5}$$

By using the coded value of the orthogonal contrast coefficients for linear and nonlinear term, the constant and coefficients with respect to the various process parameters are found out and the process model (empirical relation) is given in

$$\begin{aligned}\text{TS} = {} & -0.3714L_t^2 + 1.7483L_t + 3.19830O^2 + 0.4583O \\ & - 1.9716H_s^2 - 0.215H_s + 56.37.\end{aligned} \tag{6}$$

4.2. Empirical Relation for Flexural Strength versus Process Parameters. Similarly the process model is established between flexural strength versus process parameter (L_t, O, and H_s) as in Section 4.1 and as given in

$$\begin{aligned}\text{Fs} = {} & 1.8948L_t^2 + 0.7716L_t + 3.1848O^2 + 0.035O \\ & - 1.6149H_s^2 - 1.1983H_s + 112.7205.\end{aligned} \tag{7}$$

4.3. Empirical Relation for Impact Strength versus Process Parameters. In a similar manner, the process model is established between impact strength versus process parameter (L_t, O, and H_s) as in Section 4.1 and as given in

$$\begin{aligned}\text{IS} = {} & 1.6332L_t^2 + 0.1666L_t + 1.4983O^2 + 0.5833O \\ & - 1.5666H_s^2 - 0.2666H_s + 19.8879.\end{aligned} \tag{8}$$

5. Conclusions

Optimizing the rapid prototyping SLA process by using Taguchi method is proposed. In this paper, an attempt is made to analyze the process parameters that influence the strength aspect of the SLA parts which are useful for various applications of the prototypes in testing and tooling process. The major conclusions are as follows.

(i) Parameters L_t, O, and H_s influence much on part strength of SLA prototypes.

(ii) The optimal level combination of the process parameters is as follows:

layer thickness: 0.125 mm (level 3),
orientation: 90° V_y (level 3),
hatch spacing: 0.015 (level 2),

for tensile, flexural, and impact strength of the SLA prototypes.

(iii) Among the three process parameters, the L_t and O are major contributing parameters for the tensile strength, O and H_s are major contributing parameters for the flexural strength, and O has more significance among the parameters for the impact strength.

(iv) The empirical relationship (process model) between the part strength characteristics and the influencing parameters has been established for stereolithography process, which can predict the strength of the SLA prototypes by prior knowledge of part strength before building the prototypes.

(v) The layer thickness plays an important role since the SLA process has an inbuilt specific property of building the prototypes with layer by layer method.

The procedure is applied in order to optimize the other rapid prototyping process with different materials. The optimization is done by factorial design (Taguchi technique) to know the effect of parameter on the variables which can be determined by integrating the Taguchi method with grey relational analysis where the optimal parameter combinations of the multiple quality characteristics can be obtained. The process model may be further refined by using nonclassical optimization techniques such as genetic algorithm and neural networks.

Conflict of Interests

The authors declare that there is no conflict of interests regarding the publication of this paper.

Acknowledgments

The authors would like to thank the management of Reva Institute of Technology and Management, Bangalore, and the Director of the Gas Turbine Research Establishment, Bangalore, and Sriram Industries, Bangalore, for their immense help to carry out this work.

References

[1] J. E. Folkestad and R. L. Johnson, "Integrated rapid prototyping and rapid tooling (IRPRT)," *Integrated Manufacturing Systems*, vol. 13, no. 2, pp. 97–103, 2002.

[2] J. P. Kruth, "Material ingress manufacturing by rapid prototyping techniques," *CIRP Annals*, vol. 40, no. 2, pp. 603–614, 1991.

[3] F. L. Krause, M. Casella, C. H. Stile, and A. Ulbricht, "Enhanced RP for faster product development processes," *Journal of Rapid Prototyping*, vol. 6, no. 2, pp. 63–69, 2000.

[4] D. T. Pham and S. S. Dimov, *Rapid Manufacturing*, Springer, London, UK, 1st edition, 2001.

[5] S. H. Choy and S. Samedevan, "Modeling and optimization of RP parts," *International Journal of Machine Tools & Manufacturing*, vol. 40, no. 3, pp. 408–412, 2000.

[6] T. Luangvaranunt, C. Dhadsanadhep, J. Umeda, E. Nisaratanaporn, and K. Kondoh, "Aluminum-4 mass % Copper/Alumina composites produced from Aluminum Copper and rice husk ash silica powders by powder forging," *Materials Transactions*, vol. 51, no. 4, pp. 756–761, 2010.

[7] D. S. Prasad and A. R. Krishna, "Production and mechanical properties of A356.2/RHA composites," *International Journal of Advanced Science and Technology*, vol. 33, pp. 51–58, 2011.

[8] G. Taguchi and S. Konishi, *Taguchi Methods, Orthogonal Arrays and Linear Graphs, Tools for Quality Engineering*, American Supplier Institute, Dearborn, Mich, USA, 1997.

[9] S. Dingal, T. R. Pradhan, S. Sundar, C. A. Roy, and S. K. Roy, "Experimental investigation of selective laser sintering of iron powder by application of Taguchi method," in *Proceedings of the Laser Assisted Net Shape Engineering conference (LANE '04)*, pp. 445–456, Erlangen, Germany, 2004.

[10] S. Guharaja, A. Noorul Haq, and K. M. Karuppannan, "Optimization of green sand casting process parameters by using Taguchi's method," *International Journal of Advanced Manufacturing Technology*, vol. 30, no. 11-12, pp. 1040–1048, 2006.

[11] S. R. Rao and G. Padmanabhan, "Application of Taguchi methods and ANOVA in optimization of process parameters for metal removal rate in electrochemical machining of Al/5%SiC composites," *International Journal of Engineering Research and Applications*, vol. 2, no. 3, pp. 192–197, 2012.

[12] M. Nataraj, V. P. Arunachalam, and G. Ranganathan, "Using risk analysis and Taguchi's method to find optimal conditions of design parameters: a case study," *International Journal of Advanced Manufacturing Technology*, vol. 27, no. 5-6, pp. 445–454, 2006.

[13] P. B. Barua, P. Kumar, and J. L. Gaindhar, "Optimization of mechanical properties of V-Process castings by Taguchi method," *Indian Foundry Journal*, vol. 18, no. 3, pp. 17–25, 1997.

[14] S. O. Onuh and K. K. B. Hon, "Application of the Taguchi method and new hatch styles for quality improvement in stereolithography," *Proceedings of the Institution of Mechanical Engineers, Part B: Journal of Engineering Manufacture*, vol. 212, no. 6, pp. 461–472, 1998.

[15] J. G. Zhou, D. Herscovici, and C. C. Chen, "Parametric process optimization to improve the accuracy of rapid prototyped stereolithography parts," *International Journal of Machine Tools and Manufacture*, vol. 40, no. 3, pp. 363–379, 2000.

[16] "Standard test method for Tensile properties of plastics, D638-01," in *Annual Book of ASTM Standards*, vol. 8.02, section 8, plastics, pp. 1–15, 2001.

[17] "Flexural properties of unreinforced and reinforced plastics and electrical insulating materials, D790-03," in *Annual Book of ASTM Standards*, vol. 8.02 of *Plastics*, section 8, pp. 1–11, 2001.

[18] "Standard test methods for determining the Izod pendulum impact resistance of plastic," in *Annual Book of ASTM Standards*, vol. 8.02, pp. 1–20, Plastics, section 8, D256-04 edition, 2001.

[19] *User Manual, 3D System*, Valencia, Orlando, Fla, USa, 1988.

[20] G. Taguchi, *Introduction to Quality Engineering*, Asian Productivity Organization, 1990.

[21] L. K. Pan, C. C. Wang, Y. C. Hsiao, and K. C. Ho, "Optimization of Nd:YAG laser welding onto magnesium alloy via Taguchi analysis," *Optics & Laser Technology*, vol. 37, no. 1, pp. 33–42, 2005.

[22] R. C. Luo and J. H. Tzou, "The development of an intelligent web-based rapid prototyping manufacturing system," *IEEE Transactions on Automation Science and Engineering*, vol. 1, no. 1, pp. 4–13, 2004.

[23] D. A. Schaub, K. Rey Chu, and D. C. Montgomery, "Optimizing stereolithography throughput," *Journal of Manufacturing Systems*, vol. 19, no. 1, pp. 18–27, 2000.

[24] D. C. Montgomery, *Design and Analysis of Experiments*, John Wiley & Sons, New York, NY, USA, 3rd edition, 2001.

[25] W. H. Yang and Y. S. Tarng, "Design optimization of cutting parameters for turning operations based on the Taguchi method," *Journal of Materials Processing Technology*, vol. 84, no. 1–3, pp. 122–129, 1998.

[26] N. Tosun, C. Cogun, and G. Tosun, "A study on kerf and material removal rate in wire electrical discharge machining based on Taguchi method," *Journal of Materials Processing Technology*, vol. 152, no. 3, pp. 316–322, 2004.

[27] G. P. Syrcos, "Die casting process optimization using Taguchi methods," *Journal of Materials Processing Technology*, vol. 135, no. 1, pp. 68–74, 2003.

[28] J. A. Ghani, I. A. Choudhury, and H. H. Hassan, "Application of Taguchi method in the optimization of end milling parameters," *Journal of Materials Processing Technology*, vol. 145, no. 1, pp. 84–92, 2004.

[29] K. Chockalingam, N. Jawahar, U. Chandrasekar, and K. N. Ramanathan, "Establishment of process model for part strength in stereolithography," *Journal of Materials Processing Technology*, vol. 208, no. 1–3, pp. 348–365, 2008.

[30] C. C. Wang, T.-W. Lin, and S.-S. Hu, "Optimizing the rapid prototyping process by integrating the Taguchi method with the Gray relational analysis," *Rapid Prototyping Journal*, vol. 13, no. 5, pp. 304–315, 2007.

14

Finite Element Analysis of Residual Stress in Ti-6Al-4V Alloy Plate Induced by Deep Rolling Process under Complex Roller Path

J. J. Liou and T. I. El-Wardany

United Technologies Research Center, 411 Silver Lane, East Hartford, CT 06108, USA

Correspondence should be addressed to T. I. El-Wardany; elwardti@utrc.utc.com

Academic Editor: Archie Lockamy

The kinematics of the deep rolling tool, contact stress, and induced residual stress in the near-surface material of a flat Ti-6Al-4V alloy plate are numerically investigated. The deep rolling tool is under multiaxis nonlinear motion in the process. Unlike available deep rolling simulations in the open literature, the roller motion investigated in this study includes penetrative and slightly translational motions. A three-dimensional finite element model with dynamic explicit technique is developed to simulate the instantaneous complex roller motions during the deep rolling process. The initial motion of the rollers followed by the penetration motion to apply the load and perform the deep rolling process, the load releasing, and material recovery steps is sequentially simulated. This model is able to capture the transient characteristics of the kinematics on the roller and contacts between the roller and the plate due to variations of roller motion. The predictions show that the magnitude of roller reaction force in the penetration direction starts to decrease with time when the roller motion changes to the deep rolling step and the residual stress distributions in the near-surface material after the material recovery step varies considerably along the roller path.

1. Introduction

Deep rolling is a mechanical surface treatment method in which the workpiece surface is exposed to high local mechanical load using a spherical or cylindrical type tool to induce work hardening and compressive residual stress in the near-surface material. Depending upon the controlling parameters, this process alters the mechanical behavior of the material by cold working and enhances stability of the near-surface structure when the workpiece is exposed to a high temperature condition (exceeding 450°C). The high temperature fatigue resistance is due to the formation of a near-surface work hardened layer with a nanoscale microstructure during the deep rolling of the surface [1]. In automotive industries, fatigue damage has occurred near the crankshaft fillets due to the stress concentration. Deep rolling has been applied to the crankshaft or shaft fillets to enhance the fatigue strengths of crankshafts [2–4]. In aerospace industries, the initiation and early growth of fatigue cracks, especially in the presence of foreign-object damage, in the turbine engine components is one of the most critical challenges.

Deep rolling provides deeper case of compressive residual stress and a work hardened microstructure as well as a relatively smoother surface finish [1]. During the deep rolling process, the mechanical load is often repeatedly applied on the workpiece surface to continually increase plastic deformations in the near-surface material. In addition, the adaptability of attaching deep rolling tool to a conventional machining center to integrate this process into machining consequences has largely increased the cost effectiveness of the process.

A large number of numerical simulations have been conducted using a spherical or cylindrical type of rolling tools to understand the contact stress distributions and deformation patterns during the near-rolling process, as well as residual stress profiles after the process. Yen et al. [5] conducted 2D plain-strain and 3D finite element (FE) analyses of roller burnishing using a spherical rigid roller loaded on the hard-turned elastic-plastic material made of AISI 52100 steel to study the deformation on surface lays and residual stress in the near-surface material. Predictions from Yen et al. [5] show that the roughness peak-to-peak value (Rz) can be reduced from 11 to 1 μm. In the 2D analysis, the predicted residual stress in the tangential (out-of-plane) direction has the location of a maximum compressive residual stress in agreement with the experiment while the maximum compressive residual stress value is larger than that in the experiment. The predicted maximum compressive residual stress in the axial direction the maximum compressive residual stress is correlated well with the experimental data. However, unlike the experimental curve, the stresses drastically turned into high tensile residual stress when approaching the surface. In the 3D analysis, the predicted residual stress in the tangential direction is not in agreement with the experiment due to an insufficient thickness in the model.

Sartkulvanich et al. [6] further improved the 2D and 3D FE analyses, concluding that (i) tensile residual stresses predicted from the simulation are due to the plane strain assumption used in the 2D model since the tool is presented as a cylindrical object rather than a spherical object. Line contact is presumed in the 2D FEM simulation rather than point contact that is typically generated by spherical tool. In the 2D model, the tool geometry, which is assumed to be larger in the axial direction, pushes more material to flow aside and generates more tensile stresses in the axial direction than what is expected from a spherical tool and (ii) the FE analyses and experimental measurements reveal that the burnishing pressure is the most important factor for the amplitude of residual stress.

Ali and Pan [7] used 2D plane-strain finite element analyses to investigate the differences of residual stress and displacement distributions due to single indentation and rolling between a rigid roller and an elastic roller with a very large contact pressure. The elastic-plastic plate is modeled as a power-law strain hardening material with the nonlinear kinematic hardening rule. The residual stress distributions for both roller cases are nearly identical. The values of the residual stresses at the contact surface for both rigid and elastic rollers are about the same except a slightly (2%) higher peak near-surface compressive residual stress components in the rolling and out-of-plane directions for the rigid roller while compared to those for the elastic roller. The residual values of the displacement components in the loading direction are nearly the same for both rollers while the residual value of the displacement component in the rolling direction for the rigid roller is slightly higher than that for the elastic roller.

Other similar numerical investigation into this subject can be found in Guagliano and Vergani [8], Choi and Pan [3, 9], Majzoobi et al. [10], Fu et al. [11], Sayahi et al. [12], and Klocke et al. [13]. These analyses are modeled with a spherical or cylindrical type of roller under indentation or rolling condition. The conducted literature review indicates the need for a more detailed deep rolling FE model to take into account the multiaxis process kinematics. The present paper attempts to analyze a complex roller motion for the deep rolling which is difficult to be idealized by a simple rolling path on the elastic-plastic plate. The roller considered here is in the cylindrical shape. The roller axis is perpendicular to a mandrel axis. The mandrel rotates around its axis in a constant speed while it moves on the plate in a constant small translational speed. The motion of the roller is in such a way that the roller axis spins around the mandrel axis while the roller rotates around its axis. Figure 1 illustrates the schematic of deep rolling process under complex roller path. Such complex roller path cannot be simply represented by a two-dimensional model; therefore, a three-dimensional finite element model is proposed in this paper to better understand the reaction forces, contact stress distributions, deformation patterns, and residual stresses induced by this complex roller motion in the deep rolling process.

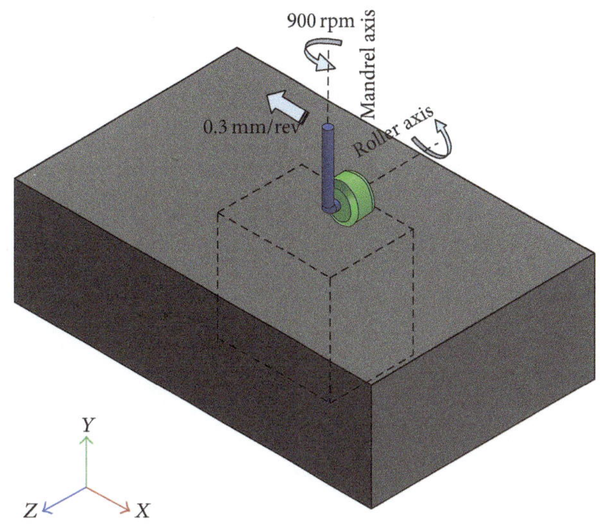

FIGURE 1: Schematic of deep rolling process on a relatively larger plate under complex roller path.

2. Modeling Procedures

As mentioned in the introduction, the main challenge in treating the deep rolling process is the multiplicity of paths that the roller involves. Thus, a holistic three-dimensional finite element model is proposed in the present paper. The model of deep rolling consists of a rigid mandrel, a rigid cylindrical roller, and a relatively larger elastic-plastic flat plate as shown in Figure 1. In order to simulate the process in an effective computational way, a finite domain (region within the dash lines in Figure 1) that contains regular continuum elements is specified. The infinite domain is arranged to surround the finite domain on four sides to represent relatively larger plate.

The multiple paths of the roller are depicted in Figure 2. The roller is free to spin around its own axis while the mandrel is rotating on the plate. The paths of the mandrel and roller are defined in the following steps.

(1) Initial Step. Initially, the roller is positioned on the starting point of the flat plate as shown in Figure 2(a). The initial condition of the plate is in a stress-free state at room temperature.

(2) Penetration Step. The mandrel rotates at a constant speed N_m for one revolution without translational motion on the plate. During this revolution, the roller starts to spin around

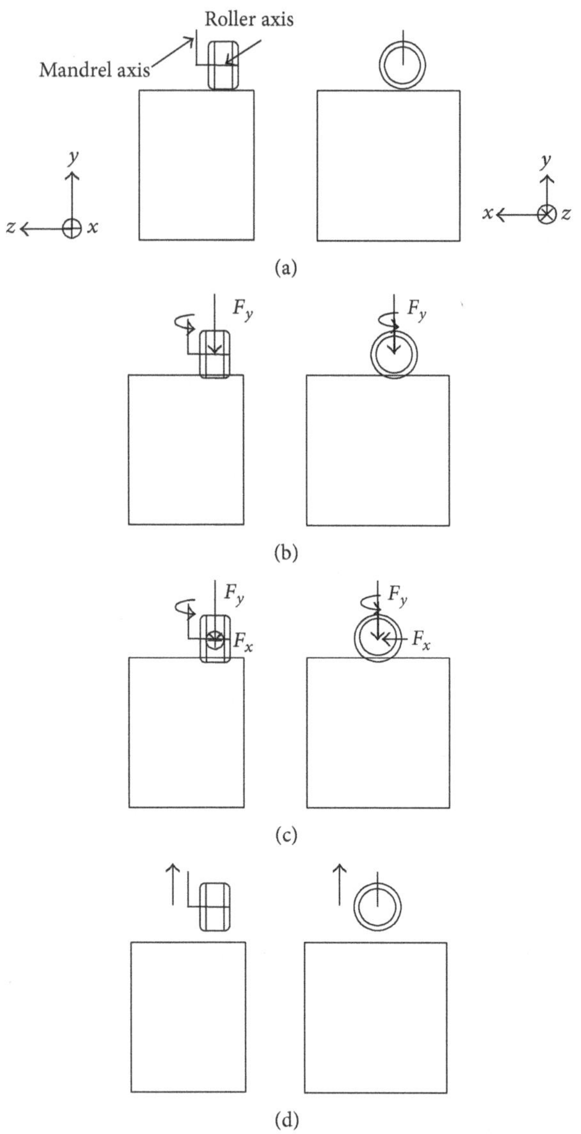

FIGURE 2: Multiple paths of the roller. (a) Initial step, (b) penetration step, (c) deep rolling step, and (d) recovery step with the left-hand column of plots being front views and the right-hand column of plots being right views.

the mandrel axis. Meanwhile, the roller is free to rotate around its own axis and linearly penetrates the flat plate at a constant speed $u_{r,y}$ so that the center of roller moves down to the full penetration δ at the end of the revolution as shown in Figure 2(b).

(3) Deep Rolling Step. The mandrel continues to rotate at the constant speed N_m and in the same time the axis of mandrel moves on the plate with a constant small translational speed $u_{r,y}$. In this step, the roller continues to spin around the mandrel axis with the fixed penetration δ. Figure 2(c) shows the deep rolling step.

(4) Releasing Step. The mandrel stops rotating and translating. The roller gradually moves up from the plate at a constant speed $u_{r,y}$.

(5) Recovery Step. The roller continues to move up to provide additional time for springback in the plate. The recovery step is shown in Figure 2(d).

The Cartesian coordinate system is used in the FE model. The selection of the sizes of the finite elements to simulate the multiple paths of roller motion follows the guidelines that are established in Ali and Pan [7] and will be reported in the next section according to the example. The bottom surface of the plate is constrained in the loading direction in all steps and this is the only constraint in the plate. The interaction behavior between the roller and the plate is modeled with Coulomb's law, taking 0.05 as the coefficient of friction for the highly lubricated rolling process [11, 14]. The cooling function of lubricant and the free-to-roll configuration in the tool design allow for the assumption of isothermal deformation.

In addition to multiple paths of roller motion, an important factor that influences the deep rolling analysis is the material model. The material model, which can adequately represent the deformation response during loading and unloading, must account for strain hardening or softening during the process. Several models have been developed, with varying degrees of accuracy, to predict the deformation response of materials. Among these models, the Johnson-Cook (JC) material model is the most widely used. The JC model has been proven to represent the plastic behavior of the material very well and has been extensively used to predict material behavior (flow) under high strain and strain rate as a function of the operating temperature. JC material model can be expressed as

$$\sigma = (A + B\varepsilon^n)(1 + C\ln\dot{\varepsilon}^*)\left[1 - (T^*)^m\right], \quad (1)$$

where σ is the effective stress, ε is the effective plastic strain, $\dot{\varepsilon}^*$ is the normalized effective plastic strain rate, n is the work hardening exponent, and T^* is the normalized temperature defined as

$$T^* = \frac{T - T_r}{T_m - T_r}, \quad (2)$$

where T, T_m, and T_r are the workpiece temperature, material melting, and room temperature, respectively. In (1), the first bracketed term represents the strain hardening of the yield stress, the next term models the increase in the yield stress at elevated strain rates, and the final bracketed term is a softening of the yield stress due to local thermal effects. Here, A, B, C, n, and m are constitutive parameters fitted to the experimental flow stress data since the formulation of JC material model is empirically based. The model assumes that the strength is isotropic.

3. Case Studies

In this section, a deep rolling process that consists of a smooth cylindrical roller and a smooth relatively larger elastic-plastic flat plate will be used as an illustrative example. Simulation results for multiple paths of roller motion will be presented to

(1) demonstrate the effectiveness of the FE model in handling multiple paths in the deep rolling process,

(2) highlight the difference between the proposed multiple paths model and the single indentation or single rolling ones,

(3) present the residual stress distributions in the near-surface material after the material recovery vary along the roller path.

Three-dimensional FE analyses involving nonlinear material behavior and nonlinear boundary effects (contact and friction) typically resulted in very large models when the "implicit" technique was used. For this reason the "explicit dynamic" technique is prevalent in FE modeling works for the similar applications, especially in many manufacturing process modeling works, for example, Fu et al. [11], Klocke et al. [13], Mohammadi et al. [15], Sayahi et al. [12] for the burnishing process, Fischer and Schleinzer [16] for the straightening process, and Taylor et al. [17] for the sheet metal forming process. In the present paper, the FE analyses were implemented using the commercially available FE code ABAQUS with the explicit dynamic analysis technique and appropriate mass scaling [18]. Details about why the explicit technique and necessary "mass scaling," that is, added artificial mass, are widely used in manufacturing applications can be found in Klocke [19]. The cylindrical roller has a diameter of 4 mm, a width of 2.5 mm, and fillets of 0.5 mm on the two edges. The roller and mandrel are modeled as analytical rigid bodies since the roller and mandrel considered in the present paper have much higher hardness than the plate. The modeled finite-domain dimensions of elastic-plastic plate are 12 × 9.5 × 12.5 mm. These dimensions are used to avoid boundary/edge effects while providing enough dimensions for establishing elastic-plastic deformation in an effective computational manner. In order to represent a relatively larger flat plate, semi-infinite elements on the four sides are employed and extrusion cuts are arranged on the four corners to accommodate the semi-infinite elements.

Figure 3 illustrates the 3D finite element mesh for deep rolling in the Cartesian coordinate (throughout this paper, the Cartesian coordinate notations x, y, and z and 1, 2, and 3 will be used interchangeably for convenience). Eight-node linear brick elements with reduced integration and hourglass control (C3D8R) were used in the finite domain and eight-node linear semi-infinite brick elements (CIN3D8) were used along the four sides. Since the semi-infinite elements provide boundary conditions by their nature, the only boundary condition utilized in the model was the constraint of plate bottom area. The fine meshes of $20 \times 20 \times 21\,\mu$m (compared to the mesh size of $50\,\mu$m for a 6-mm diameter ball in [13]) were used in the contact interactive area while coarse meshes of $250 \times 136 \times 207\,\mu$m were used in the field far away from the rolling contact so that the contact stress distribution was modeled accurately and efficiently. In total, 4,625,964 nodes and 4,386,760 elements were used for the elastic-plastic deformable plate. The complex roller paths are given as N_m = 900 rpm for the constant mandrel rotational speed, $\delta = 40\,\mu$m for the penetration step, and $u_{m,x}$ = 0.3 mm/rev for the deep rolling step. The corresponding $u_{r,y}$ can be found as 0.6 mm/s.

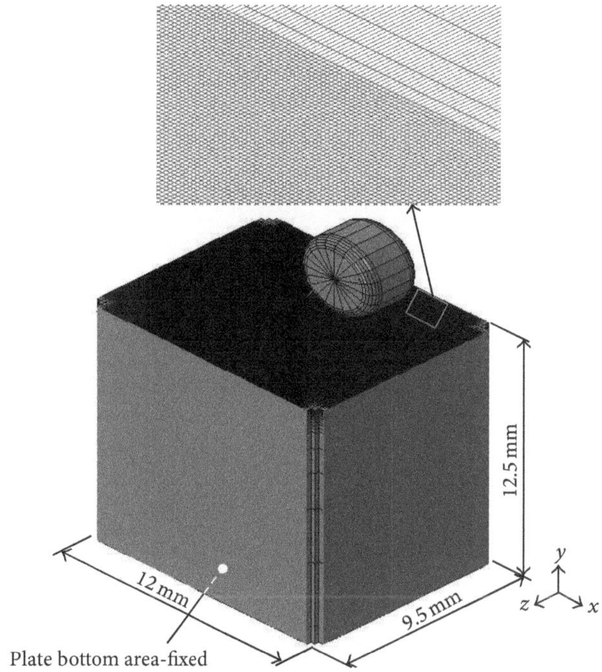

FIGURE 3: Three-dimensional finite element mesh of deep rolling.

TABLE 1: Material properties of Ti-6Al-4V, [1, 20].

Density	4430 kg/m^3
Young's modulus	110 GPa
Poisson's ratio	0.342

TABLE 2: Parameters of Johnson-Cook material model for Ti-6Al-4V [20]

Yield strength, A	968 MPa
Strain hardening coefficient, B	380 MPa
Strain rate coefficient, C	0.0197
Strain hardening exponent, n	0.421
Thermal softening exponent, m	0.577
Melting point, T_m	1605°C

The material constitutive constants for the JC model have been introduced into the finite element code in ABAQUS. The elastic-plastic flat plate considered in the present paper is Ti-6Al-4V alloy. This alloy offers a combination of high strength, light weight, formability and corrosion resistance which have made it a world standard in various applications such as aerospace, automotive, and marine equipment. The material properties of Ti-6Al-4V are listed in Table 1. The parameters of JC material for Ti-6Al-4V are shown in Table 2. These data are obtained from modified split Hopkinson pressure bar (SHPB) tests at strain rate up to $10^3\,s^{-1}$ and temperature up to 600°C [20]. Figure 4 shows the stress-strain curve of Ti-6Al-4V from the JC material model at room temperature.

The FE model has been validated through two ways. First, in an elastic contact condition—that is, the boundary conditions, element type, and mesh size-by comparison

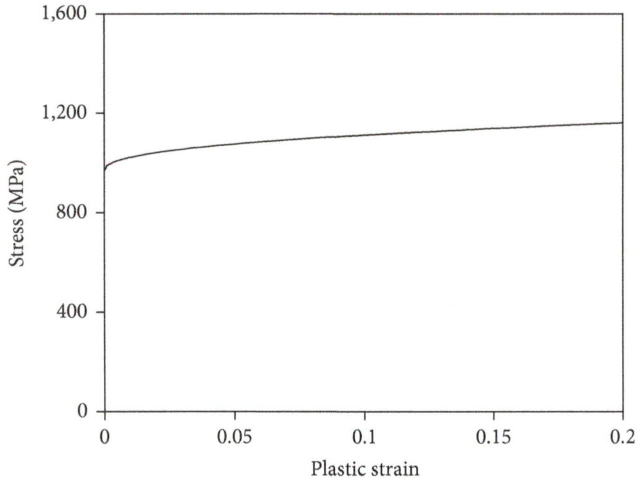

FIGURE 4: Stress-strain curve of Ti-6Al-4V from J-C model predictions at room temperature.

with the analytical solution of Hertzian line contact for an indentation simulation between the cylinder and flat plate surface when the contact is still in the elastic deformation stage, that is, for very low loads. The difference between the Hertz theory and the FE model for the contact band width, pressure, and von Mises stress was found to be smaller than 5%. The difference between the FE and analytical solution in the elastic deformation stage can be explained by the boundary conditions of the FE model that differ slightly from the half-space assumption in Hertz theory. Second, in a simple loading path of deep rolling process, details are shown in the Appendix of the present paper.

Figures 5(a) and 5(b), respectively, provide the complex paths of the reference point on the rigid mandrel for the complex roller path in the deep rolling process. In the penetration step, 0 through 0.07 seconds, $U2$ linearly reduces with time while $UR2$ linearly increases with time. In the deep rolling step, 0.07 through 0.2 seconds, $U1$ linearly declines with time due to the negative small translational displacement and $UR2$ continuously increases over the time. In the releasing step, 0.2 through 0.27 seconds, and the material recovery step, 0.27 through 0.3 seconds, only $U2$ increases with time. Note that $U3$, $UR1$, and $UR3$ are not plotted in Figure 5 since they remained zero in all time.

As mentioned previously in this section, an appropriate mass scaling has to be employed in the present FE analysis to enable a time-efficient simulation of the process. The appropriateness of mass scaling means that the artificial mass is added to the system without adulterating the global solution. One way to check the appropriateness of mass scaling is to compare the kinetic energy to the internal energy. Figure 6 shows the model kinetic energy and internal energy as a function of time, revealing the appropriateness of mass scaling.

Figure 7 shows the reaction forces with respect to the reference point on the mandrel. The magnitude of reaction force in the penetration direction, RF2, is the most significant

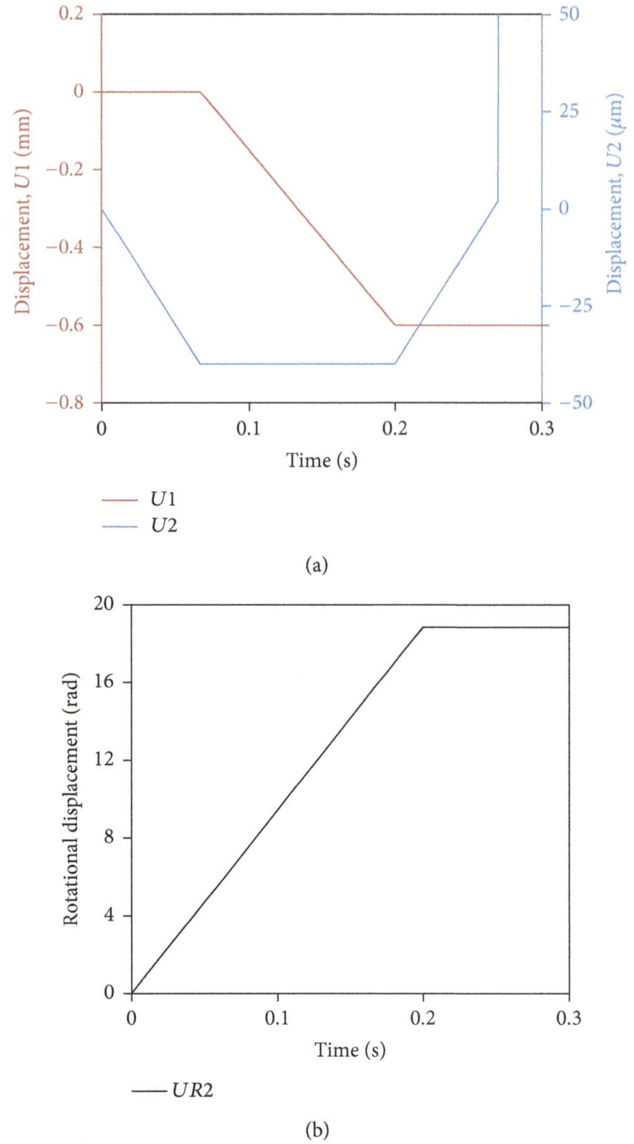

FIGURE 5: Displacements of the reference point on the rigid roller for the complex roller path in the deep rolling process. (a) Translational and (b) rotational.

one, which reaches the peak (−3291 N) at the end of penetration step. The magnitude of RF2 starts to decrease with time when the roller motion changes to the deep rolling step. The reduction of RF2 can be explained by Figure 8, which shows the historical displacement $U2$ of a point located on the plate upper surface within the rolling track and in the contact bandwidth near the end of revolution. This point is pushed down 40 μm then quickly recovers 22 μm after the roller passes the point. The near-surface material is work-hardened in this step. However, when the roller comes to pass the same point, the roller only needs to penetrate 22 μm instead of 40 μm. As a result, the RF2 magnitude is reduced. It can also be seen in Figure 6 that, with the specified roller displacement of 40 μm, the near-surface material reaches its saturated stress state after two or three times of deep rolls. The residual strain will no longer increase if no higher force

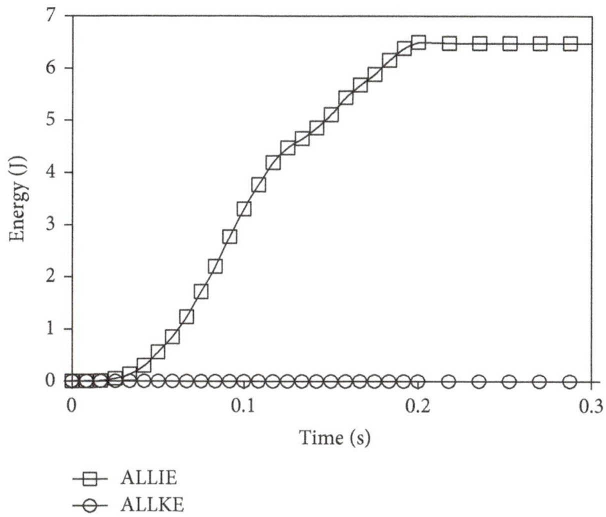

FIGURE 6: Model internal energy (ALLIE) and kinematic energy (ALLKE), showing insignificant inertia effects.

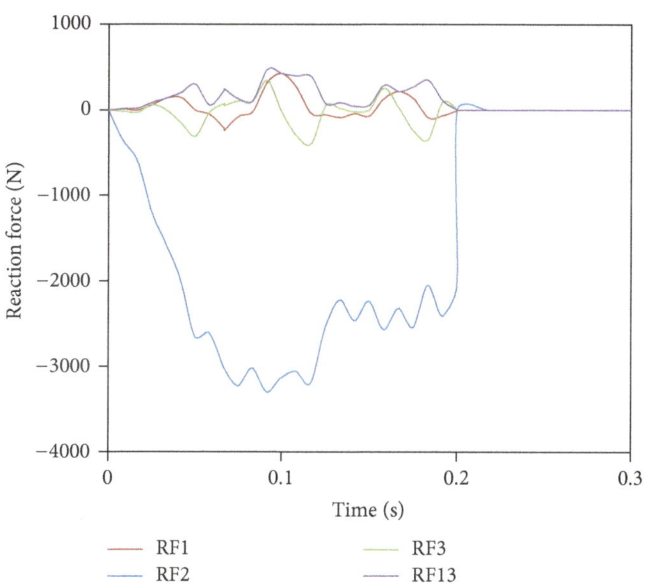

FIGURE 7: Reaction forces of the roller for the complex roller path in the deep rolling process.

or deeper roller displacement is specified in the later rolling path.

Figure 7 also shows that the magnitude of resultant reaction force (RF13) on the plate surface is approximately 10% of RF2. This indicates that the deep rolling process at the prescribed roller displacement and motion requires 5% of RF2 to overcome the material in front of the roller leading edge in addition to the 5% of RF2 for the interfacial frictional resistance. In the perspective of machine tool design, the magnitude of RF13 is most likely not a concern to the tool holder.

Figures 9(a), 9(b), 9(c), and 9(d), respectively, show the Mises stress distributions as the mandrel rotates 0.5-, 1.5-, 2.5-revolution, and at the end of material recovery step. Here,

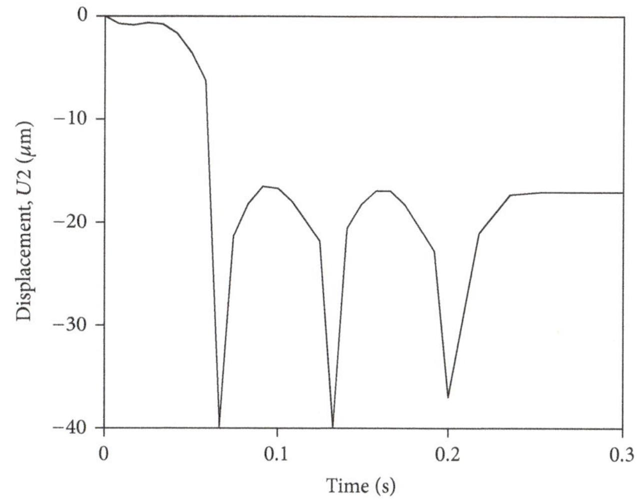

FIGURE 8: The historical displacement component $U2$ of a point on the plate upper surface within the rolling track and in the contact bandwidth near the end of revolution.

the left-hand and right-hand figures represent the top- and cut-view, respectively. Note that the roller is removed from these figures in order to see the stress distribution completely. Figure 9(a) clearly shows that the roller gradually pushes down and the material yielding has been faded in along the roller path. In the cut-view of Figure 9(a), the orange and red areas on the y-z plane are the material that is yielded and hence this area represents the plastic zone size and shape. At this instance, the maximum depth of equivalent plastic strain (PEEQ) zone is approximately 726 μm, including 20 μm of roller displacement. In the cut-view of Figures 9(b) and 9(c), the maximum depth of PEEQ zone on the y-z plane at both instances increases to 1210 μm, including 40 μm of roller displacement. At the end of material recovery step, the maximum depth of PEEQ zone on the y-z plane reduces to 260 μm.

It is interesting to compare the Mises contour shape in the plastic zone. Figure 10 shows such contours with a cutoff of minimum stress limit at 862 MPa on the y-z plane below the contacts as the roller spins 0.5-, 1.5-, and 2.5-revolution. The Mises contour shape has been changed from a symmetric form to an asymmetric form as the step changes from the penetration step to the deep rolling step.

Figure 11 shows the contact pressure profiles along the center of rolling track when the roller rotates 0.5-, 1.5-, and 2.5-revolution. These profiles indicate that the contact pressure distribution is no longer in semi-ellipsoidal shape or Hertzian pressure distribution when the plate is in the elastic-plastic deformation. The maximum contact pressure occurs at the leading edge of the roller instead of the contact center. This is due to the unpenetrated material in front of the roller, forming an obstacle to the roll motion. Similar contact pressure profiles can also be found in Ali and Pan [7] by using an elastic-plastic power-law strain hardening material with a nonlinear kinematic hardening rule. In the present paper, the peak pressures are 3.3, 3.7, and 4.0 GPa when

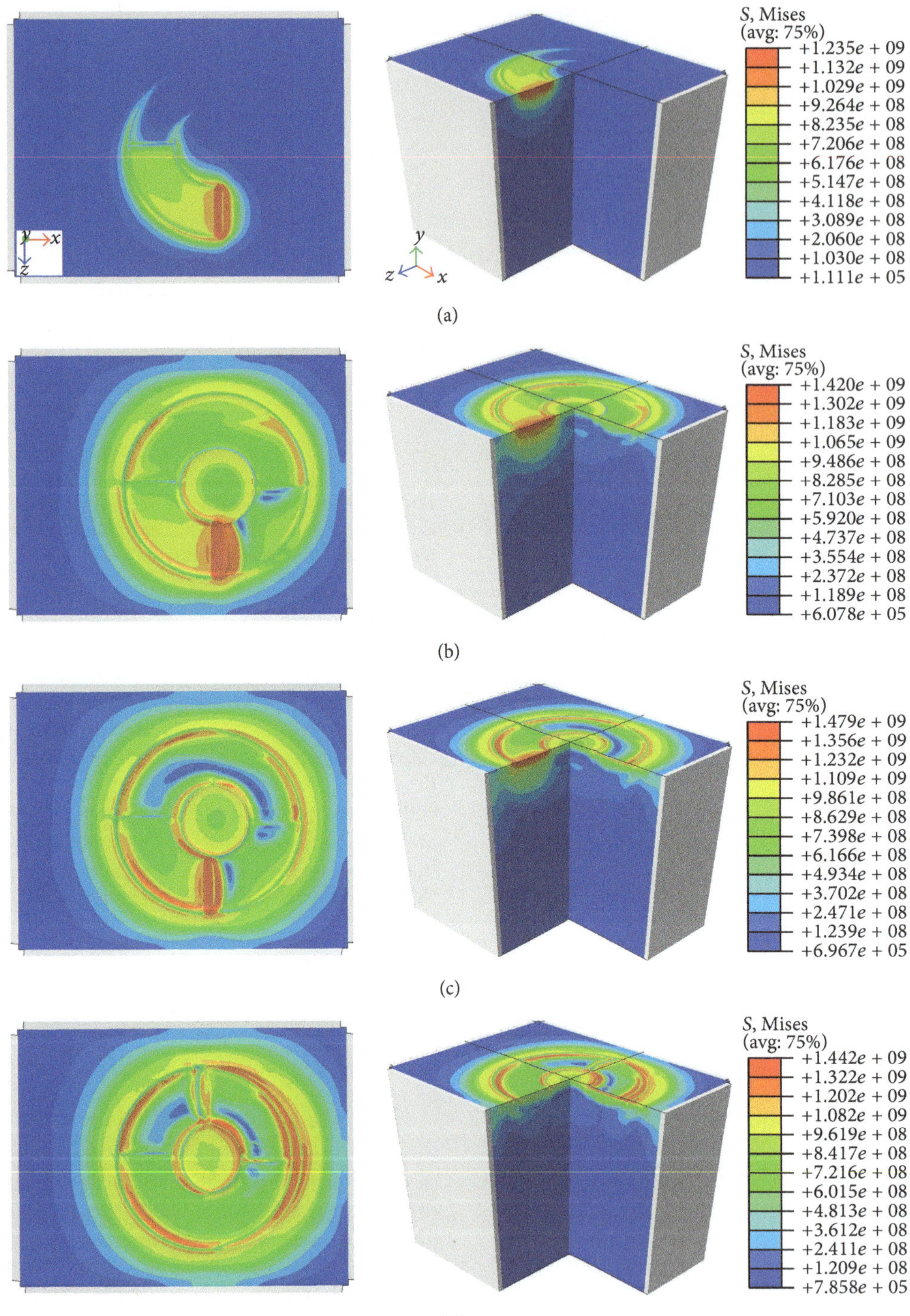

FIGURE 9: Mises stress distributions on the plate upper surface (top-view in the left-hand column) and in the plate (cut-view in the right-hand column) when the roller spins (a) 0.5-revolution, (b) 1.5-revolution, (c) 2.5-revolution, and (d) the roller is at the end of recovery step. Note that the roller is removed from the figure in order to see the stress distribution completely and all units are in Pa.

the roller rotates 0.5-, 1.5-, and 2.5-revolution, respectively. The corresponding contact widths are 0.452, 0.819, and 0.530 mm for these roller locations. The results reflect the complexity of material penetration and strain-hardening in this present paper. For example, at 0.5-revolution, the plate is penetrated approximately 20 μm. At 1.5-revolution, the plate is approximately penetrated 35 μm and in the same time the material that the roller contacts with has been strain-hardened; therefore, a higher peak pressure and wider contact width are expected. At 2.5-revolution, although the roller penetrates only 23 μm of material, the material has much higher strain-hardening compared to that at 1.5-revolution;

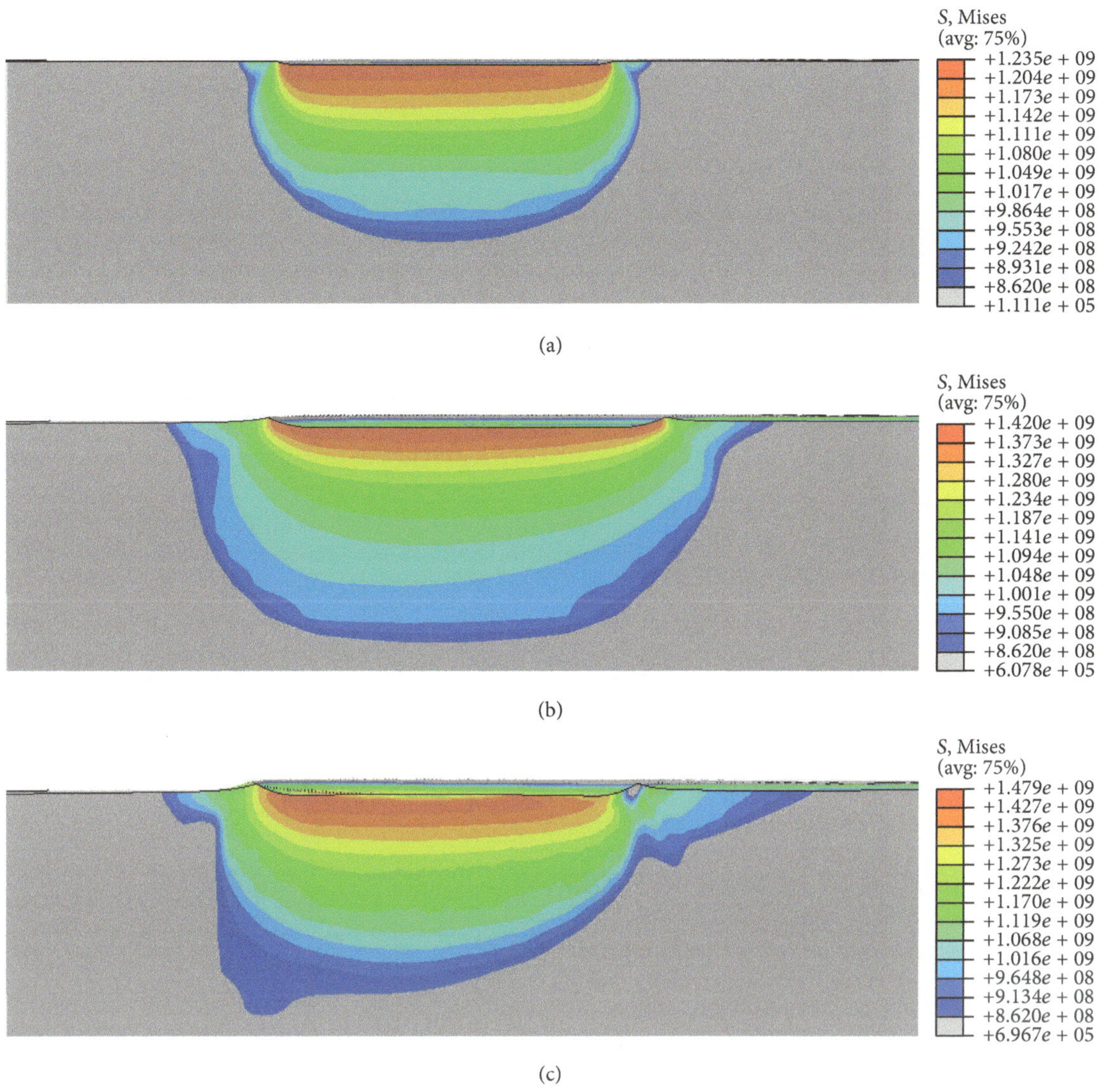

FIGURE 10: Mises contours with a cutoff of minimum stress limit at 862 MPa, the yield strength of Ti-6Al-4V on the y-z plane below the contacts as the roller spins (a) 0.5-revolution, (b) 1.5-revolution, and (c) 2.5-revolution. All units are in Pa.

hence, a higher peak pressure is generated but narrower contact width between the roller and the plate.

Figure 12 shows a comparison of the residual stresses distributions starting from the plate top surface when the roller is at the end of recovery step. Note that the residual stresses diminish as the depth increases from the plate top surface. In order to clearly show the residual stress profiles in the near-surface material, only the residual stresses within 2 mm depth are depicted. Four spots on the plate top surfaces are selected to review the residual stress distributions: 12, 9, 6, and 3 o'clock positions. These positions are corresponding to the 0-, 0.25-, 0.5-, and 0.75-revolution of the roller around the mandrel axis, respectively. These positions are located within the deep rolling track. It can be obviously seen in Figure 12 that the complex roller path induces compressive residual stress into the near-surface of the material. Residual stress components $S11$ and $S33$, which align with the two orthogonal directions of the deep-rolling plane, are considerable while the other four components, $S22$, $S12$, $S13$, and $S23$, are relatively insignificant and can be considered negligible.

In Figure 12, the maximum magnitudes of $S11$ and $S33$ range approximately from 900 to 1,500 MPa and the induced compressive residual stresses $S11$ and $S33$ are about 1 mm deep. Figure 11 shows the $S11$ and $S33$ contours of residual stress on the plate top surface. It can be seen from Figure 13 that the residual stress distributions within the deep rolling track are not uniform. It is also found from Figure 13 that $S11$ in the regions near the outer boundary of the roller track and near 12 and 6 o'clock positions drastically turns into tensile stress when approaching the surface. Similar conditions also occur for $S33$ in the regions near the outer boundary of the roller track and near 9 and 3 o'clock positions. Some regions near the inner boundary of the roller track also have tensile stress on the surface.

There is little information available in the literature regarding these tensile residual stresses near the inner and

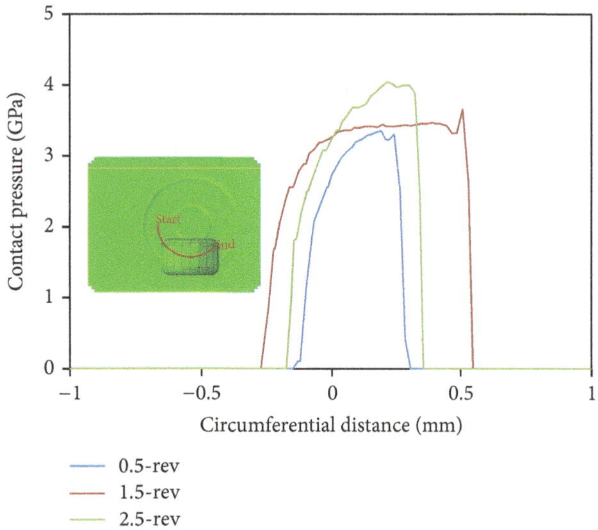

FIGURE 11: Contact pressure distributions at different steps.

outer boundaries of the roller track since most of the deep rolling processes are simulated by using simplified 2D models. However, the authors in the present paper find that these positive stress regions could be generated by the combination of microslip and small translational motion of roller for the complex roller paths specified in the present paper. The microslip phenomenon was first observed by Heathcote [21] in 1921 for a ball rolling in a groove. When a ball is rolling in a groove with a cross-sectional radius of curvature close to that of the ball itself, the contact area is no longer planar but shares the transverse curvature of the ball and groove; surface points in different transverse positions on the ball have different peripheral speeds which lead to microslip. More details about the Heathcote microslip can be found in Johnson [22]. In the present paper, the roller has rounded fillets of 0.5 mm on the two edges. When the roller is in the deep rolling step, the rolling track boundaries and the edges of the roller have very close conformity, this can lead to the microslip on the interface near the edges of the roller. The small translational motion in the deep rolling step also contributes an offset effect so that the high tensile stresses of $S11$ do not occur exactly at the 12 and 6 o'clock positions but offset toward the 11 and 5 o'clock positions. Similarly, the high tensile stresses of $S33$ offset a little bit further from the 3 and 9 o'clock positions.

Although the tensile stress appears in some regions in the roller track, the authors find that the neighboring surfaces surrounding the inner and outer track boundaries tend to have significant compressive stress levels, especially in $S11$ that aligns with the mandrel translational direction. Residual stress profiles for a depth path near the outer track at the 3 o'clock position as shown in Figure 12(e) is plotted in Figure 14. This figure shows that residual stress distributions out of the deep rolling track can be different. A strip or a band with significant compress residual stress is expected to form if the deep rolling continues this complex path further to the left side of the plate top shown in Figure 12(e).

4. Comparison to Published Experimental Studies of Deep Rolling under Simple Roller Path Using a Ball Tool

So far, there is no documented experimental result using a cylindrical roller under complex roller path. There are, however, limited numbers of well-documented experimental studies available in the literature for the deep rolling process under simple roller path using a ball tool with limited normal load due to the usage of hydraulic pressure system. All of the experimental measurements used the ball tools to deep roll the specimen on a bar-stock or plate structure. Among these experimental measurements, Nalla et al. [1] and Tsuji et al. [23] used the ball tool to deep roll Ti-6Al-4V alloy bars around the circumferential direction. Residual stresses induced by the deep rolling like process under simple roller path in [1, 23] were measured using X-ray diffraction method. In Figure 15, the predicted residual stress $S11$ from Figure 14 for the penetration-controlled deep rolling process using cylindrical roller under complex roller path are compared to the measurements of Nalla et al. [1] and Tsuji et al. [23] under simple roller path. Note that the residual stress profiles from experiments [1, 23] are in the axial direction of bar-stock specimen. The experimental results available in [1, 23] are used here to illustrate the benefit of using high depth of penetration deep rolling process to achieve higher compressive residual stress in comparison with the limited normal loading system. It is observed that (i) the predicted $S11$ on the surface and in the near-surface material for the complex roller path using a cylindrical tool has significant higher compressive residual stress than those obtained from experiments [1, 23] for the simple roller path using a ball tool, (ii) both the predicted and measured maximum compressive residual stress from experiments [1, 23] occur in the depth of 0.02 to 0.07 mm from the surface, (iii) the predicted $S11$ and measured residual stresses from experiments [1, 23] are in the same range at 0.5 mm below the surface, and (iv) the compressive residual stress affected depth is approximately in the same range of 0.6 mm for the model predictions and measurements from experiments [1, 23].

It should be noted that Figure 15 has been extracted for a comparison of residual stress on the surface, at their maximum values and at 500 μm below the surface in the absence of exact and direct experimental measurements of deep rolling process under complex roller path using a cylindrical tool. However, as more closed experimental data of complex roller path becomes available it may be incorporated readily into a quantitative comparison and model validation.

5. Conclusions

A comprehensive 3D finite element dynamic analysis with considering a complex roller path is conducted to simulate the deep rolling process. In this analysis, the roller is modeled as a rigid body and is in contact with a relatively larger plate under highly lubricated condition. The plate material is modeled as an elastic-plastic material by using the JC material

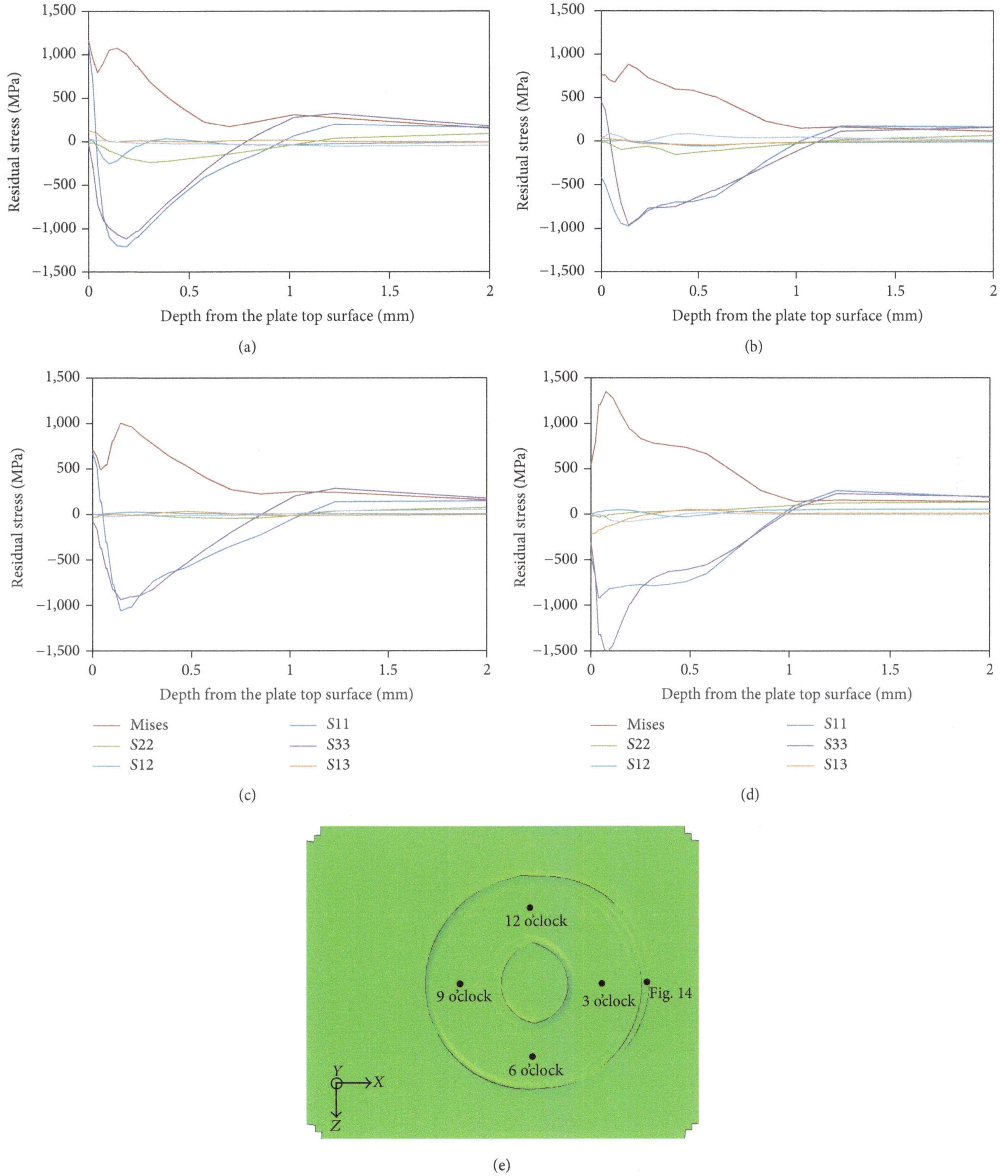

FIGURE 12: Residual stress distributions within the rolling track at (a) 12 o'clock, (b) 9 o'clock, (c) 6 o'clock, (d) 3 o'clock position when the roller is at the end of recovery step, and (e) the positions for paths of residual stress.

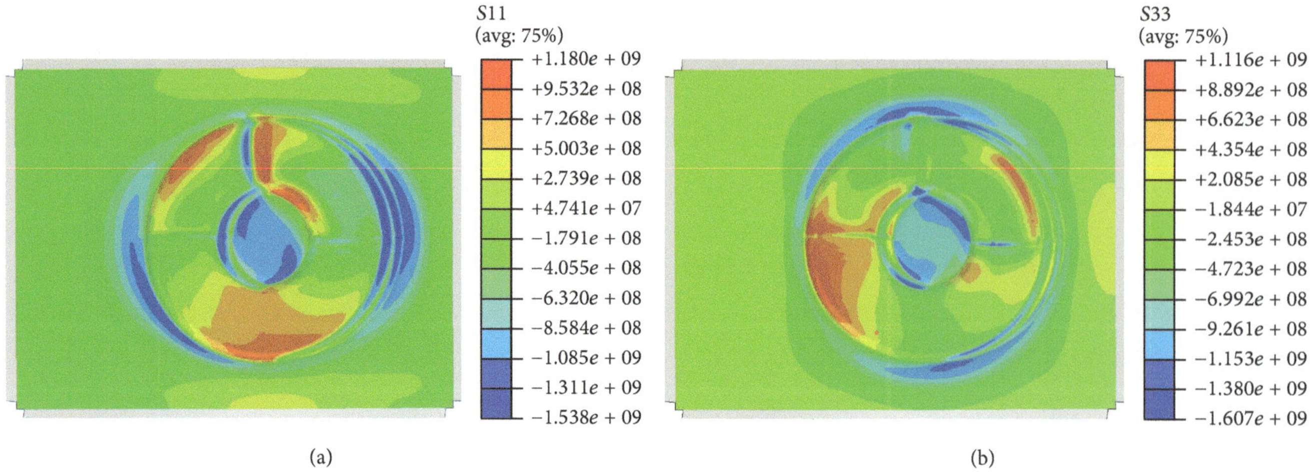

FIGURE 13: Contours of residual stress on the plate top surface at the end of recovery step: (a) $S11$ and (b) $S33$. All units are in Pa.

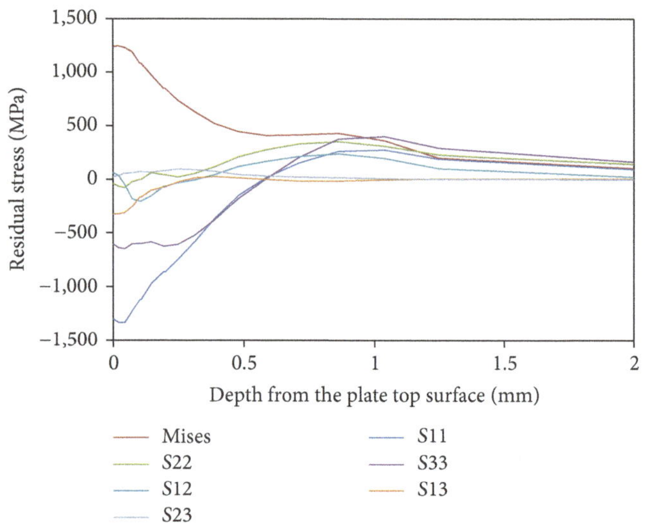

FIGURE 14: Residual stress distributions in the material neighboring the outer track at the 3 o'clock position shown in Figure 12(e).

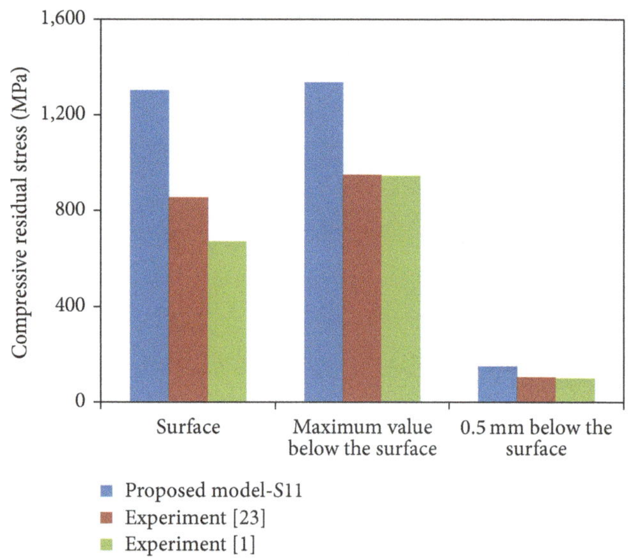

FIGURE 15: Comparison of the predicted residual stress $S11$ from Figure 14 for the complex roller path to the measured residual stresses from [1, 23] for the simple roller path.

model for the strain hardening and deformation response in the process. The results can be summarized as follows.

(1) The magnitude of resultant resistant force on the plate surface is approximately 10% of the loading force, which consists of 5% of the loading force on the prescribed interfacial friction and 5% of loading force on the material penetration in front of the roller. This helps a designer in implementing the high speed spindle tool design.

(2) The magnitude of loading force in the penetration direction (depth direction) starts to decrease with time when the roller path changes to the deep rolling step and the residual stress distributions in the near-surface material after the material recovery step vary along the roller path.

(3) With a specified roller displacement, the subsurface material reaches its saturated stress state after two or three times of deep rolls. The residual strain will not increase if no deeper roller displacement is performed.

(4) The contact pressure distributions are no longer in semiellipsoidal shape or elastic Hertzian pressure distribution when the plate is in the deep rolling process.

(5) The residual stress components $S11$ and $S33$ in the near-surface material within the deep rolling track are greater than the other components. The peak values of compressive $S11$ and $S33$ are in the range between 900 and 1,500 MPa for the proposed complex roller path in the deep rolling process.

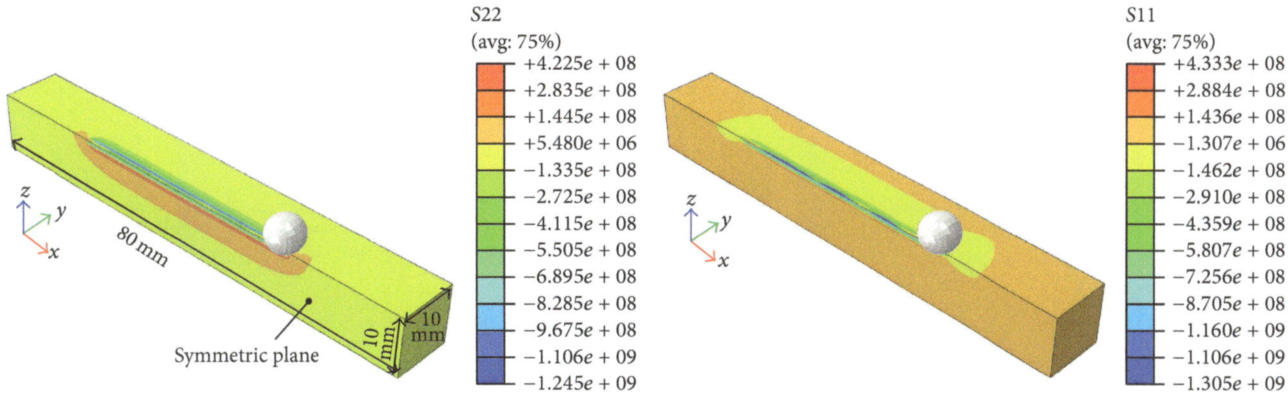

FIGURE 16: Predicted distribution of residual stresses $S11$ and $S22$ for the simple roller path. All units are in Pa.

(6) The residual stress components $S11$ and $S33$ on the deep rolled surface vary with the roller path. $S11$ near 12 and 6 o'clock positions and $S33$ near 9 and 3 o'clock positions drastically turned into tensile stress levels.

(7) The microslip and small translational motion of the roller can cause regional tensile stress on the plate top surface near the roller edges in the roller track.

(8) The residual stress distributions in the neighboring subsurface material surrounding the inner and outer track boundaries are quite different from those within the deep rolling track. Significant compressive residual stresses are found on these surfaces.

(9) While the proposed FE model is shown to have significantly higher compressive residual stress compared to the published experiments representing simple roller path conditions, its qualitative aspects have not been validated directly. The model must be validated further in the future for its capability in predicting residual stress in the complex roll path conditions.

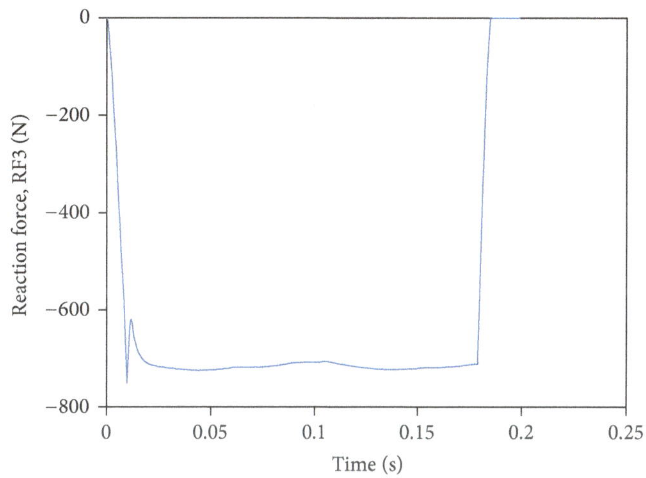

FIGURE 17: Predicted reaction force on the roller for the simple roller path.

Appendix

Model Validation in Simple Loading Path

To validate the material model, a three-dimensional finite element model was built to simulate the deep rolling on Ti-6Al-4V workpiece under simple loading path. The workpiece has a dimension of $10 \times 10 \times 80$ mm. The roller is in spherical shape as a common tool for the deep rolling under simple loading path. The diameter of roller is 6 mm. The Johnson-Cook material model and constitutive parameters for Ti-6Al-4V described previously in the present paper was employed to simulate the material behavior under loading and unload conditions. The roller was assumed to be rigid to reduce the computational time since the roller is much harder than the workpiece. The model is designed to be symmetric in the later direction to simplify the simulation process and decrease the computational time. Eight-node linear brick finite elements with reduced integration and hourglass control (C3D8R) were used for the workpiece.

The simulation consists of three steps. Initially, the roller sits on the starting position of the workpiece. The first step is to load the roller at its center to penetrate the workpience with a depth of 40 μm, the second step is to move the roller horizontally by assigning both the rotational speed of 750 rpm and translation speed of 236 mm/s. The third step is to lift the roller after the full rolling distance of 40 mm is completed.

Figure 16 shows the residual stress distribution in the workpiece from the three-dimensional finite element model at the end of third step. Figure 17 shows the predicted reaction force on the roller. The average reaction force during the second step is 714 N. Figure 18 shows the residual stress profiles ($S11$ and $S22$) in the depth below the middle point of rolling path. Figure 19 shows a comparison of predicted residual stress $S11$ in Figure 18 for the simple roller path to the measured residual stresses from Nalla et al. [1] and Tsuji et al. [23] for the simple roller path. The model predictions have, in terms of residual stress range, satisfactory agreement with experimental results, although there is a slight difference

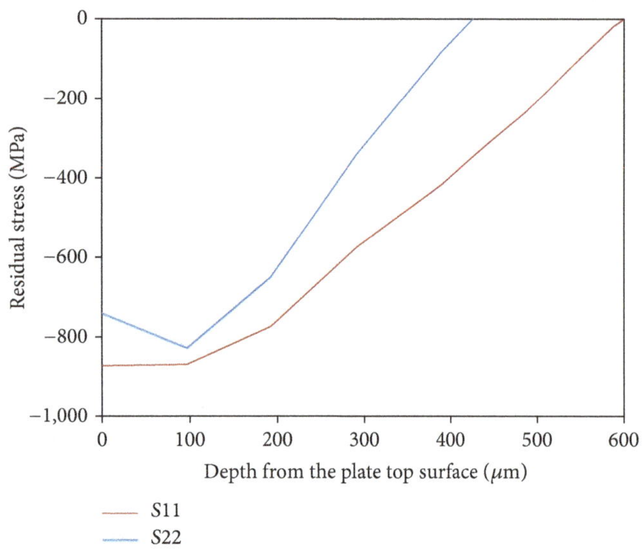

FIGURE 18: Predicted near-surface compressive residual stress for the simple roller path.

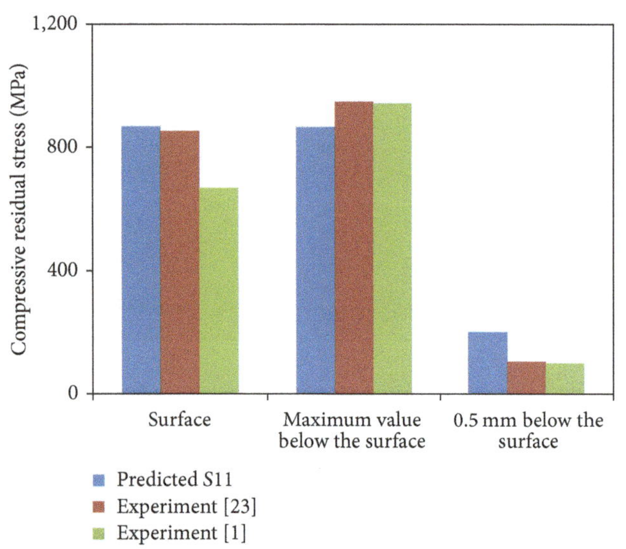

FIGURE 19: Comparison of the predicted residual stress S11 from Figure 18 for the simple roller path to the measured residual stresses from [1, 23] for the simple roller path.

in the residual stresses between the model predictions and experimental measurements.

Nomenclature

$U1, U2, U3$:	Translational displacements in the x, y, and z directions, respectively
$UR1, UR2, UR3$:	Rotational displacements along the x, y, and z axes, respectively
RF1, RF2, RF3:	Reaction forces in the x, y, and z directions, respectively
$S11, S22, S33$:	Normal stresses in the x, y, and z directions, respectively
$S12, S13, S23$:	Equivalent to shear stresses τ_{xy}, τ_{xz}, and τ_{yz}.

Conflict of Interests

The authors declare that there is no conflict of interests regarding the publication of this paper.

Acknowledgment

The authors would like to thank United Technologies Research Center for permission to publish this paper.

References

[1] R. K. Nalla, I. Altenberger, U. Noster, G. Y. Liu, B. Scholtes, and R. O. Ritchie, "On the influence of mechanical surface treatments-deep rolling and laser shock peening-on the fatigue behavior of Ti-6Al-4V at ambient and elevated temperatures," *Materials Science and Engineering A*, vol. 355, no. 1-2, pp. 216–230, 2003.

[2] K. S. Choi, J. Pan, and S. Ho, "Fatigue failures of rollers in crankshaft fillet rolling," SAE Technical Paper 2004-01-1498, 2004.

[3] K. S. Choi and J. Pan, "Simulations of stress distributions in crankshaft sections under fillet rolling and bending fatigue tests," *International Journal of Fatigue*, vol. 31, no. 3, pp. 544–557, 2009.

[4] K. H. Kloos and J. Adelmann, "Schwingfestigkeitssteigerung durch Festwalzen," *Materialwissenschaft und Werkstofftechnik*, vol. 19, no. 1, pp. 15–23, 1998.

[5] Y. C. Yen, P. Sartkulvanich, and T. Altan, "Finite element modeling of roller burnishing process," *CIRP Annals—Manufacturing Technology*, vol. 54, no. 1, pp. 237–240, 2005.

[6] P. Sartkulvanich, T. Altan, F. Jasso, and C. Rodriguez, "Finite element modeling of hard roller burnishing: an analysis on the effects of process parameters upon surface finish and residual stresses," *Journal of Manufacturing Science and Engineering, Transactions of the ASME*, vol. 129, no. 4, pp. 705–716, 2007.

[7] M. Y. Ali and J. Pan, "Effect of a deformable roller on residual stress distribution for elastic-plastic flat plate rolling under plane strain conditions," *SAE International Journal of Materials and Manufacturing*, vol. 5, no. 1, pp. 129–142, 2012.

[8] M. Guagliano and L. Vergani, "Residual stresses induced by deep rolling in notched components," *Metallurgical Science and Technology*, vol. 14, no. 2, pp. 61–68, 1996.

[9] K. S. Choi and J. Pan, "Effects of pressure-sensitive yielding on stress distributions in crankshaft sections under fillet rolling and bending fatigue tests," *International Journal of Fatigue*, vol. 31, no. 10, pp. 1588–1597, 2009.

[10] G. H. Majzoobi, K. Azadikhah, and J. Nemati, "The effects of deep rolling and shot peening on fretting fatigue resistance of Aluminum-7075-T6," *Materials Science and Engineering A*, vol. 516, no. 1-2, pp. 235–247, 2009.

[11] C. H. Fu, Y. B. Guo, J. McKinney, and X. T. Wei, "Process mechanics of low plasticity burnishing of nitinol alloy," *Journal of Materials Engineering and Performance*, vol. 21, no. 12, pp. 2607–2617, 2012.

[12] M. Sayahi, S. Sghaier, and H. Belhadjsalah, "Finite element analysis of ball burnishing process: comparisons between numerical results and experiments," *International Journal of Advanced Manufacturing Technology*, vol. 67, no. 5–8, pp. 1665–1673, 2013.

[13] F. Klocke, V. Bäcker, H. Wegner, and M. Zimmermann, "Finite element analysis of the roller burnishing process for fatigue resistance increase of engine components," *Proceedings of the Institution of Mechanical Engineers, Part B: Journal of Engineering Manufacture*, vol. 225, no. 1, pp. 2–11, 2011.

[14] R. M. Guo, "Investigation of roll bite behavior with various cold rolling conditions using semi-analytic solutions of von Karman's rolling equation," *La Metallurgia Italiana*, vol. 2014, no. 4, pp. 29–38, 2014.

[15] F. Mohammadi, R. Sedaghati, and A. Bonakdar, "Finite element analysis and design optimization of low plasticity burnishing process," *International Journal of Advanced Manufacturing Technology*, vol. 70, no. 5–8, pp. 1337–1354, 2014.

[16] F. D. Fischer and G. Schleinzer, "Residual stress formation and distortion of rail steel," in *Handbook of Residual Stress and Deformation of Steel*, G. E. Totten, M. Howes, and T. Inoue, Eds., pp. 424–436, ASM International, Material Park, Ohio, USA, 2002.

[17] L. Taylor, J. Cao, A. P. Karafillis, and M. C. Boyce, "Numerical simulations of sheet-metal forming," *Journal of Materials Processing Technology*, vol. 50, no. 1–4, pp. 168–179, 1995.

[18] SIMULIA, *ABAQUS User's Manual, Version 6.12*, SIMULIA, Providence, RI, USA, 2012.

[19] F. Klocke, *Manufacturing Processes, 1, Cutting*, Springer, Berlin, Germany, 2011.

[20] L. Li and N. He, "A FEA study on mechanisms of saw-tooth chip deformation in high speed cutting of Ti–6Al–4V alloy," in *Proceedings of the 5th International Conference on High Speed Machining*, pp. 759–767, 2006.

[21] H. L. Heathcote, "The ball bearing: in the making, under test and on service," *Proceedings of the Institute of Automobile Engineers*, vol. 15, no. 1, pp. 569–702, 1921.

[22] K. L. Johnson, *Contact Mechanics*, Cambridge University Press, New York, NY, USA, 1987.

[23] N. Tsuji, S. Tanaka, and T. Takasugi, "Evaluation of surface-modified Ti-6Al-4V alloy by combination of plasma-carburizing and deep-rolling," *Materials Science and Engineering A*, vol. 488, no. 1-2, pp. 139–145, 2008.

15

Experimental Investigation and Multiobjective Optimization of Turning Duplex Stainless Steels

Rastee D. Koyee,[1] Uwe Heisel,[1] Siegfried Schmauder,[2] and Rocco Eisseler[1]

[1] *Institute for Machine Tools, University of Stuttgart, Holzgartenstraße 17, 70174 Stuttgart, Germany*
[2] *IMWF, University of Stuttgart, Pfaffenwaldring 32, 70569 Stuttgart, Germany*

Correspondence should be addressed to Rastee D. Koyee; rastee.ali@ifw.uni-stuttgart.de

Academic Editor: Godfrey C. Onwubolu

This paper addresses experimental investigations of turning EN 1.4462 and EN 1.4410 duplex stainless steel grades with multilayer coated carbide inserts. Single-point wet and dry longitudinal turning tests of cylindrical bars are conducted; cutting forces, effective cutting powers, and tool wear are measured. The parametric influences of cutting speed, feed rate, and process conditions on the cutting performances such as resultant cutting force, specific effective cutting power, and flank wear are analyzed and proper conclusions are drawn. Nature-inspired metaheuristic bat algorithm is employed to handle the multiobjective optimization of the conflicting performances. Finally, the optimum cutting condition for each process condition can be selected from calculated Pareto optimal fronts by the user according to the planning requirements.

1. Introduction

Duplex stainless steels (DSSs) can be defined as a family of stainless steels whose structures are approximately 50% austenite and 50% ferrite, and its physical properties are a combination of the ferritic and the austenitic grades. In addition to their relatively low cost, they combine the best attributes of both austenitic and ferritic stainless steels which provide high strength and ductility with good resistance to corrosion (including stress corrosion cracking). Therefore, they are most commonly used when a combination of high mechanical strength and high corrosion resistance is required and are increasingly seen as an attractive alternative to the conventional stainless steels. However, owing to their high tensile and yield strength (roughly twice the yield strength of their counterpart austenitic grades, see Table 1), high work hardening rate, low thermal conductivity, high fracture toughness, strong tendency to form the built-up edge (BUE), and relatively high austenite and nitrogen content, modern duplex stainless steels are regarded as poorly machinable materials [1].

Over the past several years, few researchers have investigated the machining of duplex stainless steels. For instance, Bordinassi et al. studied the main effects of the turning in the superficial integrity of the duplex stainless steel ASTM A890-Gr6A. Their findings have shown that the smaller feed rate, smaller cutting speed, and the greater cutting depth provided the smaller values for the tensile residual stress, the smaller surface roughness, and the greater microhardness [2]. Królczyk et al. examined the influence of cutting parameters on surface roughness after DSS turning process. Their results have clearly showed that the feed rate was the main influencing factor on the surface roughness [3]. Królczyk et al. determined the coated carbide tool life and drew the tool wear curve when machining DSS. Their results have confirmed no effect of cutting speed and cooling on metallographic structure and designated the optimum cutting speed between 130 and 150 m/min and [4]. Nomani et al. have conducted machinability tests on duplex alloys SAF 2205 and SAF 2507, while employing austenite stainless steel 316L as a benchmark during drilling. Both duplex alloys displayed poorer machinability responses, with 2507 being worst [5]. Oliveira Jr. et al. have studied the turning operation of SAF 2507 and its influence on the alloy's corrosion resistance in practical applications. Their results have indicated that turning with PVD-coated inserts under high-pressure cooling resulted in

TABLE 1: Workpiece materials properties.

Chemical composition %weight	EN 1.4462	EN 1.4410
C	0.018	0.015
Cr	22.42	24.92
Ni	5.44	6.91
Mo	3.12	4.06
Mn	0.84	0.75
Si	0.37	0.25
N	0.18	0.3
P	0.025	0.021
S	0.0033	0.0007
Mechanical properties		
Yield strength (MPa)	514	579
Tensile (MPa)	737	826
Hardness (BHN)	212	236
Elongation (%)	41	40

long tool lives, good workpiece roughness, and high corrosion resistance of the material after machining. The most frequent wear mechanism found during the tests was notch wear, while the main tool wear mechanism was attrition [6]. Philip Selvaraj et al. have optimized dry turning parameters of two different grades of nitrogen alloyed duplex stainless steel by using Taguchi method. Their results revealed that the feed rate is the most significant parameter influencing the surface roughness and cutting force. On the other hand, the cutting speed was identified as the most significant parameter influencing the tool wear [7].

In this paper, multiobjective bat algorithm (MOBA) is applied to optimize the conflicting turning performances such as resultant cutting force, specific effective cutting power, and maximum flank wear. The algorithm suggests sets of optimum solutions for the parameter setting of DSS turning process. The models of conflicting performances were obtained from experimental data. The rest of this paper is organized as follows. The description of MOBA is introduced in Section 2. Discussions of the obtained experimental results are presented in Section 3. Modeling of performance characteristics is described in Section 4. Results of MOBA are shown in Section 5 and some conclusions are given in Section 6.

2. Multiobjective Bat Algorithm (MOBA)

Nature-inspired metaheuristic algorithms are based on strategies that try to imitate the behavior observed in species found in nature to update a population of candidates to solve optimization problems. They have contributed significantly to the development of new optimization techniques. These algorithms can be classified as population-based and trajectory-based. BA is a population-based swarm intelligence algorithm which is inspired by the echolocation of microbats. Echolocation is an advanced hearing based navigation system used by bats and some other animals to detect objects in their surroundings by emitting a sound to the environment. In general echolocation calls are characterized by three features, namely, pulse frequency, pulse emission rate, and loudness (intensity). In general the frequency φ in a range $[\varphi_{\min}, \varphi_{\max}]$ corresponds to a range of wavelengths $[\Omega_{\min}, \Omega_{\max}]$. For example, a frequency range of [20 kHz, 500 kHz] corresponds to a range of wavelengths from 0.7 mm to 17 mm in reality. Obviously, one can choose the ranges freely to suit different applications. In order to develop a standard bat algorithm, the following approximate or idealized rules are applied.

(a) All bats use echolocation to sense distance, and they also "know" the difference between food/prey and background barriers in some magical way.

(b) Bats fly randomly with velocity v_i at position x_i with a frequency $\varphi_{\min}$, varying wavelength, and loudness A_0 to search for prey. They can automatically adjust the wavelength (or frequency) of their emitted pulses and adjust the rate of pulse emission $r \in [0, 1]$, depending on the proximity of their target.

(c) Although the loudness can vary in many ways, we assume that the loudness varies from a large (positive) A_0 to a minimum constant value $A_{\min}$.

For the bats in simulations, we have to define the rules how their positions x_i and velocities v_i in a d-dimensional search space are updated. The new solutions x_{ti} and velocities v_{ti} at time step t are given by

$$\begin{aligned} f_i &= f_{\min} + (f_{\max} - f_{\min}) \beta, \\ v_i^{t+1} &= v_i^t + (x_i^t - x_*) f_i, \\ x_i^{t+1} &= x_i^t + v_i^t, \end{aligned} \tag{1}$$

where $\beta \in [0, 1]$ is a random vector drawn from a uniform distribution. Here x_* is the current global best location (solution) which is located after comparing all the solutions among all the n bats at each iteration t. Based on the above approximations and idealization, the basic steps of the multiobjective bat algorithm (MOBA) can be summarized as the pseudocode shown in Algorithm 1.

2.1. Pareto Optimality. The concept of Pareto optimum was formulated by Vilfredo Pareto in the XIX century and constitutes by itself the origin of research in multiobjective optimization. A solution vector $u = (u_1, \dots, u_n)^T \in F$ is said to dominate another vector $v = (v_1, \dots, v_n)^T$ if and only if $u_i \le v_i$ for all $i \in \{1, \dots, n\}$ and $\exists i \in \{1, \dots, n\} : u_i < v_i$. In other words, no component of u is larger than the corresponding component of v, and at least one component is smaller. Similarly, we can define another dominance relationship $\preccurlyeq$ by

$$u \preceq v \Longleftrightarrow u \prec v \vee u = v. \tag{2}$$

It is worth pointing out that, for maximization problems, the dominance can be defined by replacing $\prec$ with $\succ$. Therefore, a point $x_* \in F$ is called a nondominated solution if no solution can be found that dominates it [8]. The Pareto front PF of

```
Objective functions f_1(x), ..., f_K(x), x = (x_1, ..., x_d)^T
Initialize the bat population x_i (i = 1, 2, ..., n) and v_i
for j = 1 to N (points on Pareto fronts)
    while (t < Max number of iterations)
        Generate new solutions and update by (1)
        if (rand > r_i)
            Random walk around a selected best solution
        end if
        Generate a new solution by flying randomly
        if (rand < A_i & f(x_i) < f(x_*))
            Accept the new solutions,
            and increase r_i & reduce A_i
        end if
        Rank the bats and find the current best x_*
    end while
    Record x_* as a non-dominated solution
end
Postprocess results and visualization
```

Algorithm 1: Multiobjective bat algorithm (MOBA) [10].

a multiobjective can be defined as the set of nondominated solutions so that

$$\mathrm{PF} = \{s \in S \mid \nexists s' \in S : s' \prec s\} \tag{3}$$

or in terms of the Pareto optimal set in the search space

$$\mathrm{PF}^* = \{x \in F \mid \nexists x' \in F : G(x') \prec G(x)\}, \tag{4}$$

where $G = (G_1, \ldots, G_K)^T$. To obtain a good approximation to Pareto front, a diverse range of solutions should be generated using efficient techniques [9].

3. Materials and Methods

Full-factorial cutting tests are carried out on a CNC lathe (CTX 420 Linear V5) with maximum drive power 25 kW and a speed range of 35–7000 rpm. The workpiece materials for turning tests were standard DSS EN 1.4462 and super DSS EN 1.4410. Their chemical compositions and mechanical properties are given in Table 1.

The cutting tools used in the tests were made by Sandvik Coromant AB, Sweden. The solid carbide inserts were of type CNMG 120408-MM 2025. The coating consists of 5.5 μm CVD TiCN-Al_2O_3-TiN layers on a substrate which features excellent resistance to both mechanical and thermal shock. The triple coating consists of a 2.5 μm thick TiCN at the bottom followed by a 2 μm multilayer TiN/Al_2O_3 at the middle and finally a thin (~1 μm) outer coating of TiN on top. The inserts were mounted on a right hand style PCLNL-2525M-12 ISO type tool holder with tool geometry as follows: including angle = 80°, back rake angle = −6°, clearance angle = 5°, approach angle = 95°, and cutting edge radius ≈ 40 μm. The cutting and process conditions are given in Table 2. It has been designed to study the effects of the workpiece materials,

Table 2: Cutting and process conditions.

Cutting speed, m/min	100, 180
Feed rate, mm/rev	0.15, 0.2, 0.25, 0.3, 0.35, 0.4
Cutting depth, mm	1
Process condition	Dry, wet

process, and cutting conditions on the constants in Kienzle's cutting force model.

The width of maximum flank wear (VB_{max}) was periodically measured during progressive tool wear experiments using an optical microscope after every cutting pass of machining operation. The criterion for tool life was VB_{max} = 0.6 mm or catastrophic tool failure of the tool edge. During the turning tests, the main cutting force (F_c), axial force (F_a), and radial force (F_r) were measured using a Kistler type 9129A three-component piezoelectric dynamometer, which was connected to a charged Kistler type 5070A amplifier and personal computer through an analog to digital converter card. The resultant cutting force is calculated using

$$R = \sqrt{F_c^2 + F_a^2 + F_r^2}. \tag{5}$$

Thereafter, specific cutting pressures such as main k_c, feed k_f, and radial k_r are calculated using the following equations:

$$k_c = \frac{F_c}{fa_p}, \quad k_a = \frac{F_a}{fa_p}, \quad k_r = \frac{F_r}{fa_p}, \tag{6}$$

where f is the feed rate and a_p is the cutting depth. The effective cutting power ($P_{\text{eff.}}$) component of the motor is always proportional to the torque emitted by the motor; therefore it is often used as a signal within the control system for quantifying the motor load. During the course of each cutting test, the electrical current consumed by the machine tool was measured from the external effective-power modules which were installed between the frequency convertor and the motor. In the present study, the range of energy requirements at the drive motor of the machine tool while turning DSS is estimated using the specific effective cutting power which can be calculated using the following formula:

$$P_s = \frac{P_{\text{eff.}}}{v_c f a_p} = \frac{\sqrt{3} UI \cos\varphi}{v_c f a_p}, \tag{7}$$

where U is the line voltage in volts, I is the line current in ampere, $\cos\varphi$ is the power factor, and v_c is the cutting speed. The experimental setup is shown in Figure 1. In the present research, high-resolution thermographic camera of brand VarioCam head Hires 640 by InfraTec is used to film the chip formation and measure the chip temperature at a resolution of 640 × 480 pixels.

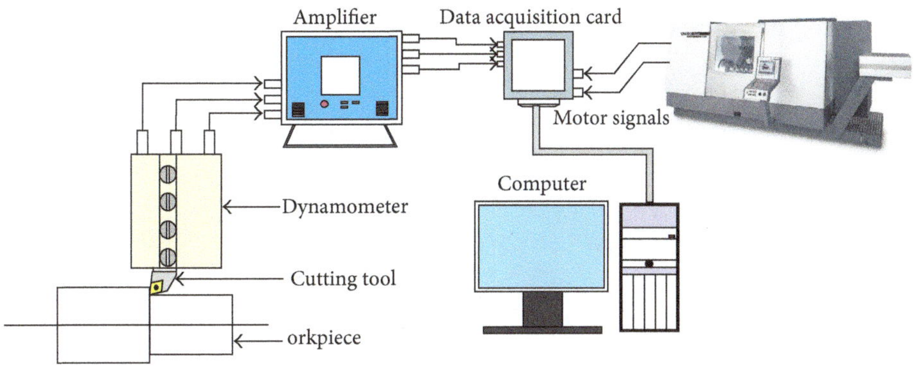

FIGURE 1: Diagram of the experimental setup.

TABLE 3: Summary of the models coefficients.

Model	Proc. cond.	EN 1.4462			EN 1.4410		
		$k_{1.1}$	a	b	$k_{1.1}$	a	b
F_c	Dry	2367	−0.28	0.1	2920	−0.29	0.08
	Wet	2215	−0.26	0.13	2471	−0.26	0.19
F_f	Dry	852	−0.48	0.47	1182	−0.41	0.47
	Wet	737	−0.49	0.53	1126	−0.64	0.5
F_r	Dry	741	−0.27	0.23	921	−0.22	0.32
	Wet	899	−0.46	0.18	908	−0.23	0.3
R	Dry	2550	−0.32	0.2	3145	−0.31	0.21
	Wet	2431	−0.32	0.22	2811	−0.35	0.28

4. Results and Discussion

4.1. Specific Cutting Pressure. The specific cutting pressure is often considered as an indication of the machinability of a given work material. The higher the specific cutting pressure is, the lower the machinability of the work material is. The specific cutting pressures for turning EN 1.4462 and EN 1.4410 were calculated from the cutting data employing (6) and the results are shown in Figure 2. Based on a quick review of the results,

(1) no drastic difference between the cutting pressures of dry and wet conditions was observed. However, overall wet machining shows an improvement in the machining performance through lower cutting pressures;

(2) specific cutting pressures when machining EN 1.4462 DSS were seen lower than the specific cutting pressures when machining EN 1.4410. Therefore, it can be predicted that the machinability index EN 1.4410 will be smaller than the machinability index of EN 1.4462;

(3) the cutting pressures generally showed a decreasing trend with increasing cutting speed and feed rate (see Figure 2). As it can be seen in this figure, the maximum cutting pressures occur at low cutting speeds and low feed rates. Thus, in order to achieve low cutting pressures, machining process should be done in high cutting speeds and high feed rates;

(4) modified Kienzle formula can be applied to evaluate the parametric effects of cutting conditions on the cutting forces as follows:

$$F = k_{1.1} \left(\frac{v_c}{100} \right)^a (f)^{(1-b)}, \quad (8)$$

where F represents the described cutting and resultant forces, $k_{1.1}$ is the specific cutting pressure for $1\,\text{mm}^2$ cross-sectional area of the cut, and a and b are constants. The model fitting has employed the least square method. Summary of models coefficients is listed in Table 3.

To avoid misleading conclusions, the adequacy of the fitted models should be checked. Therefore, analysis of variance (ANOVA) for 95% a level of confidence was performed in order to estimate the predictive accuracy of the models and to determine the relative significances of the different factors using Matlab. From the ANOVA shown in Table 4, it is apparent that almost all correlation coefficients are near to 1, showing that significant terms have been included in the model and that the model is capable of predicting the responses, P value estimators are all close to zero (much lower than 0.05 corresponding to the confidence interval) which show the significant effect of factors on the corresponding response, and F-values are much larger than 1, which indicate that factors have a significant effect on the response. Moreover, variations of cutting force, feed force, and radial force towards cutting speed and feed rate are studied in dry and wet machining

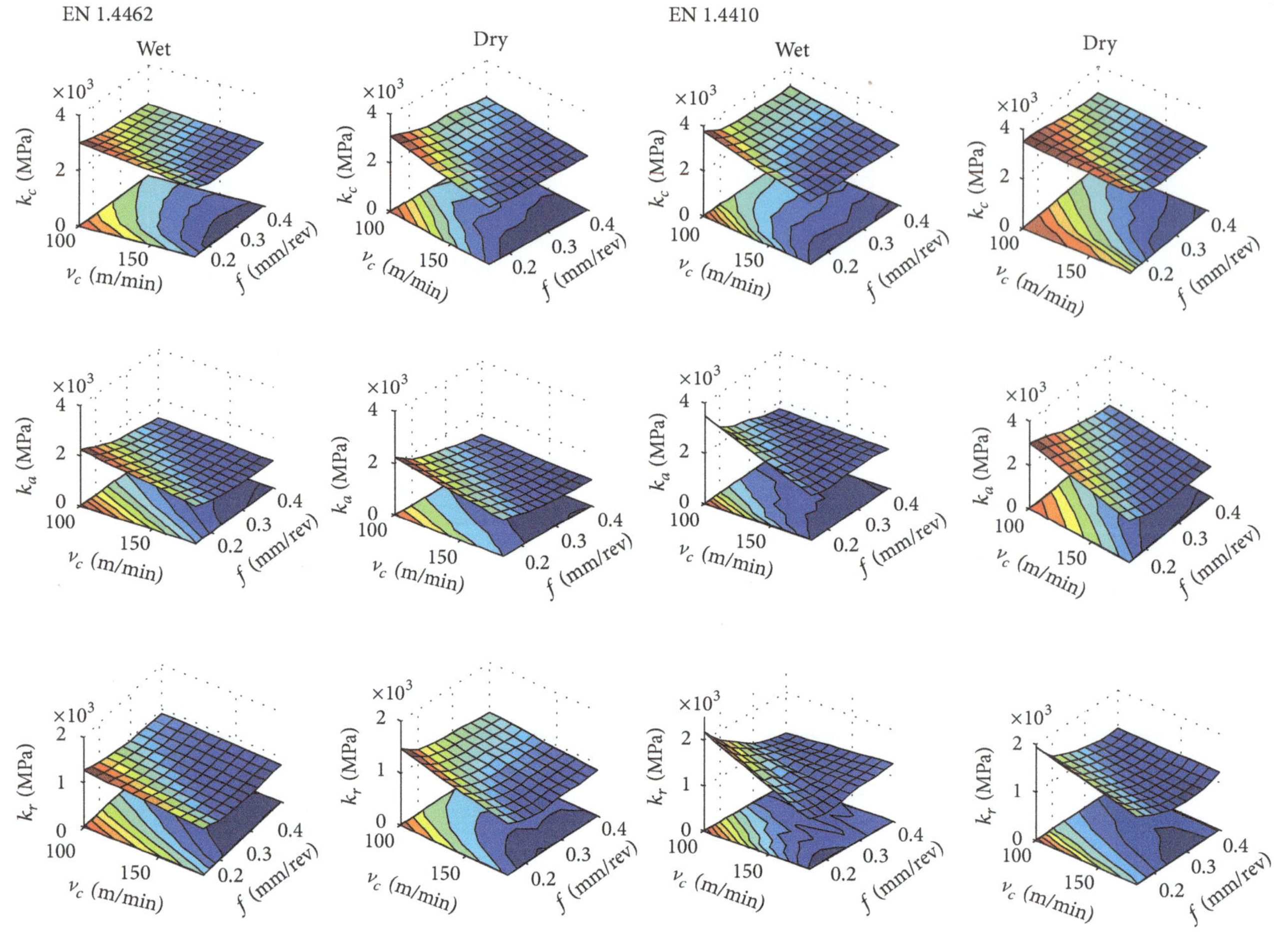

FIGURE 2: Computed specific cutting pressures.

TABLE 4: Analysis of variance for cutting force models.

Proc. cond.	EN 1.4462				EN 1.4410			
	R^2	$R^2_{adj.}$	F-value	P value	R^2	$R^2_{adj.}$	F-value	P value
Dry	0.99	0.99	8430	$7.8E-16$	0.99	0.99	5310	$6.25E-15$
Wet	0.99	0.98	4010	$2.2E-14$	0.99	0.99	7900	$1.05E-15$
Dry	0.96	0.95	1660	$1.2E-12$	0.84	0.80	350	$1.24E-09$
Wet	0.94	0.92	1130	$6.6E-12$	0.95	0.93	949	$1.43E-11$
Dry	0.97	0.96	1720	$9.9E-13$	0.73	0.67	159	$4.14E-08$
Wet	0.96	0.95	1060	$8.8E-12$	0.88	0.85	457	$3.76E-10$
Dry	0.98	0.98	3870	$2.6E-14$	0.96	0.95	1120	$6.73E-12$
Wet	0.98	0.97	2250	$3E-13$	0.98	0.97	2850	$1.02E-13$

separately. It should be noted that the average $k_{1.1}$ at wet cutting is generally seen 5% lower than $k_{1.1}$ at dry cutting and $k_{1.1}$ of turning EN 1.4410 is generally 25.281% higher than $k_{1.1}$ of turning EN 1.4462. These results clearly indicate the impact of process conditions and workpiece materials on the amount of energy required to cut 1 mm^2 chip cross-sections at 100 m/min.

4.2. Specific Effective Cutting Power (P_s). It has been suggested that more than 90% of environmental impact from machine tools is due to electrical energy consumption [11]. Therefore, minimizing this consumption is supposed to significantly improve the sustainability performance of manufacturing systems. To achieve this, the estimation of the energy consumed in machining operation should be accomplished through calculation of the specific effective cutting power (see (7)), which can be defined as the amount of the machine-tool energy required to remove a certain volume of the material. Beside the consumed cutting energy in chip formation, it is comprised of the energy consumed by varieties of machine-tool functions such as workpiece handling and movement, cutting fluid system, chip removal, and tool changing.

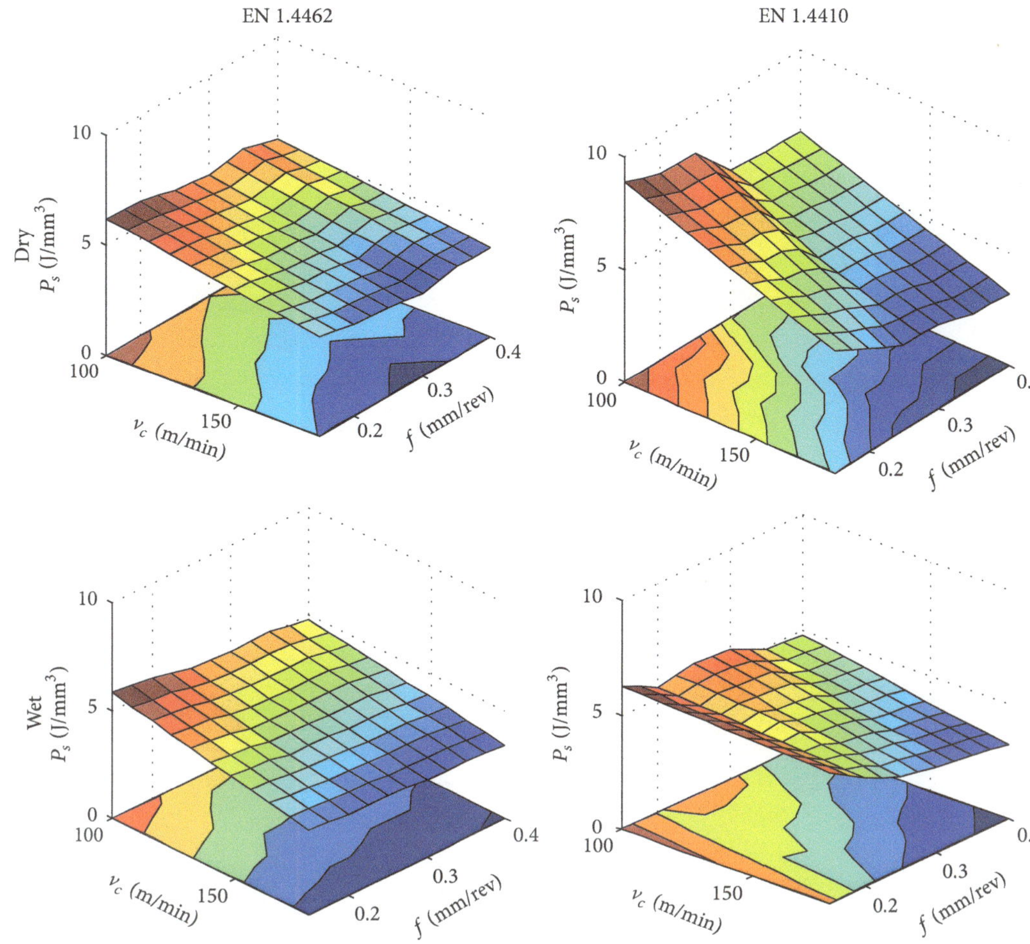

FIGURE 3: Specific effective cutting power.

The specific effective cutting power is supposed to depend on the machinability of the work material, which takes into account the material properties, process conditions, and cutting conditions. The influences of cutting parameters v_c and f and process conditions on the specific effective cutting power are shown in Figure 3, from which, the following observations can be made.

(1) Specific effective cutting power is inversely proportional to the cutting speed and feed rate.

(2) Compared to the dry cutting, the wet cutting of EN 1.4462 and EN 1.4410 has shown lower values of P_s by 10.82% and 18.81%, respectively. This is mainly attributed to the lower cutting forces in wet cutting. Therefore, the advantage of employing wet cutting when machining DSSs overtakes its disadvantage of being more energy consumable due to the operation of the coolant pump.

(3) Average P_s when machining EN 1.4410 is 12.3% higher than the average P_s when machining EN 1.4462. This result is expected as the latter has inferior mechanical properties and higher percentage of chip breaking constituents such as sulfur and phosphorus (see Table 1).

(4) The effect of the process parameters on the P_s results can be modelled using the nonlinear model:

$$P_s = C_1 \left(\frac{v_c}{100}\right)^{C_2} (f)^{(1-C_3)}, \quad (9)$$

where C_{1-3} are the model coefficients. The values of model coefficients and the check of adequacy of the models for each turning case are shown in Table 5. It can be seen from this table that the R^2 values are high and close to 1, which are usually desirable. The large model F-values and small mode P values imply that the models are significant. Therefore, results from the statistical analyses indicate that the developed mathematical models can be successfully applied for predicting the P_s.

(5) The mean specific cutting pressure can be estimated by the equation:

$$E_c = \frac{k_c + k_a + k_r}{1000} \text{ in (J/mm}^3\text{)} \quad (10)$$

when cutting EN 1.4462 and EN 1.4410 were found to be 79.7% and 60.72% higher than corresponding P_s values (see Figure 4).

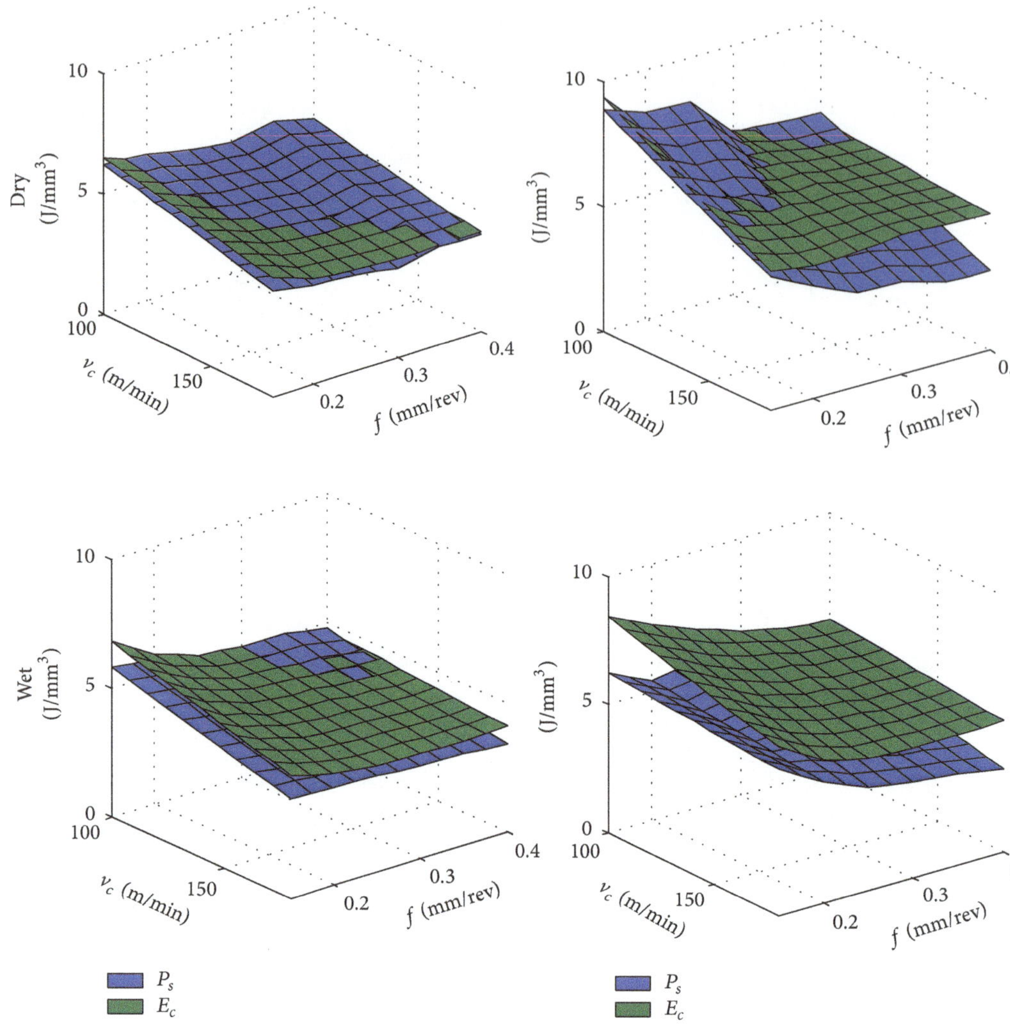

FIGURE 4: Comparison between P_s and E_c.

(6) The comparison between the effective cutting power ($P_{eff.}$) and the overall consumed cutting power (P_c) (see Figure 5) calculated by (11) has revealed that the average $P_{eff.}$ is 8.965% and 33.3195% higher than average P_c for cutting both EN 1.4462 and EN 1.4410, respectively. Results have also showed that the average P_c in wet cutting EN 1.4462 and EN 1.4410 were 2.9931% and 8.27% lower than the corresponding dry cutting:

$$P_c = \left(\frac{v_c}{60}\right)\left(F_c + \frac{f}{\pi D}F_a\right) \text{ in (J/mm}^3\text{)}. \quad (11)$$

4.3. The Width of Maximum Flank Wear (VB_{max}). The impacts of cutting parameters v_c and f and process conditions on the VB_{max} are shown in Figure 6 and the summary of the findings is presented below.

(1) Based on the observations of tool wear behavior with respect to time during dry machining of DSSs, EN 1.4410 is considered more difficult-to-machine than EN 1.4462. For instance, machining EN 1.4410 at cutting speed value of 180 m/min and feed rate value of 0.35 mm/rev has resulted in significantly higher wear rate and thus shorter tool life, higher cutting temperature, higher cutting forces, and higher energy consumption than machining EN 1.4462 (see Figure 7).

TABLE 5: P_s model coefficients and check of the models adequacy.

Coeff.	EN 1.4462		EN 1.4410	
	Dry	Wet	Dry	Wet
C_1	4.859	4.217	4.183	3.329
C_2	−0.595	−0.631	−1.055	−0.445
C_3	1.128	1.163	1.421	1.360
Check of models adequacy				
R^2	0.972	0.989	0.973	0.874
$R^2_{adj.}$	0.966	0.987	0.968	0.845
F-value	$3.35E+03$	$7.70E+03$	$1.06E+03$	$6.29E+04$
P value	$4.99E-14$	$1.17E-15$	$8.58E-12$	$9.05E-11$

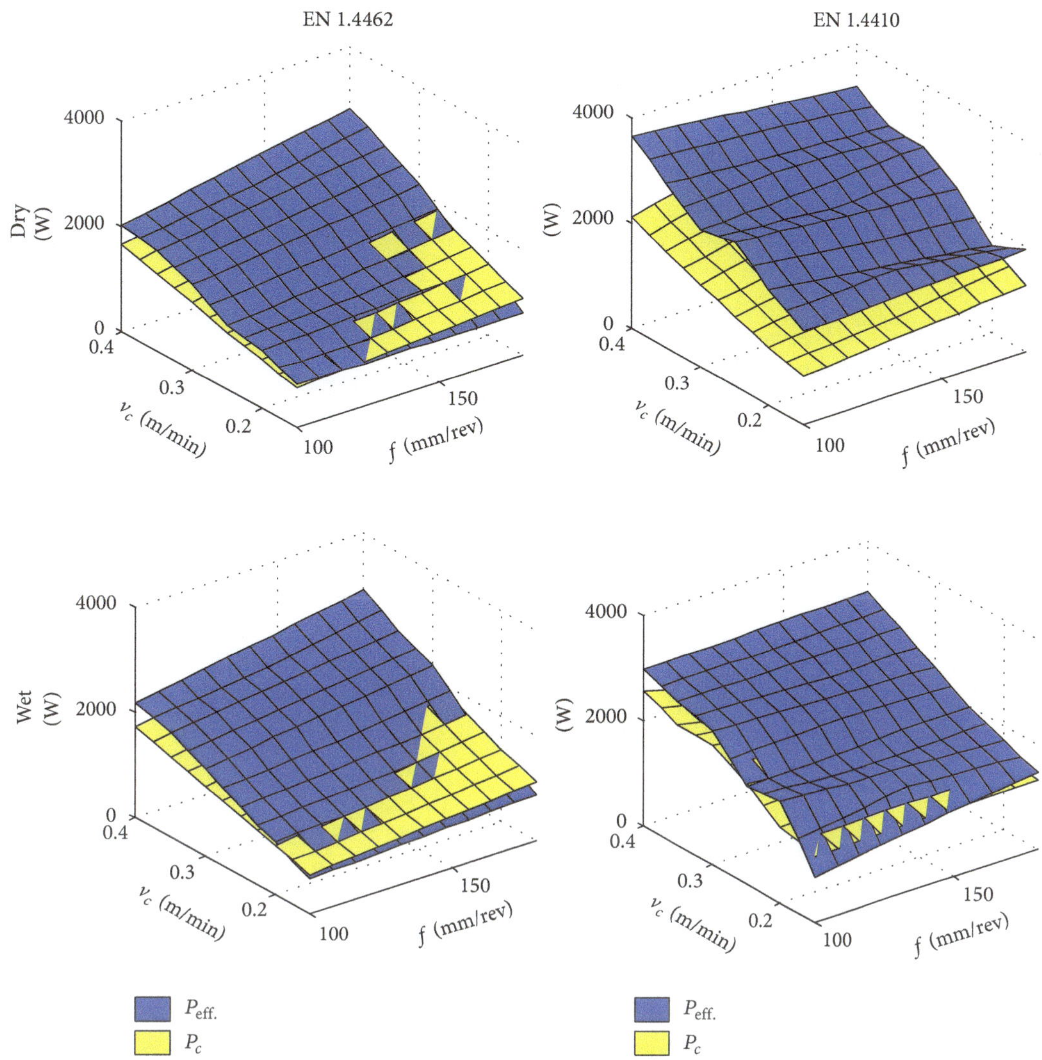

FIGURE 5: Comparison between P_c and $P_{eff.}$.

Figure 8 shows the average value of recorded cutting forces in dry conditions over 60 sec and 320 sec of machining EN 1.4462 and 20 sec and 40 sec of machining EN 1.4410.

An increasing trend in all cutting forces (main, feed, and radial) over time was observed. However, the largest percentage of cutting force increase was recorded in radial cutting force with 32.933% and 164.34% for machining EN 1.4462 and EN 1.4410, respectively. From this perspective, the most suitable parameter to monitor and correlate with progressive tool wear during machining is the radial cutting force component.

(2) The illustration of typical wear forms when machining DSSs are shown in Figure 9. The following can be seen.

(a) Severe adhesion between the chip and the rake face of the tool (BUE) was visible throughout the worn crater area. This is believed to be attributed to the high pressure and temperature encountered in machining ductile materials such as DSSs.

(b) Regardless of the adopted process conditions, chipping of the cutting edge at high feed rate was often observed, which has significantly contributed to acceleration of tool failure. Heavy load and impact when the tool entered and/or extracted the workpiece are considered to be the main reasons behind this phenomenon.

(c) The most dominant tool wear mode under low cutting speed conditions was the notch wear which was typically located near the depth of cut line. These notches act as stress and temperature raisers of the already high mechanical strength, strain hardening rate, fracture toughness, and low conductivity DSSs chips. Flaking occurrence because of the concentrated thermal load

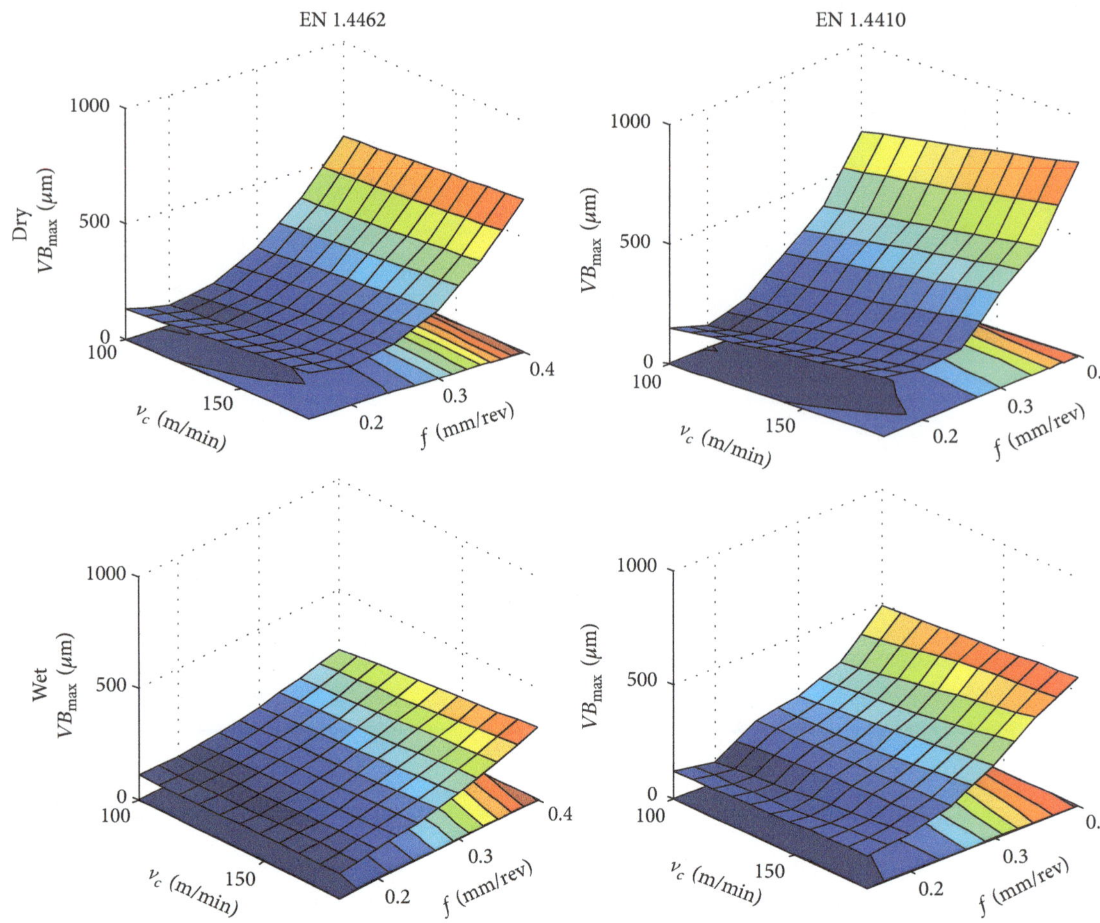

FIGURE 6: The width of maximum flank wear (VB_{max}).

is more possible there. Ultimately, the combination of notch wear and flaking caused the cutting edge to fail abruptly.

(d) Another form of the main cutting edge damage is caused by the unfavorable chip morphology and chip flow (see Figure 10). Cutting DSSs at low cutting speed and feed rate has contributed to the formation of strong ribbon and snarled chips with dominant side-curl flow. The chips were entangled around the cutting tool, tool post, and workpiece and damaged to the cutting edge and the workpiece surface. The damage is often propagated along the main cutting edge of the tool with cutting time and had exceeded 5 mm length of damage in feed rate ranges of 0.15–0.20 mm/rev.

(e) In addition to the combined notching and flaking effects, nearly equal proportions of soft ferrite and hard austenite grains in the DSS structure make the cutting tool alternate cutting between soft and hard grains; this leads to an automatic tendency to initiate chatter in the cutting system and promote the catastrophic failure of the cutting tools.

(3) To model VB_{max}, an empirical formula described by (8) is applied as follows:

$$VB_{max} = \lambda_1 \left(\frac{v_c}{100}\right)^{\lambda_2} (f)^{(1-\lambda_3)}, \quad (12)$$

where λ_{1-3} are model constants. To check the adequacy of derived models, ANOVA for 95% confidence interval has been applied. Table 6 summarizes the values of coefficients and adequacy criterion. It can be seen that high correlation coefficients are existing between the experimental and predicted VB_{max} values and high model F-values, and low model P values ($\ll 0.05$) confirm the significance of the models.

5. Multiobjective Optimization of Turning DSSs

The objective of the present study is to simultaneously minimize the resultant cutting force, the specific effective cutting power, and the width of maximum flank wear. The mathematical formulation of the current optimization problem can be stated as follows:

$$\text{Min} : R(v_c, f),$$

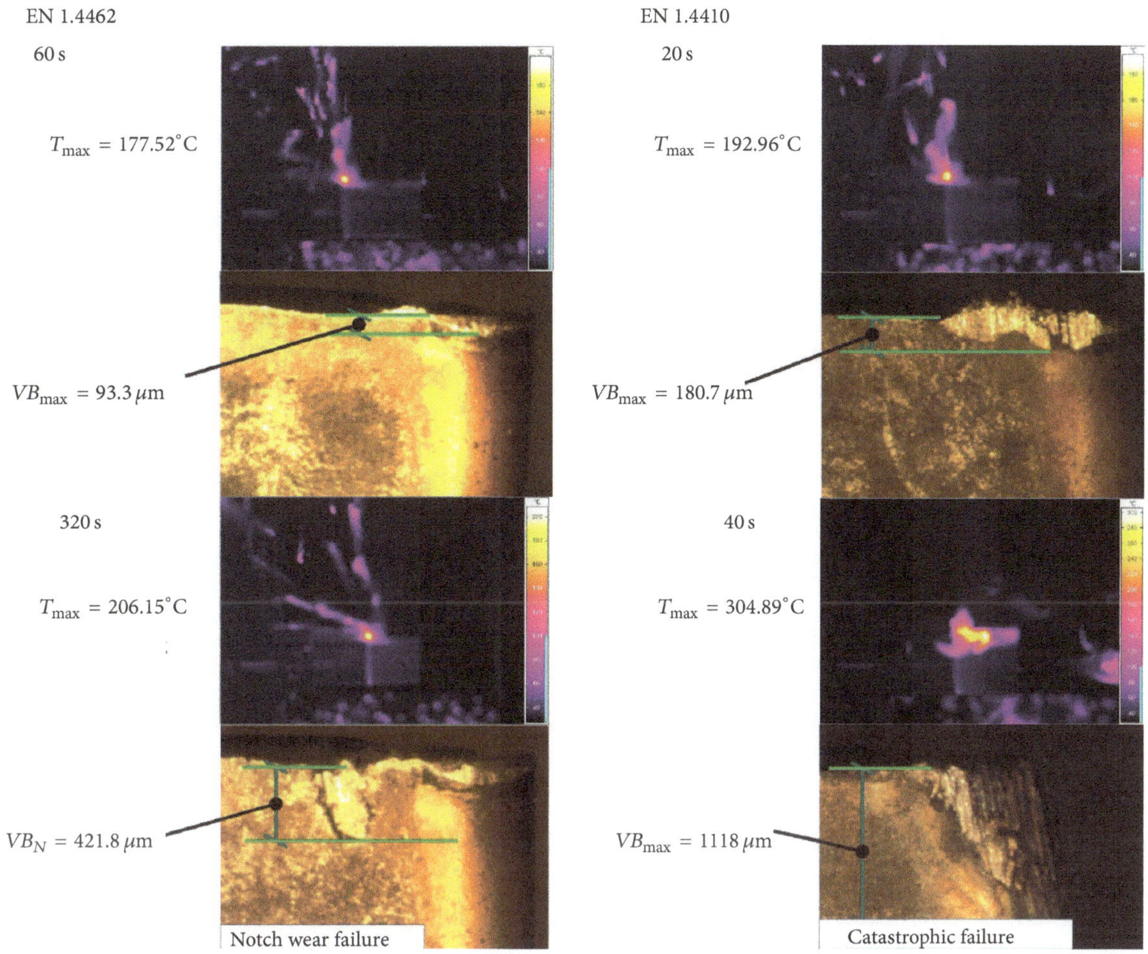

FIGURE 7: Progression of the maximum chip temperature and tool wear pattern with cutting time in dry cutting of DSSs.

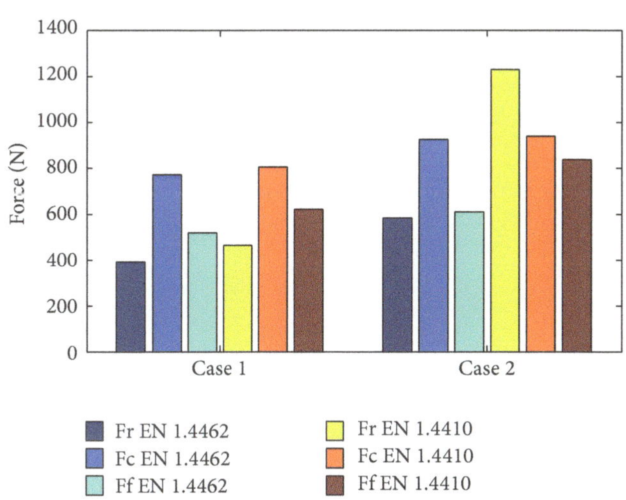

FIGURE 8: Variation of cutting forces with respect to the cutting time.

$$\text{Min}: P_s(v_c, f),$$
$$\text{Min}: VB_{max}(v_c, f), \quad (13)$$

TABLE 6: VB_{max} models coefficients and check of their adequacies.

Coeff.	EN 1.4462		EN 1.4410	
	Dry	Wet	Dry	Wet
λ_1	4618.4	654.091	4473.3	2369.6
λ_2	0.29156	0.49853	0.50425	0.40716
λ_3	−1.335	−0.15483	−1.236	−0.8276
Check of models adequacy				
R^2	0.908	0.947	0.904	0.957
$R^2_{adj.}$	0.888	0.936	0.882	0.948
F-value	$1.17E+02$	$4.55E+02$	$1.10E+02$	$3.15E+02$
P value	$1.56E-07$	$3.84E-10$	$2.04E-07$	$1.98E-09$

subjected to the following constraints:

(1) arithmetic average roughness (R_a)

$$R_a = \frac{0.032 f^2}{r_\varepsilon} \leq 2\,\mu\text{m}, \quad (14)$$

where r_ε is the tool nose radius (r_ε = 0.8 mm);

TABLE 7: Optimal solutions obtained by MOBA.

EN 1.4462										EN 1.4410									
Dry					Wet					Dry					Wet				
v_c (m/min)	f (mm/rev)	R (N)	P_s (J/mm^3)	VB_{max} (μm)	v_c (m/min)	f (mm/rev)	R (N)	P_s (J/mm^3)	VB_{max} (μm)	v_c (m/min)	f (mm/rev)	R (N)	P_s (J/mm^3)	VB_{max} (μm)	v_c (m/min)	f (mm/rev)	R (N)	P_s (J/mm^3)	VB_{max} (μm)
100.000	0.150	555.092	6.240	55.038	100.002	0.150	552.772	5.745	73.142	100.000	0.150	707.119	9.297	64.323	100.000	0.150	723.222	6.591	73.943
180.000	0.150	459.186	4.399	65.327	178.925	0.150	457.782	3.980	97.786	180.000	0.150	588.395	5.001	86.423	180.000	0.150	590.268	5.074	93.937
179.809	0.227	640.756	4.174	171.766	100.002	0.150	552.772	5.745	73.142	142.126	0.150	633.924	6.414	76.894	180.000	0.227	793.979	4.371	200.299
118.334	0.150	526.421	5.644	58.024	179.999	0.227	631.188	3.706	158.178	179.642	0.227	815.592	4.209	218.033	109.167	0.151	706.220	6.318	77.919
100.000	0.150	555.092	6.240	55.038	179.842	0.163	487.654	3.913	107.908	118.837	0.150	670.395	7.747	70.277	115.621	0.151	690.684	6.165	79.277
179.729	0.212	606.666	4.212	146.460	111.331	0.151	535.995	5.364	77.620	179.329	0.166	638.255	4.810	108.341	134.366	0.150	654.234	5.773	83.786
170.492	0.150	467.298	4.543	64.301	179.874	0.227	631.330	3.708	158.123	179.568	0.226	813.246	4.218	216.130	100.000	0.150	723.222	6.591	73.943
179.832	0.169	504.905	4.335	85.972	179.672	0.175	515.706	3.871	117.106	179.173	0.221	800.248	4.266	205.824	179.955	0.213	759.780	4.469	178.940
157.628	0.150	480.225	4.759	63.208	179.251	0.218	612.239	3.741	150.590	179.268	0.207	757.641	4.390	176.307	179.747	0.191	702.116	4.654	146.053
179.888	0.205	590.144	4.228	135.316	165.634	0.151	471.164	4.175	94.621	151.814	0.150	620.582	5.985	79.337	179.967	0.159	616.636	4.964	105.004
179.839	0.156	473.085	4.326	71.161	178.925	0.154	467.061	3.963	100.732	179.280	0.202	745.058	4.429	168.127	131.197	0.151	660.861	5.830	83.364
101.749	0.151	553.565	6.174	55.776	132.912	0.153	512.025	4.785	86.274	135.041	0.151	645.543	6.762	75.408	178.007	0.199	725.385	4.606	156.756
114.216	0.150	532.589	5.764	57.465	102.988	0.150	547.537	5.639	74.226	173.716	0.150	595.388	5.190	85.064	180.000	0.158	613.579	4.976	103.704
179.929	0.176	523.372	4.309	95.492	179.916	0.220	616.011	3.726	152.516	159.154	0.150	611.486	5.694	81.241	124.713	0.150	671.418	5.968	81.315
179.923	0.215	612.819	4.202	151.017	179.477	0.198	568.806	3.795	135.229	114.730	0.150	677.709	8.040	69.017	179.931	0.223	783.550	4.401	193.553
145.718	0.151	493.871	4.984	62.257	172.328	0.151	465.442	4.071	96.599	179.226	0.192	714.582	4.531	149.243	179.955	0.221	779.580	4.412	191.091
179.907	0.227	640.697	4.173	171.836	114.327	0.152	533.722	5.270	79.164	179.368	0.214	779.816	4.320	191.524	179.893	0.218	771.685	4.435	186.105
179.808	0.199	577.399	4.244	126.930	179.547	0.159	479.353	3.932	105.034	179.642	0.227	815.592	4.209	218.033	179.976	0.207	743.816	4.517	169.519
134.048	0.151	506.443	5.240	60.444	179.453	0.195	561.494	3.806	132.648	179.632	0.158	613.573	4.902	97.078	179.966	0.194	709.180	4.627	150.071
179.903	0.152	463.419	4.394	67.048	178.956	0.172	508.174	3.893	114.138	179.835	0.172	656.062	4.723	117.616	179.498	0.201	729.354	4.569	160.676
179.768	0.211	603.537	4.215	144.314	153.385	0.152	486.718	4.375	92.087	127.549	0.150	656.158	7.187	72.956	179.544	0.225	788.856	4.392	196.372
123.923	0.150	518.011	5.493	58.604	139.497	0.152	500.678	4.648	87.505	131.156	0.150	650.430	6.979	73.975	178.354	0.216	769.510	4.465	182.732
128.672	0.151	514.614	5.366	60.214	179.737	0.184	535.372	3.840	123.815	178.196	0.199	738.674	4.482	162.685	179.794	0.199	722.278	4.587	157.058
179.729	0.196	570.192	4.254	122.316	179.272	0.151	459.954	3.970	98.660	178.573	0.188	704.297	4.586	142.492	179.070	0.212	757.201	4.491	176.267
179.925	0.204	588.777	4.229	134.442	130.574	0.151	510.221	4.849	84.347	178.995	0.204	749.654	4.423	170.712	155.898	0.150	620.335	5.409	88.596
109.829	0.150	538.550	5.902	56.563	122.953	0.152	521.943	5.033	82.244	109.399	0.150	687.638	8.456	67.322	118.672	0.151	684.428	6.095	80.102
179.924	0.188	551.210	4.274	111.005	179.727	0.209	592.598	3.759	143.873	179.112	0.182	687.429	4.629	133.563	175.593	0.169	649.518	4.910	116.157
179.444	0.172	512.779	4.331	89.687	174.712	0.151	462.912	4.037	97.120	154.531	0.150	617.407	5.873	80.142	180.000	0.153	597.737	5.042	97.002
179.933	0.202	582.598	4.236	130.391	178.781	0.192	556.182	3.823	130.316	100.000	0.150	707.119	9.297	64.323	179.989	0.173	652.387	4.825	121.279
179.551	0.198	574.023	4.252	124.567	179.838	0.202	575.613	3.780	137.899	179.742	0.218	789.326	4.281	198.813	167.869	0.150	604.674	5.234	91.306
179.810	0.220	625.221	4.191	159.945	179.481	0.167	496.658	3.904	110.650	106.371	0.151	695.876	8.695	66.973	104.371	0.150	713.438	6.462	75.467
179.920	0.163	491.388	4.353	79.502	179.821	0.179	524.095	3.855	120.031	178.411	0.179	678.856	4.683	128.181	179.955	0.206	739.773	4.530	167.150
179.923	0.215	613.379	4.202	151.418	114.521	0.150	529.447	5.273	78.358	179.461	0.209	765.346	4.360	181.724	180.000	0.163	627.234	4.921	109.701
179.954	0.187	548.298	4.277	109.332	179.889	0.214	603.031	3.743	147.766	179.298	0.190	709.018	4.547	146.046	179.722	0.188	694.634	4.679	142.086

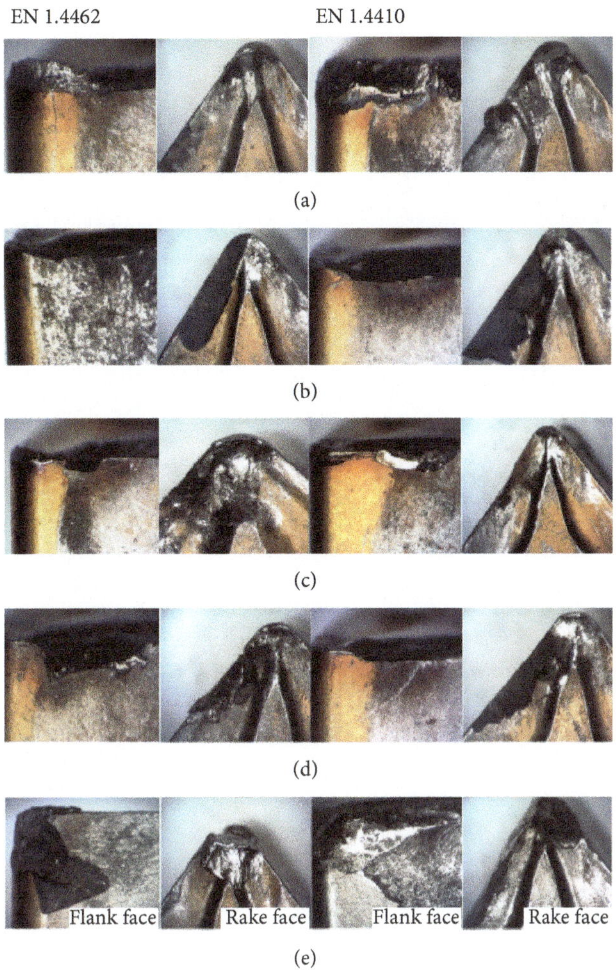

FIGURE 9: Wear patterns observed on the flank and rake face of cutting tools under dry and wet machining of DSSs.

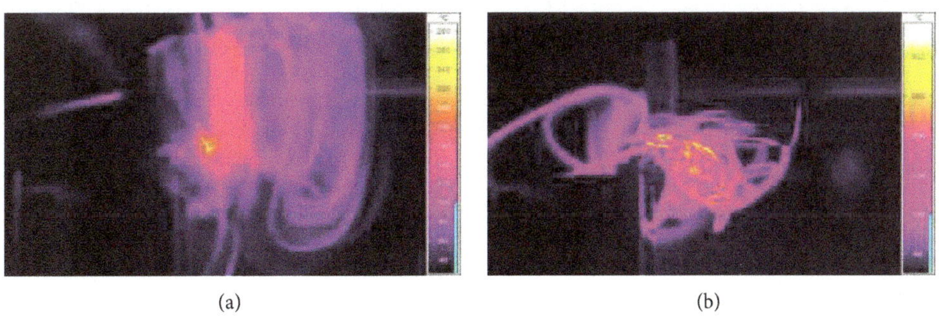

FIGURE 10: Entanglement of chips around the cutting tool and workpiece at $v_c = 100\,\text{m/min}$ and $f = 0.15\,\text{mm/rev}$: (a) EN 1.4462, (b) EN 1.4410.

(2) cutting parameters lower and upper bounds:

$$100 \le v_c \le 180\,\text{m/min},$$
$$0.15 \le f \le 0.4\,\text{mm/rev}. \tag{15}$$

The proposed MOBA is implemented in Matlab. The computing time was less than a minute, depending on the iteration number and population size. It was found that the best population size (n), loudness reduction (α), and pulse reduction rate (γ) were 40, 0.8, and 0.8, respectively. Optimization results have shown that MOBA is very efficient and consistently converges to the sets of optimal solutions. Figure 11 shows the Pareto optimal frontier points for different process conditions, at which the designers can determine the final solutions depending on their preferences. The final optimum v_c and f and their corresponding R, P_s, and $VB_{\max}$ are shown in Table 7. These process parameters are the Pareto optimal process parameters that should be shown to the decision

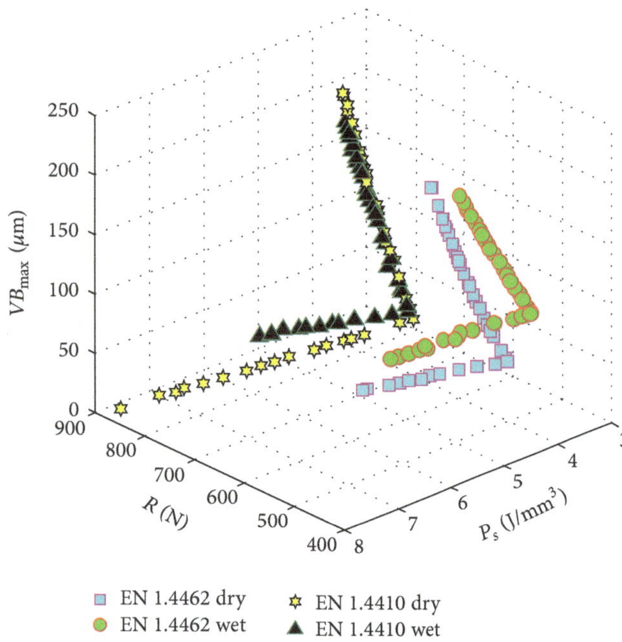

FIGURE 11: Pareto front points.

maker to simultaneously achieve the desired objectives in turning of the DSSs.

6. Conclusions

An experimental investigation on cutting of EN 1.4462 and EN 1.4410 duplex stainless steels was presented. A modelling technique based on the modified Kienzle's equation was adopted to model the performances and ANOVA tests were performed to check the models adequacies. With the aid of three-dimensional surface plots, the effects of workpiece materials, process, and cutting conditions on the different cutting performances have been analysed and proper conclusion points have been drawn. Results of the early stages of this study have shown that the values of no-beneficial performances when cutting EN 1.4410 were generally higher than those of cutting EN 1.4462, and in comparison to the dry cutting, wet cutting has shown a general improvement in the machining performance.

This paper also presented multiobjective optimization of machining duplex stainless steels based on the nature-inspired metaheuristic bat algorithm (MOBA). Three objectives are minimized simultaneously: resultant cutting force, specific effective cutting power, and maximum flank wear. Arithmetic average roughness has been included in the formulation of the optimization problem as a constraint. Results of optimization have shown that MOBA is very efficient and consistently converges to the sets of optimal solutions. It has provided Pareto frontiers of nondominated solution sets for optimum cutting conditions, providing decision makers with a resourceful and efficient means of achieving it. In addition to Pareto's front graph, the paper is complemented with numerical outcomes.

Conflict of Interests

The authors declare that there is no conflict of interests regarding the publication of this paper.

References

[1] I. A. Armas and S. M. Degalaix, *Duplex Stainless Steels*, ISTE Ltd, London, UK, 2009.

[2] E. C. Bordinassi, G. F. Batalha, S. Delijaicov, N. B. de Lima, and S. Paulo, "Superficial integrity analysis in a super duplex stainless steel after turning," *Journal of Achievements in Materials and Manufacturing Engineering*, vol. 18, pp. 335–338, 2006.

[3] G. Królczyk, S. Legutko, and M. Gajek, "Predicting the surface roughness in the dry machining of duplex stainless steel (DSS)," *Metalurgija*, vol. 52, no. 2, pp. 259–262, 2013.

[4] G. Królczyk, M. Gajek, and S. Legutko, "Effect of the cutting parameters impact on tool life in duplex stainless steel turning process," *Tehnicki Vjesnik*, vol. 20, no. 4, pp. 587–592, 2013.

[5] J. Nomani, A. Pramanik, T. Hilditch, and G. Littlefair, "Machinability study of first generation duplex (2205), second generation duplex (2507) and austenite stainless steel during drilling process," *Wear*, vol. 304, no. 1-2, pp. 20–28, 2013.

[6] C. D. Oliveira Jr., A. Diniz, and R. Bertazzoli, "Correlating tool wear, surface roughness and corrosion resistance in the turning process of super duplex stainless steel," *Journal of the Brazilian Society of Mechanical Sciences and Engineering*, vol. 36, no. 4, pp. 775–785, 2013.

[7] D. Philip Selvaraj, P. Chandramohan, and M. Mohanraj, "Optimization of surface roughness, cutting force and tool wear of nitrogen alloyed duplex stainless steel in a dry turning process using Taguchi method," *Measurement*, vol. 49, no. 1, pp. 205–215, 2014.

[8] Coello CAC, "An updated survey of evolutionary multiobjective optimization techniques: state of the art and future trends," in *Proceedings of the Congress on Evolutionary Computation (CEC '99)*, pp. 3–13, Washington, DC, USA, July 1999.

[9] A. Konak, D. W. Coit, and A. E. Smith, "Multi-objective optimization using genetic algorithms: a tutorial," *Reliability Engineering and System Safety*, vol. 91, no. 9, pp. 992–1007, 2006.

[10] X.-S. Yang, "Multi-objective optimization," in *Nature-Inspired Optimization Algorithms*, X.-S. Yang, Ed., chapter 14, pp. 197–211, Elsevier, Oxford, UK, 2014.

[11] CECIMO, Concept Description for CECIMO's Self -Regulatory Initiative (SRI) for the Sector Specific Implementation of the Directive 2005/32/EC (EuP Directive) Goal of the SRI Scope of the SRI, 2011.

16

3D Point Cloud Data Basis Shape Management for Assembly of Modularized Large and Complicated Marine Structures

Deok-Hyun Yoon,[1,2] Hee Jin Kang,[3] Jin Choi,[3] Deok Eun Kim,[2] and Xin Huang[1]

[1] *College of Civil Engineering, Nanjing Forestry University, Jiangsu, Nanjing 210037, China*
[2] *Samin Information Co., Ltd., Busan 612-020, Republic of Korea*
[3] *Advanced Ship Research Division, Korea Research Institute of Ships & Ocean Engineering, Daejeon 305-343, Republic of Korea*

Correspondence should be addressed to Xin Huang; huangxin@njfu.com.cn

Academic Editor: Konstantinos Salonitis

As global competition heats up, in order to improve the productivity, simulation-based methods are becoming increasingly dominant in shipyards. The advancement of the CAD-based production management process even allows verification of installability and functionality before beginning the actual construction. However, whether the ship has been exactly constructed as designed can still and only be manually verified for a limited area. Therefore, significant interblock and intermodule errors are inevitably present in assembly, resulting in costly, time-consuming inspections and modifications. If the construction errors and defects can be investigated and controlled in each shop before assembly of modules, the productivity will be considerably improved. In the installation simulation of large structures, early detection and correction of the errors in junction allow fast and efficient assembly and provide better quality product development even with distributed construction yards. This technique can promote interindustrial collaboration among companies of different sizes, resulting in a significant improvement in overall productivity. In this paper, 3D point cloud data basis shape management framework has been studied with several case studies in a shipyard.

1. Introduction

Incorporating IT into the design process, 3-dimensional computer aided design (3D CAD) models are rapidly evolving from the "Does it fit?" model, which verifies the geometry and installability of structures, to the "Does it work?" model, which investigates the functionality of the assembled structures [1]. Through the contribution of 3D CAD modeling, the design of complicated structures and shapes has become efficient, and shipbuilding industries are now able to design and construct megablocks at a high preerection rate [2]. Currently 3D product model-based designs and simulation-based process management are implemented in marine plant industries for the design and installation of large structures, such as turrets. Despite advancements in 3D CAD technologies, the industries still rely on limited manual inspections to assess whether the structure has been constructed according to the original design or the 3D CAD product model [3]. The current design and construction process for a block, such as an engine room, in which complicated pipelines and valves are placed is an example. Once it is confirmed that the designed equipment module can function as intended, the module is then configured and constructed. The constructed module needs to be connected to the other blocks. Generally, the specific locations for connection work, such as junctions of pipes, are marked on the panel sides of the adjacent blocks [4]. However, such markings cannot be labeled on an open block, which does not have a connection panel. In such a case, the connection work refers to the installability-verified 3D design drawings. Furthermore, the accuracy control (AC) of the outline and quality control (QC) of the equipment modules are conducted during the construction of each block to allow connectivity to the other blocks.

2. Existing Methodology for Shape Management

The QC for an equipment module in a shipyard is divided into the equipment test, installation test, and module test [5, 6].

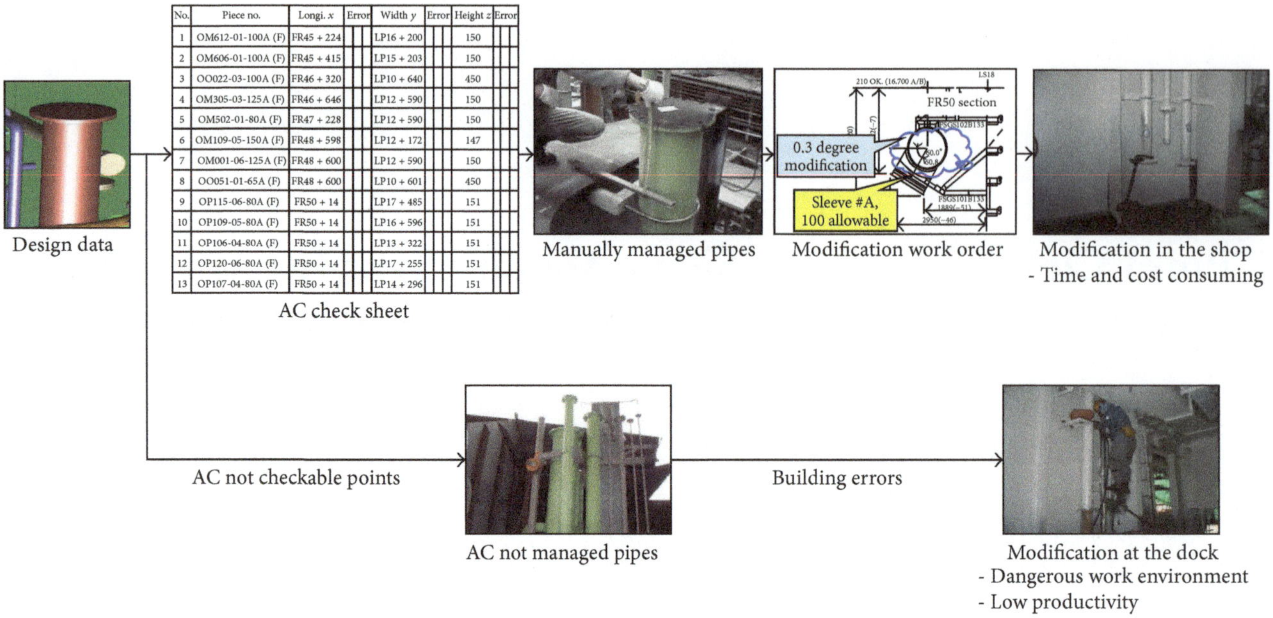

No.	Piece no.	Longi. x	Error	Width y	Error	Height z	Error
1	OM612-01-100A (F)	FR45 + 224		LP16 + 200		150	
2	OM606-01-100A (F)	FR45 + 415		LP15 + 203		150	
3	OO022-03-100A (F)	FR46 + 320		LP10 + 640		450	
4	OM305-03-125A (F)	FR46 + 646		LP12 + 590		150	
5	OM502-01-80A (F)	FR47 + 228		LP12 + 590		150	
6	OM109-05-150A (F)	FR48 + 598		LP12 + 172		147	
7	OM001-06-125A (F)	FR48 + 600		LP12 + 590		150	
8	OO051-01-65A (F)	FR48 + 600		LP10 + 601		450	
9	OP115-06-80A (F)	FR50 + 14		LP17 + 485		151	
10	OP109-05-80A (F)	FR50 + 14		LP16 + 596		151	
11	OP106-04-80A (F)	FR50 + 14		LP13 + 322		151	
12	OP120-06-80A (F)	FR50 + 14		LP17 + 255		151	
13	OP107-04-80A (F)	FR50 + 14		LP14 + 296		151	

AC check sheet

Figure 1: Manual AC check and modification process for preoutfitting.

In each of the tests, the connectivity, such as the location and angle of the pipe terminals, is manually inspected for the adjacent planes before actual assembly [6]. However, as the number of pipe members aligned to the connection plane of the block increases or the architecture of the pipe assembly becomes more complex, the manual inspection and modification will be conducted less accurately and at a slower pace [7, 8]. Furthermore, manual tasks experience additional difficulties due to thermal transformation of engagement surfaces. This is because the expansion or contraction of the engagement surfaces of a module affects the location or shape of the junction. Therefore, this aspect brings efforts to either install complex equipment modules in a single block or simplify the connection architecture between the adjacent blocks. Actually, the percentage of preoutfitting for ships with complicated upper structures, such as the Floating Production Storage and Offloading (FPSO), is about half of that of relatively simple ships [9]. Therefore, when developing high value-added ships, increasing preoutfitting and reducing correction works, which are caused by the mistakes in block assembly at the dock, have significant impacts on cost and time. Table 1 shows some example of QC items for modular building process.

In order to correct the difference in the level of the ship outline or pipelines, which is discovered during the block assembly, additional work, such as hot processing, cutting, bending, or even reinstallation, has to be conducted. This results in delay and degradation in quality. Figure 1 illustrates the present control flow of major shipyards for errors during the block assembly process. In Figure 1, it is shown that the inspection and modification are conducted manually, and the management fails to properly inspect hard-to-detect items, resulting in costly long period modifications in the module assembly site which is the main cause of low dock turnover ratio at the block assembly stage.

Table 1: Example of QC items for equipment modules [6].

Equipment test	Pipe length, pipe bending angle, flange perpendicularity, and so forth.
Installation test	Location of pipe terminal, angle of pipe terminal, and so forth.
Module test	Length of primary panel, level of primary panel, location of internal member, angle of internal member, length, width, level of module, and so forth.

Inconcinnity of engagement surfaces actually happens frequently throughout the preoutfitting and block installation process. As the correction work for the equipment often requires dangerous high place work and welding operations, the inconcinnity is one of the major factors that threaten safety and productivity. Furthermore, because it is difficult in the field process planning to estimate the potential modifications, the experience of field managers generally plays the primary role, resulting in a high level of uncertainty in the process management [10]. In light of this fact, and considering that marine industries practice block and module assemblies, the improvement of construction productivity heavily relies on the effectiveness of error detection and modification for large blocks or equipment modules before actual assembly at the dock. Ultimately, the "Does it work?" and "Does it fit?" simulation concepts that advanced 3D CAD technology brings are concluded in the "Constructed as designed?" concept, which describes whether the outcome of the construction agrees with the designed product. In other words, the agreement of inspection between the constructed product and the designed product completes the 3D CAD product model-based production simulation technology. If fast and accurate inspection can be conducted on the location and shape of the pipeline junctions of the modules between

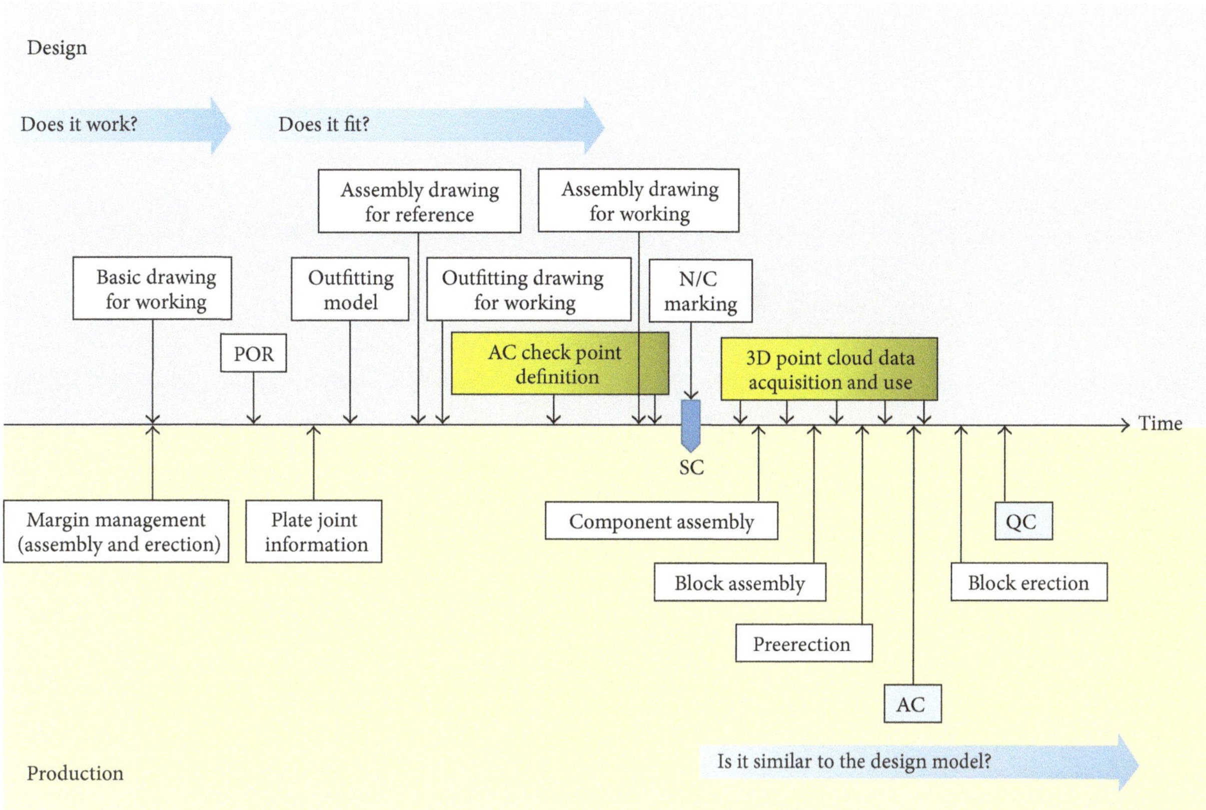

FIGURE 2: Application of the 3D point cloud data in design and construction process.

the blocks, considers the temperature effects, and perfectly finishes the modifications before the block assembly, the attendant modifications in the block installation and module assembly at dock will be significantly reduced and will ultimately result in a much higher dock turnover.

3. 3D Point Cloud Data Basis Shape Management

A wide variety of commercial 3D laser scanners is utilized in various fields and can be used to generate point cloud data. A point cloud refers to a set of x, y, and z coordinates that are constructed using the phase difference between a laser beam emitted from a laser scanner and the returning beam reflected by an object [11]. Although a diverse array of hardware that is capable of constructing a point cloud is available, the software that is able to utilize and implement them for ships and marine plants is only at the elementary stages of development. In the current marine industries, shape management for product against the design drawings is conducted in two ways: QC and AC of these two inspection options; the point cloud data can be applied to AC experimentally for equipment, pipe, vent, and so forth at the preerection and block erection phases as shown in Figure 2. Through these AC activities at the construction phases, the compliance of the product with the design is inspected. Furthermore, the installability, functionality, and maintainability are investigated under the design phase before the steel cutting process.

The following information from the data generated during the design process is utilized for the automation of panel machining at the construction phase [8]:

(i) panel list of blocks and hull,

(ii) boundary of each block and panel,

(iii) information of equipment pipe pads and angle support models,

(iv) information of hull panel and angle support attachment,

(v) information of pads and angles contained on hull panels.

As shown in Figure 3, this information is utilized through equipment allocation using the hull panel, which is generated in the design process, as the background model. As the 3D point cloud data are available in this task, any structure in 3D space that cannot be marked on hull panels is compared with the design data and modified during the preoutfitting process before block installation [12]. Furthermore, it is possible for the equipment constructed at a distant location to be checked before the delivery. At that point, it can be assessed to verify whether it can be installed to the destination block, whether it causes interference, or how much modification will be required once it is delivered with using attribute data of ship

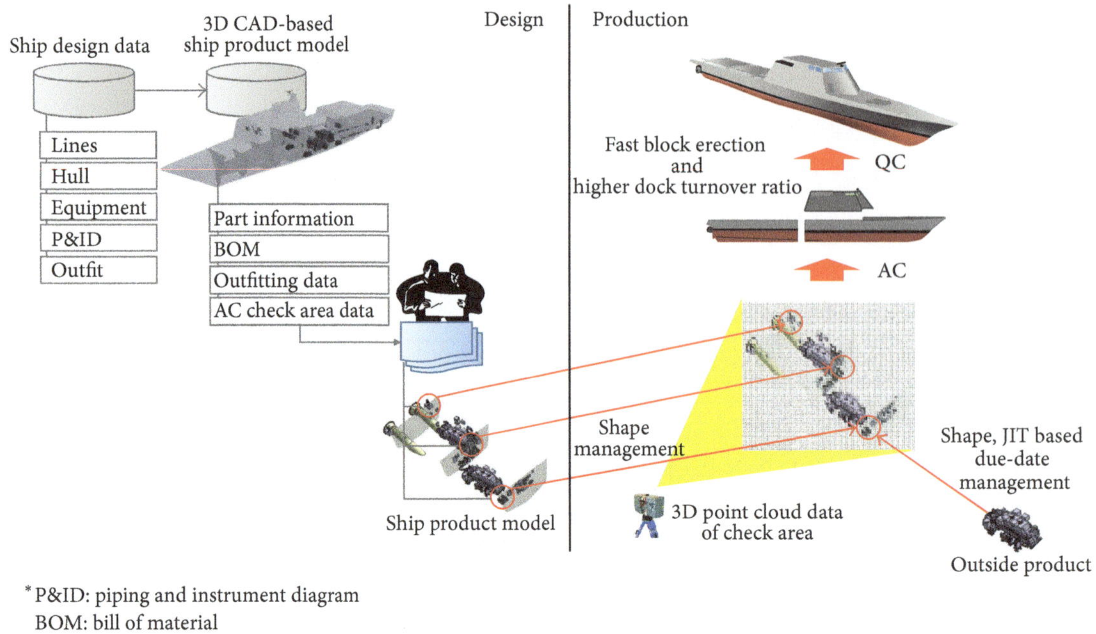

Figure 3: Shape management framework on the basis of the 3D point cloud data.

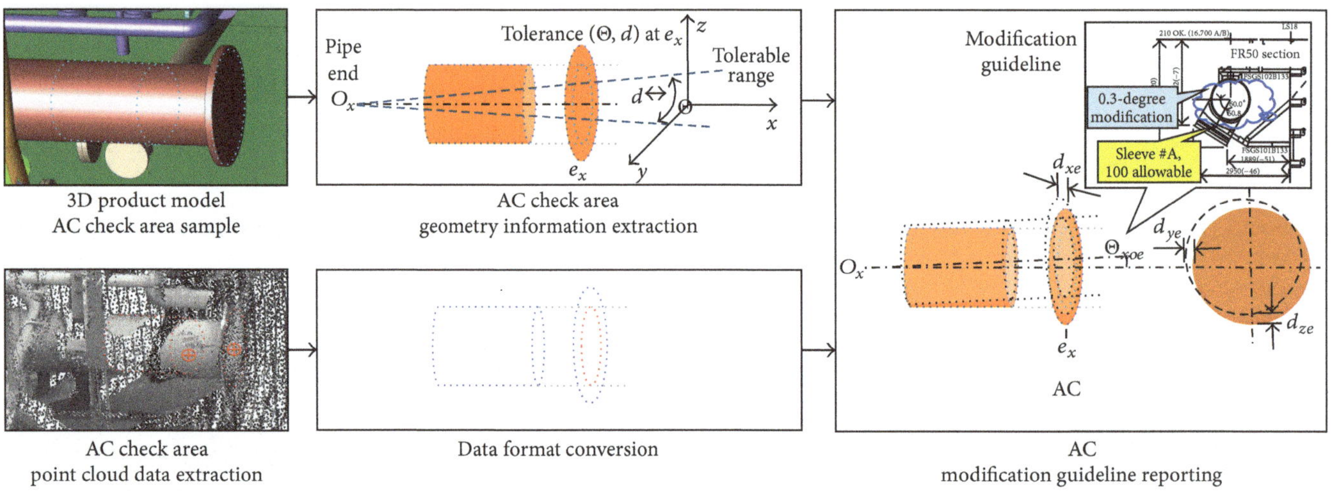

Figure 4: An example of the compliance check and modification order using 3D point cloud data.

product model [12, 13]. Therefore, a higher preoutfitting rate is achievable and the storage period after delivery can be reduced. This results in the implementation of the just in time (JIT) inventory management and provides efficient overall process control.

If the collected 3D point cloud data of the AC check area exhibit disagreement with the design data, considering the interference with surrounding structures, a decision is made to either accept expecting minor modification work at the block installation site or to reject the demand for further modification. In this case the 3D point cloud data becomes a primary basis of the modification job order. Figure 4 illustrates how a modification job order is generated using the 3D point cloud data.

4. Data Management for Mobile Device-Basis Site Availability

The point cloud data can be transformed into mesh models or surface models [14–16]. This is so that they can be incorporated into the 3D CAD data that is utilized in the design and construction stages [17]. However, it is important to note that the 3D point cloud data collected at the field occupies anywhere from hundreds of megabytes to dozens of gigabytes, depending on the data usage [17, 18]. Therefore, it is not easy to manage the data on the shipyard network, as the construction fields and design fields are separate from each other. Consequently, a ship manufacturer's design system requires smart data load management to utilize the 3D point cloud data received from laser scanners. In this study, the data

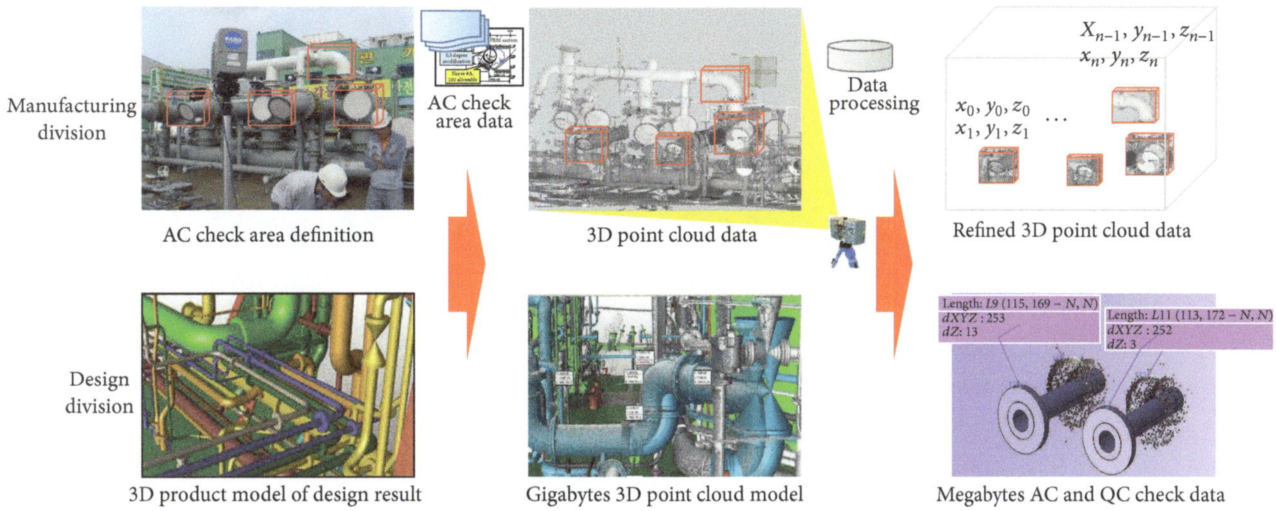

FIGURE 5: Data processing for online application of 3D point cloud data with small data volume.

load is relieved of dozens of megabytes experimentally as the intensive point data is collected for important regions of machinery room, such as boundaries, pipe couplings, and flanges, with which the design data is compared. Low density or no data is recorded for ignorable regions. In order to implement this method, the data collection boundaries should be predefined based on the AC check items that were identified in the design stages as shown in Figure 5. Furthermore, the collected point data should be refined to better focus on the predefined regions. Furthermore, the quality of the line and surface information can be significantly improved using density functions and thresholds. This can be seen when the point density is increased for high-density regions and decreased for low-density regions with respect to the threshold.

5. Case Study

The ability to inspect and modify various assembly methods of 3D point cloud data, including vertically moving assembly, laterally moving assembly, and unit module shapes, is adoptable to various fields of large and complicated marine structures assembly [6, 19]. It is also expected that the cooperation of designers and field mangers at the shop allows the detection and correction of errors of each module before it is delivered to the dock. This results in precise schedule management as shown in Figure 6 and Tables 2 and 3. In the three cases of leg block assembly, point cloud data basis preverified module's assembly shortened installation time in the dock from several days to an hour. Though a couple of days are required for 3D laser scanning, it experimentally does not affect whole process time and cost for delivery.

As shown in Figure 7, 3D point cloud data are effective especially for inspection of large and complicated structures' engagement surfaces [19]. The scanned point cloud data are easily transformed into a form that the CAD systems of each shipyard [14–17]. The detected errors are reflected on the modification job order and corrected at the construction site regardless of the distance of each module's work place. As a result, it is acknowledged that the unnecessary risky modifications, which degrade productivity (e.g., dock turnover), can be reduced innovatively. Through a further collaboration with the design data, the modification job order issuance, delivery, and storage can be automated. This results in a significant reduction in demand of modification during the block assembly.

In summary, the use of the 3D point cloud data aids in relieving the limits of the current construction method and allows a higher preoutfitting rate. This ultimately results in precise scheduling of product delivery and is accompanied by the reduction of production time, cost, and risk. Shin et al. [4] reported that the improvements in the construction method resulted in a higher preoutfitting rate followed by a more efficient scheduling and faster production. In their study, the reduction in the FPSO production time only reached 44%. As they achieved a preoutfitting rate of 75%, it was more than doubled compared to what it had been by 32%. This is due to the increase in the collateral modification work at the block installation site for the increased preoutfitting. This result leads to the conclusion that, in order to achieve full reduction in production time, increments in preoutfitting have to be supported by the elimination of the modification work through inspection and verification before the block assembly. Since utilization of the 3D point cloud data enables online remote checkup for installability of long lead time equipment, the construction can have a margin of time for installation and modification, but not for shipping. In addition, the just in time (JIT) management of the material and equipment is possible for the construction process of individual blocks and the installation schedule of blocks and large modules.

6. Conclusions and Future Work

This paper discussed the application method of the 3D point cloud data to investigate whether large and complicated

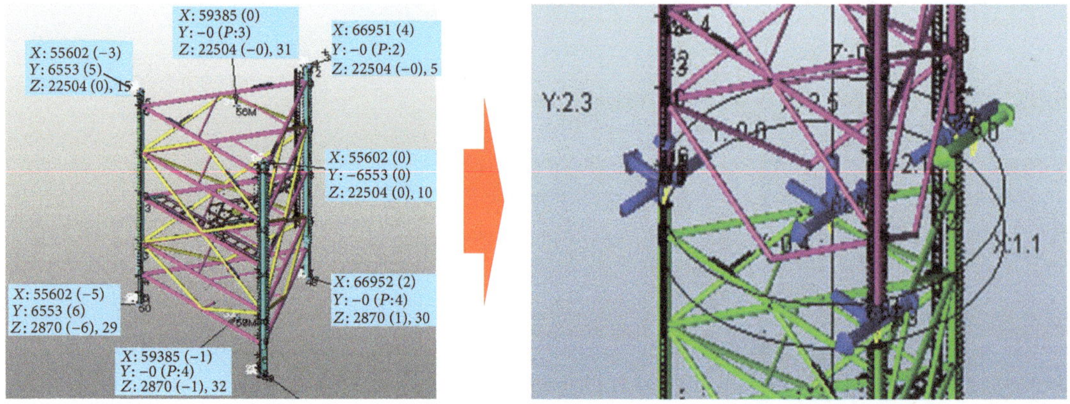

AC with 3D point cloud data and product model of leg block

Interference check and modification

FIGURE 6: 3D point cloud data basis simulation for the leg block assembly of a jackup rig.

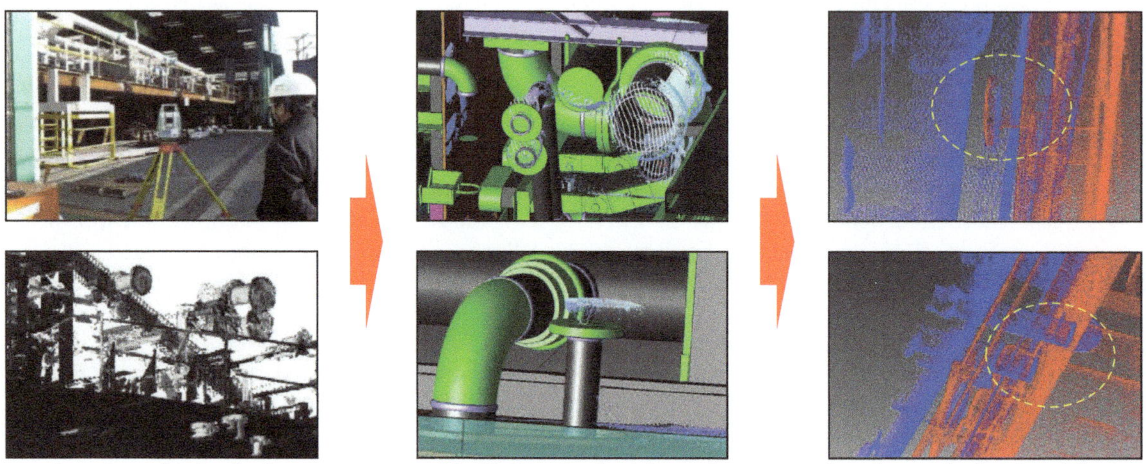

Scanning and point cloud data generation

AC with 3D point cloud data and product model of leg block

Interference check and modification

FIGURE 7: 3D point cloud data basis simulation for assembly enables long distance dispersion cooperation.

TABLE 2: Cost burden examples of existing methods for the leg block assembly.

	Dimensions	Weight (abt.)	Crane capacity	Additional cost caused by construction delay in the dock			
				Crane	Man-hour	Construction delay	Additional cost
Case 1	65 m × 32 m	7,000 ton	9,000 ton	14 hours	37 m/h × 14 h	1.5 days	268,000 US dollars
Case 2	37 m × 34 m	4,500 ton	9,000 ton	10 hours	37 m/h × 10 h	1 day	250,000 US dollars
Case 3	37 m × 34 m	4,500 ton	9,000 ton	12 hours	37 m/h × 12 h	1 day	223,000 US dollars

TABLE 3: Application results of 3D point cloud data basis simulation for the leg block assembly.

	Dimensions	Weight (abt.)	Crane capacity	Additional cost caused by construction delay in the dock			
				Crane	Man-hour	Construction delay	Additional cost
Case 1	65 m × 32 m	7,000 ton	9,000 ton	1 hour	37 m/h × 1 h	—	24,000 US dollars
Case 2	37 m × 34 m	4,500 ton	9,000 ton	1 hour	37 m/h × 1 h	—	24,000 US dollars
Case 3	37 m × 34 m	4,500 ton	9,000 ton	1 hour	37 m/h × 1 h	—	24,000 US dollars

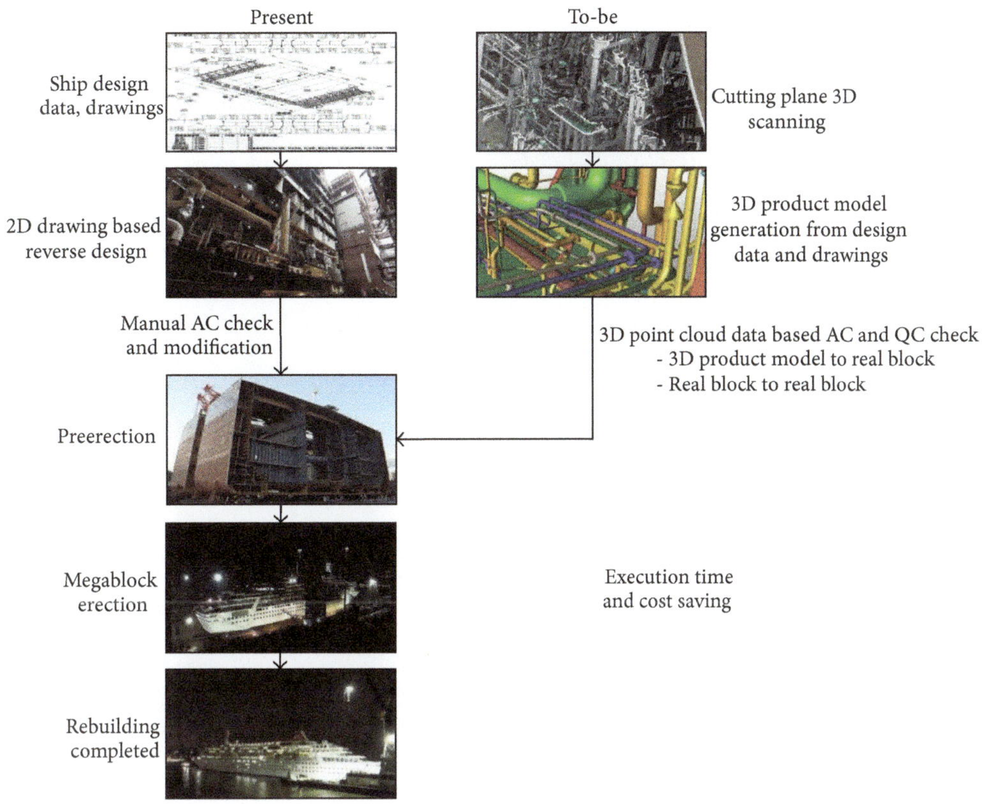

FIGURE 8: Reverse engineering concept using the proposed method [17, 20].

marine structures such as block modules are constructed as originally designed and to inspect the compatibility with other blocks and installability on the platform. Although the proposed method still contains some technical issues that it must resolve, it will significantly contribute to the increase in the level of productivity in marine industries, where most of the inspections are manually performed. These industries focus on 3D CAD-based design and construction. As the products are compared with the design data and other products in real time and the errors are corrected before installation, the risky, delay-causing modification work can be significantly reduced. Furthermore, the distributed production process can be precisely scheduled for installation of the modular products, and the JIT management of equipment and materials can be possible. Ultimately, this method can lead to stronger interindustrial collaboration within the industry clusters, resulting in higher operation rate of middle class shipbuilders and equipment manufacturers as well as higher dock turnover for major shipbuilders. Further investigation is required in the field of data transfer security during the application of this method.

Despite the limited usage of the current 3D point cloud data method, its capability of operating with 3D information of complicated shapes makes the method extremely valuable for large-scale applications, such as open-space remodeling shown in Figure 8. "Present" of the figure is a scene of remodeling that was disclosed in 2014 for the Norwegian Crown from Blohm + Voss, Hamburg, Germany [20]. This remodeling process includes cutting the body into two pieces and then filling in a new block to elongate the ship with many additional features. If the 3D point cloud data were utilized for this remodeling, the design information of the complicated cut section would be more easily achievable. Furthermore, reverse engineering-based rebuilding can be feasible for not only the middle of the ship, but also any complex portion of the ship.

Even the concurrent distributed work can yield a high preoutfitting rate and requires less modification work when the 3D point cloud data is utilized for the shape management of the remote construction of complicated structures. Because the productivity of the distributed collaboration is improved, the operation rate of the small and middle scale shipyards and equipment manufacturers can be effectively maintained, resulting in higher productivity for all. To realize this concept before long, further investigation and discussion are needed for some issues, such as security, to prevent leakage of design knowledge, know-how, data, and other information.

Conflict of Interests

The authors declare that there is no conflict of interests regarding the publication of this paper.

Acknowledgment

This study is partially supported by Research Projects of KRISO (PES1990 and PNS2251).

References

[1] ITE, *Design Tools to Support NAVSEA Ship Design & Program Management & NAVSEA-CNET HSI for Training Systems*, 2007.

[2] J. H. Park, G. J. Jin, J. M. Kim, D. S. You, M. K. Oh, and D. S. Lim, "Ship building IT status and prospects," *Electronics and Telecommunications Trends*, vol. 25, no. 4, pp. 19–54, 2010.

[3] M. l. Roh, W. Y. Choi, and K. Y. Lee, "Rapid pipe modeling method considering the relationship with a hull structure," *Journal of the Society of Naval Architects of Korea*, vol. 43, no. 2, pp. 258–267, 2006.

[4] S.-C. Shin, J.-B. Cho, K.-Y. Shin, and S.-Y. Kim, "Process improvements for elevating pre-outfitting rate of FPSO," *Journal of the Society of Naval Architects of Korea*, vol. 46, no. 3, pp. 325–334, 2009.

[5] M. P. Tedesco, *An approach to standardization of naval equipment and components [Master of Science Thesis]*, Massachusetts Institute of Technology, 1994.

[6] Y.-T. Choi, H.-W. Suh, and S.-S. Lee, "Establishment of information interface technology between hull and outfitting designs," *Journal of the Society of Naval Architects of Korea*, vol. 50, no. 6, pp. 458–465, 2013.

[7] The Waste Reduction Resource Center, *Shipbuilding and Ship Repair Surface Coating NESHAP Compliance Inspection: How to Perform the Inspection*, 2003.

[8] H. Y. Park, H. G. Kang, N. S. Park, and M. S. Lee, "Development of the E/R insulation modeling automation system using structural hull model information," *Special Issued of the Society of Naval Architects of Korea*, pp. 112–117, 2008.

[9] Y. Watanabe and N. Yamaga, "Increasing speed of ship repair through three-dimensional laser measurement in compliance with environmental regulations," *Mitsubishi Heavy Industries Technical Review*, vol. 50, no. 2, pp. 66–69, 2013.

[10] H. S. Kim, D. H. Lee, K. W. Bang, H. K. Cheon, and I. S. Kwak, "Establishing execution schedule system in ship building industry," in *Proceedings of the Korean Institute of Industrial Engineers*, Seoul, Republic of Korea, 2010.

[11] F. Bernardini and H. Rushmeier, "The 3D model acquisition pipeline," *Computer Graphics Forum*, vol. 21, no. 2, pp. 149–172, 2002.

[12] H. J. Kang, Y.-S. Yang, J. Choi, J.-K. Lee, and D. Lee, "Time basis ship safety assessment model for a novel ship design," *Ocean Engineering*, vol. 59, pp. 179–189, 2013.

[13] H. J. Kang, J.-G. Shin, and J. K. Lee, "A business model-based design of a damage control support system for naval ships," *Systems Engineering*, vol. 15, no. 1, pp. 14–27, 2012.

[14] S. Z. A. Abdul-Rahman and V. Coors, "Improving the realism of existing 3d city models," *Sensors*, vol. 8, no. 11, pp. 7323–7343, 2008.

[15] M. Y. Yang and W. Förstner, "Plane detection in point cloud data," Tech. Rep. TR-IGG-P-2010-01, 2010, http://www.ipb.uni-bonn.de/technicalreports/.

[16] A. Adamson and M. Alexa, "Ray tracing point set surfaces," in *Proceedings of the Shape Modeling International (SMI '03)*, pp. 272–279, May 2003.

[17] SAMIN, SAMIN Information System, 2014, http://www.saminis.com/.

[18] D. Lichti, M. P. Stewart, M. Tsakiri, and A. J. Snow, "Calibration and testing of a terrestrial laser scanner," *International Archives of the Photogrammetry, Remote Sensing and Spatial Information Sciences*, vol. 33, no. B5/2, pp. 485–492, 2000.

[19] W. Boehler, M. Bordas, and A. Marbs, "Investigating laser scanner accuracy," in *Proceedings of the 19th CIPA International Symposium*, Antalya, Turkey, 2003.

[20] The Mail on Sunday & Metro Media Group, Associated Newspapers Limited Part of the Daily Mail, 2014, http://www.dailymail.co.uk.

17

Observation of Drilling Burr and Finding out the Condition for Minimum Burr Formation

Nripen Mondal, Biswajit Sing Sardar, Ranendra Nath Halder, and Santanu Das

Department of Mechanical Engineering, Kalyani Government Engineering College, Kalyani, West Bengal 741235, India

Correspondence should be addressed to Santanu Das; sdas.me@gmail.com

Academic Editor: Godfrey Onwubolu

Suppression or elimination of burr formation at the exit edge of the workpiece during drilling is essential to make quality products. In this work, low alloy steel specimens have been drilled to observe burr height under different machining conditions. Taper shank, uncoated 14 mm diameter HSS twist drills are used in these experiments. Dry environment is maintained in experiment set I. Water is applied as cutting fluid in experiment set II. In the next four sets of experiments, the effect of providing back-up support material and exit edge bevel is observed on formation of burr at the exit edge of specimens under dry and wet conditions. It is revealed that an exit edge bevel of 31 degrees with water as the cutting fluid gives negligible burr at the exit edge of the drilled hole at certain machining conditions. Use of a back-up support can also reduce drill burr to a large extent. In this paper, artificial neural networks (ANN) are developed for modeling experimental results, and modeled values show close matching with the experimental results with small deviations.

1. Introduction

Drilling is a common hole-making operation, and the majority of workpieces are subject to hole-drilling before they leave a machine shop. However, presence of burr on the drilled workpieces creates problems not only in handling, but also in the assembly line. Burrs are undesired projections attached to the edge of drilled holes. These are found to be substantially greater at exit side than entry side. Hence, elimination, or large-scale reduction, of exit burr is the necessary requirement of an industry [1–5].

Many researchers worked on burr-related problems associated with drilling and other processes and also investigated the mechanism behind burr formation. Control charts were applied by Min [6] and Kim et al. [7] for control and prediction of burr height during drilling different steels. The same technique was also employed by Lee and Dornfeld [8] for estimating burr size during microdrilling to some success. Finite element method (FEM) was used by some others [9–11] to observe stress and deformation patterns analytically to understand the reason behind burr formation. Guo and Dornfeld [9] made finite element analysis to assess the substantial reduction of drilling burr with back-up support. They also successfully did [10] this analysis to understand burr formation in stainless steels. FEM was also used by Park and Dornfeld [11] to find out the influence of exit edge angle of a specimen, tool rake angle, and back-up support on burr formation. The estimates made showed remarkable results.

An overview on different strategies to control burr was presented by some researchers [1–5, 12, 13]. Investigations on the effect of drill size [14, 15], use of drills with special geometry [16–19], effect of using different coolants, exit edge modification, and providing back-up support [20–22] were explored by other research groups. Lee and Dornfeld [8] carried out experiments on four different materials using HSS and carbide drills of varying shapes and geometry. They concluded that step drills with less step angle and step size produced quite small burr. Min et al. [20] performed experimental and analytical investigation on drilling burr formation by varying interaction angle at the exit edge. It is the angle by which work exit surface is inclined. They observed that a large interaction angle results in quite less burr due to less bending of job material to form a burr. Another group of researchers found out [19] variable feed drilling with a suitable amplitude to give high tool life with quite low burr height. Although variable feed was reported

Table 1: Experimental conditions.

Machine tool	Radial drilling machine, Make: Energy Limited, India, Model: RDH-32/930 Main motor power: 1.5 kW
Tool holder	R/L 265 ME-20 AL, Make: Sandvik Asia Limited, India
Cutting tool	Taper shank, uncoated 14 mm diameter HSS twist drill, Make: Miranda (India)
Workpiece material	Low alloy steel, hardness: 225 HB Composition: C (0.17%), Si (0.21%), Mn (0.63%), P (0.09%)
Workpiece size	100 mm × 50 mm × 6 mm
Cutting velocity	20, 25 and 31 m/min
Feed	0.032, 0.05 and 0.08 mm/rev
Environment Experiment sets	Dry and water-cooled I—Drilling in dry condition without any back-up support or edge bevel II—Drilling with water cooling without any back-up support or edge bevel III—Drilling in dry condition with a back-up support IV—Drilling in water-cooled condition with a back-up support V—Drilling in dry condition with 31° exit edge bevel VI—Drilling in water-cooled condition with 31° exit edge bevel

to reduce burr remarkably compared to that of constant feed drilling, implementation of this method needs special facility to be provided on a drilling machine.

During tool/cutter exit from the workpiece, cutting edge of the tool/cutter was observed [5–12] to have been chipped off or broken beside formation of burr/foot. This problem also was seen to reduce with the introduction of special tool geometry, suitable exit order sequence, and work edge bevel. Provision of a beveled work edge was reported to lower burr height noticeably in milling and drilling operations [12–15, 21–23] due to gradually decreasing depth of cut and the reducing need of back-up support. Suitable tool path selection also reduces burr height [24]. Significant effect of different size of drills on burr formation was observed in some other works [14, 15] under varying cutting conditions. Introduction to different shape and geometry of cutting tool was also reported [8, 17, 18, 25] to reduce burr significantly for specific applications. A typical stepped drill was tested to restrict or to remove burr effectively. For laminated composites, delamination and burr formation at interface layers were studied [26] and some conditions were reported to reduce formation of burr and to avoid delamination.

In another work, mechanism of exit burr formation during drilling of aluminium alloys was closely studied [27], and an analytical model was proposed considering the effect of temperature that matched well with experimental results. Burr size was also modeled by other researchers [28] analytically for ductile metals to make good prediction of burr formation. The nature of burr formed during drilling of small size drills under dry and minimum quantity lubrication (MQL) with water soluble oil condition was discussed [29]. They also tried to obtain good quality holes with minimum burr by using optimized tool selection. Investigation of drill burr during hole-making on low carbon steel specimens with 40 Brinell Hardness Number was carried out by Roy et al. [23]. Utility of providing a back-up support or beveling the exit edge by 31° was observed in that work. However, effect of water-cooled condition was not explored suitably with or without a bevel in this work. Providing bevel in dry condition was reported not to show remarkable reduction in burr height.

Karnik et al. [30] tried to minimize burr size in drilling stainless steel workpieces with the use of genetic algorithm (GA) and Taguchi method of designing experiments. An innovative measurement technique to find out burr height was also reported [31]. Gaitonde and Karnik [32] utilized artificial neural networks (ANN) and particle swarm optimisation (PSO) approaches for optimal selection of drilling parameters to achieve minimum burr size (height and thickness) during drilling within the domain of experiments.

Even as a lot of works have been carried out on burr, still a lot of scope remains to find out appropriate strategy to minimize burr for a particular process of machining different workpiece materials. The obbjective of the present investigation is to explore effects of different machining conditions and strategies on drilling burr formation and selection of the optimal condition to suppress burr formation significantly. 14 mm diameter holes are drilled with or without using a back-up support and bevel at the exit edge to notice formation of burr in low alloy steel specimens under dry and wet with water cooled environment.

2. Experimental Setup and Machining Condition

Experiments are conducted in this work on a radial drilling machine (Make-Energy Limited, India) under dry and wet (water cooled) conditions. Low alloy steel specimens are taken for drilling experiments using 14 mm diameter drills.

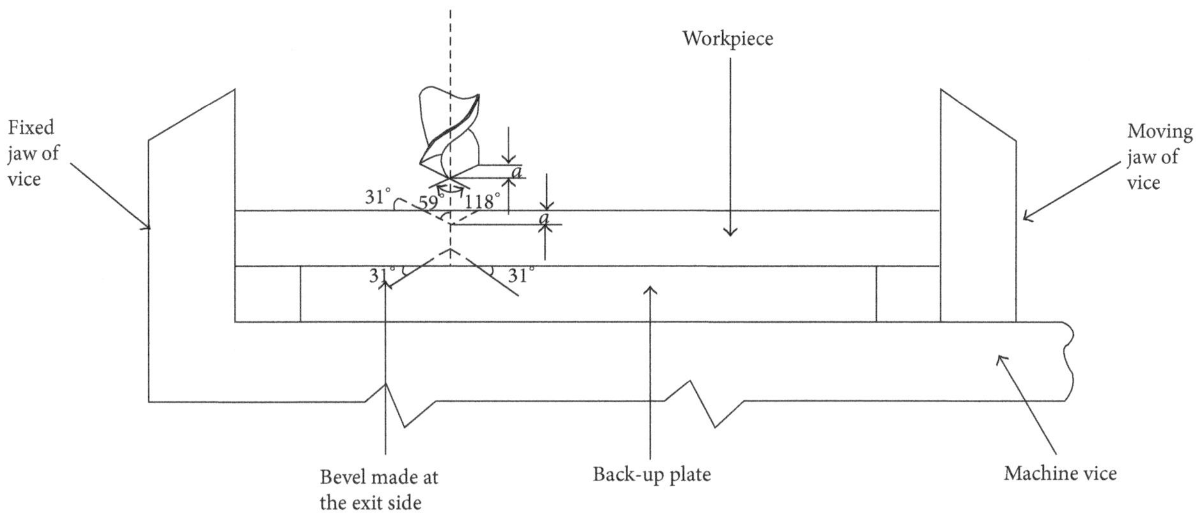

FIGURE 1: Schematic representation of drilling using a back-up plate and/or exit edge bevel.

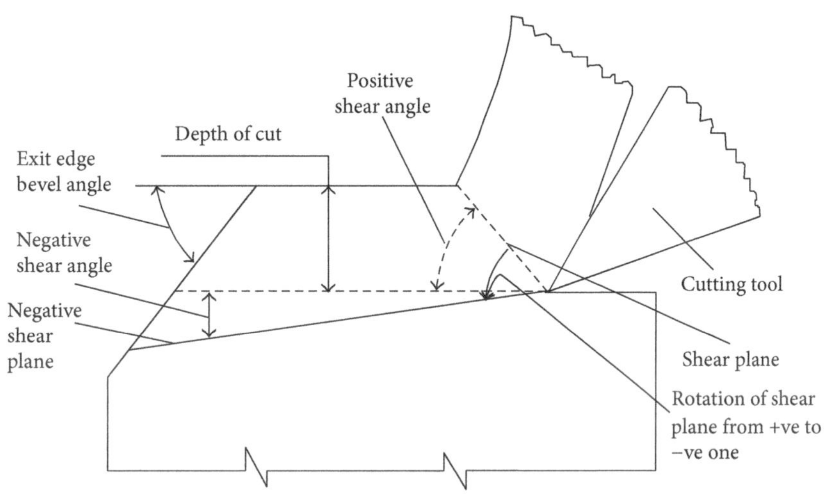

FIGURE 2: Schematic diagram showing orientation of positive shear plane to a negative one and exit edge bevel.

Experimental conditions are given in Table 1. Cutting velocity, V_c, and feed, S_z, are chosen within 20–31 m/min and 0.032–0.08 mm/rev, respectively. Ranges of V_c and S_z are selected considering usual industrial practice for drilling low alloy steels with tungsten grade HSS drill bit. Back-up support is provided to explore its influence on reduction of burr formation in some sets of experiments. It is done by placing similar specimens as that of the test piece with slightly less width below the test piece as shown in Figure 1 and without providing any edge bevel. Less width of the support facilitates easy clamping of test piece in a machine vice. Drilling is continued up to the middle of back-up support thickness.

Back-up plate is used to provide the required support during the tool exit and intends not to allow rotation of shear plane to a negative shear plane and suppresses bending of chip to form a burr. The shear plane rotation was reported to result in burr formation in a number of works [4, 5, 9, 13, 21]. Possible rotation of a positive shear plane to a negative one is shown schematically in Figure 2. When the tool or cutter approaches the emerging edge of a workpiece, no back-up material exits at that position to support the cutting forces exerted by the cutting tool onto the part of the workpiece. As a result, the shear plane gets oriented to the negative direction. If a back-up material is placed below the workpiece, then required support is likely to be provided by it resulting in expected reduction in the formation of burr. The effectiveness of suppressing formation of burr by using a back-up support is compared to that of using edge beveling in this work.

A bevel of 31° is provided at the rear side of the workpiece as shown in Figure 1 for experiment sets V and VI. The bevel is made with the help of drilling with the same 14 mm drill bit up to the drill point length, a, as shown in Figure 1. In this way, 31° exit edge bevel angle is made, as it is complementary to 59° which is half the point angle, 118°. Thus, no other processes are needed to produce the bevel. Only appropriate marking at the location of the bevel making with the drill is to be done. Bevel portion of the specimen is placed at the rear side of the workpiece aligning with the hole to make with suitable marking. When this exit edge bevel is used, no back-up support is provided.

Number of other parameters, such as lip clearance angle and point angle, may have an influence on burr formation, and this may be explored experimentally. However, investigation is restricted in this work to find out the effect on burr height by varying cutting velocity, feed, machining environment, and with and without the use of back-up and exit edge bevel of 31°.

Along the bevel provided at the exit edge of the workpiece, depth of cut gets gradually reduced. When the drill bit goes down along the bevel position, chip width gets reduced gradually. Correspondingly, the influence of chisel edge to deliver axial thrust causing bulging out of the thinning work material at the exit end is eliminated. When peripheral point of cutting edge of the drill bit approaches the rear end of the specimen, depth of cut reduces gradually needing correspondingly decreasing cutting force (thrust) and torque. Hence, at some bevel angle, it is expected that no substantial requirement of back-up support will be there leading to suppression of burr formation at the exit edge [4, 5, 7, 13, 20–22]. In the present work, 31° exit edge bevel is provided as it is easily achievable by using the twist drill point portion having a point angle of 118° as detailed in Figure 1. Influence of this is explored on the extent of suppressing burr formation.

Table 1 shows the detail of experimental conditions. Process parameters are selected in-line with that normally practiced in industry using standard HSS twist drills. Three levels of feed (S_z) and cutting velocity (V_c) are employed following full factorial design of experiments at each experiment set as detailed in Table 2. Hence, at each experiment set, $3^2 = 9$ tests are performed. Although a burr size may be characterized by its thickness and height, in the present work, burr height is considered to characterize a burr in line with many other works reported earlier [7, 18, 19, 27]. Burr height is measured using a vernier height gauge. Six different sets of experiments are conducted as detailed in Table 1 to find out the condition giving minimum drilling burr within the domain of experiments. Experiment sets I and II are performed under dry and wet with water cooling condition, respectively, without using any backing plate or edge bevel. Experiments with the provision of backing support are done under dry and water-cooled conditions in experiment sets III and IV, respectively. In experiment sets V and VI, drilling is carried out at respective dry and wet environments with an exit edge bevel of 31°.

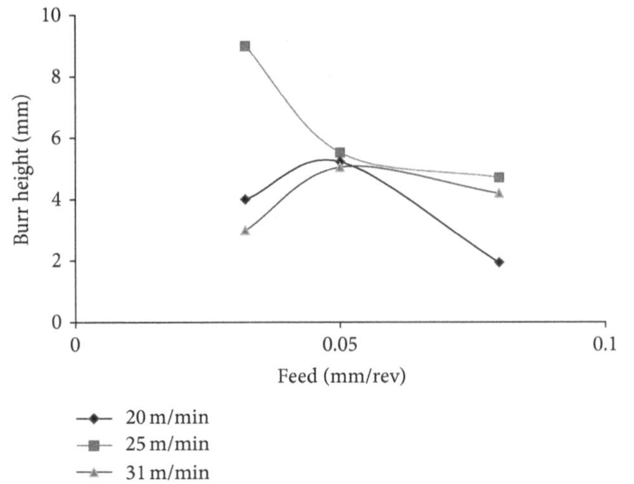

FIGURE 3: Plot of variation of burr height at different cutting velocity and feed for experiment set I.

TABLE 2: Test conditions in each experiment set.

Sl. no.	Feed (S_z) (mm/rev)	Cutting velocity (V_c) (m/min)
1	0.032	20
2	0.032	25
3	0.032	31
4	0.050	20
5	0.050	25
6	0.050	31
7	0.080	20
8	0.080	25
9	0.080	31

3. Results and Discussion

3.1. Discussion on the Observation of Drilling Burr at Dry Condition without Any Back-Up Support or Beveled Edge (Experiment Set I). Observation on experiments in drilling at dry condition without using any back-up support material, or edge beveling, is presented in Figure 3. Large burrs are found in these tests. Variation of burr height is noticed at different machining conditions, although no clear trend is seen with the individual machining parameters. Large size burrs are expectedly formed in these tests as no back-up support material is available at the exit edge of the drilled hole. For this, shear plane is likely to have oriented to a negative direction facilitating formation of burr as shown in Figure 2 [21, 24]. When shear plane is oriented towards a negative plane, the chip is bent downwards and can be attached to the exit edge of workpiece forming a burr. Large extent of plastic deformation under dry condition with high temperature rise may have resulted in large burrs. Corresponding photographs of burrs observed are shown in Figure 4. For experiment set I, only at a cutting velocity (V_c) of 20 m/min and a feed (S_z) of 0.08 mm/rev (test No. 7), burr height of slightly less than 2 mm is found. In all the nine tests in experiment set I, transient, nonuniform burrs are observed. Classification of drilling burr was discussed by Kim et al. [7]. They classified drilling burr as uniform, crown, transient, and uniform with drill cap.

3.2. Discussion on the Observation of Drilling Burr under Water-Cooled Condition without Back-Up Support or Edge Bevel (Experiment Set II). Experiment set II is conducted in water cooled condition without using a back-up material and

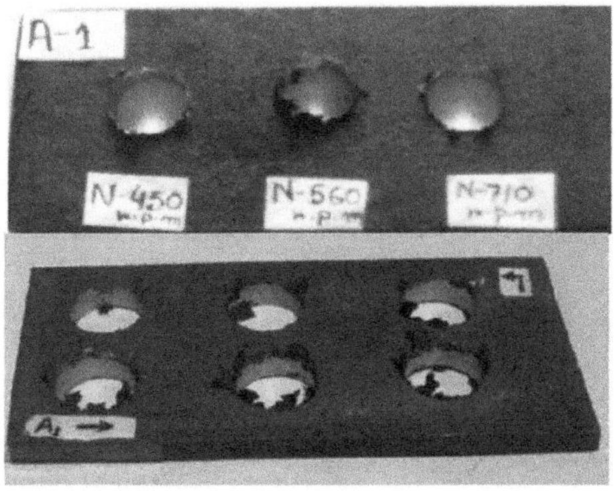

FIGURE 4: Photographic view of burr formed in experiment set I (left-right order: top for Sl. numbers 1–3, bottom for Sl. numbers 4–6, middle for Sl. numbers 7–9).

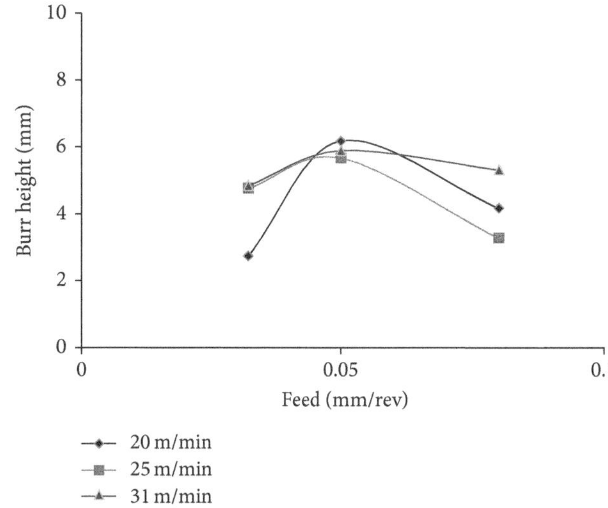

FIGURE 5: Plot of variation of burr height at different cutting velocity and feeds for experiment set II.

edge bevel. The experimental results are plotted in Figure 5 showing the variation of burr height observed under different cutting conditions. No remarkable reduction in burr height is found to be there at water-cooled condition compared to that at dry environment. At certain machining conditions, burr height is lesser with water cooling than dry condition. However, at other conditions, larger burr height is observed with water cooled condition than that with dry condition. Figure 6 shows typical burrs seen at the exit edge of the drilled holes. In this experiment set II, mainly transient burrs are observed. Water cooling is expected to reduce drilling temperature and thereby may reduce the extent of plastic deformation during drilling. However, marginal contribution of water in lubricating tool-chip-workpiece interface regions may be the possible reason behind having no significant effect of it on burr reduction. In this experiment set also, no definite trend of variation of burr height with machining parameters is noticed as that of experiment set I.

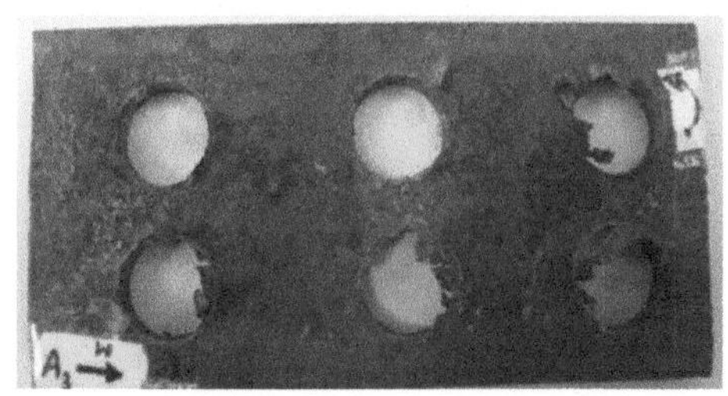

FIGURE 6: Photographic view of typical burr formed in experiment set II (left-right order: bottom for Sl. numbers 1–3, top for Sl. numbers 4–6).

3.3. Discussion on Drilling Burr Formation with a Back-Up Support at Dry Condition (Experiment Set III). Experiment set III is performed in dry condition by providing a low alloy steel back-up support. This backing plate is similar to that of the test specimen with slightly less width to facilitate easy clamping of the test piece in a machine vice (Figure 1). This is done to render a support during tool exit, so that rotation of shear plane from positive value to a negative one about a pivot point is suppressed. With this, burr formation is expected to be restricted. Burrs formed at these conditions are observed, and the burr heights measured at different cutting velocity and feed are shown in Figure 7. Burr height observed is found to be substantially reduced with the use of back-up support at dry condition compared to that at experiment set I. The largest burr height seen in experiment set III is 0.64 mm. Figure 8 depicts the photographic view of burr formed using a back-up plate in dry condition. Transient burrs are seen in these tests around the hole exit end at few locations only.

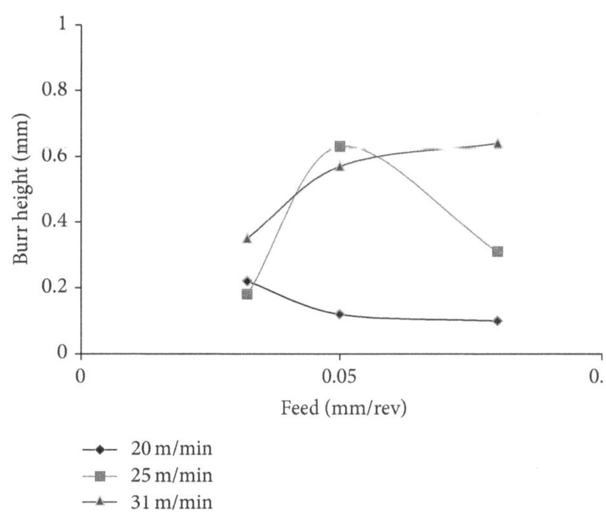

FIGURE 7: Plot of variation of burr height at different cutting velocity and feeds for experiment set III.

It is seen that burr height is quite less at low cutting velocity at all the three feeds. At low cutting velocity (V_c), when rise in machining temperature, and hence, plastic deformation is less, possibility of large burr through sustained

FIGURE 8: Photographic view of burr formed in experiment set III (left-right order: middle for Sl. numbers 1–3, top for Sl. numbers 4–6, bottom for Sl. numbers 7–9).

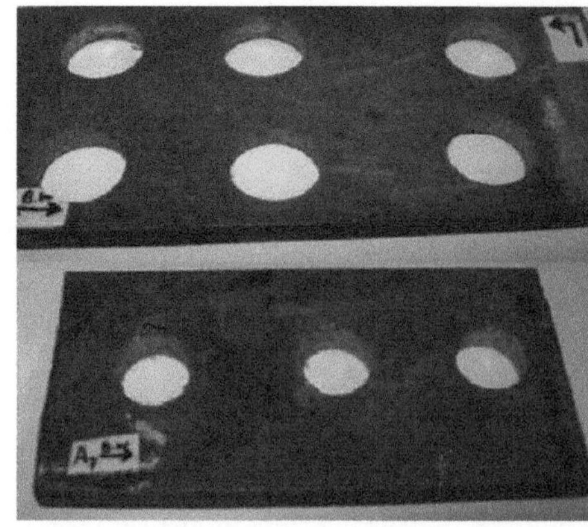

FIGURE 10: Photographic view of burr formed in experiment set IV (left-right order: middle for Sl. numbers 1–3, top for Sl. numbers 4–6, bottom for Sl. numbers 7–9).

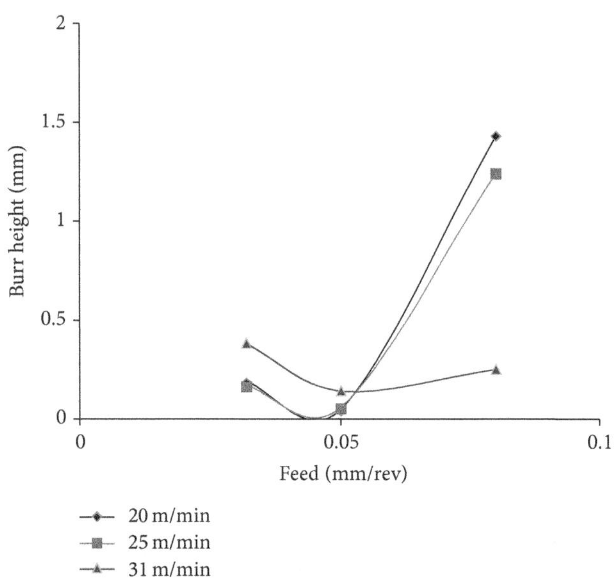

FIGURE 9: Plot of variation of burr height at different cutting velocity and feeds for experiment set IV.

bending is also expected to be less. Minimum burr height of only 0.10 mm is observed with the use of back-up support at dry condition at a feed of 0.08 mm/rev and cutting velocity of 20 m/min. Burr height more than 0.5 mm but less than 0.65 mm is found at some other higher cutting velocity conditions.

3.4. Discussion on Drilling Burr Formation Using Water-Cooled Condition with a Back-Up Support (Experiment Set IV). Experiment set IV is carried out with a back-up support material under water-cooled condition. Burrs formed at these conditions are noted, and the result is shown in Figure 9. Photographic views of burrs present in this experiment are shown in Figure 10. It is found that burrs are reduced considerably under this water-cooled condition using a back-up support at 0.032 and 0.05 mm/rev feed conditions. Quite less burrs as low as 0.04 mm and 0.05 mm in height are obtained at a feed of 0.05 mm/rev and cutting velocity (V_c) of 20 and 25 m/min, respectively. However, at a higher feed of 0.08 mm/rev and cutting velocity (V_c) of 20 and 25 m/min, large burr of more than 1 mm is detected. Only at one or two points, small size burrs are found in this experiment set.

Combined effect of attaching the back-up support and reduction in cutting temperature due to water cooling may have caused the formation of less burr height. The back-up support is expected to restrict the rotation of shear plane to a negative orientation as indicated in Figure 2, and to reduce burr formation significantly. Water cooling reduces cutting temperature and may have not allowed large increase in plasticity of the workpiece during machining causing less tendency of burr formation.

3.5. Discussion on Drilling Burr Formation Using Exit Edge Beveling under Dry Condition (Experiment Set V). Variation of burr height with the change in feed at three cutting velocities is plotted in Figure 11 with the provision of 31° edge beveling in dry condition. The burr formed is observed, and its photograph is shown in Figure 12. Moderate burr height less than 1 mm is formed at all the machining conditions undertaken in experiment set V. Burr size in this case is lower than that without using a back-up support or an edge bevel but higher than that using a back-up support. Gradual reduction of depth of cut along the bevel needs reducing force values and tends to reduce burr formation. At some points around the exit edge, few burrs are found to be attached. Further experiments are next performed using water-cooled condition to reduce temperature maintaining the exit edge bevel of 31° to investigate its effectiveness. This observation is also supported by somewhat similar reports made previously [15, 23, 27] under varied experimental conditions.

3.6. Discussion on Drilling Burr Formation Using an Exit Edge Bevel with Water Cooling (Experiment Set VI). Experiment

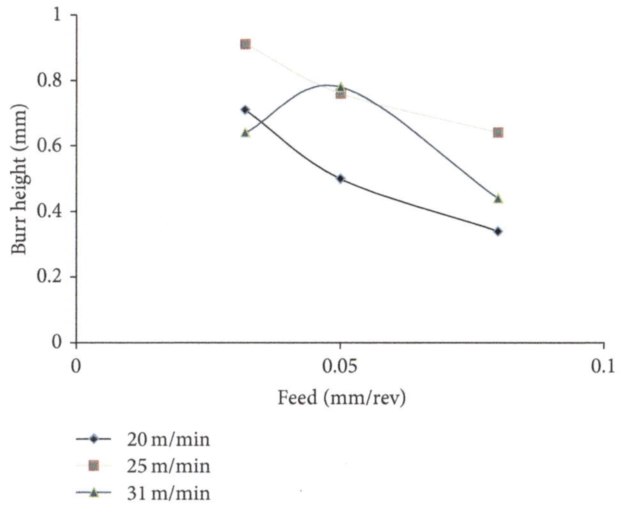

FIGURE 11: Plot of variation of burr height at different cutting velocity and feeds for experiment set V.

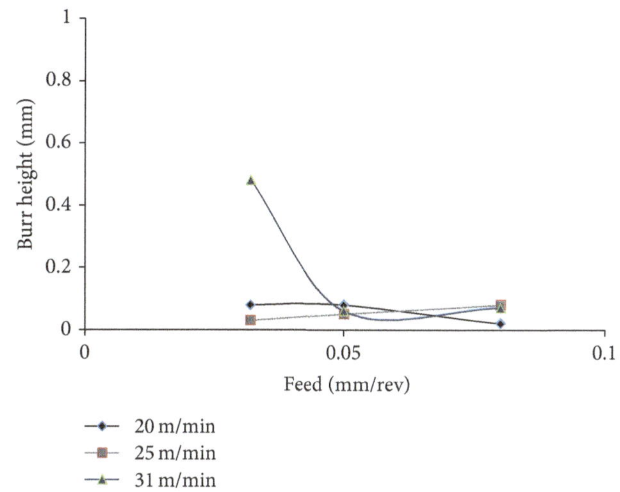

FIGURE 13: Plot of variation of burr height at different cutting velocity and feeds for experiment set VI.

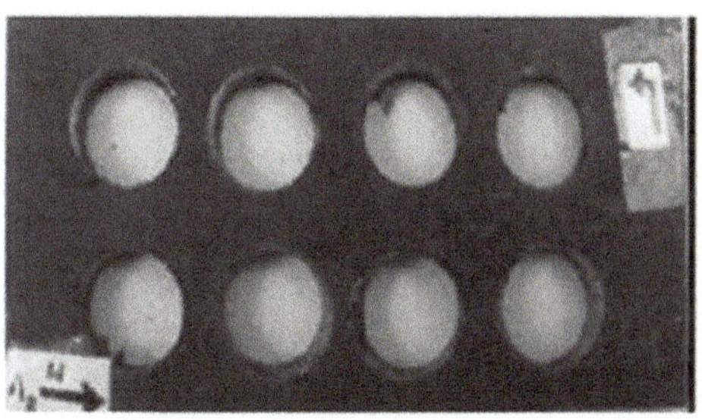

FIGURE 12: Photographic view of burr formed in experiment set V (left-right order: bottom for Sl. numbers 1–4, top for Sl. numbers 5–8, right side for Sl. number 9).

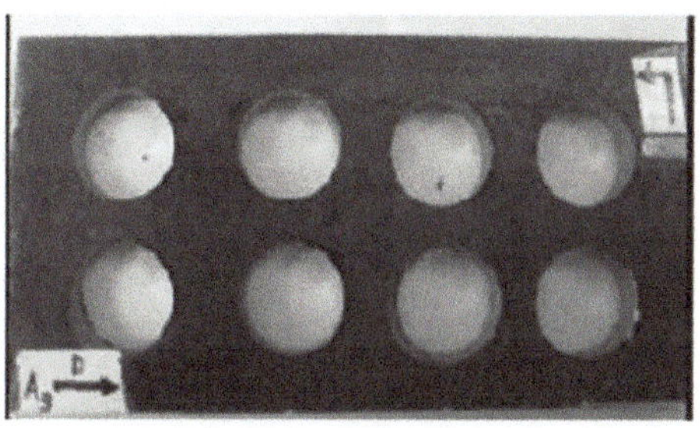

FIGURE 14: Photographic view of burr formed in experiment set VI (left-right order: bottom for Sl. numbers 1–4, top for Sl. numbers 5–8, right side for Sl. number 9).

set VI is carried out in water-cooled condition with three values of cutting velocity and feed with the provision of an exit edge bevel of 31°. Burr formation in these conditions is also observed to be similar to the other five experiment sets. Results are presented in the form of Figure 13. Quite less burr height is observed from the plot except at 31 m/min cutting velocity and 0.032 mm/rev feed condition. The photographic view of drilling burr corresponding to experiment set VI is shown in Figure 14. Tiny infrequent burr is seen around the exit edge of the drilled hole in this experiment set.

Figure 14 reveals that burr is reduced substantially by using an exit edge bevel angle of 31° at water-cooled condition. Only at a cutting velocity (V_c) of 31 m/min and feed of 0.032 mm/rev, large burr height of 0.48 mm is observed. Under other experimental conditions, low burr height of up to 0.08 mm is observed. At a cutting velocity (V_c) of 20 m/min and a feed of 0.08 mm/rev, a minimum burr height of 0.02 mm is obtained. This is the smallest height of burr seen among different sets of the present experimental work, and hence, the condition corresponding to this quite low burr can be recommended to adopt.

This small burr height may have occurred mainly due to the provision of the 31° exit edge bevel that causes gradual reduction of depth of cut when the drill approaches the rear surface of the hole. This results in gradually less requirement of cutting force during tool exiting, and therefore, needing no additional back-up support during emergence of the drill from the rear side of the job. Consequently burr formation may have been suppressed significantly as negative shear plane formation becomes less likely. Water-cooled machining conditions further help in reducing plastic deformation by taking away the heat generated and reducing the extent of burr formation. This is in line with the report on the temperature effect on burr formation [23, 27].

Burr height at the exit edge is next modeled as a function of cutting conditions with a complex nonlinear algorithm, namely, artificial neural networks (ANN), or simply neural networks (NN). Detail of the NN applied in this work is given in the following sections.

4. The Neural Networks (NN) Model Used

There are several algorithms used in a neural network. In the present work, Levenberg-Marquardt multilayer Neural Networks (NN) with back propagation training algorithm and feed forward system [32–34] are used to model burr height using the data observed. Matlab software package with neural network toolbox is utilized in this work. In this algorithm, an iterative gradient method is employed to compute connection weights corresponding to minimum total mean-square error between the obtained output of the multilayer network and the target output. Multilayer NN consists of an input layer, one or more hidden layer(s) and an output layer [32, 33]. The neural network structure used in the present work is shown in Figure 15, where there are one input layer having five input nodes, a single hidden layer, and an output layer with one output node. Cutting velocity, feed, use of coolant, back-up support, and edge bevel comprise of the input layer, and burr height is there as the output node. The NN algorithm used in this work is detailed elsewhere [34].

5. Using the NN for Testing and Experimental Validation

The neural networks (NN) are first trained with a training data set. There are total 54 experimental datasets. These datasets are shown in Table 3. Train ratio chosen is 80/100 which means that 44 datasets (i.e., 80% of 54 datasets) are used for training (1st columns Sl. number 1 to 44). A validation ratio of 10/100 is used, meaning the use of 5 samples (10% of datasets) for validation (1st columns Sl. numbers 45 to 49). Testing ratio of 10/100 is selected in this work that means 5 sample data (10% of datasets) are to be used for testing (1st columns Sl. numbers 50 to 54). Training sample data are required for determining weights of the network during training. Validation of sample data is used to measure network generalization and to halt training when generalization stops improving. The testing sample data have no effect on training. They provide an independent measure of network performance during and after training. The training dataset consists of 44 sets of sample data consisting of normalized datasets of input data and the corresponding output data. The normalizing factor considered is (x_{in}/x_{max}) where x_{in} is input data and x_{max} is the maximum value data. The five input variables are cutting velocity, feed, use of coolant, use of a back-up support, and provision of exit edge bevel, and the output is experimentally observed burr height at the exit edge. For use of NN training, dry condition, use of no back-up, and no edge bevel of the work piece are assigned 0 values as the input, and water-cooled condition, use of back-up support, and 31° edge bevel angle of the workpiece are assigned a value of 1 each.

ANN-based burr size modeling is done in MATLAB software package using neural network toolbox. Levenberg-Marquardt multilayer neural networks (NN) with back propagation training algorithm are employed in this work. Maximum number of epochs chosen is 1000. Initial and maximum values of mu (a factor promoting convergence of a network by a typical iterative method) are 0.001 and 10^{10} with the decreasing and increasing factors of 0.1 and 10, respectively. Minimum performance gradient (MSE) considered is 10^{-7}.

The number of hidden neurons chosen in each hidden layer is (2 × input + output). Considering 1 hidden layer, the total number of hidden neurons becomes 11 for an optimum structure of NN [33, 34]. During network analysis, it is found that in general, increasing the hidden layer from 1 to 3 results in little change; the network becomes too complex to solve the problem, and so, the number of hidden layer is usually taken as 1 and the number of hidden neurons can be chosen to be 11 as an optimum one. After training of the NN, all of the 54 sample datasets are used for getting the output. The estimated burr height, thus obtained, and the error of estimation are shown in Table 3 in the last two columns beside the input dataset.

The comparison of experimental findings and neural network estimates of burr height is displayed in Figure 16, and percentages of error between experimental values and model estimates are shown in Figure 17. It is found that the NN model estimates are having quite close matching with the experimental data, barring few deviations, and showing the effectiveness of the NN algorithm for modeling the input-output system to outline the possibility of estimation of burr height within the experimental domain. Occasionally, only small deviations are observed between the estimates and measured burr height. This may be due to the inherent experimental variability of the machining system and possible existence of high degree of nonlinearity in the system. It is clear from Figures 16 and 17 that model estimates deviate more when burr height is noticeably high. In the case of experiments using back-up support, or edge bevel with the use of water cooling environment that gives remarkably less burr height, estimation error is quite less. ANN was employed in earlier works [30, 32] successfully to model and estimate burr size. In line with this, finding out the condition to reduce burr height significantly is likely to be facilitated by the proposed modeling technique.

6. Conclusions

From the present work with 14 mm diameter HSS twist drills for making holes in low alloy steel specimens, conclusions drawn are given below.

(i) Usual drilling under dry condition shows large burr formation at the exit edge. Applying water as the cutting fluid does not yield remarkable results in reducing burr height. Absence of back-up support or bevel during tool exit may have resulted in formation of large burr both in dry and wet conditions.

(ii) Use of a back-up support reduces burr height substantially within the domain of experiments conducted due to less possibility of negative orientation of the shear plane. Use of water is found to reduce burr height further at these experimental conditions. This may be due to lowering of temperature, and hence, lessening plasticity of workpiece.

(iii) Use of 31° exit edge bevel, made with a standard twist drill of 118° point angle, causes significant reduction in burr height in most of the cutting conditions. Use of water cooling minimizes burr height to a great extent. Slow gradual reduction of depth of cut along the bevel causes decreasing requirement of cutting

TABLE 3: Training dataset for neural networks (NN) and estimated burr height.

Training data number	Cutting velocity, V_c (m/min)	Feed (mm/rev)	Cooling applied	Use of back up plate	Edge beveliing	Measured burr height (ΔM) (mm)	NN estimated burr height (ΔS) (mm)	Percentage of predication error $\frac{\Delta M - \Delta S}{\Delta M} \times 100$
1	20	0.032	0	0	0	4	4.178	−4.45
2	20	0.032	1	0	0	2.73	2.794	−2.34432
3	20	0.032	0	1	0	0.22	0.214	2.727273
4	20	0.032	1	1	0	0.18	0.192	−6.66667
5	20	0.032	0	0	1	0.71	0.818	−15.2113
6	20	0.032	1	0	1	0.08	0.074	7.5
7	20	0.05	0	0	0	5.22	5.047	3.314176
8	20	0.05	1	0	0	6.17	5.938	3.76013
9	20	0.05	0	1	0	0.12	0.132	−10
10	20	0.05	1	1	0	0.04	0.0393	1.75
11	20	0.05	0	0	1	0.5	0.49	2
12	20	0.05	1	0	1	0.08	0.076	5
13	20	0.08	0	0	0	1.93	2.092	−8.39378
14	20	0.08	1	0	0	4.17	3.075	26.25899
15	20	0.08	0	1	0	0.1	0.109	−9
16	20	0.08	1	1	0	1.43	1.373	3.986014
17	20	0.08	0	0	1	0.34	0.429	−26.1765
18	20	0.08	1	0	1	0.02	0.024	−20
19	25	0.032	0	0	0	7	6.625	5.357143
20	25	0.032	1	0	0	4.76	4.894	−2.81513
21	25	0.032	0	1	0	0.18	0.219	−21.6667
22	25	0.032	1	1	0	0.16	0.159	0.625
23	25	0.032	0	0	1	0.91	0.791	13.07692
24	25	0.032	1	0	1	0.03	0.027	10
25	25	0.05	0	0	0	5.52	5.66	−2.53623
26	25	0.05	1	0	0	5.66	6.044	−6.78445
27	25	0.05	0	1	0	0.63	0.568	9.84127
28	25	0.05	1	1	0	0.05	0.041	18
29	25	0.05	0	0	1	0.76	0.664	12.63158
30	25	0.05	1	0	1	0.05	0.047	6
31	25	0.08	0	0	0	4.71	4.246	9.85138
32	25	0.08	1	0	0	3.27	3.718	−13.7003
33	25	0.08	0	1	0	0.31	0.362	−16.7742
34	25	0.08	1	1	0	1.24	1.25	−0.80645
35	25	0.08	0	0	1	0.64	0.523	18.28125
36	25	0.08	1	0	1	0.08	0.079	1.25
37	31	0.032	0	0	0	3	3.823	−27.4333
38	31	0.032	1	0	0	4.84	5.052	−4.38017
39	31	0.032	0	1	0	0.35	0.368	−5.14286
40	31	0.032	1	1	0	0.38	0.331	12.89474
41	31	0.032	0	0	1	0.64	0.732	−14.375
42	31	0.032	1	0	1	0.48	0.471	1.875
43	31	0.05	0	0	0	5.05	4.479	11.30693
44	31	0.05	1	0	0	5.88	6.255	−6.37755
45	31	0.05	0	1	0	0.57	0.472	17.19298

Table 3: Continued.

Training data number	Cutting velocity, V_c (m/min)	Feed (mm/rev)	Cooling applied	Use of back up plate	Edge beveliing	Measured burr height (ΔM) (mm)	NN estimated burr height (ΔS) (mm)	Percentage of predication error $\frac{\Delta M - \Delta S}{\Delta M} \times 100$
46	31	0.05	1	1	0	0.14	0.139	0.714286
47	31	0.05	0	0	1	0.78	0.734	5.897436
48	31	0.05	1	0	1	0.48	0.358	25.41667
49	31	0.08	0	0	0	4.2	4.64	−10.4762
50	31	0.08	1	0	0	5.31	5.27	0.753296
51	31	0.08	0	1	0	0.64	0.452	29.375
52	31	0.08	1	1	0	0.35	0.449	−28.2857
53	31	0.08	0	0	1	0.44	0.423	3.863636
54	31	0.08	1	0	1	0.07	0.079	−12.8571

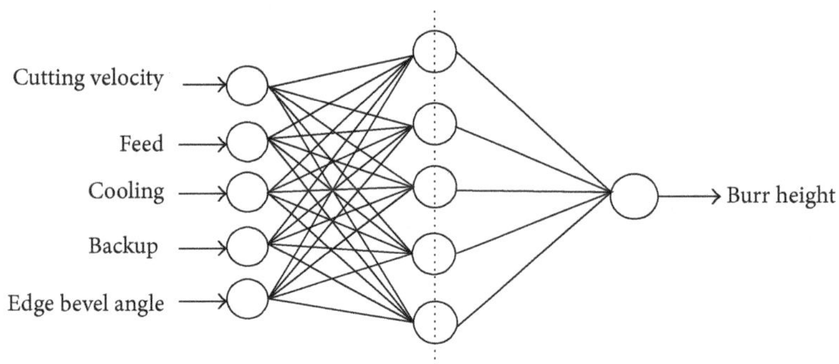

Figure 15: Architecture of artificial neural networks used.

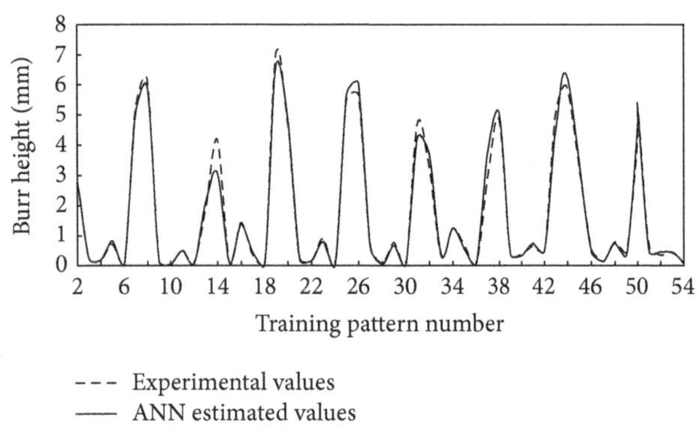

Figure 16: Experimental and ANN estimates of burr height for training patterns.

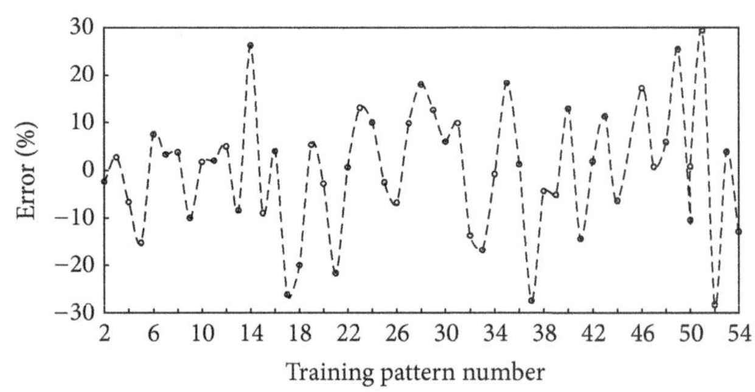

Figure 17: Experimental and ANN estimates of burr height in percentage of error.

force/torque needing reduced back-up support. This possibly causes suppression of burr formation to a large extent. Application of water cools down the tool and workpiece. This reduces plastic deformation resulting in further lowering of the chance of formation of large burrs.

(iv) Within the experimental domain, hole-making at cutting velocity 20 m/min and feed 0.08 mm/rev using 14 mm drills and 31° exit edge bevel gives minimum burr under water-cooled condition. In this case, burr height as low as 0.02 mm is achieved. Hence, this condition may be recommended to obtain negligible burr.

(v) The three-layer neural networks algorithm is applied to model the experimental data, and the model estimates are seen to have close matching with the observed burr height with small deviations, thereby showing the possibility of using the model for estimation of burr height within the domain of experimentation.

Conflict of Interests

The authors declare that they have no conflict of interests regarding the publication of this paper.

Acknowledgment

This research received no specific grant from any funding agency in the public, commercial, or not-for-profit sectors. The laboratory facilities available at Kalyani Government Engineering College are utilized for doing the work.

References

[1] L. K. Gillespie, "Burrs produced by drilling," Tech. Rep. BDX-613-1248, Bendix Corporation, 1975.

[2] L. K. Gillespie and P. T. Blotter, "The formation and properties of machining burrs," *Journal of Engineering for Industry, Transactions of the ASME*, vol. 98, no. 1, pp. 66–74, 1976.

[3] K. Nakayama and M. Arai, "Burr formation in metal cutting," *CIRP Annals, Manufacturing Technology*, vol. 36, no. 1, pp. 33–36, 1987.

[4] I. W. Park and D. A. Dornfeld, "A study of burr formation mechanism," *Journal of Engineering Materials and Technology, Transactions of the ASME*, vol. 133, pp. 75–87, 1991.

[5] G.-L. Chern and D. A. Dornfeld, "Burr/breakout model development and experimental verification," *Journal of Engineering Materials and Technology, Transactions of the ASME*, vol. 118, no. 2, pp. 201–206, 1996.

[6] S. Min, "Control chart of drilling exit burr in low carbon steel," LMA Report AISI4118, 2001.

[7] J. Kim, S. Min, and D. A. Dornfeld, "Optimization and control of drilling burr formation of AISI 304L and AISI 4118 based on drilling burr control charts," *International Journal of Machine Tools and Manufacture*, vol. 41, no. 7, pp. 923–936, 2001.

[8] K. Lee and D. A. Dornfeld, "Micro-burr formation and minimization through process control," An e-Scholarship Respiratory, Barkclcy, Calif, USA, Univcrsity of California, 2004.

[9] Y. B. Guo and D. A. Dornfeld, "Finite element analysis of drilling burr minimization with a backup material," *Transactions of NAMRI/ SME*, vol. 26, pp. 207–212, 1998.

[10] Y. B. Guo and D. A. Dornfeld, "Finite element modeling of burr formation process in drilling 304 stainless steel," *Journal of Manufacturing Science and Engineering, Transactions of the ASME*, vol. 122, no. 4, pp. 612–619, 2000.

[11] I. W. Park and D. A. Dornfeld, "A study of burr formation processes using the finite element method—part I," *Journal of Engineering Materials and Technology, Transactions of the ASME*, vol. 122, no. 2, pp. 221–228, 2000.

[12] D. A. Dornfeld, "Strategies for preventing and minimizing burr formation," An e-Scholarship Respiratory, Berkeley, Calif, USA, University of California, 2000.

[13] J. C. Aurich, D. Dornfeld, P. J. Arrazola, V. Franke, L. Leitz, and S. Min, "Burrs-analysis, control and removal," *CIRP Annals, Manufacturing Technology*, vol. 58, no. 2, pp. 519–542, 2009.

[14] R. Neugebauer, G. Schmidt, and M. Dix, "Size effects in drilling burr formation," in *Proceedings of the CIRP International Conference on Burrs*, pp. 117–128, Kaiserslautern, Germany, 2009.

[15] S. Kundu, *Experimental investigation on the effect of different drilling conditions on burr formation towards its minimization [Masters Dissertation]*, Kalyani Government, Engineering College, Kalyani, India, 2011.

[16] S.-L. Ko and J.-K. Lee, "Analysis of burr formation in drilling with a new-concept drill," *Journal of Materials Processing Technology*, vol. 113, no. 1–3, pp. 392–398, 2001.

[17] B. Shyu, "Burr reduction by tool design," LMA Research Reports, University of California, Berkeley, Calif, USA, 2001.

[18] S. L. Ko and J. E. Chang, "Development of drill geometry for burr minimization in drilling," *CIRP Annals, Manufacturing Technology*, vol. 52, no. 1, pp. 45–48, 2003.

[19] T.-R. Lin and R.-F. Shyu, "Improvement of tool life and exit burr using variable feeds when drilling stainless steel with coated drills," *International Journal of Advanced Manufacturing Technology*, vol. 16, no. 5, pp. 308–313, 2000.

[20] S. Min, D. A. Dornfeld, and Y. Nakao, "Influence of exit surface angle on drilling burr formation," *Journal of Manufacturing Science and Engineering, Transactions of the ASME*, vol. 125, no. 4, pp. 637–644, 2003.

[21] S. P. Pratim and S. Das, "Burr minimization in face milling: an edge bevelling approach," *Proceedings of the Institution of Mechanical Engineers B: Journal of Engineering Manufacture*, vol. 225, no. 9, pp. 1528–1534, 2011.

[22] P. P. Saha and S. Das, "An investigation on the effect of machining parameters and exit edge beveling on burr formation in milling," *Journal of Mechatronics and Intelligent Manufacturing*, vol. 2, no. 1/2, pp. 73–84, 2011.

[23] K. Roy, P. Mukherjee, and U. K. Hansda, "An experimental investigation on drilling burr formation," in *Proceedings of Poster Presentation of the 3rd International and 24th AIMTDR Conference*, pp. 211–216, Visakhapatnam, India, 2010.

[24] S. Tripathi and D. A. Dornfeld, "Review of geometric solutions for milling burr prediction and minimization," *Proceedings of the Institution of Mechanical Engineers B: Journal of Engineering Manufacture*, vol. 220, no. 4, pp. 459–466, 2006.

[25] K. H. Kim, C. H. Cho, S. Y. Jeon, K. Lee, and D. A. Dornfeld, "Drilling and deburring in a single process," *Proceedings of the Institution of Mechanical Engineers Part B: Journal of Engineering Manufacture*, vol. 217, no. 9, pp. 1327–1331, 2003.

[26] J. Choi, "Formation of Burr when drilling multi-layer materials," in *Proceedings of the CODEF Annual Meeting*, University of California, Berkeley, Calif, USA, 2003.

[27] L. K. Lauderbaugh Saunders and C. A. Mauch, "An exit burr model for drilling of metals," *Journal of Manufacturing Science and Engineering, Transactions of the ASME*, vol. 123, no. 4, pp. 562–566, 2001.

[28] J. Kim and D. A. Dornfeld, "Development of an analytical model for drilling burr formation in ductile materials," *Journal of Engineering Materials and Technology, Transactions of the ASME*, vol. 124, no. 2, pp. 192–198, 2002.

[29] U. Heisel and M. Schaal, "Burr formation in short hole drilling with minimum quantity lubrication," *Production Engineering*, vol. 3, no. 2, pp. 157–163, 2009.

[30] S. R. Karnik, V. Gaitonde, and J. P. Davim, "Integrating Taguchi principle with genetic algorithm to minimize burr size in drilling of AISI 316L stainless steel using an artificial neural network model," *Proceedings of the Institution of Mechanical*

Engineers B: Journal of Engineering Manufacture, vol. 221, no. 12, pp. 1695–1704, 2007.

[31] Y. Nakao and Y. Watanabe, "Measurements and evaluations of drilling burr profile," *Proceedings of the Institution of Mechanical Engineers B: Journal of Engineering Manufacture*, vol. 220, no. 4, pp. 513–523, 2006.

[32] V. N. Gaitonde and S. R. Karnik, "Minimizing burr size in drilling using artificial neural network (ANN)-particle swarm optimization (PSO) approach," *Journal of Intelligent Manufacturing*, vol. 23, pp. 1783–1793, 2012.

[33] R. P. Lippmann, "An introduction to computing with neural nets," *IEEE ASSP Magazine*, vol. 4, no. 2, pp. 4–22, 1987.

[34] J. J. Moré, "The Levenberg-Marquardt algorithm: implementation and theory," *Numerical Analysis: Lecture Notes in Mathematics*, vol. 630, pp. 105–116, 1978.

18

Analyzing the Effectiveness of Microlubrication Using a Vegetable Oil-Based Metal Working Fluid during End Milling AISI 1018 Steel

Vasim Shaikh,[1] Nourredine Boubekri,[2] and Thomas W. Scharf[3]

[1] *Department of Applied Engineering, Safety and Technology, Millersville University of Pennsylvania, Millersville, PA 17551, USA*
[2] *Department of Engineering Technology, University of North Texas, Denton, TX 76207, USA*
[3] *Department of Materials Science and Engineering, University of North Texas, Denton, TX 76207, USA*

Correspondence should be addressed to Nourredine Boubekri; boubekri@unt.edu

Academic Editor: Gilles Dessein

Microlubrication minimizes the exposure of metal working fluids to the machining operators leading to an economical, safer, and healthier workplace environment. In this study, a vegetable oil-based lubricant was used to conduct wear analysis and to analyze the effectiveness of microlubrication during end milling AISI 1018 steel. A solid carbide cutting tool with bright oxide finish was used with varying cutting speed and feed rate having a constant depth of cut. Abrasion was the dominant wear mechanism for all the cutting tools under consideration. Other than abrasion, sliding adhesive wear of the workpiece materials was also observed. The scanning electron microscope investigation of the used cutting tools revealed microfatigue cracks, welded microchips, and unusual built-up edges on the cutting tools flank and rake side. A full factorial experiment was conducted and regression models were generated for both the sides of tool flank wear. The study shows that with a proper selection of the cutting parameters it is possible to obtain higher tool life.

1. Introduction

Metal working fluids (MWFs) are used to cool and lubricate the tool/workpiece interface during machining. The MWFs perform several important functions including reducing the friction-heat generation and dissipating generated heat at tool/workpiece interface which results in the reduction of tool wear. MWFs also flush the chips away from the tool and clean the workpiece causing less built-up edge (BUE). Therefore, MWFs cannot be completely avoided; however, their exposure to machine operators is a cause of growing occupational health hazards. U.S. National Institute for Occupational Safety and Health (NIOSH) recommends that occupational permissible exposure limits to MWF aerosols be limited to 0.4 mg/m^3 of thoracic particulate mass which corresponds to approximately 0.5 mg/m^3 of total particulate mass as a time-weighted average (TWA) concentration for up to 10 hours per day during a 40-hour workweek [1]. However, the oil mist level in the U.S. automotive parts manufacturing facilities has been estimated to be 20–90 mg/m^3 with the use of conventional lubrication by flood coolant [2]. The exposure to such amounts of metal working fluid may contribute to adverse health effects and safety issues, including toxicity, dermatitis, respiratory disorders, and cancer [3]. Also, the costs related to the use of MWFs range from 7 to 17% of the total costs of the manufactured workpiece [4] as compared to the tool cost which is only about 2–4% [5]. Microlubrication is also known as "minimum quantity lubrication" (MQL) and "near-dry machining." In microlubrication, a small amount of cutting fluid, usually 2–100 milliliters per hour mL/hr [6–8] in the form of aerosol is delivered to the cutting tool/workpiece interface and does not recirculate through the lubrication system.

It is almost all evaporated at the point of application. Hence no recirculation is required. It is important however to ensure an efficient extraction of aerosol from the machine. In a flood application, the same coolant is recirculated through the system, filtered, and used again [9]. The lubrication is obtained via the lubricant, and the cooling is achieved by the pressurized air that reaches the cutting surface in a microlubrication application. Further, microlubrication reduces induced thermal shock and helps to increase the workpiece surface integrity in situations of high tool pressure [10]. The efficiency of a machining process depends on the tribological conditions which lower the tool wear rate obtained by the cooling and lubricating effectiveness of the cutting fluid. The main objective of this work is to investigate the effectiveness of microlubrication during end milling AISI 1018 steel with a solid carbide cutter under varying cutting speed, feed rate levels, and a constant depth of cut. A full factorial experiment and regression models for the tool flank wear are generated.

2. Experimental Methods

End milling experiments were carried out on a Mori Seiki computer numerical control (CNC) Dura vertical 5060 milling machine with a spindle power of 15 HP and a maximum spindle speed of 10,000 revolutions per minute (RPM). Figure 1 shows the machining set-up. The nozzle was positioned approximately 15 cms from the cutting tool tip at an angle of 45°. AISI 1018 steel was used as workpiece material. The alloy can be easily formed, machined, welded, and fabricated. The alloy is a free machining grade that is often employed in high volume screw machine parts applications and is commonly employed in shafts, spindles, pins, rods, sprocket assemblies, and a wide variety of component parts.

A Kuroda Ecosaver KEP3 microlubrication unit was installed on the milling machine to provide a constant aerosol flow rate of 12 mL/hr through one external nozzle as shown in Figure 2. Accu-Lube LB-2000 was used as minimum quantity lubricant manufactured by ITW Rocol North America. The lubricant was specially recommended for microlubrication of ferrous and nonferrous materials. It is manufactured using renewable raw materials and is nontoxic and biodegradable. It is not diluted in water before use.

Vegetable oils are by their chemical nature long chain fatty acid triesters of glycerol and provide most of the desirable lubricant properties such as good boundary lubrication, high viscosity index, high flash point, and low volatility. These advantages are mainly due to their polar ester structure and higher molecular weight in comparison to petroleum derived hydrocarbons [11]. Vegetable oil-based MWFs are highly attractive substitutes for petroleum oil-based MWFs because they are environmentally friendly, renewable, less toxic, and readily biodegradable. As a result, vegetable oil-based MWFs are more potential candidates for use in metal working industry as MWFs [12].

It is suggested that vegetable oil-based MWFs perform better than mineral oil-based MWFs because of their lubricity. Vegetable oil carries a slight polar charge. This charge draws the vegetable oil-based MWFs molecule to a metallic surface and is tenacious enough to resist being wiped off. Mineral oil-based MWFs have no charge and, therefore, adhere less tightly to a metal surface. Also, mineral oil-based MWFs are just a straight hydrocarbon. The vegetable oil-based MWFs have functional groups containing oxygen which makes them more attractive to the metal surface and they bond more tightly making them better MWFs [13]. It is reported that vegetable oil-based MWFs increase the tool life 30% more and give better lubricity as compared to mineral oil-based MWFs.

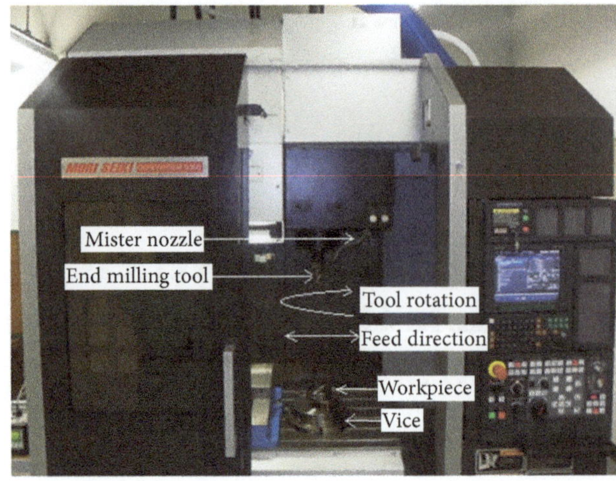

Figure 1: Machining set-up.

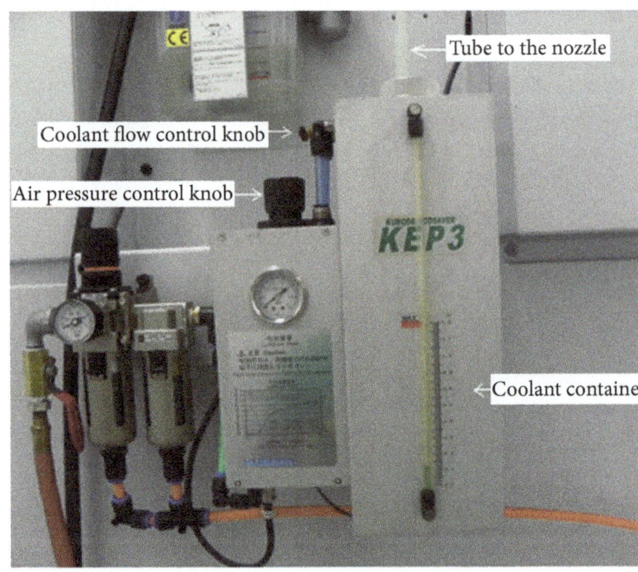

Figure 2: Microlubrication unit.

Vegetable oil-based MWFs have higher flashpoint and are more compatible with human skin than mineral oil-based MWFs. Also, vegetable oil-based MWFs are able to machine chlorine-free. Disposal costs dramatically decrease if MWFs are used without chlorine which is an inherent part of most MWFs. Vegetable oil-based MWFs also producing less mist improve the tool life and surface finish of the machined surface [14].

An exhaust pump was used to collect the mist generated during machining to replicate the exact factory production environment. End milling was carried out with cutting speeds of 24, 30, and 36 meters per minute (m/min) and feed rates of 0.15 and 0.25 millimeters per revolution (mm/rev).

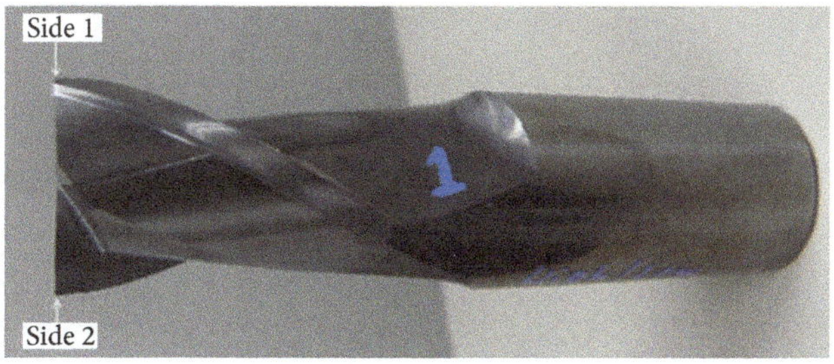

FIGURE 3: Cutting tool.

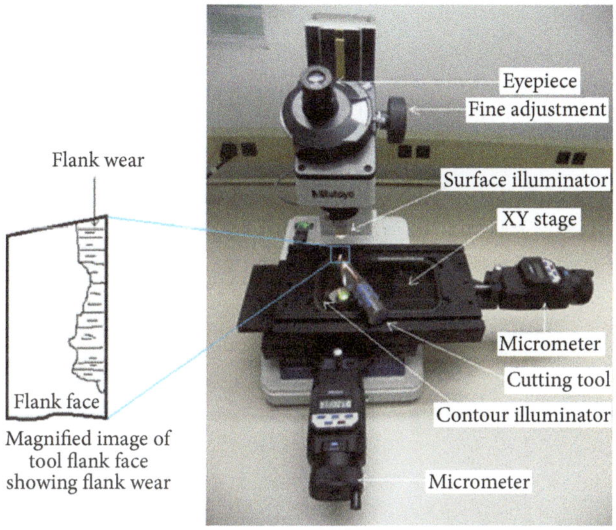

FIGURE 4: The Mitutoyo toolmakers microscope.

The cutting parameters were decided based on the recommendation in the ASM metal handbook for machining [15]. The axial depth of cut was 3.175 mm and the radial depth of cut was 6.35 mm. The cutting tool used in the experiment was 25.4 mm diameter of solid carbide end mill having square ends and two flutes with bright oxide finish manufactured by Guhring Inc. Figure 3 shows the cutting tool with clearly marked sides 1 and 2. The helix angle of the flute was 30° and the flute length was 38.1 mm. The radial rake angle is 9°. The radial relief is 10°.

The flank wear on both the flanks was measured at regular interval by the Mitutoyo toolmakers microscope. The microscope and a magnified inset image of the tool flank face showing the flank wear as seen from the microscope are shown in Figure 4.

The cutting tool was set on the XY stage with its flank aligned with the y-axis as seen from the eyepiece. The microscope has a contour illumination from the bottom suited for measurement and inspection of specimen contour and a surface illuminator for observation and inspection of specimen surfaces. A digital micrometer head was attached to the x- and y-axis and was set to zero. The micrometer was equipped with digital data display which gave live readings. The y-axis was carefully moved in $+y$-directions and was aligned with the maximum flank wear valley as seen on the tool from the eyepiece. Reading from the micrometer was then collected and recorded in manually.

TABLE 1: Factorial experiment layout of cutting speed and feed rate combination.

Solid carbide end mill	Cutting speeds (m/min)		
	24	30	36
Feed rates (mm/rev)			
0.15	24, 0.15	30, 0.15	36, 0.15
0.25	24, 0.25	30, 0.25	36, 0.25

The tool was declared failed if the flank wear of any one of the two flanks went above 0.5 mm [15]. A dual beam FEI Nova 200 NanoLab scanning electron microscope (SEM) was used to conduct the wear analysis of the cutting tool. The FEI Quanta environmental SEM (ESEM) was used to do the wear analysis of the workpiece. The secondary electron detector (ETD) was used to acquire all the SEM images. The toolmakers microscope has a maximum magnification of 40x. It is only suitable to measure the flank wear and to have an overall impression of the tool failure. However, SEM provides magnification which is way beyond the capability of the toolmakers microscope making it suitable to conduct the wear analysis of the cutting tool and the workpiece. The Vickers microhardness measurements were carried out using the Shimadzu dynamic ultramicrohardness tester model number DUH-211S. The hardness tester used a 115° standard pyramid indenter. The normal load used was 300 mN. The indentation depth is automatically measured as the indenter is pressed against the specimen eliminating human error. This allows dynamic measurement of changes that occur in the specimen's resistance to deformation during the indentation process. The hardness tester measures dynamic hardness and evaluates the hardness that corresponds to both plastic and elastic deformation. Indents were made at an increment of 10 μm up to the bulk of the material as shown in Figure 5. Adequate spacing of 50 μm was kept between two indents to avoid any possible alteration in readings due to the deformation caused by the earlier indent.

3. Design of Experiments

The study was conducted using a randomized factorial design. Full factorial design was used in the current study because the numbers of runs are uncertain as they were based on the tool failure which cannot be predetermined. Table 1 shows the factorial experiment with six cutting speed and feed rate combinations. The purpose to conduct a full factorial experiment was to effectively investigate the main effects as well as the interaction among the independent variables (i.e., cutting speed and feed rate). Each combination (i.e., treatment) was carried out only once, and there were no replicates. The experiment was fully randomized. An analysis of variance (ANOVA) was carried out for flank wear and F-ratios were calculated. All collected data were recorded using Microsoft Excel and transferred to Design Expert 8.0 analytical software for the ANOVA analyses.

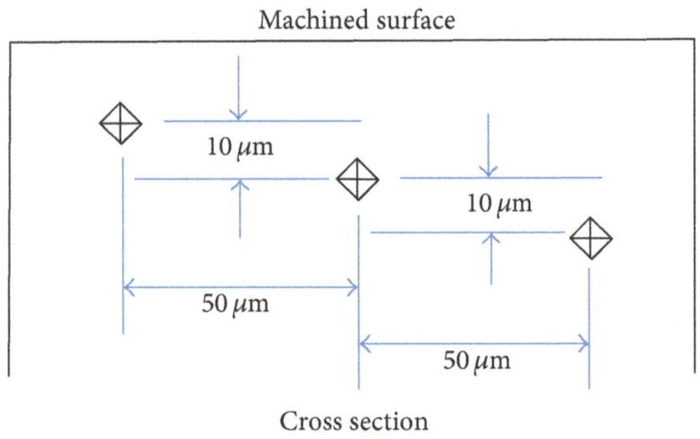

FIGURE 5: Schematics of the Vickers microhardness indenter.

3.1. ANOVA Assumptions

(1) Individual differences and errors of measurement are normally distributed within each group.

(2) Size of the variance, distribution of individual differences, and random errors are identical in each group.

(3) Individual differences and errors of measurement are independent from group to group.

3.2. Hypothesis

(1) *Null Hypothesis.* There is no significant difference between the responses (i.e., tool wear) obtained by varying the individual input variables levels (i.e., cutting speed and feed rate).

(2) *Alternate Hypothesis.* There is a significant difference between the responses (i.e., tool wear) obtained by varying the individual input variables levels (i.e., cutting speed and feed rate).

4. Results and Discussions

A standard technique used to check the normality assumption is to construct a normal probability plot [16]. The normal probability plot determines if the distribution of data approximates a normal distribution. In the ANOVA, the normal probability plot is usually more effective and straightforward when it is done with residuals.

The normal probability plot of residuals for both the flank wear sides is shown in Figures 6 and 7. The residual is the difference between the actual and predicted response values. The normal probability plot with a general straight line pattern indicates that the residuals follow a normal distribution. The far points which are slightly scattered do not follow any significant pattern. The analysis of variance is supposed to be robust for the normality assumption.

Figures 8 and 9 show plots of residuals versus predicted values which were used to test the assumption of constant variance for both sides 1 and 2 of the flank wear, respectively. To fulfill the test, these plots should be a random scatter. The constant variance in the data demonstrates that the data does not follow a particular pattern, indicating constant variance.

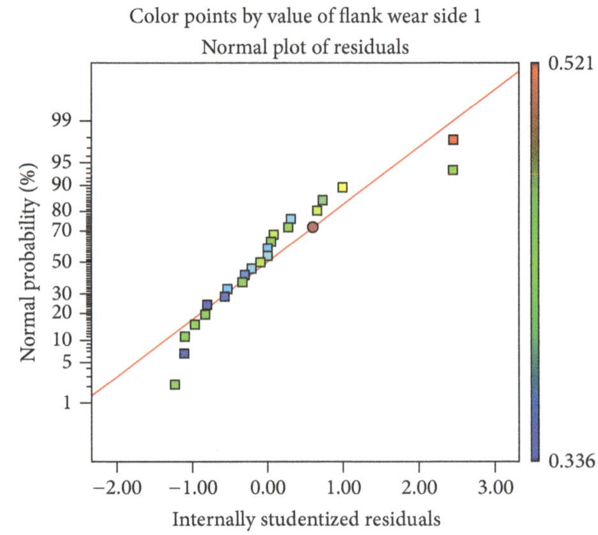

FIGURE 6: Normal plots of residual in data for flank wear side 1.

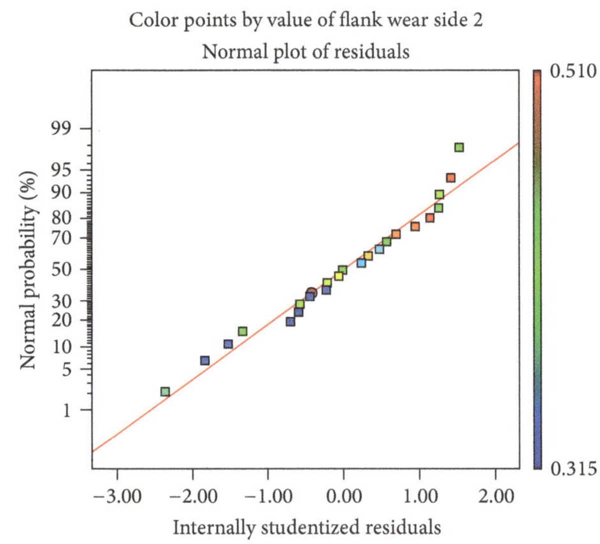

FIGURE 7: Normal plots of residual in data for flank wear side 2.

The ANOVA for flank wear side 1 is shown in Table 2. The model provided an F value of 10.57 which indicates that the model is statistically significant based on a 95% confidence level. The resulting R-squared value indicates that the model is able to predict 61.32 percent of the variation in the data. The remaining percent is considered noise and may not be predicted by this model, the sources of which may be manual variations while taking flank wear readings, machine vibrations, variations within the tools, and workpiece materials. The "Adeq Precision" measures the signal-to-noise ratio. The ratio of 7.85 indicates an adequate signal. Hence, the model can be used as a good predictor of the flank wear side 1 as a result of changing the cutting speed and feed rate levels. The resulting regression model is shown below.

Flank wear side 1 $= +0.5857 - (9.64E - 03 \times \text{speed}) - (0.2434 \times \text{feed}) + (0.0302 \times \text{speed} \times \text{feed})$.

TABLE 2: Analysis of variance for flank wear side 1.

ANOVA for selected factorial model [partial sum of squares—type III]						
Source	Sum of squares	df	Mean square	F value	P value Probability > F	
Model	0.037	3	0.012	10.57	0.0002	Significant
A speed	$6.99E-03$	1	$6.99E-03$	6.01	0.0236	
B feed	0.020	1	0.020	17.21	0.0005	
AB	$1.23E-03$	1	$1.23E-03$	1.06	0.3162	
Pure error	0.018	18	$1.02E-03$			
Cor total	0.060	23				

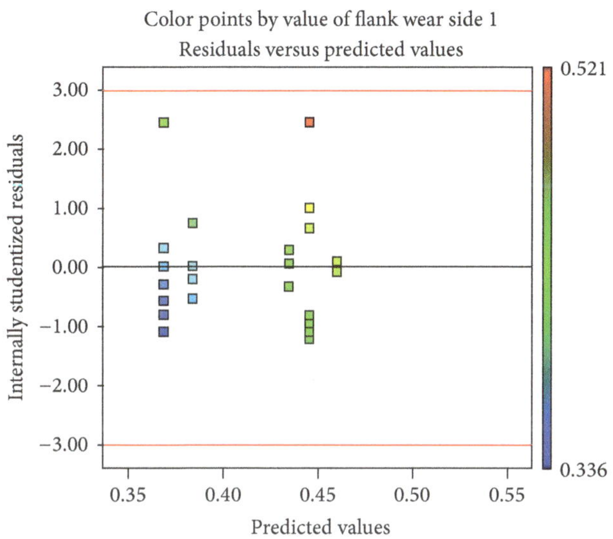

FIGURE 8: Constant variance in data for flank wear side 1.

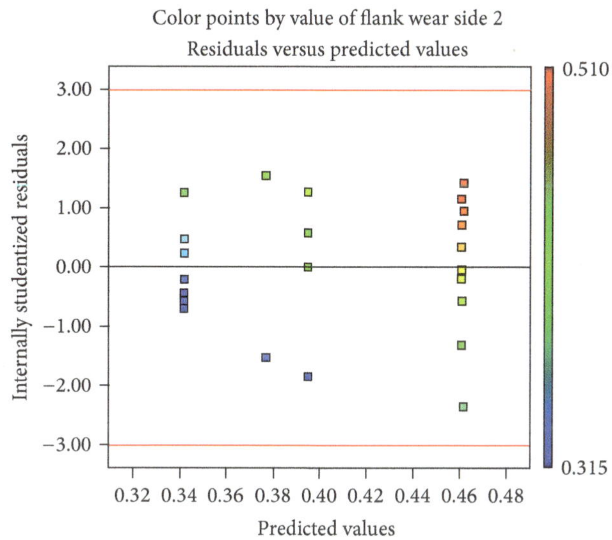

FIGURE 9: Constant variance in data for flank wear side 2.

Similarly, the ANOVA for flank wear side 2 is shown in Table 3. The model provided an F value of 8.27 which indicates that the model is statistically significant based on a 95% confidence level. The resulting R-squared value indicates that the model is able to predict 55.38 percent of the variation in the data. The remaining percent is considered noise and may not be predicted by this model, the sources of which may be the same as given for flank wear side 1. The "Adeq Precision" measures the signal-to-noise ratio. The ratio of 5.67 indicates an adequate signal. Hence, the model can be used as a good predictor of the flank wear side 2 as a result of changing the cutting speed and feed rate levels. The resulting regression model is shown below.

Flank wear side 2 $= +0.3447 + (7.36E-03 \times \text{speed}) + (0.3654 \times \text{feed}) - (0.0400 \times \text{speed} \times \text{feed})$.

The robustness of the regression models was validated by comparing the computed model values to the actual experimental values as shown in Table 4.

Figures 10 and 11 show the tool life in minutes and material removal volume (MRV) in mm^3 for different cutting speed and feed rate combinations, respectively. The MRV is the product of the length of cut and axial and radial depth of cut multiplied by the number of workpieces taken by the cutting tool to fail. Cutting with lower feed rate resulted in prolonged tool life and the highest MRV whereas cutting with higher feed rates decreased the tool life and MRV significantly. The cutting performance under microlubrication was approximately five times better in terms of tool life

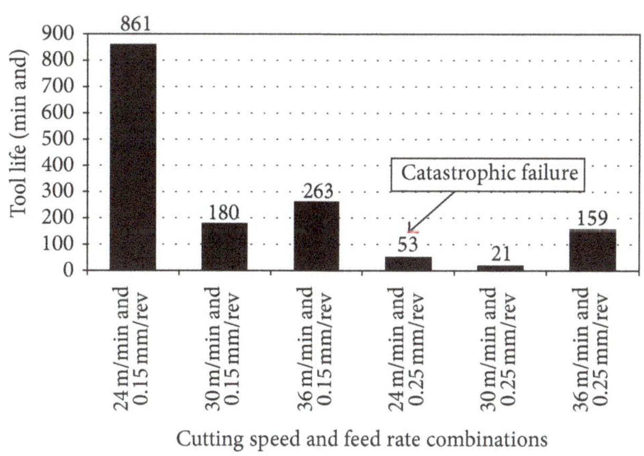

FIGURE 10: Tool life for different cutting speed and feed rate combinations.

TABLE 3: Analysis of variance for flank wear side 2.

ANOVA for selected factorial model [partial sum of squares—type III]						
Source	Sum of squares	df	Mean square	F value	P value Probability > F	
Model	0.054	3	0.018	8.27	0.0009	Significant
A speed	$2.27E-04$	1	$2.27E-04$	0.11	0.7491	
B feed	0.032	1	0.032	14.76	0.0010	
AB	$2.16E-03$	1	$2.16E-03$	1.00	0.3290	
Pure error	0.031	18	$1.72E-03$			
Cor total	0.097	23				

TABLE 4: Comparison of experimental and predicted flank wear values.

Flank wear side	Cutting speed and feed rate combination	Experimental mean value (mm)	Predicted value (mm)	% error
1	24 m/min and 0.15 mm/rev	0.43	0.42	+2.3
	36 m/min and 0.25 mm/rev	0.44	0.45	−2.2
2	24 m/min and 0.15 mm/rev	0.46	0.43	+6.9
	36 m/min and 0.25 mm/rev	0.34	0.34	0.0

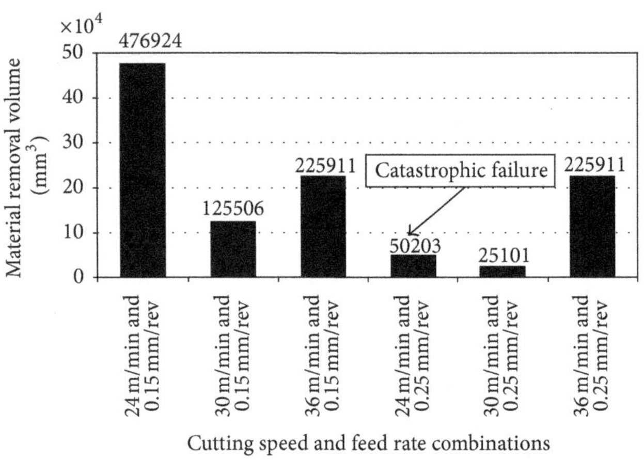

FIGURE 11: Material removal volume for different cutting speed and feed rate combinations.

and two times better in terms of MRV under low cutting speed and feed rate combination (i.e., 24 m/min and 0.15 mm/rev) as compared to high cutting speed and feed rate combination (i.e., 36 m/min and 0.25 mm/rev). Lower cutting speed produces lower revolutions per minute (RPM) of the spindle. Lower spindle RPM combined with lower feed rates produces more chips and higher MRV and vice versa. The highest tool life was realized using treatment at 24 m/min and 0.15 mm/rev. The tool failed after 861 minutes of machining at this treatment. Due to the lower cutting speed and feed rate [17], it is hypothesized that the aerosol particle produced by microlubrication easily penetrated through to the tool/workpiece interface [18], lubrication occurs by diffusion through the workpiece, and the cutting fluid reacts to form a boundary layer of solid-film lubricant [19]. This led to the cooling of the cutting tool and simultaneously the lubrication of the tool/workpiece interface. At low cutting speed, the cutting temperature is also expected to be low as compared to the other cutting speed and feed rate combinations [19–21]. This also contributes to the increase in tool life.

But it was observed that the MRV for cutting speed of 36 m/min was identical for both the feed rates of 0.15 and 0.25 mm/rev used in the study. This is because feed rate is more significant parameter than the cutting speed in controlling the MRV [22]. Also, high spindle speed in combination with high feed rate causes evaporative cooling and enhances the cutting tool performance resulting in higher MRV [6]. One other fact to consider is that both the cutting combinations machined exactly the same number of workpieces. However, the feed rate of 0.15 mm/rev being slower exhibited a higher tool life in minutes than at 0.25 mm/rev.

The lowest tool life was realized using treatment at 30 m/min and 0.25 mm/rev. The tool failed only after 21 minutes of machining. Several reasons are hypothesized to have contributed to the lower tool life. Due to high feed rate, the cutting fluid cannot effectively cool and lubricate the cutting zone leading to higher temperatures. The tip of the tool is highly sensitive at high temperature [23]. Also, under high feed rate, the subsurface of the workpiece is harder, as shown in Figure 12, due in part to work hardening resulting in earlier tool failure.

It was also observed that the tool life when machining at the cutting speed of 30 m/min and feed rate of 0.25 mm/rev was lower than when machining at the highest speed of 36 m/min for the same feed rate. It is hypothesized that machining at the cutting speed of 36 m/min causes evaporative cooling of the fluid droplets that impact the target surface enhancing the heat transfer from the tools cutting edge surface as compared to the cutting speed of 30 m/min. The vapor phase near the vicinity of the cutting edge is then continuously removed by the air stream carrying the MQL fluid along with the higher cutting speed of 36 m/min, allowing access for fresh cutting fluid droplets to hit the

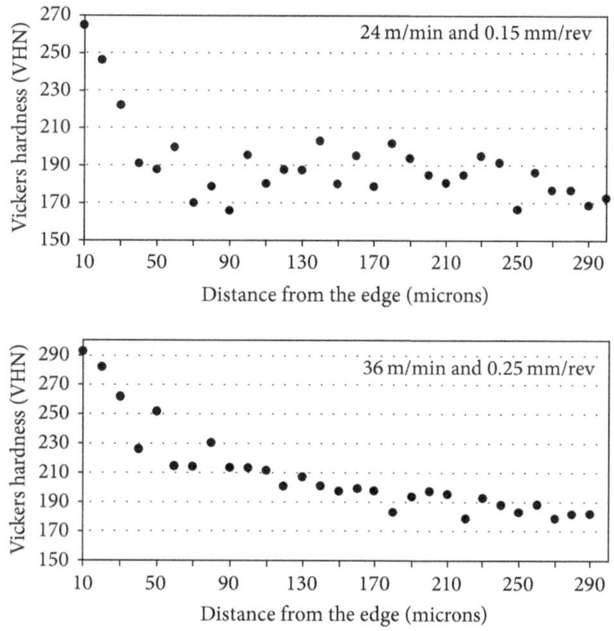

FIGURE 12: Microhardness depth profile of cross section for treatment at 24 m/min and 0.15 mm/rev and 36 m/min and 0.25 mm/rev.

surface required to be cooled [6]. This phenomenon of evaporative cooling is responsible for the higher tool life at the cutting speed of 36 m/min as compared to 30 m/min. This cooling effect helps the cutting tool to perform with a tool life of 159 min. But at the cutting speed of 30 m/min, this cooling effect is less dominant giving a tool life of only 21 min.

All the cutting tools under consideration failed with gradual abrasive wear except for treatment at 24 m/min and 0.25 mm/rev, which underwent a catastrophic failure. Figures 13(a)–13(f) show the cutting tool at failure for treatment at 24 m/min and 0.15 mm/rev, treatment at 36 m/min and 0.15 mm/rev, treatment at 30 m/min and 0.25 mm/rev, treatment at 36 m/min and 0.25 mm/rev, treatment at 30 m/min and 0.15 mm/rev, and treatment at 24 m/min and 0.25 mm/rev, respectively. Gradual abrasion of the cutting tool edge has been reported in other studies related to machining steels [24–26]. However, no study has been reported to date on end milling AISI 1018 steel under microlubrication.

Figure 14 shows the cutting tool for treatment at 24 m/min and 0.25 mm/rev, which exhibited catastrophic failure. The massive chipping shown in the figure is the reason for the failure of the tool. Two main conditions or causes of chipping are mechanical shock (i.e., impact by interrupted cutting as in milling) and thermal fatigue (i.e., cyclic variation in temperature of the tool in interrupted cutting) [27]. Chipping can also be caused by high temperatures at higher speed or higher feed rate conditions. Failure is linked to the fatigue crack promoted by cyclic variation of the tool temperature during machining [28].

SEM examination enabled the cutting edge of the flank and the rake side of the cutting tool to be analyzed and revealed that various other microwear mechanisms were activated at different cutting speed and feed rate levels. The cutting speed and feed rate level of 36 m/min and 0.25 mm/rev as well as 24 m/min and 0.15 mm/rev were selected for further SEM analysis. Based on SEM observation, two-body abrasion wear mode is likely dominant as shown in Figures 15(a) and 15(b). Microgrooves are formed which expose the tool substrate. This is due to the high temperature resulting from the high cutting speed and feed rate levels of 36 m/min and 0.25 mm/rev [29].

In addition, chips form a foreign body and slide between the tool rake wear and the workpiece activating sliding wear as shown in Figure 16 and causing the tool tip to gradually wear out at the contact zone. In low carbon steel, the softer ferrite phase coexist with the patches of harder pearlite phase. During machining, these pearlite patches "break off" from the workpiece and work as an abrasive particle, abrading the tool surface [30].

Figure 17 shows development of microfatigue crack near to the cutting edge resulting in the weakening of the bright oxide finish coating and decreasing the local hardness. This may finally result in the removal of the oxide finish from the tool and exposure of the tool substrate [31]. These results indicate that wear mechanism started by severe microabrasion leading to sliding wear, progressed to microfatigue at finish layer, and eventually caused the oxide finish layer to spall of the surface on the cutting tool.

Figures 18(a) and 18(b) show microchips being welded on rake and flank face, respectively. A higher magnification inset picture is also shown to clarify the view. It is likely that chips at higher speed and feed rate level get welded to the cutting tool because of inadequate time for the MWF to reach the cutting tool/workpiece interface [23].

Figures 19, 20, 21, and 22 show SEM images of cutting tool edges for the cutting speed and feed rate level of 24 m/min and 0.15 mm/rev. Sliding wear and nonuniform microabrasion similar to high speed and feed rate level were

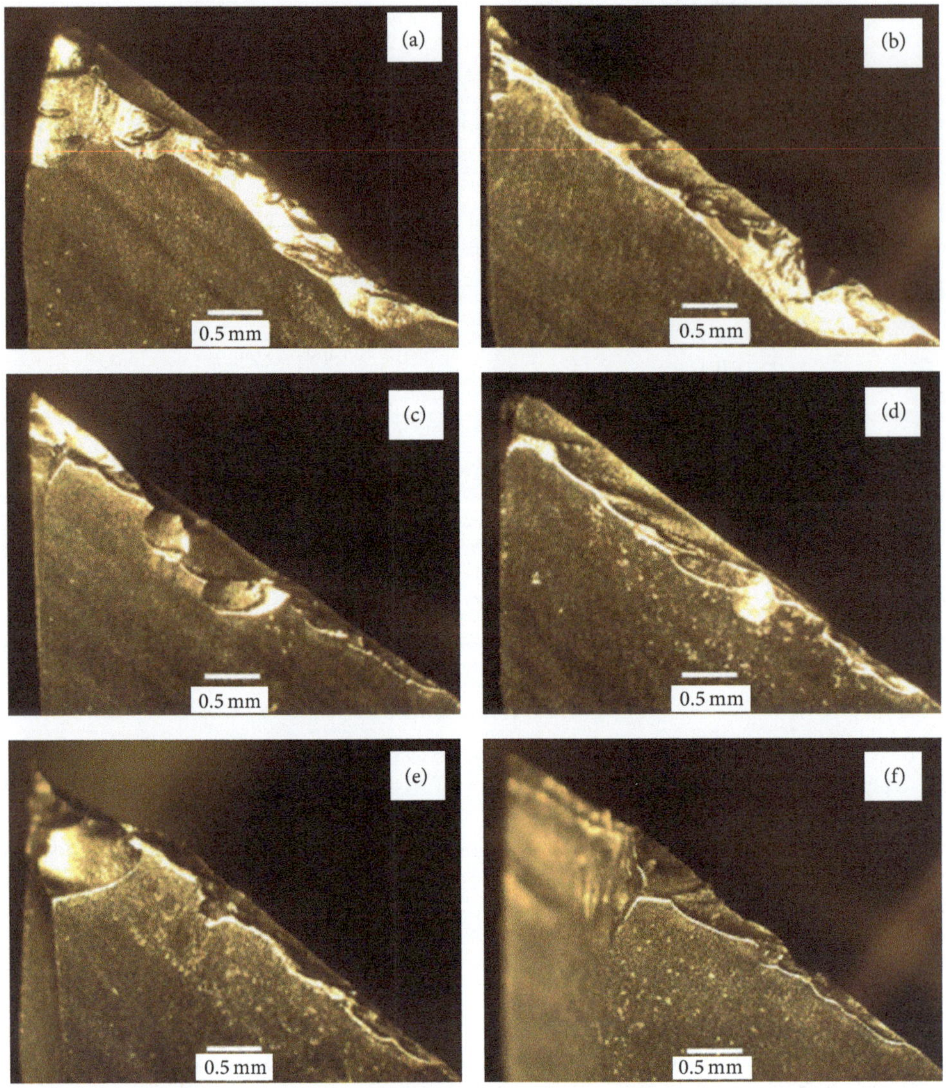

FIGURE 13: Optical micrograph of tool flank wear for treatment at (a) 24 m/min and 0.15 mm/rev, (b) 36 m/min and 0.15 mm/rev, (c) 30 m/min and 0.25 mm/rev, (d) 36 m/min and 0.25 mm/rev, (e) 30 m/min and 0.15 mm/rev, and (f) 24 m/min and 0.25 mm/rev.

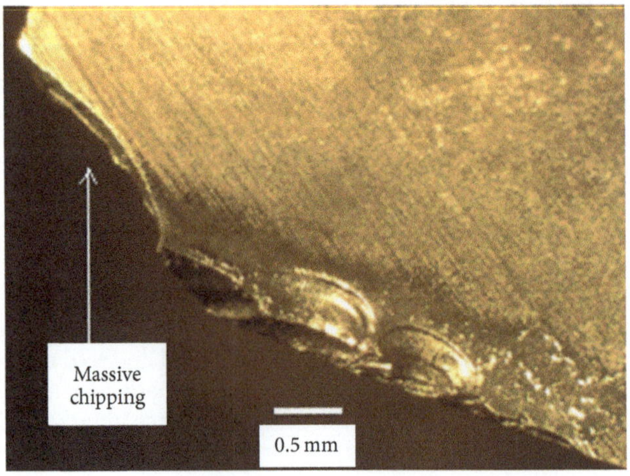

FIGURE 14: Optical micrograph of tool rake for treatment at 24 m/min and 0.25 mm/rev.

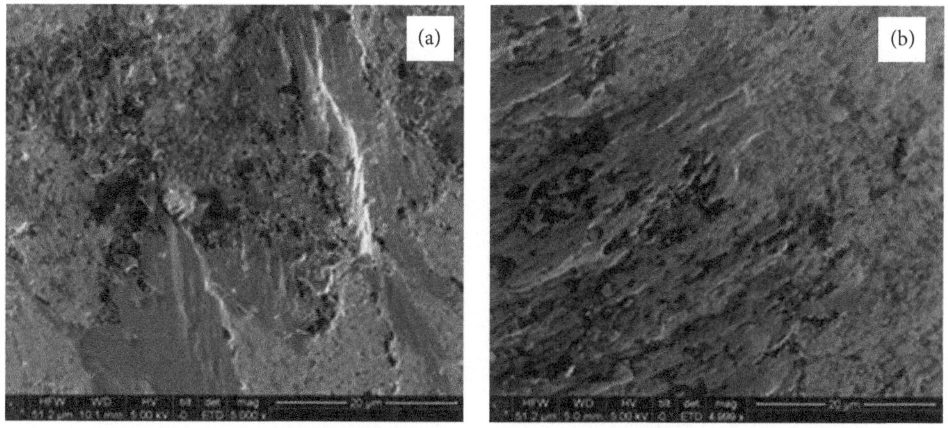

FIGURE 15: SEM micrograph of tool for treatment at 36 m/min and 0.25 mm/rev showing nonuniform microabrasion. (a) Rake and (b) flank.

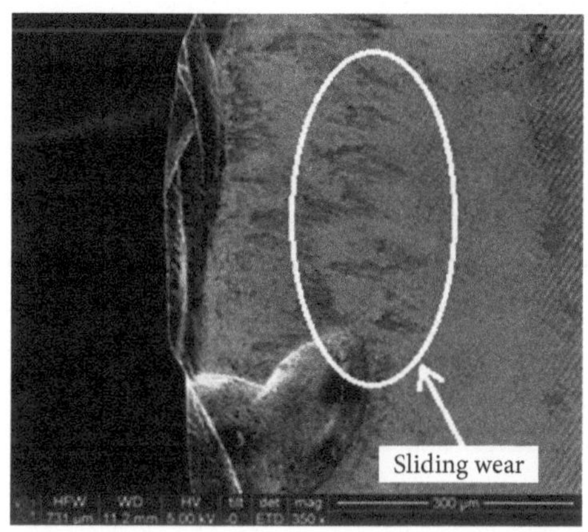

FIGURE 16: SEM micrograph of tool rake for treatment at 36 m/min and 0.25 mm/rev showing sliding wear.

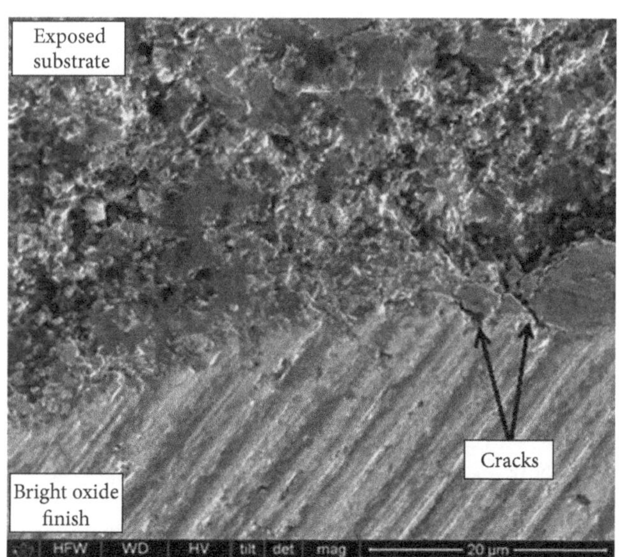

FIGURE 17: SEM micrograph of tool rake for treatment at 24 m/min and 0.15 mm/rev showing microfatigue crack.

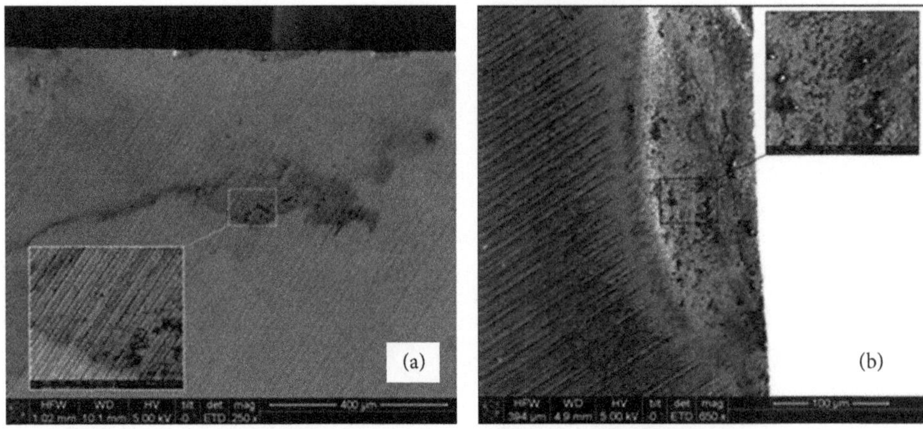

FIGURE 18: SEM micrograph of tool for treatment at 36 m/min and 0.25 mm/rev showing welded microchips. (a) Rake and (b) flank.

FIGURE 19: SEM micrograph of tool rake for treatment at 24 m/min and 0.15 mm/rev showing sliding wear.

observed for low speed and low feed rate levels (i.e., 24 m/min and 0.15 mm/rev).

Figure 19 shows the occurrence of sliding wear having grooves parallel to the metal flow direction. Figures 20(a) and 20(b) show nonuniform microabrasion for rake and the flank side, respectively. Microfatigue cracks were not observed at this treatment. The tool flank and rake faces showed unusual build-up edge (BUE) and adhesion of workpiece material to the cutting edges for treatment at 24 m/min and 0.15 mm/rev, as shown in Figures 21 and 22, respectively. The BUE shown in Figure 22 was not a stable kind of BUE.

Figure 23 shows the SEM image of the workpiece surface for treatment at 24 m/min and 0.15 mm/rev. The images clearly show abrasive wear through the mechanism of plowing. It is hypothesized that these small fragments of workpiece materials have adhered to the cutting tool faces causing adhesive wear. However, it is to be noted that plowing did not happen consistently throughout the machined workpiece surface.

5. Conclusion

The focus of this study was to investigate the effects of microlubrication during end milling American Iron and Steel Institute (AISI) 1018 steel with a 1-inch solid carbide cutter under varying cutting speed and feed rate levels and a constant depth of cut using Accu-Lube 2000 vegetable oil-based lubricant at a constant flow rate, an air pressure of 12 mL/h, and 0.1 megapascals (MPa), respectively. Machining experiments were carried out for a total of six combinations having cutting speeds of 24, 30, and 36 meters per minute (m/min) and feed rates of 0.15 and 0.25 millimeters per revolution (mm/rev). The cutting performance under microlubrication is five times better in terms of tool life and two times better in terms of MRV under low cutting speed and feed rate combination (i.e., 24 m/min and 0.15 mm/rev) as compared to high cutting speed and feed rate combination (i.e., 36 m/min and 0.25 mm/rev) for AISI 1018 steel using a vegetable oil-based cutting fluid. Therefore, the best combination of cutting parameters that led to lower tool life was the cutting speed of 24 m/min and feed rate of 0.15 mm/rev. Massive chipping was observed at the cutting tool edge for treatment at 24 m/min and 0.25 mm/rev causing catastrophic tool failure. Gradual two-body abrasion was the dominant wear mechanism except for treatment at 24 m/min and 0.25 mm/rev where the tool underwent catastrophic failure due to massive chipping. Similar results on gradual abrasive

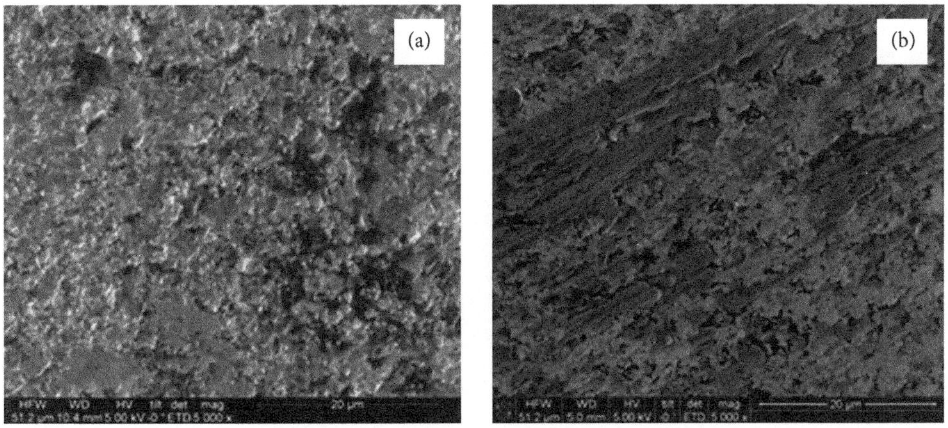

Figure 20: SEM micrograph of tool for treatment at 24 m/min and 0.15 mm/rev showing nonuniform microabrasion. (a) Rake and (b) flank.

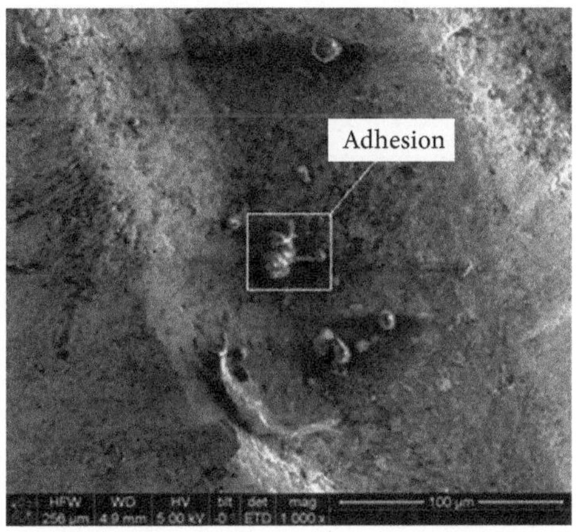

Figure 21: SEM micrograph of tool flank for treatment at 24 m/min and 0.15 mm/rev showing adhesion.

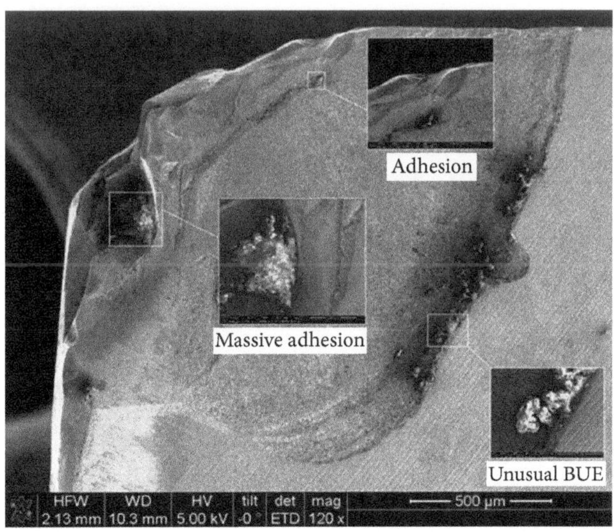

Figure 22: SEM micrograph of tool rake for treatment at 24 m/min and 0.15 mm/rev showing adhesion and built-up edge (BUE).

wear mechanism were reported in the literature [24–26] on other steels than the one investigated in this study. Exact comparison with similar grade steel cannot be done because no studies have been reported in the literature to date on end milling AISI 1018 steel under microlubrication. Other than abrasion, sliding adhesive wear of the workpiece materials was also observed. The SEM investigation of the used cutting tools revealed microfatigue cracks, welded microchips, and

Figure 23: SEM micrograph of workpiece for treatment at 24 m/min and 0.15 mm/rev showing abrasive wear through plowing mechanism.

unusual built-up edges on the cutting tools flank and rake side. The ANOVA clearly indicated that both the cutting speed and feed rate are statistically significant factors based on a 95% confidence level for both the flank wear sides. It can be concluded that the flank wear can be predicted using the obtained regression models, under the conditions investigated in this study.

Conflict of Interests

The authors declare that there is no conflict of interests regarding the publication of this paper.

References

[1] U.S. Department of Health and Human Services, *Occupational Exposure to Metal Working Fluid*, 98-102, NIOSH Publication, 1998.

[2] E. O. Bennett and D. L. Bennett, "Occupational airway diseases in the metalworking industry. Part 1: respiratory infections, pneumonia, chronic bronchitis and emphysema," *Tribology International*, vol. 18, no. 3, pp. 169–176, 1985.

[3] N. Boubekri and V. Shaikh, "Machining using minimum quantity lubrication: a technology for sustainability," *International Journal of Applied Science and Technology*, vol. 2, pp. 111–115, 2012.

[4] K. Weinert, I. Inasaki, J. W. Sutherland, and T. Wakabayashi, "Dry machining and minimum quantity lubrication," *CIRP Annals—Manufacturing Technology*, vol. 53, no. 2, pp. 511–537, 2004.

[5] S. Zhang, J. F. Li, and Y. W. Wang, "Tool life and cutting forces in end milling Inconel 718 under dry and minimum quantity cooling lubrication cutting conditions," *Journal of Cleaner Production*, vol. 32, pp. 81–87, 2012.

[6] A. D. Jayal and A. K. Balaji, "Effects of cutting fluid application on tool wear in machining: interactions with tool-coatings and tool surface features," *Wear*, vol. 267, no. 9-10, pp. 1723–1730, 2009.

[7] S. Kurgin, G. Barber, and Q. Zou, "Cutting insert and work piece materials for minimum quantity lubrication," in *International Seminar on Modern Cutting and Measurement Engineering*, vol. 7997 of *Proceedings of the SPIE*, pp. 1–10, Beijing, China, December 2010.

[8] M. H. Sadeghi, M. J. Hadad, T. Tawakoli, A. Vesali, and M. Emami, "An investigation on surface grinding of AISI 4140 hardened steel using minimum quantity lubrication-MQL technique," *International Journal of Material Forming*, vol. 3, no. 4, pp. 241–251, 2010.

[9] V. Shaikh and N. Boubekri, "Effects of minimum quantity lubrication in drilling 1018 steel," *Journal of Manufacturing Technology Research*, vol. 1, pp. 1–14, 2010.

[10] A. Attanasio, M. Gelfi, C. Giardini, and C. Remino, "Minimal quantity lubrication in turning: effect on tool wear," *Wear*, vol. 260, no. 3, pp. 333–338, 2006.

[11] A. K. Singh and A. K. Gupta, "Metalworking fluids from vegetable oils," *Journal of Synthetic Lubrication*, vol. 23, no. 4, pp. 167–176, 2006.

[12] S. A. Lawal, I. A. Choudhury, and Y. Nukman, "Application of vegetable oil-based metalworking fluids in machining ferrous metals—a review," *International Journal of Machine Tools and Manufacture*, vol. 52, no. 1, pp. 1–12, 2012.

[13] S. Woods, "Going green," *Cutting Tool Engineering*, vol. 57, no. 2, pp. 11–13, 2005.

[14] W. R. Stott, "Environmentally friendly cutting fluids," *Gear Technology*, vol. 22, no. 2, pp. 16–18, 2005.

[15] J. R. Davis, *ASM Metals Handbook, Machining*, vol. 16, 19th edition, 1989.

[16] D. C. Montgomery, *Design and Analysis of Experiments*, John Wiley & Sons, New York, NY, USA, 5th edition, 2001.

[17] T. H. C. Childs, K. Maekawa, T. Obikawa, and Y. Yamane, *Metal Machining, Theory and Applications*, Oxford, UK, 2000.

[18] S. Min, I. Inasaki, S. Fujimura, T. Wakabayashi, and S. Suda, "Investigation of adsorption behaviour of lubricants in near-dry machining," *Proceedings of the Institution of Mechanical Engineers B: Journal of Engineering Manufacture*, vol. 219, no. 9, pp. 665–671, 2005.

[19] D. P. Adler, W. W.-S. Hii, D. J. Michalek, and J. W. Sutherland, "Examining the role of cutting fluids in machining and efforts to address associated environmental/health concerns," *Machining Science and Technology*, vol. 10, no. 1, pp. 23–58, 2006.

[20] S. M. Ali, N. R. Dhar, and S. K. Dey, "Effect of minimum quantity lubrication (MQL) on cutting performance in turning medium carbon steel by uncoated carbide insert at different speed-feed combinations," *Advances in Production Engineering & Management*, vol. 6, pp. 185–198, 2011.

[21] M. Shaw, J. Pigott, and L. Richardson, "The effect of cutting fluid upon chip-tool interface temperature," *Transaction of ASME*, vol. 73, pp. 45–56, 1951.

[22] H. Yanda, J. A. Ghani, M. N. A. M. Rodzi, K. Othman, and C. H. C. Haron, "Optimization of material removal rate, surface roughness and tool life on conventional dry turning of FCD700," *International Journal of Mechanical and Materials Engineering*, vol. 5, no. 2, pp. 182–190, 2010.

[23] N. R. Dhar, M. T. Ahmed, and S. Islam, "An experimental investigation on effect of minimum quantity lubrication in machining AISI 1040 steel," *International Journal of Machine Tools and Manufacture*, vol. 47, no. 5, pp. 748–753, 2007.

[24] H. Caliskan, C. Kurbanoglu, P. Panjan, M. Cekad, and D. Kramar, "Wear behavior and cutting performance of nanostructured hard coatings on cemented carbide cutting tools in hard milling," *Tribology International*, vol. 62, pp. 215–222, 2013.

[25] R. Suresh, S. Basavarajappa, V. N. Gaitonde, and G. L. Samuel, "Machinability investigations on hardened AISI 4340 steel using coated carbide insert," *International Journal of Refractory Metals and Hard Materials*, vol. 38, pp. 124–133, 2013.

[26] A. K. Sahoo and B. Sahoo, "Experimental investigations on machinability aspects in finish hard turning of AISI, 4340 steel using uncoated and multilayer coated carbide inserts," *Measurement*, vol. 45, pp. 2153–2165, 2012.

[27] S. Kalpakjian, *Manufacturing Engineering and Technology*, Addison-Wesley Publishing Company, New York, NY, USA, 3rd edition, 1995.

[28] R. B. Da Silva, J. M. Vieira, R. N. Cardoso et al., "Tool wear analysis in milling of medium carbon steel with coated cemented carbide inserts using different machining lubrication/cooling systems," *Wear*, vol. 271, no. 9-10, pp. 2459–2465, 2011.

[29] S. Dhiman, R. Sehgal, S. K. Sharma, and V. S. Sharma, "Machining behavior of AlSI 1018 steel during turning," *Journal of Scientific and Industrial Research*, vol. 67, no. 5, pp. 355–360, 2008.

[30] P. Kwon, "Predictive models for flank wear on coated inserts," *Journal of Tribology*, vol. 122, no. 1, pp. 340–347, 2000.

[31] K. Abu-Shgair, M. Al-Hasan, A. K. Abdul Jawwad, A. Al-Bashir, H. H. Abu-Safe, and M. H. Gordon, "Characterizing (Ti, AI)N film coating produced by inverted cylindrical magnetron sputtering for metal machining applications," *Reviews on Advanced Materials Science*, vol. 24, no. 1-2, pp. 48–55, 2010.

19

Integrated Inventory Model for Deteriorating Items with Price-Dependent Demand under Quantity-Dependent Trade Credit

Karuppuchamy Annadurai

Department of Mathematics, Government Arts College, Udumalpet, Tirupur District, Tamil Nadu 642126, India

Correspondence should be addressed to Karuppuchamy Annadurai; anna.vadivu@gmail.com

Academic Editors: A. Che and G. Onwubolu

This paper explores an integrated inventory model when the deterioration rate follows exponential distribution under trade credit. Here, it is assumed that demand rate is a function of selling price and the permissible delay in payment depends on the order quantity. In the model shortages are completely backlogged. The maximization of the total profit per unit of time is taken as the objective function to study the retailer's optimal ordering policy. This paper also presents a practical application example where the proposed inventory model is utilized to support business decision making. Particularly, the model developed in the paper could be useful in the area of supply chain management. Finally, sensitivity analysis of the optimal solution with respect to major parameters is carried out. Our result illustrates that this model can be quite useful in determining the optimal ordering policy when the trade credit period is being analyzed.

1. Introduction

Both in deterministic and probabilistic inventory models of classical type, it is observed that payments are made to the supplier immediately after receiving the items. In practice, the supplier will offer the retailer a delay period in payment for the amount of purchase to increase the demand known as trade credit period. Offering such a credit period to the retailer will encourage the supplier's selling and reduce on-hand stock level. Simultaneously, without a primary payment the retailer can take the advantages of a credit period to reduce cost and increase profit. The customer does not have to pay any interest during the fixed period but if the payment gets delayed beyond the period interest will be charged by the supplier. This arrangement comes out to be very advantageous to the customer as he may delay the payment till the end of the permissible delay period. During this credit period the retailer can start to accumulate revenues on the sales and earn interest on that revenue. Thus, the delay in the payment offered by the supplier is a kind of price discount since paying later indirectly reduces the cost of holding and it encourages the retailers to increase their order quantity. Moreover, paying later indirectly reduces the cost of holding stock. Hence, trade credit can play an important role in inventory model for both the suppliers as well as the retailers in integrated inventory model. Goyal [1] first derived an EOQ model under the conditions of permissible delay in payments. Chung and Huang [2] extended Goyal's [1] model to consider the case that the units are replenished at a finite rate under permissible delay in payments. Teng [3] assumed that selling price is not equal to the purchasing price to modify Goyal's model [1]. Salameh et al. [4] extended this issue to the continuous review inventory model.

The control and maintenance of inventories for deteriorating items with shortages have received much attention of several researchers in the recent years because most of the physical goods are deteriorated over time. Deterioration refers to decay, damage, or spoilage. In respect of foods, films, drugs, chemicals, electronic components, and radioactive substances, deterioration may happen during normal period of storage. Therefore, the loss due to deterioration can not be neglected. So, decay or deterioration of physical goods in stock is a very realistic feature and researchers felt the necessity to take this factor into consideration in developing inventory models. Goyal and Giri [5] studied recent trends in modeling of deteriorating inventory. Various types of inventory models with constant deterioration rate and a linearly time varying demand rate were discussed by Goswami and Chaudhuri [6], Chakrabarti and Chaudhuri [7], Giri et al. [8], and others. Jalan and Chaudhuri [9] discussed inventory models with constant demand, constant deterioration, and instantaneous replenishment. Inventory models with a fixed deteriorating rate and an exponentially time varying demand were studied by Hariga and Benkherouf [10], Wee [11], and others. Jamal et al. [12] and Chang and Dye [13] extended Goyal's [1] model to the case for deterioration and allowable shortages. Annadurai and Uthayakumar [14] analyzed the partial trade credit financing in a supply chain by EOQ-based model for decaying items by allowing shortages. They assumed that the supplier may offer a partial permissible delay in payments even if the order quantity is less than predetermined quantity. From the sensitivity analysis, it was proved that the deterioration rate, ordering cost, and purchasing cost affect the total profit of the retailer. Recently, Annadurai [15] developed a mathematical model for obtaining the optimal purchase quantity for exponentially distributed deteriorating items with some salvage value associated with the deteriorated items and partial backlogging.

At the beginning, demand rate was assumed to be constant which is in general likely to be time dependent and stock dependent. Burwell et al. [16] developed economic lot size model for price-dependent demand under quantity and freight discounts. Inventory model for ameliorating items for price-dependent demand rate was proposed by Mondal et al. [17] and inventory model with price- and time-dependent demand was developed by You [18]. Wee [19] developed a replenishment policy for items with a price-dependent demand and varying rate of deterioration. Lin et al. [20] developed an economic order quantity model with time varying demand, partial back ordering, and linearly time-dependent deterioration rate. Annadurai and Uthayakumar [21] developed an inventory model under two levels of trade credit policy for deteriorating items by assuming that the demand is a function of credit period offered by the retailer to the customers. In a recent paper, Annadurai [22] presented an integrated inventory model for exponentially deteriorated items with back orders under two levels of trade credit. He considered the impact of a replenishment policy on the timing of the cash flows associated with payments to suppliers and revenue streams from customers. That is, the same cash amount will possess different money value at different future time.

There are some other research articles dealing with deteriorating inventory model under trade credit. Aggarwal and Jaggi [23] developed ordering policies of deteriorating items under permissible delay in payments. Chang et al. [24] established an inventory model for deteriorating items with linear trend demand under conditions of permissible delay in payments. Dye [25] proposed inventory model by considering the stock-dependent demand for deteriorating items for partial backlogging and conditions of permissible delay in payments. He used initial stock-dependent demand function. In the existing literature about permissible delay in payments, it is assumed that the demand is either mostly a function of time or a function of the retail price. Soni and Shah [26] established optimal ordering policy for stock-dependent demand under progressive payment scheme. In a recent paper, Tripathy and Mishra [27] dealt with development of an inventory model when the deterioration rate follows Weibull two-parameter distributions. It was assumed that demand rate is a function of selling price and holding cost is time dependent. From their model, one can calculate the optimum average profit margins for with shortage case and without shortage case for the deterministic inventory model with varying demand rate and holding cost subject to the conditions. In a recent paper, Annadurai and Uthayakumar [28] formulated a two-echelon inventory model for deteriorating items with credit-period-dependent demand including shortages under two-level trade credit financing and determining the retailer's optimal replenishment policy when both the supplier as well as the retailer offers the credit period to stimulate customer demand.

In real life, the occurrence of shortage in an inventory system is a phenomenon. Shortages are of great importance especially in a model that considers a delay in payment due to the fact that shortages can affect the quantity ordered to benefit from the delay in payment. Now there arises a natural question whether the length of the permissible delay period gets influenced by the volume of the quantity ordered. Intuition leads to the fact that the volume of the ordered quantity should have a direct impact on the length of this period. More precisely, we can say that the more we order the longer the delay period is likely to be allotted.

The present paper incorporates this fact in an integrated inventory model allowing for shortages and obtains the optimal replenishment policy for deteriorating items with price-dependent demand under quantity-dependent trade credit. That is, if the order quantity is less than the predetermined quantity, the payment for the items must be made immediately. Otherwise, a fixed trade credit period is permitted. Next, we provide the fundamental notations and assumptions for the proposed model in Section 2. In Section 3, a mathematical model is established and solution procedure is discussed for maximizing the total profit, based on which an algorithm for finding the optimal policy is suggested. Numerical examples are provided to illustrate the proposed model in Section 4. In Section 5, sensitivity analysis of the optimal solution with respect to major parameters is carried out. Lastly, in Section 6, we draw the conclusions and give suggestions for future research.

2. Notations and Assumptions

The proposed model is developed based on the following assumptions and notations.

2.1. Notations. The notation is summarized in the following:

A: ordering cost per order,

Q: ordering quantity,

Q_0: a specified value of Q,

T: the inventory cycle length of each cycle,

t_1: time when the inventory level comes down to zero,

θ: the deterioration rate, where $0 \le \theta < 1$,

C: unit purchase cost of the item,

C_1: shortage cost for backlogged item,

p: selling price per unit item,

M: permissible delay period in settling the accounts, $0 \le M \le T$,

h: the inventory holding rate per unit time excluding interest charges,

I_e: interest which can be earned per \$ per unit time,

I_k: interest payable per \$ per unit time,

$I(t)$: the inventory level at time t,

TP: the total profit per unit time.

2.2. Assumptions. The following assumptions are made in the model.

(1) Replenishment occurs instantaneously at an infinite rate. That is, lead time is zero.

(2) The demand rate is a function of selling price.

(3) Shortages are allowed to occur and completely backlogged.

(4) The length of the permissible delay period M for repaying the supplier is given by

$$M = \begin{cases} M_1 & \text{if } 0 < Q < Q_0, \\ M_2 & \text{if } Q \ge Q_0, \end{cases} \tag{1}$$

where Q is the ordered quantity, Q_0 is a specified value of Q, and $M_2 > M_1$.

(5) No payment to the supplier is outstanding at the time of placing an order; that is, $M < T$.

(6) The retailer can accumulate revenue and earn interest from the very beginning that his customer pays for the amount of purchasing cost to the retailer until the end of the trade credit period offered by the supplier. That is, the retailer can accumulate revenue and earn interest during the period $t = M$ with rate I_e under the condition of trade credit.

(7) There is no repair or replacement of deteriorated units during the planning horizon. The item will be withdrawn from warehouse immediately as they become deteriorated.

(8) The inventory carrying cost is proportional to the inventory level and the deterioration rate is proportional to time.

(9) Selling price p follows an increasing trend, and demand rate possesses the negative derivative throughout its domain where demand rate is $f(p) = (a - p) > 0$.

3. Mathematical Formulation and Solution

In this section, a mathematical model is developed to determine the optimal replenishment cycle time that maximizes the total annual profit in an inventory system for exponential deteriorating items under quantity-dependent permissible delay in payments including shortages. Let $I(t)$ be the inventory level at time $0 \le t \le T$. Inventory level decreases due to both the demand and deterioration of the item during the period $[0, t_1]$ and ultimately falls to zero at $t = t_1$. Thereafter, shortages are allowed to occur and all of the demands during the period $[t_1, T]$ are completely backlogged.

As described above, the differential equation representing the inventory status is given by

$$\frac{dI(t)}{dt} = \begin{cases} -(a-p) - \theta I(t) & \text{if } 0 \le t \le t_1, \\ -(a-p) & \text{if } t_1 \le t \le T \end{cases} \tag{2}$$

with boundary condition $I(t) = 0$ at $t = t_1$. The solution of (2) is given by

$$I(t) = \begin{cases} \dfrac{(a-p)}{\theta}\left[e^{\theta(t_1 - t)} - 1\right] & \text{if } 0 \le t \le t_1, \\ (a-p)(t - t_1) & \text{if } t_1 \le t \le T. \end{cases} \tag{3}$$

The beginning inventory level for each cycle can be obtained as

$$S = I(0) = \frac{(a-p)}{\theta}\left(e^{\theta t_1} - 1\right). \tag{4}$$

The total number of items that become deteriorated in the interval $[0, t_1]$, say D_τ, is given by

$$D_\tau = S - \int_0^{t_1} (a-p)\,dt = \frac{(a-p)}{\theta}\left(e^{\theta t_1} - \theta t_1 - 1\right). \tag{5}$$

Hence, the order quantity per cycle is given by

$$\begin{aligned} Q &= D_\tau + \int_0^T (a-p)\,dt \\ &= \frac{(a-p)}{\theta}\left(e^{\theta t_1} - \theta t_1 - 1\right) + (a-p)T. \end{aligned} \tag{6}$$

Next, the total profit per unit of time of the inventory system is computed using the following various components.

(1) Ordering cost $= A$.

(2) Holding cost (excluding interest charges) per cycle $= h\int_0^{t_1} I(t)dt = (h(a-p)/\theta^2)(e^{\theta t_1} - \theta t_1 - 1)$.

(3) Purchase cost $= CQ = (C(a-p)/\theta)(e^{\theta t_1} - \theta t_1 - 1) + C(a-p)T$.

(4) Shortage cost $= C_1 \int_{t_1}^{T} I(t)dt = (C_1/2)(a-p)(T-t_1)^2$.

There are two major cases to occur in the interest charged and earned in each order cycle and we discuss each case in detail as follows.

Case 1 ($M \leq t_1$). In this case, since the length of period with positive stock is larger than the credit period, the buyer can use the revenue to earn interest at the rate I_e per dollar per year in $(0, t_1)$. Hence, the interest earned per unit time is calculated and is given by $I_e p \int_0^{t_1} -(a-p)(t-t_1)dt = I_e p(a-p)(t_1^2/2)$. On the other hand, beyond the credit period, the unsold stock is supposed to be financed with rate I_k and the interest payable in each order cycle is given by $I_k C \int_M^{t_1}(a-p)(t_1-t)dt = (I_k C(a-p)/2)(t_1-M)^2$.

Hence, the total profit per unit time is given by

$$\begin{aligned} \mathrm{TP}_1(t_1,T,p) &= p(a-p) - \frac{a-p}{T} \\ &\times \left\{ \frac{A}{a-p} + \frac{(h+C\theta)}{\theta^2}\left(e^{\theta t_1} - \theta t_1 - 1\right) + CT \right. \\ &\left. + \frac{1}{2}\left[C_1(T-t_1)^2 + I_k C(t_1-M)^2 - I_e p t_1^2\right] \right\}. \end{aligned} \tag{7}$$

Let $t_1 = \gamma T$, $0 < \gamma < 1$. Hence, we get the profit function

$$\begin{aligned} \mathrm{TP}_1(T,p) &= p(a-p) - \frac{a-p}{T} \\ &\times \left\{ \frac{A}{a-p} + \frac{(h+C\theta)}{\theta^2}\left(e^{\theta\gamma T} - \theta\gamma T - 1\right) + CT \right. \\ &+ \frac{1}{2}\left[C_1(1-\gamma)^2 - (I_e p - I_k C)\gamma^2\right]T^2 \\ &\left. + \frac{I_k C}{2}M(M-2\gamma T) \right\}. \end{aligned} \tag{8}$$

Our objective is to maximize the profit function $\mathrm{TP}_1(T,p)$. The necessary conditions for maximizing the profit are $\partial \mathrm{TP}_1(T,p)/\partial T = 0$ and $\partial \mathrm{TP}_1(T,p)/\partial p = 0$ which yield

$$\begin{aligned} \frac{\partial \mathrm{TP}_1(T,p)}{\partial T} &= \frac{a-p}{T^2}\left\{ \frac{A}{a-p} + \frac{(h+C\theta)}{\theta^2}\left(e^{\theta\gamma T} - \theta\gamma T - 1\right) + CT \right. \\ &+ \frac{1}{2}\left[C_1(1-\gamma)^2 - (I_e p - I_k C)\gamma^2\right]T^2 \\ &\left. + \frac{I_k C}{2}M(M-2\gamma T) \right\} \\ &- \frac{a-p}{T}\left[\frac{(h+C\theta)}{\theta}\gamma\left(e^{\theta\gamma T} - 1\right) \right. \\ &+ \left[C_1(1-\gamma)^2 - (I_e p - I_k C)\gamma^2\right]T \\ &\left. + C - I_k C M\gamma \right] = 0, \end{aligned}$$

$$\begin{aligned} \frac{\partial \mathrm{TP}_1(T,p)}{\partial p} &= a - 2p \\ &+ \frac{1}{T}\left\{ \frac{(h+C\theta)}{\theta^2}\left(e^{\theta\gamma T} - \theta\gamma T - 1\right) - CT \right. \\ &- \frac{1}{2}\left[C_1(1-\gamma)^2 - (I_e p - I_k C)\gamma^2\right]T^2 \\ &\left. - \frac{I_k C}{2}M(M-2\gamma T) \right\} = 0. \end{aligned} \tag{9}$$

Using the software MATLab 7.0, from (9), we can calculate the optimum values of T^* and p^* simultaneously. From (9), the optimal value $\mathrm{TP}_1^*(T,p)$ of the average net profit can be determined provided T^* and p^* satisfy the conditions $\partial^2 \mathrm{TP}_1(T,p)/\partial T^2 < 0$, $\partial^2 \mathrm{TP}_1(T,p)/\partial p^2 < 0$ and $(\partial^2 \mathrm{TP}_1(T,p)/\partial T^2)(\partial^2 \mathrm{TP}_1(T,p)/\partial p^2) - (\partial^2 \mathrm{TP}_1(T,p)/\partial T \partial p)^2 > 0$ at $p = p^*$ and $T = T^*$.

If the solutions obtained from (9) do not satisfy the sufficiency conditions, then we conclude that no feasible solution will be optimal for the set of parameters taken to solve (9). Such a situation will imply that the parameter values are inconsistent and there are some errors in their estimation.

Case 2 ($M > t_1$). In this case, since $M > t_1$, the buyer pays no interest but earns interest at the rate I_e during the period $[0, M]$. Therefore, the interest earned in this case is calculated and is given by $(a-p)I_e P t_1(M - t_1/2)$ and the interest payable in each order cycle = 0.

Then, in this case, the total profit per unit time is given by

$$\begin{aligned} \mathrm{TP}_2(t_1,T,p) &= p(a-p) \\ &- \frac{a-p}{T}\left\{ \frac{A}{a-p} + \frac{(h+C\theta)}{\theta^2}\left(e^{\theta t_1} - \theta t_1 - 1\right) \right. \\ &\left. + CT + \frac{C_1}{2}(T-t_1)^2 - I_e p t_1\left(M - \frac{t_1}{2}\right) \right\}. \end{aligned} \tag{10}$$

By letting $t_1 = \gamma T$, $0 < \gamma < 1$, the profit function becomes

$$\begin{aligned} \mathrm{TP}_2(T,p) &= p(a-p) - \frac{a-p}{T} \\ &\times \left\{ \frac{A}{a-p} + \frac{(h+C\theta)}{\theta^2}\left(e^{\theta\gamma T} - \theta\gamma T - 1\right) + CT \right. \\ &\left. + \frac{C_1}{2}(1-\gamma)^2 T^2 - I_e P\gamma T\left(M - \frac{\gamma T}{2}\right) \right\}. \end{aligned} \tag{11}$$

Our objective is to maximize the profit function $\mathrm{TP}_2(T,p)$. The necessary conditions for maximizing the profit are $\partial \mathrm{TP}_2(T,p)/\partial T = 0$ and $\partial \mathrm{TP}_2(T,p)/\partial p = 0$ which yield

$$\frac{\partial \mathrm{TP}_2(T,p)}{\partial T} = \frac{a-p}{T^2}\left\{\frac{A}{a-p} + \frac{(h+C\theta)}{\theta^2}\left(e^{\theta\gamma T} - \theta\gamma T - 1\right) + CT + \frac{C_1}{2}(1-\gamma)^2 T^2 - I_e p\gamma T\left(M - \frac{\gamma T}{2}\right)\right\} - \frac{a-p}{T}\left[\frac{(h+C\theta)}{\theta}\gamma\left(e^{\theta\gamma T}-1\right) + C_1(1-\gamma)^2 T + C + I_e p\gamma(\gamma T - M)\right] = 0,$$

$$\frac{\partial \mathrm{TP}_2(T,p)}{\partial p} = a - 2p + \frac{1}{T}\times\left\{\frac{(h+C\theta)}{\theta^2}\left(e^{\theta\gamma T} - \theta\gamma T - 1\right) + CT + \frac{C_1}{2}(1-\gamma)^2 - I_e p\gamma T\left(M - \frac{\gamma T}{2}\right)\right\} = 0. \tag{12}$$

Using the software MATLab 7.0, from (12), we can calculate the optimum values of T^* and p^* simultaneously. From (11), the optimal value $\mathrm{TP}_2^*(T,p)$ of the average net profit can be determined provided T^* and p^* satisfy the sufficient conditions $\partial^2\mathrm{TP}_2(T,p)/\partial T^2 < 0$, $\partial^2\mathrm{TP}_2(T,p)/\partial p^2 < 0$ and $(\partial^2\mathrm{TP}_2(T,p)/\partial T^2)(\partial^2\mathrm{TP}_2(T,p)/\partial p^2) - (\partial^2\mathrm{TP}_2(T,p)/\partial T\partial p)^2 > 0$ at $p = p^*$ and $T = T^*$.

If the solutions obtained from (12) do not satisfy the sufficiency conditions, then we conclude that no feasible solution will be optimal for the set of parameters taken to solve (12). Such a situation will imply that the parameter values are inconsistent and there are some errors in their estimation.

We observe the following property of optimal t_1, T, and p.

Result 1. $\mathrm{TP}^M(t_1,T,p)$ is an increasing function of M.

Proof. We observe that for given (t_1,T,p), $\mathrm{TP}_1^M(t_1,T,p)$ and $\mathrm{TP}_2^M(t_1,T,p)$ are increasing functions of M. Now, consider any M_1 and M_2 such that $M_2 > M_1$.

(i) Suppose $M_1 < M_2 < t_1$. Then, $\mathrm{TP}^{M_1}(t_1,T,p) = \mathrm{TP}_1^{M_1}(t_1,T,p) > \mathrm{TP}_1^{M_2}(t_1,T,p) = \mathrm{TP}^{M_2}(t_1,T,p)$.

(ii) Suppose $M_1 < t_1 < M_2$. Then, $\mathrm{TP}^{M_1}(t_1,T,p) = \mathrm{TP}_1^{M_1}(t_1,T,p) > \mathrm{TP}_1^{M_2}(t_1,T,p) > \mathrm{TP}_2^{M_2}(t_1,T,p) = \mathrm{TP}^{M_2}(t_1,T,p)$.

(iii) Suppose $t_1 < M_1 < M_2$. Then, $\mathrm{TP}^{M_1}(t_1,T,p) = \mathrm{TP}_2^{M_1}(t_1,T,p) > \mathrm{TP}_2^{M_2}(t_1,T,p) = \mathrm{TP}^{M_2}(t_1,T,p)$.

Hence, the result follows. □

TABLE 1: Optimal solutions under different parametric values of Example 1.

θ	t_1^*	T^*	Q^*	p^*	$\mathrm{TP}(t_1^*, T^*, p^*)$
0.03	0.041	0.082	161.568	28.157	1284.232
0.06	0.042	0.085	160.356	29.348	1281.513
0.09	0.044	0.089	159.314	30.523	1279.354

TABLE 2: Optimal solutions under different parametric values of Example 2.

θ	t_1^*	T^*	Q^*	p^*	$\mathrm{TP}(t_1^*, T^*, p^*)$
0.03	0.054	0.109	162.382	30.451	1286.364
0.06	0.056	0.112	161.351	31.343	1285.348
0.09	0.058	0.115	160.131	32.013	1283.814

Based on the above result, we develop the following algorithm to find the optimal values of T and p and then, after knowing optimal T and p, we can determine the optimal t_1.

Algorithm

Step 1. Find (T^*, p^*) maximizing $\mathrm{TP}^{M_2}(T,p)$.

Step 2. If $(a-p)T^* \geq Q_0$, then (T^*, p^*) is optimal. Otherwise, go to Step 3.

Step 3. Find (T^{**}, p^{**}) maximizing $\mathrm{TP}^{M_1}(T,p)$.

Step 4. Compute $\mathrm{TP}^{M_1}(T^{**}, p^{**})$ and $\mathrm{TP}^{M_2}(T^0, p^0)$, where $(a-p)T^0 = Q_0$ and T^0 is the optimal value of T for given $T = T^0$.

Step 5. If $\mathrm{TP}^{M_1}(T^{**}, p^{**}) \geq \mathrm{TP}^{M_2}(T^0, p^0)$, then (T^{**}, p^{**}) is optimal; otherwise, (T^0, p^0) is optimal.

Once we have optimal solution (T^*, p^*), then we can find the optimal $t_1^* = \gamma T^*$.

4. Numerical Examples

4.1. Example 1. In order to evaluate the proposed model in the presented paper, we designed numerical example for different values of θ : A = \$250 per order, $a = 100$, h = \$1.5 per unit per year, C = \$15 per unit, C_1 = \$1.5 per unit per year, $I_k = 0.1$ and $I_e = 0.2$, $\gamma = 0.5$, and

$$M = \begin{cases} M_1 & \text{if } 0 < Q < Q_0, \\ M_2 & \text{if } Q \geq Q_0, \end{cases} \tag{13}$$

where $M_1 = 0.055$ year, $M_2 = 0.082$ year, and $Q_0 = 15000$.

With the help of the software MATLab 7.0 we apply the proposed algorithm to solve the problem for the values θ = 0.03, 0.06, and 0.09. The results are summarized in Table 1.

Table 3: Effect of change in various parameters of Example 1.

Changing parameter	% Change in system	Change in t_1^*	T^*	Q^*	p^*	$\mathrm{TP}(t_1^*, T^*, p^*)$
h	−50	0.029	0.052	155.312	31.536	1288.642
	−25	0.038	0.061	157.583	30.324	1281.521
	+25	0.042	0.083	168.358	29.467	1280.323
	+50	0.044	0.084	174.347	27.358	1272.514
a	−50	0.044	0.085	158.357	24.325	1202.512
	−25	0.041	0.082	162.356	27.524	1231.321
	+25	0.024	0.061	171.581	28.512	1283.245
	+50	0.016	0.053	173.384	29.518	1292.511
C_1	−50	0.051	0.089	163.348	26.523	1271.412
	−25	0.048	0.082	162.365	27.341	1275.241
	+25	0.031	0.053	155.381	31.621	1282.523
	+50	0.027	0.042	155.219	36.325	1291.345
A	−50	0.039	0.080	156.581	28.025	1282.632
	−25	0.040	0.081	158.357	27.625	1278.782
	+25	0.043	0.086	162.132	31.618	1280.654
	+50	0.045	0.088	163.245	30.812	1278.713
C	−50	0.046	0.088	163.413	26.514	1278.612
	−25	0.045	0.087	162.156	27.681	1279.221
	+25	0.039	0.082	158.913	31.625	1284.912
	+50	0.038	0.081	157.682	32.918	1285.613

4.2. Example 2. Given $A = \$500$ per order, $a = 100$, $h = \$2$ per unit per year, $C = \$30$ per unit, $C_1 = \$2.5$ per unit per year, $I_k = 0.1$ and $I_e = 0.2$, $\gamma = 0.5$ and

$$M = \begin{cases} M_1 & \text{if } 0 < Q < Q_0, \\ M_2 & \text{if } Q \geq Q_0, \end{cases} \tag{14}$$

where $M_1 = 0.082$ year, $M_2 = 0.019$ year and $Q_0 = 9000$.

With the help of the software MATLab 7.0 and apply the proposed Algorithm to solve the problem for the values $\theta = 0.03, 0.06$, and 0.09. The results are summarized in Table 2.

5. Sensitivity Analysis

Here we study the effects of changes in the system parameters h, a, C_1, A, and C on the optimal length of inventory interval with positive inventory t_1^*, the optimal length of order cycle T^*, the optimal order quantity per cycle Q^*, the optimal selling price p^*, and the maximum total profit TP^* of Example 1. A sensitivity analysis is performed by changing each of the parameters by +50%, +25%, −25%, and −50%, taking one parameter at a time and keeping the remaining parameters unchanged. The results are shown in Table 3.

Based on the results of Table 3, we can obtain the following managerial insights.

(1) Increase in the values of the parameter h will result in decrease of p^* and $\mathrm{TP}(t_1^*, T^*, p^*)$ but increase of t_1^*, T^*, and Q^*.

(2) Decrease in the values of the parameter h will result in increase of p^* and $\mathrm{TP}(t_1^*, T^*, p^*)$ but decrease of t_1^*, T^*, and Q^*.

(3) Increase in the values of the parameter a will result in increase of Q^*, p^*, and $\mathrm{TP}(t_1^*, T^*, p^*)$ but decrease of t_1^* and T^*.

(4) Decrease in the values of the parameter a will result in decrease of Q^*, p^*, and $\mathrm{TP}(t_1^*, T^*, p^*)$ but increase of t_1^* and T^*.

(5) Increase in the values of the parameter C_1 will result in increase of p^* and $\mathrm{TP}(t_1^*, T^*, p^*)$ but decrease of t_1^*, T^*, and Q^*.

(6) Decrease in the values of the parameter C_1 will result in decrease of p^* and $\mathrm{TP}(t_1^*, T^*, p^*)$ but increase of t_1^*, T^*, and Q^*.

(7) Increase in the values of the parameter A will result in decrease of p^* and $\mathrm{TP}(t_1^*, T^*, p^*)$ but increase of t_1^*, T^*, and Q^*.

(8) Decrease in the values of the parameter A will result in increase of p^* and $\mathrm{TP}(t_1^*, T^*, p^*)$ but decrease of t_1^*, T^*, and Q^*.

(9) Increase in the values of the parameter C will result in decrease of t_1^*, T^*, and Q^* but increase of p^* and $\mathrm{TP}(t_1^*, T^*, p^*)$.

(10) Decrease in the values of the parameter C will result in increase of t_1^*, T^*, and Q^* but decrease of p^* and $\mathrm{TP}(t_1^*, T^*, p^*)$.

6. Concluding Remarks

In this paper, we formulated an integrated inventory model for deteriorating items where shortage is completely backlogged and the permissible delay in payment depends on the order quantity. The deterministic demand rate is assumed to be a function of selling price. The objective is to determine the retailer's optimal ordering quantity and the optimal length of order cycle for maximizing the profit. An algorithm is suggested to find the optimal ordering policy, which helps the inventory manager to decide whether it would be worthwhile to take advantage of a longer credit period for repaying the supplies by ordering a larger amount of the commodity. Numerical examples are provided to demonstrate the applicability of the proposed models. The sensitivity analysis of the optimal solution with respect to the parameters is also included. The outcome shows that a higher value of deterioration rate causes lower value of order quantity and lower value of inventory cost. Hence, it suggests that the retailer should offer higher credit period to customers as the deterioration rate increases. This implies that credit period offered to customers has positive impact.

This paper also presented a practical application example where the proposed inventory model is utilized to support business decision making. Particularly, the model developed in the paper could be useful in the area of supply chain management. Our result illustrates that this model can be quite useful in determining the optimal ordering policy when the trade credit period is being analyzed. The proposed model can be used in inventory control of certain decaying items such as photographic film, electronic components, and radioactive materials which exhibit stock-dependent consumption.

A natural extension of the model would be to study that the case of n breaks in the permissible delay period. Additionally, this work can be extended for deteriorating items with a two-parameter Weibull distribution. The proposed model can further incorporate more realistic assumptions, such as probabilistic demand and trapezoidal fuzzy deterioration.

Acknowledgments

The author greatly appreciates the six anonymous reviewers for their very valuable and insightful comments, which helped in improving the paper. The author also wishes to thank the Journal Publishing Editor for her helpful suggestions, which have led to a significant improvement in the earlier version of the paper. Best efforts have been made by the author to revise the paper abiding by the constructive comments of the referees and all their suggestions were incorporated directly into the text.

References

[1] S. K. Goyal, "Economic order quantity under conditions of permissible delay in payments," *Journal of the Operational Research Society*, vol. 36, no. 4, pp. 335–338, 1985.

[2] K.-J. Chung and Y.-F. Huang, "The optimal cycle time for EPQ inventory model under permissible delay in payments," *International Journal of Production Economics*, vol. 84, no. 3, pp. 307–318, 2003.

[3] J.-T. Teng, "On the economic order quantity under conditions of permissible delay in payments," *Journal of the Operational Research Society*, vol. 53, no. 8, pp. 915–918, 2002.

[4] M. K. Salameh, N. E. Abboud, A. N. El-Kassar, and R. E. Ghattas, "Continuous review inventory model with delay in payments," *International Journal of Production Economics*, vol. 85, no. 1, pp. 91–95, 2003.

[5] S. K. Goyal and B. C. Giri, "Recent trends in modeling of deteriorating inventory," *European Journal of Operational Research*, vol. 134, no. 1, pp. 1–16, 2001.

[6] A. Goswami and K. Chaudhuri, "EOQ model for an inventory with a linear trend in demand and finite rate of replenishment considering shortages," *International Journal of Systems Science*, vol. 22, no. 1, pp. 181–187, 1991.

[7] T. Chakrabarti and K. S. Chaudhuri, "An EOQ model for deteriorating items with a linear trend in demand and shortages in all cycles," *International Journal of Production Economics*, vol. 49, no. 3, pp. 205–213, 1997.

[8] B. C. Giri, T. Chakrabarty, and K. S. Chaudhuri, "A note on a lot sizing heuristic for deteriorating items with time-varying demands and shortages," *Computers and Operations Research*, vol. 27, no. 6, pp. 495–505, 2000.

[9] A. K. Jalan and K. S. Chaudhuri, "Structural properties of an inventory system with deterioration and trended demand," *International Journal of Systems Science*, vol. 30, no. 6, pp. 627–633, 1999.

[10] M. A. Hariga and L. Benkherouf, "Optimal and heuristic inventory replenishment models for deteriorating items with exponential time-varying demand," *European Journal of Operational Research*, vol. 79, no. 1, pp. 123–137, 1994.

[11] H. M. Wee, "A deterministic lot-size inventory model for deteriorating items with shortages and a declining market," *Computers and Operations Research*, vol. 22, no. 3, pp. 345–356, 1995.

[12] A. M. M. Jamal, B. R. Sarker, and S. Wang, "An ordering policy for deteriorating items with allowable shortage and permissible delay in payment," *Journal of the Operational Research Society*, vol. 48, no. 8, pp. 826–833, 1997.

[13] H.-J. Chang and C.-Y. Dye, "An inventory model for deteriorating items with partial backlogging and permissible delay in payments," *International Journal of Systems Science*, vol. 32, no. 3, pp. 345–352, 2001.

[14] K. Annadurai and R. Uthayakumar, "Analysis of partial trade credit financing in a supply chain by EOQ-based model for decaying items with shortages," *International Journal of Advanced Manufacturing Technology*, vol. 61, pp. 1139–1159, 2012.

[15] K. Annadurai, "An optimal replenishment policy for decaying items with shortages and salvage value," *International Journal of Management Science and Engineering Management*, vol. 8, pp. 38–46, 2013.

[16] T. H. Burwell, D. S. Dave, K. E. Fitzpatrick, and M. R. Roy, "Economic lot size model for price-dependent demand under quantity and freight discounts," *International Journal of Production Economics*, vol. 48, no. 2, pp. 141–155, 1997.

[17] B. Mondal, A. K. Bhunia, and M. Maiti, "An inventory system of ameliorating items for price dependent demand rate," *Computers and Industrial Engineering*, vol. 45, no. 3, pp. 443–456, 2003.

[18] P.-S. You, "Inventory policy for products with price and time-dependent demands," *Journal of the Operational Research Society*, vol. 56, no. 7, pp. 870–873, 2005.

[19] H.-M. Wee, "A replenishment policy for items with a price-dependent demand and a varying rate of deterioration," *Production Planning and Control*, vol. 8, no. 5, pp. 494–499, 1997.

[20] C. Lin, B. Tan, and W.-C. Lee, "An EOQ model for deteriorating items with time-varying demand and shortages," *International Journal of Systems Science*, vol. 31, no. 3, pp. 391–400, 2000.

[21] K. Annadurai and R. Uthayakumar, "An inventory model for deteriorating items with credit period dependent demand under two-level trade credit financing," *Journal of Advanced Research in Applied Mathematics*, vol. 3, pp. 24–46, 2011.

[22] K. Annadurai, "An integrated inventory model for decaying items with time discounting and shortages under two levels of trade credit," *Bulletin of Kerala Mathematics Association*, vol. 10, pp. 49–66, 2013.

[23] S. P. Aggarwal and C. K. Jaggi, "Ordering policies of deteriorating items under permissible delay in payments," *Journal of the Operational Research Society*, vol. 46, pp. 658–662, 1995.

[24] H.-J. Chang, C.-H. Hung, and C.-Y. Dye, "An inventory model for deteriorating items with linear trend demand under the condition of permissible delay in payments," *Production Planning and Control*, vol. 12, no. 3, pp. 274–282, 2001.

[25] C. Y. Dye, "A deteriorating inventory model with stock dependent rate and partial backlogging under conditions of permissible delay in payments," *Opsearch*, vol. 39, pp. 189–198, 2002.

[26] H. Soni and N. H. Shah, "Optimal ordering policy for stock-dependent demand under progressive payment scheme," *European Journal of Operational Research*, vol. 184, no. 1, pp. 91–100, 2008.

[27] C. K. Tripathy and U. Mishra, "An inventory model for Weibull deteriorating items with price dependent demand and time-varying holding cost," *Applied Mathematical Sciences*, vol. 4, no. 41-44, pp. 2171–2179, 2010.

[28] K. Annadurai and R. Uthayakumar, "Two-echelon inventory model for deteriorating items with credit period dependent demand including shortages under trade credit," *Optimization Letters*, vol. 7, pp. 1227–1249, 2013.

20

Material Removal Rate, Electrode Wear Rate, and Surface Roughness Evaluation in Die Sinking EDM with Hollow Tool through Response Surface Methodology

Teepu Sultan, Anish Kumar, and Rahul Dev Gupta

Department of Mechanical Engineering, M. M. University, Mullana, Ambala, Haryana 133203, India

Correspondence should be addressed to Anish Kumar; anish_kaushik@rediffmail.com

Academic Editor: Gilles Dessein

Electrical discharge machining is one of the earliest nontraditional machining, extensively used in industry for processing of parts having unusual profiles with reasonable precision. In the present work, an attempt has been made to model material removal rate, electrode wear rate, and surface roughness through response surface methodology in a die sinking EDM process. The optimization was performed in two steps using one factor at a time for preliminary evaluation and a Box-Behnken design involving three variables with three levels for determination of the critical experimental conditions. Pulse on time, pulse off time, and peak current were changed during the tests, while a copper electrode having tubular cross section was employed to machine through holes on EN 353 steel alloy workpiece. The results of analysis of variance indicated that the proposed mathematical models obtained can adequately describe the performances within the limits of factors being studied. The experimental and predicted values were in a good agreement. Surface topography is revealed with the help of scanning electron microscope micrographs.

1. Introduction

Electrodischarge machining is a nonconventional machining process extensively used in industry for processing of parts having unusual profiles with reasonable precision [1]. Steel is a widely used engineering material. There are a variety of steels used for numerous applications. The steel is being divided into low carbon, medium carbon, and high carbon steel on the basis of carbon content. Low carbon steel contains a carbon content from 0.15% to 0.45%. It is the most common form of steel as it provides material properties that are acceptable for many applications [2]. EN 353 steel is cheaply available and widely used alloy. EN 353 is low alloy case carburized steel, predominantly used for manufacturing heavy-duty gears, shafts, pinions, and especially crown wheel [3].

Discharge current was the most significant controlling parameter in machining Al-15% SiC MMC using multi-hole electrode by grey relational analysis, while Taguchi method was employed to determine the relations between the machining parameters and process characteristics like MRR (material removal rate), EWR (electrode wear rate), and SR (surface roughness) [4]. The combination of maximum pulse on time and minimum pulse off time gives maximum MRR on Al-7075 alloy with brass electrode using Taguchi approach [5]. Amorim and Weingaertner concluded that the best results of MRR and surface texture for duty factor of 0.5 were obtained with tungsten-copper electrodes, no matter the machining conditions while machining copper-beryllium ASTM C17200 alloy [6]. Single-channel electrodes have comparatively better MRR and lower EW during fast hole drilling of Inconel 718 and Ti-6Al-4V with tubular electrodes of brass and copper, while copper electrode undergoes lower wear than brass electrode [7]. Discharge current, pulse on time and pulse off time, had a significant influence on the machining characteristics such as MRR, EWR, and overcut while machining Inconel 718 [8] and on gap current while machining high strength low alloy steel (HSLA), respectively [9].

Kerosene decomposes at higher temperature due to larger discharge energy and produces carbon particles that adhere to the electrode surface and this phenomenon restricts rapid tool wear during machining than deionized water [10]. Globules of debris, pockmarks, and melted drops were observed on the surfaces of blind holes which were more dispersed and pronounced. Most significant parameters like voltage and current in the optimization of a single quality characteristic are not being significant as in multiple quality characteristics [11]. Patel et al. done the multiobjective optimization with copper electrode and concluded that lower value of parameters reduces the tool wear ratio substantially, while their higher values increase the MRR drastically on AISI tool steel [12]. MRR not only increases with increase in discharge current but also increases the microcracks density on machining tungsten carbide with graphite electrode [13]. Increasing discharge energy increases instability and, due to which, the quality of the workpiece surface becomes rougher and the white layer thickness increases. Also the amount of particles in the gap becomes too large and forms electrically conducting paths between the tool electrode and the workpiece causing unwanted discharges, which becoming electric arcs (arcing). These electric arcs damage the electrodes surfaces (tool and workpiece surfaces) and can create microcracks [14]. During EDM processes, a significant amount of workpiece material was found to be transferred from workpiece surface to tool surface and vice versa. Also, a continuous burning of cutting fluid gives out a carbon residue, a visual black layer on machined surface, which decreases the EWR [15]. Kansal et al. reported that surface roughness of EN-31 tool steel with copper electrode gives the best surface finish if obtained at the lower level of peak current and pulse on time [16]. While increasing the pulse current or reducing the pulse on duration suppresses the formation of surface cracks in the SKD11 machined surface and hence improves the fatigue life [17]. Kumar et al. [18] revealed that intense heat conditions in the machining zone yielded higher crack densities and surface roughness while machining titanium at higher peak current along with a higher frequency of the globules of debris than on steel. Mathematical modeling of process using response surface methodology shows that the developed model can achieve reliable prediction of experimental results within acceptable accuracy [19]. Çaydaş and Hasçalik also added that RSM is an economical way for obtaining information for any system with the fewest numbers of experiments [20]. Many researchers used RSM to quantify the effect of control factors on their problems [21–24]. Box-Behnken design gives us robustness to the unavailability of data over central composite design [25]. Reversing the polarity of sparking alters the material removal phenomenon with an appreciable amount of electrode material depositing on the workpiece surface [26]. Due to its structural integrity, copper has ability to produce very fine surface finishes, even without special polishing circuits. It also makes copper electrodes highly resistant to DC arcing in poor flushing situations [27]. Rajesha et al. [1] demonstrated that hollow tool is particularly useful for drilling holes with low tool wear rate. It was found that while machining the same length of the Inconel 718 with a solid tool, it takes approximately 40% more machining time than that taken by a hollow tool. It also helps in minimizing the dielectric fluid degradation. Consequently, the approach light is cost effective with higher yield and reduced material and energy loss.

By scrutiny of the published research work, this paper addresses the issue that an alternate type of tool like hollow tube electrode may have a positive impact on MRR with low tool wear rate due to improved flushing conditions. Also there is a little work that has been carried out regarding the modelling and analysis of the EDM process on EN 353 steel alloy material with the MRR, EWR, and SR as the machining performances. In this study, mathematical models have been developed using response surface methodology, while analysis of variance (ANOVA) is used to check the validity of the models. Surface topography is revealed with the help of scanning electron microscope (SEM) micrographs.

2. Experimental Procedure

Experiments were conducted on a die sinking EDM machine, model Electronica Xpert 1 CNC as shown in Figure 1(a). In this study, EN 353 steel is selected as the work material. Chemical composition of the workpiece material is given in Table 1. A cylindrical electrode with tubular section having 12 mm external and 9 mm internal diameter cooper was used as the electrode (tool). Figures 1(b) and 1(c) show a typical copper electrode used in the experimentation. Kerosene oil was used as a dielectric.

The process parameters and their levels for the main experiments were decided on the basis of the pilot experiments conducted using one-factor-at-a-time approach as shown in Table 2.

The Box-Behnken design (BBD) was used for planning and executing the subsequent main experimentation, as shown in Table 3. In this study, an effort has been made to model the empirical relationship between machining parameters by using response surface methodology. The workpiece was connected to the positive polarity while the tool electrode was maintained at negative polarity. Side flushing method was employed for the dielectric fluid. A hole depth of 8 mm and diameter of 12 mm was machined throughout, for each run. The process parameters and depth of cut were programmed in the NC controlled unit. Once the experimentation was completed, the workpieces were cleaned thoroughly using acetone and the final individual weight of electrode was measured. Material removal rate was calculated by using the following formula:

$$\text{MRR} = \frac{V\left(\text{mm}^3\right)}{t\,(\text{min})}\left(\text{mm}^3/\text{min}\right), \quad (1)$$

where V is volume of material removed and t is the machining time.

Electrode wear rate was calculated by using the following formula:

$$\text{EWR} = \frac{E_b - E_a\,(\text{g})}{t\,(\text{min}) * \text{density}\left(\text{g/mm}^3\right)}\left(\text{mm}^3/\text{min}\right), \quad (2)$$

where E_b and E_a are the weights of electrode material before and after machining, respectively, and t is the machining

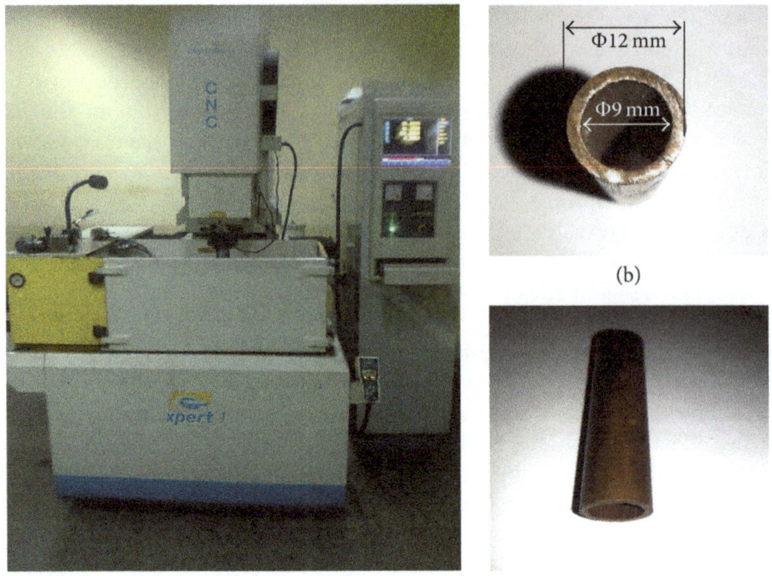

(a) (b) (c)

FIGURE 1: (a) CNC die sinking EDM; ((b), (c)) cross-sectional view of electrode.

TABLE 1: The composition of EN 353.

Composition									
Contents	C	Si	Mn	Cr	Mo	S	P	Al	Ni
Percentage	0.18	0.22	0.66	0.89	0.13	0.033	0.034	0.026	1.06

TABLE 2: Factors and their levels.

Serial number	Symbols	Input factors	Levels			Units
			1	2	3	
1	A	Pulse on time	100	340	580	μs
2	B	Pulse off time	4	16	28	μs
3	C	Peak current	5	25	45	A

time. The weights of the electrodes were measured using a balance. Surface roughness measurements were carried out at the sidewall of the holes using a Brand-Mitutoyo Surftest, Model-SJ-301.

3. Results and Discussion

3.1. Effect of Process Parameters on Material Removal Rate. The mathematical relationship for correlating the machining rate and the considered process variables is obtained as follows:

$$\begin{aligned} \mathrm{MRR} = {} & 20.12185 + 0.015014 * T_{\mathrm{ON}} \\ & - 0.23106 * T_{\mathrm{OFF}} - 0.42890 * I_P \\ & + 2.88377E - 005 * T_{\mathrm{ON}}^2 + 4.52138E \\ & - 003 * I_P^2 - 8.47222E - 004 * T_{\mathrm{ON}} * T_{\mathrm{OFF}} \\ & + 0.013052 * T_{\mathrm{OFF}} * I_P. \end{aligned} \quad (3)$$

The fit summary recommended that the quadratic model is statistically significant for analysis of MRR. The results of the quadratic model for MRR in the form of ANOVA are given in Table 4. This model was developed for 95% confidence level. The model F value of 43.10 implied that the model is significant for MRR. There is only a 0.01% chance that a "model F value" this large could occur due to noise.

The values of "Prob > F" less than 0.05 indicate that the model terms are significant. In this case, A, B, A^2, C^2 AB, and BC are significant model terms. The other model terms are said to be nonsignificant. The lack of fit F value of 0.0801 implies that it is not significant relative to the pure error. The "Pred. R^2" of 0.8179 is in reasonable agreement with the "Adj. R^2" of 0.9404. "Adeq Precision" measures the signal-to-noise ratio. A ratio greater than 4 is desirable. Your ratio of 14.843 indicates an adequate signal. Figures 2(a), 2(b), and 2(c) show the effects of different input factors (pulse on time, pulse off time, and peak current) on the response parameter MRR. Figure 2(d) shows the contour plot, while

TABLE 3: Factors and their responses.

Run number	Factors			Responses		
	Pulse off time T_{OFF} (μs)	Pulse on time T_{ON} (μs)	Pulse current I_P (A)	Material removal rate MRR (mm^3/min)	Electrode wear rate EWR (mm^3/min)	Surface roughness SR (μm)
1	28	340	5	14.6	4.7	5.39
2	4	340	5	23.37	11.6	6.3
3	16	340	25	18.25	7.4	7.22
4	28	340	45	20.96	12.9	6.25
5	16	340	25	18.85	5.7	7.45
6	16	340	25	19.55	7.2	7.57
7	4	580	25	28.99	13.8	10.12
8	4	100	25	14.53	3.5	5.42
9	16	340	25	18.86	5.8	7.23
10	16	580	45	29.03	10.2	10.16
11	16	100	5	14	3.9	4.28
12	16	580	5	26.42	10.9	9.15
13	16	100	45	16.16	4.1	5.34
14	28	100	25	15.66	3.5	4.33
15	16	340	25	17.8	5.9	7.43
16	4	340	45	22.2	7.2	7.73
17	28	580	25	24.36	10.4	9.2

TABLE 4: The analysis of variance for main and interaction effects of parameters on MRR.

Source	SS	DOF	MS	F value	Prob > F	At 95% CI	% contribution
Model	364.25	6	60.71	43.10	<0.0001	Significant	
A	293.43	1	293.43	208.33	<0.0001	Significant	57.19
B	22.82	1	22.82	16.20	0.0024	Significant	11.97
C	12.40	1	12.40	8.80	0.0141	Significant	0.03
A^2	13.14	1	13.14	9.33	0.0122	Significant	3.25
AB	8.29	1	8.29	5.89	0.0356	Significant	6.65
BC	14.18	1	14.18	10.06	0.0099	Significant	10.97
Lack of fit	12.31	6	2.05	4.62	0.0801	Nonsignificant	
R^2	0.9628						
Adjusted R^2	0.9404						
Predicted R^2	0.8179						

CI: confidence interval; DOF: degrees of freedom; MS: mean sum of squares; SS: sum of squares.

Figures 2(e) and 2(f) show three-dimensional interaction response surfaces for MRR in relation to the input parameters of peak current and pulse off time. It can be seen from the figure that MRR increases with increase in peak current. Maximum MRR is obtained at high peak current 45 A and pulse on time at 580 μs. This can be explained on the basis of increase in the rate of discharge energy, as high concentration of discharge energy in the spark gap leads to rapid melting and vaporization of metal, which leads to rises in the MRR. It was observed from the plots that MRR increases (from 14 to 29.03 mm^3/min) with increase in peak current and pulse on time. Figures 2(g), 2(h), and 2(i) display the normal probability plot of residuals and residual versus predicted plots for MRR. It was observed that the residuals generally fall on a straight line implying that the errors are normally distributed.

3.2. Effect on Electrode Wear Rate. The mathematical relationship for correlating the electrode wear rate and the considered process variables is obtained as follows:

$$\begin{aligned}
\text{Electrode Wear Rate} \\
&= +1.51042 - 7.34375E - 003 * T_{ON} \\
&\quad + 0.16875 * T_{OFF} + 0.028125 * I_P + 7.13976E \\
&\quad - 005 * T_{ON}^2 - 2.17014E - 003 * T_{OFF}^2 \\
&\quad - 1.46875E - 003 * I_P^2 - 8.15972E \\
&\quad - 004 * T_{ON} * T_{OFF} - 9.89583E \\
&\quad - 005 * T_{ON} * I_P + 2.70833E - 003 * T_{OFF} * I_P.
\end{aligned} \tag{4}$$

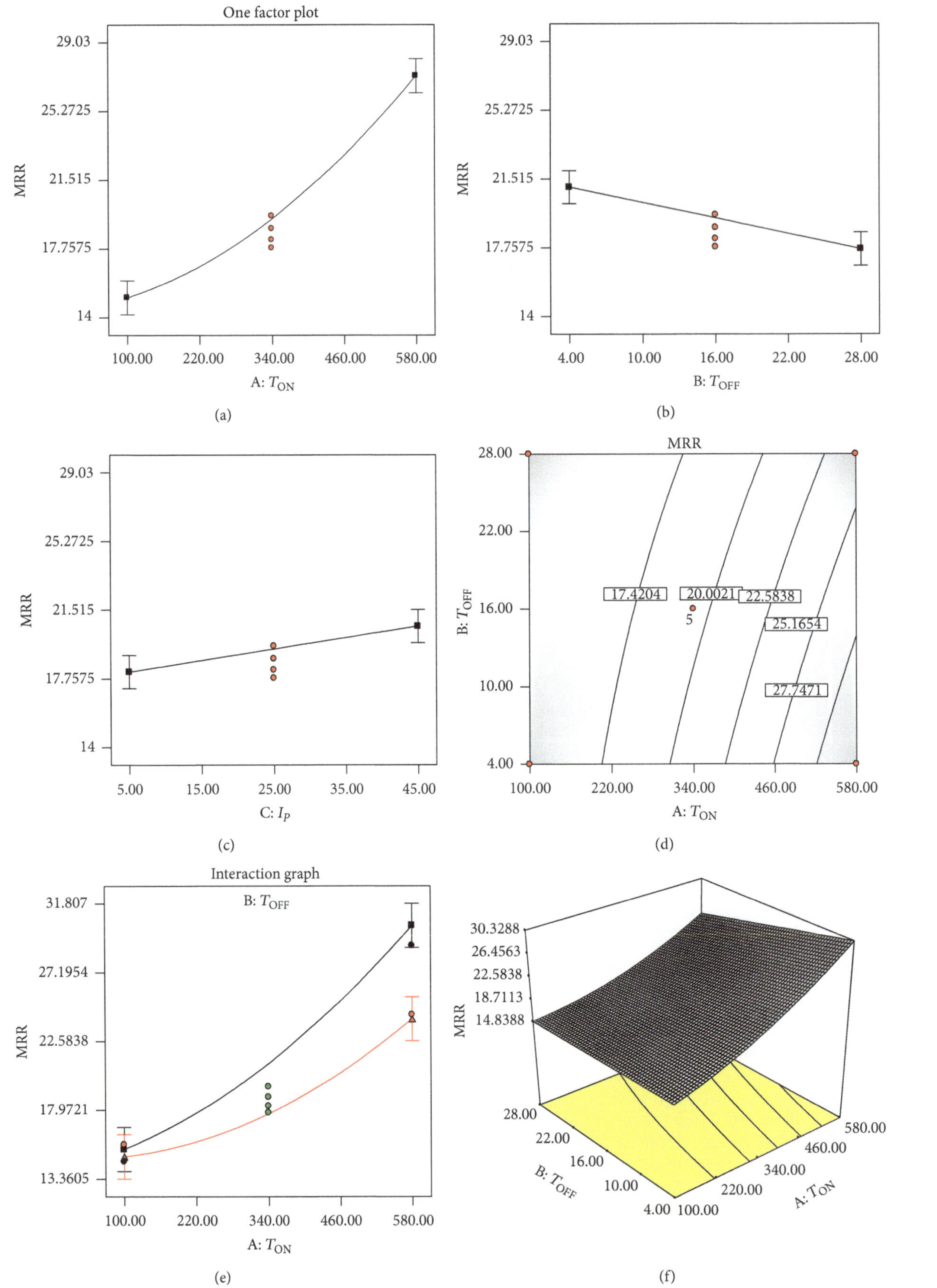

FIGURE 2: Continued.

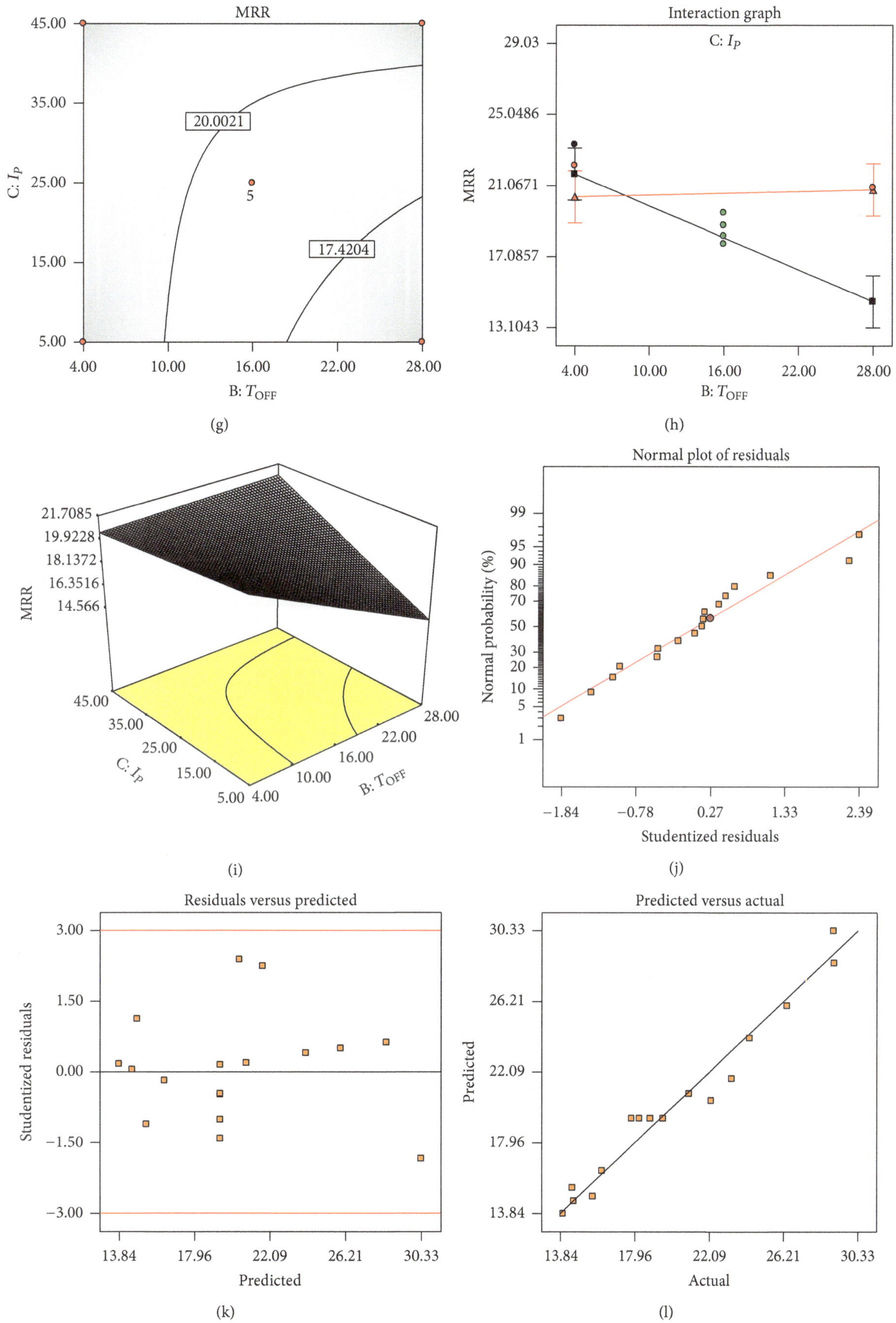

FIGURE 2: Response surface plot for MRR: (a) T_{ON} factor plot, (b) T_{OFF} factor plot, (c) I_P factor plot, (d) contour plot between T_{ON} and T_{OFF}, ((e) and (f)) interactions plot between T_{ON} and T_{OFF}, (g) contour plot between T_{OFF} and I_P, ((h) and (i)) interactions plot between T_{OFF} and I_P, (j) normal plot of residuals, (k) residuals versus predicted values, and (l) predicted versus actual values.

Table 5: The analysis of variance for main and interaction effects of parameters on EWR.

Source	SS	DOF	MS	F value	Prob > F	At 95% CI	% contribution
Model	170.14	5	34.03	23.50	<0.0001	Significant	
A	114.76	1	114.76	79.24	<0.0001	Significant	70.39
B	2.64	1	2.64	1.83	0.2037		3.25
C	1.36	1	1.36	0.94	0.3531		0.94
B^2	11.69	1	11.69	8.07	0.0161		0.09
BC	39.69	1	39.69	27.40	0.0003	Significant	0.39
Lack of fit	13.19	7	1.88	2.75	0.1725	Nonsignificant	
R^2	0.9144						
Adjusted R^2	0.8755						
Predicted R^2	0.7301						

CI: confidence interval; DOF: degrees of freedom; MS: mean sum of squares; SS: sum of squares.

Table 6: The analysis of variance for main and interaction effects of parameters on SR.

Source	SS	DOF	MS	F value	Prob > F	At 95% CI	% contribution
Model	53.68	7	7.67	550.62	<0.0001	Significant	
A	46.37	1	46.37	3329.21	<0.0001	Significant	86.17
B	2.42	1	2.42	173.75	<0.0001	Significant	4.49
C	2.38	1	2.38	170.61	<0.0001	Significant	4.42
A^2	0.52	1	0.52	37.30	0.0002	Significant	0.96
B^2	0.91	1	0.91	65.02	<0.0001	Significant	1.69
C^2	1.05	1	1.05	75.20	<0.0001	Significant	1.95
BC	0.081	1	0.081	5.83	0.0389	Significant	0.15
Lack of fit	0.034	5	$6.75E-003$	0.29	0.8937	Nonsignificant	
R^2	0.9977						
Adjusted R^2	0.9959						
Predicted R^2	0.9921						

CI: confidence interval; DOF: degrees of freedom; MS: mean sum of squares; SS: sum of squares.

The fit summary recommended that the quadratic model is statistically significant for analysis of EWR. The results of the quadratic model for EWR in the form of ANOVA are given in Table 5. This model was developed for 95% confidence level. The model F value of 23.50 implies that the model is significant. There is only a 0.02% chance that "model F value" this large could occur due to noise. Values of "Prob > F" less than 0.05 indicate that model terms are significant. In this case, A, B, A^2, and AB are significant model terms. The "lack of fit F value" of 0.1725 implies the lack of fit is not significant relative to the pure error. The "Pred. R^2" of 0.7301 is in reasonable agreement with the "Adj. R^2" of 0.8755. "Adeq. Precision" measures the signal-to-noise ratio. A ratio greater than 4 is desirable. The ratio of 16.222 indicates an adequate signal. This ratio of 16.222 indicates an adequate signal. Figures 3(a), 3(b), and 3(c) show the effects of different input factors (pulse on time, pulse off time, and peak current) on the response parameter EWR. Figure 3(d) shows the contour plot, while Figures 3(e) and 3(f) show three-dimensional interaction response surfaces for EWR in relation to the input parameters of peak current and pulse off time. From the contour plot and response surface, it was observed that minimum EWR 3.5 mm^3/min with the low value of peak current at 4 A and pulse off time at 4 μs. In addition to this, Figures 2(g), 2(h), and 2(i) show that the normal plot of residuals data is normally distributed.

3.3. Effect of Process Parameters on Surface Roughness. The mathematical relationship for correlating the surface roughness and the considered process variables is obtained as follows:

$$\begin{aligned} \text{Surface Roughness} &= +2.88516 + 5.88455E-003 * T_{\text{ON}} \\ &\quad + 0.072066 * T_{\text{OFF}} + 0.099094 * I_P \\ &\quad + 6.09809E-006 * T_{\text{ON}}^2 - 3.22049E \\ &\quad - 003 * T_{\text{OFF}}^2 - 1.24687E-003 * I_P^2 \\ &\quad - 5.93750E-004 * T_{\text{OFF}} * I_P. \end{aligned} \tag{5}$$

Based on ANOVA as shown in Table 6, the model F value of 550.62 implies that the model is significant. There is only

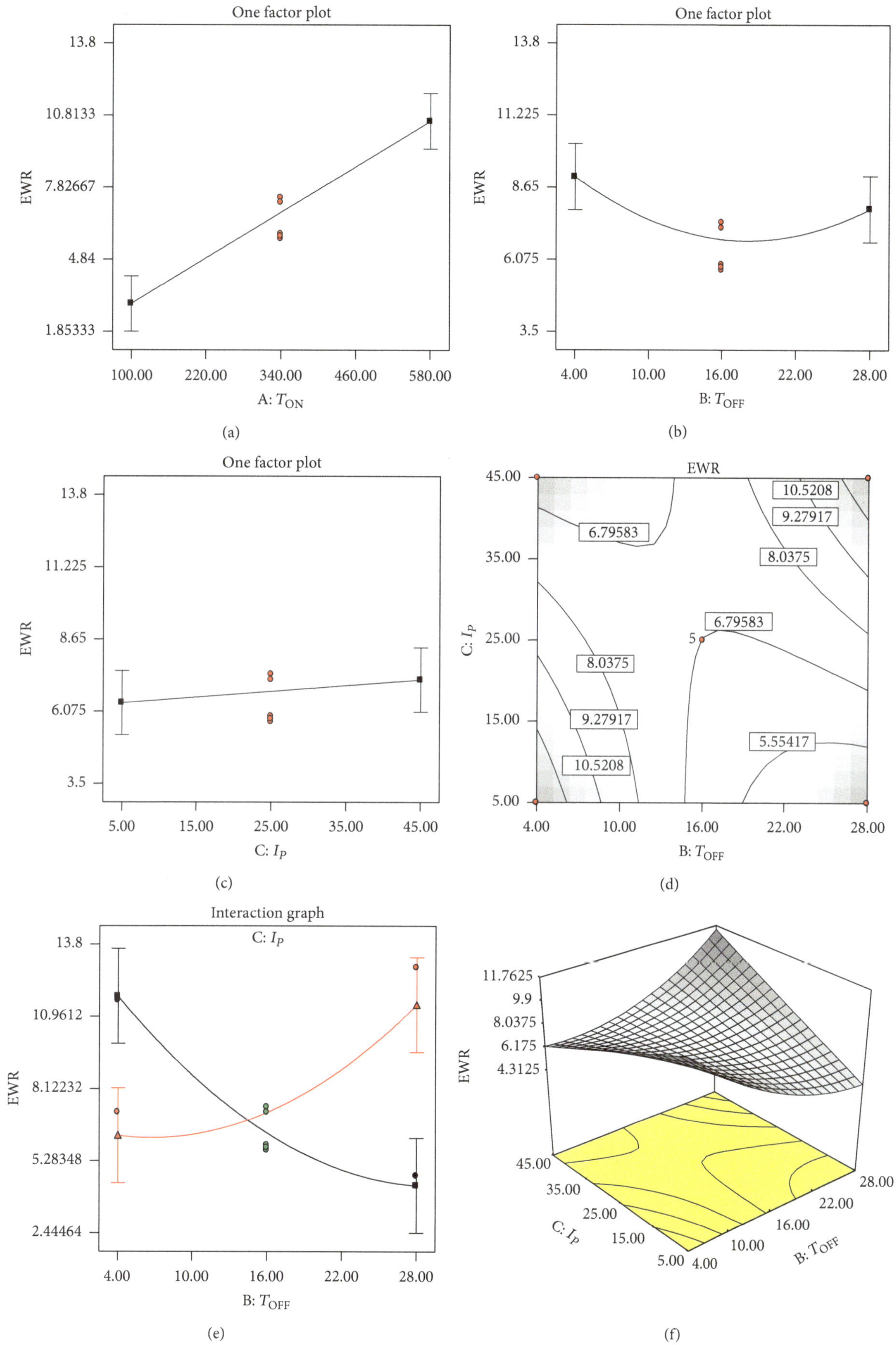

Figure 3: Continued.

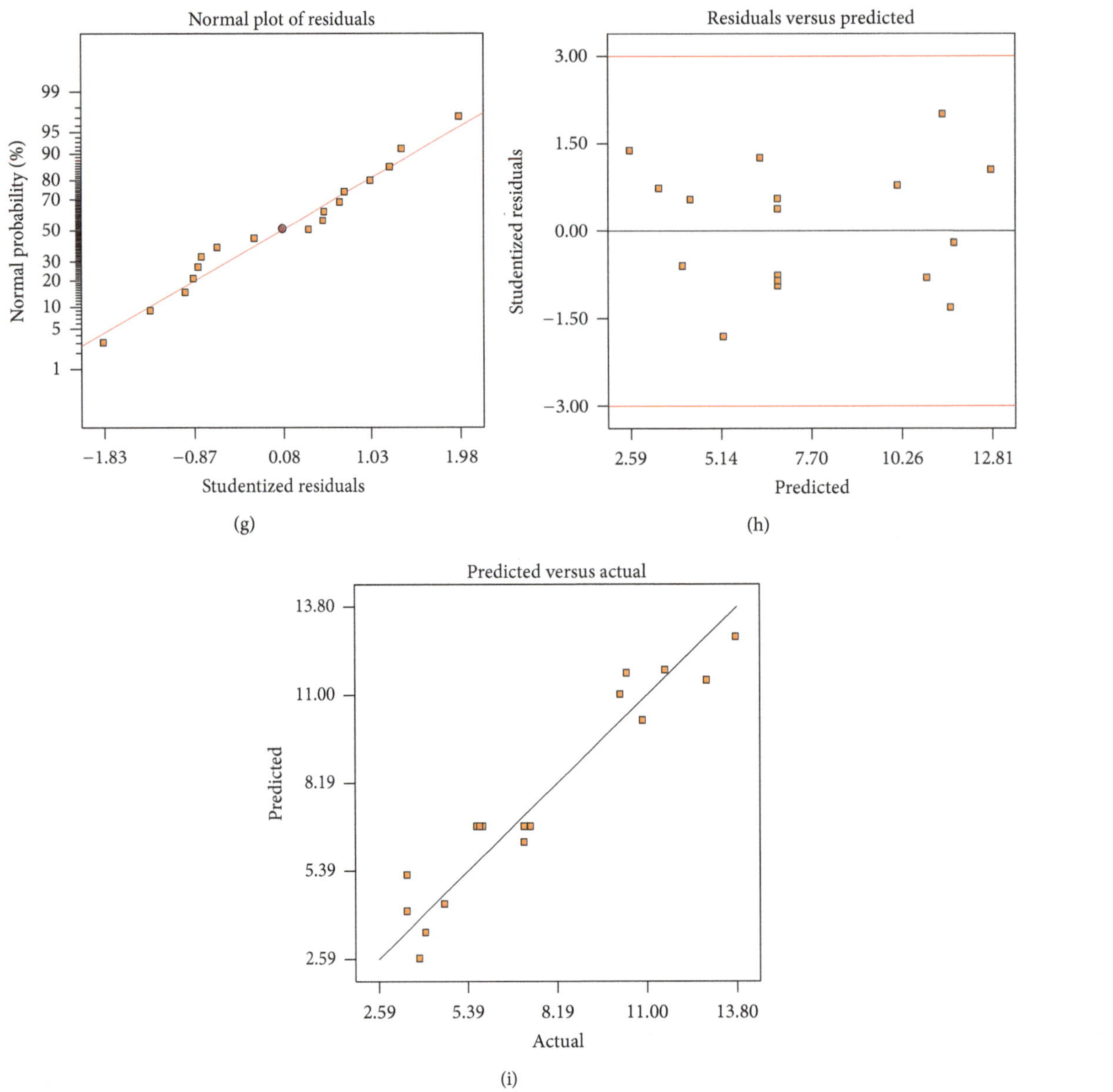

FIGURE 3: Response surface plot for EWR: (a) T_{ON} factor plot, (b) T_{OFF} factor plot, (c) I_P factor plot, (d) contour plot between T_{OFF} and I_P, ((e) and (f)) interactions plot between T_{OFF} and I_P, (g) normal plot of residuals, (h) residuals versus predicted values, and (i) predicted versus actual values.

a 0.01% chance that a "model F value" this large could occur due to noise. Values of "Prob > F" less than 0.0500 indicate model terms are significant. In this case A, B, C, A^2, B^2, C^2, and BC are significant model terms. The "lack of fit F value" of 0.8937 implies that the lack of fit is not significant relative to the pure error. There is a 89.37% chance that a "lack of fit F value" this large could occur due to noise. The "Pred. R^2" of 0.9921 is in reasonable agreement with the "Adj. R^2" of 0.9959. "Adeq Precision" measures the signal-to-noise ratio. A ratio greater than 4 is desirable. Your ratio of 73.433 indicates an adequate signal. This model can be used to navigate around the design space. Figure 4(a) shows the contour plot, while Figures 4(b) and 4(c) show three-dimensional interaction response surfaces for SR in relation to the input parameters of peak current and pulse off time.

From the contour plot and response surface, it was observed that minimum SR 4.29 μm with the value of peak current at 16 A and pulse off time at 100 μs. In addition to this, Figures 4(d), 4(e), and 4(f) show that the normal plot of residuals data is normally distributed and all the experimental results are in the region very near to the predicted values, and, hence, the developed model can be effectively used to predict the surface roughness in EDM of EN 353. At higher peak current, the discharge energy per pulse increases, which

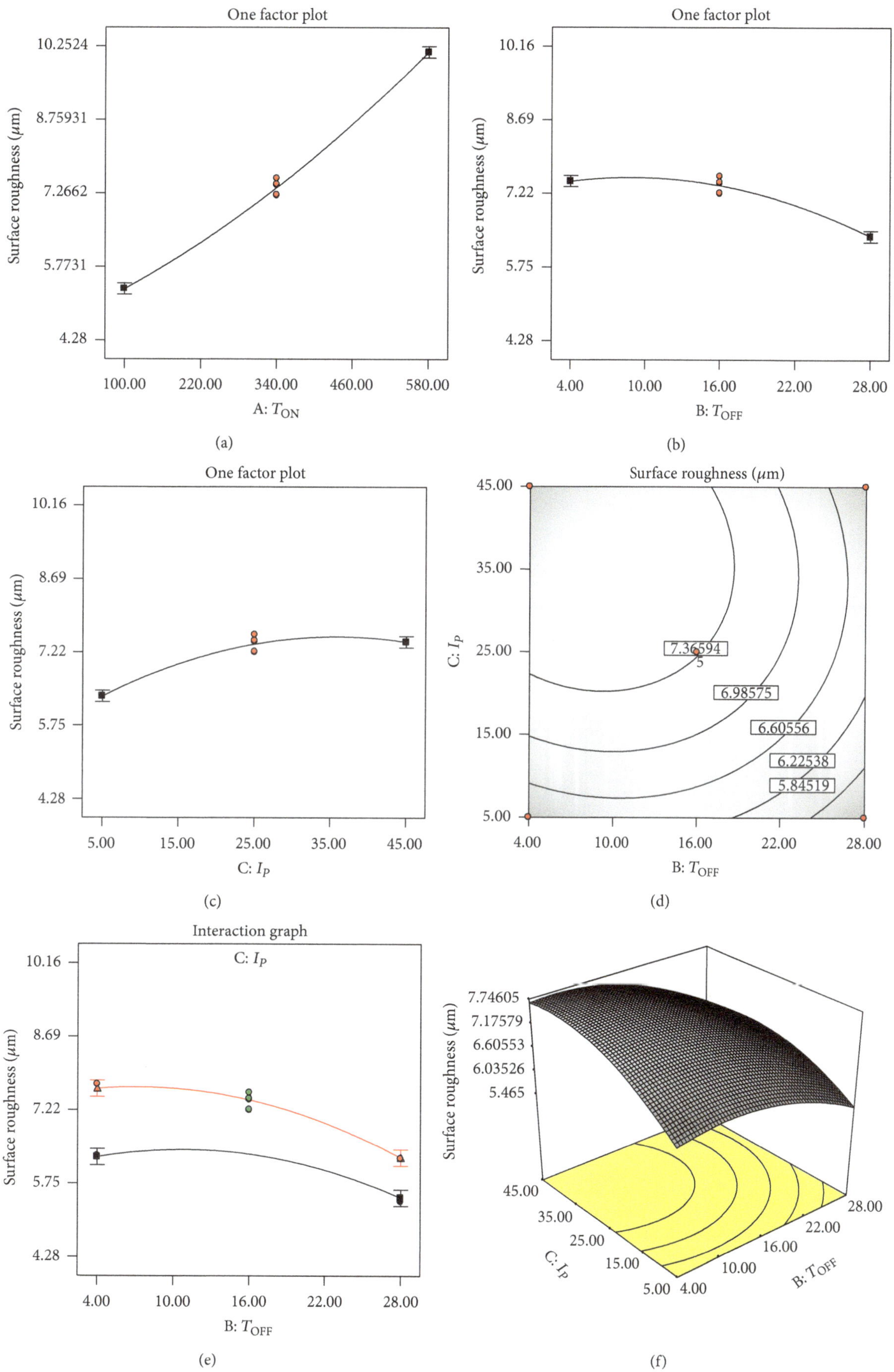

FIGURE 4: Continued.

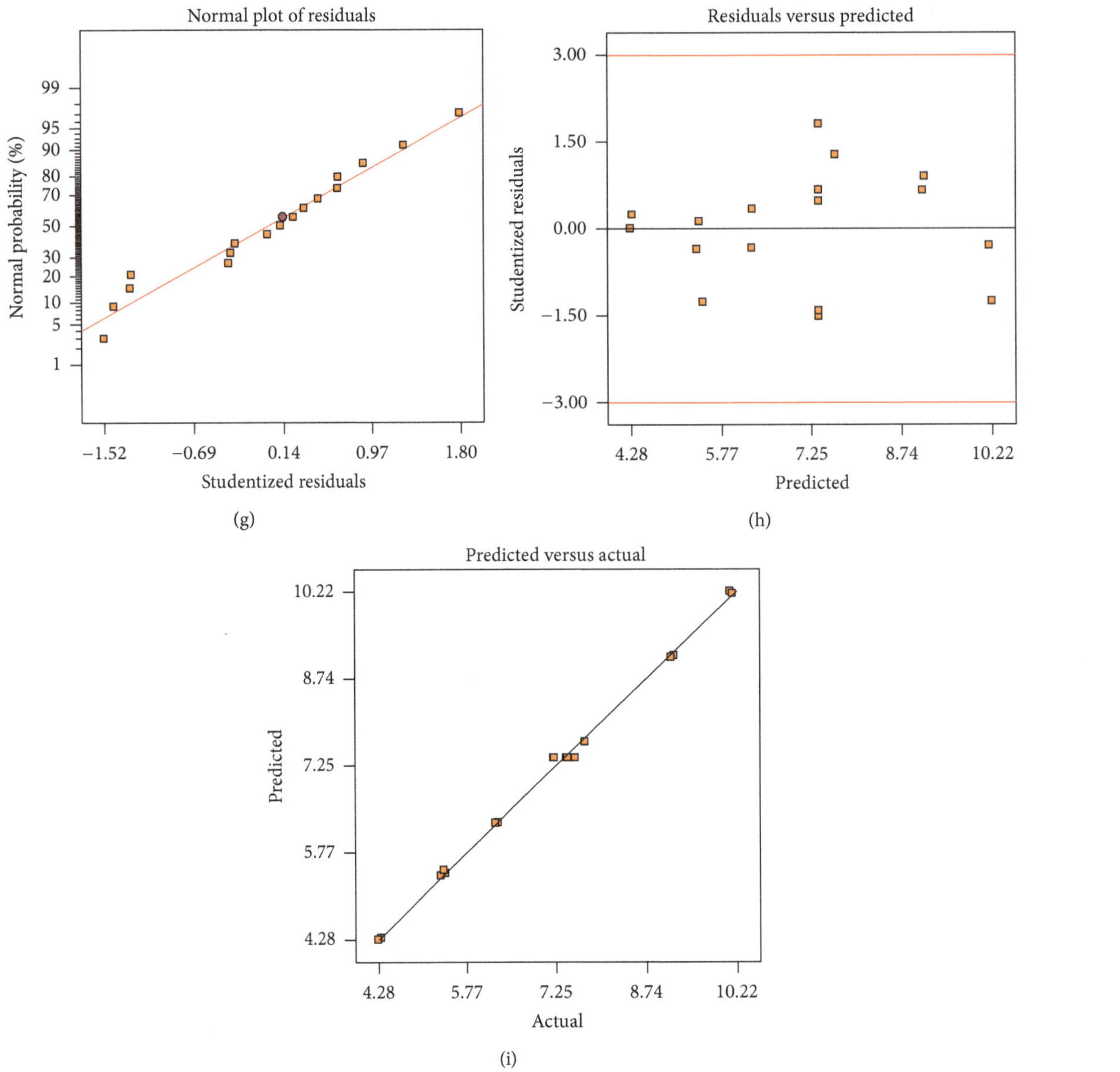

FIGURE 4: Response surface plot for SR: (a) T_{ON} factor plot, (b) T_{OFF} factor plot, (c) I_P factor plot, (d) contour plot between T_{OFF} and I_P, ((e) and (f)) interactions plot between T_{OFF} and I_P, (g) normal plot of residuals, (h) residuals versus predicted values, and (i) predicted versus actual values.

produces the deeper and wider overlapping craters, globules of debris, and microcracks on the machined samples as seen in SEM micrographs (Figure 5).

3.4. Microstructure Analysis. During EDM process, the discharged energy produces very high temperatures at the point of the spark, causing a minute part of the sample to melt and vaporizes. With each discharge, a crater was formed on the machined surface. It was observed from Figure 5. SEM micrographs that, EDM surface produces irregular topography and defects included globules of debris, spherical particles, varying size craters, and microcracks [28]. The surface topography was altered owing to significant electrical parameters such as pulse on time, pulse off time, and peak current. The pulse on time and peak current are the most significant parameters that lead to deterioration of the surface texture. When pulse on time was increased, the surface texture of the machined surface is composed of varying sizes of deep craters. These deep and overlapping craters were formed owing to successive electrical discharge, intense heat, and local melting or vaporization of work material. Some of the molten materials produced by the discharge was carried away by the kerosene. The remaining molten material resolidifies to form lumps of debris. Under shorter pulse on time, electrical sparks generate smaller craters on the work surface. The high pulsed current caused frequent cracking

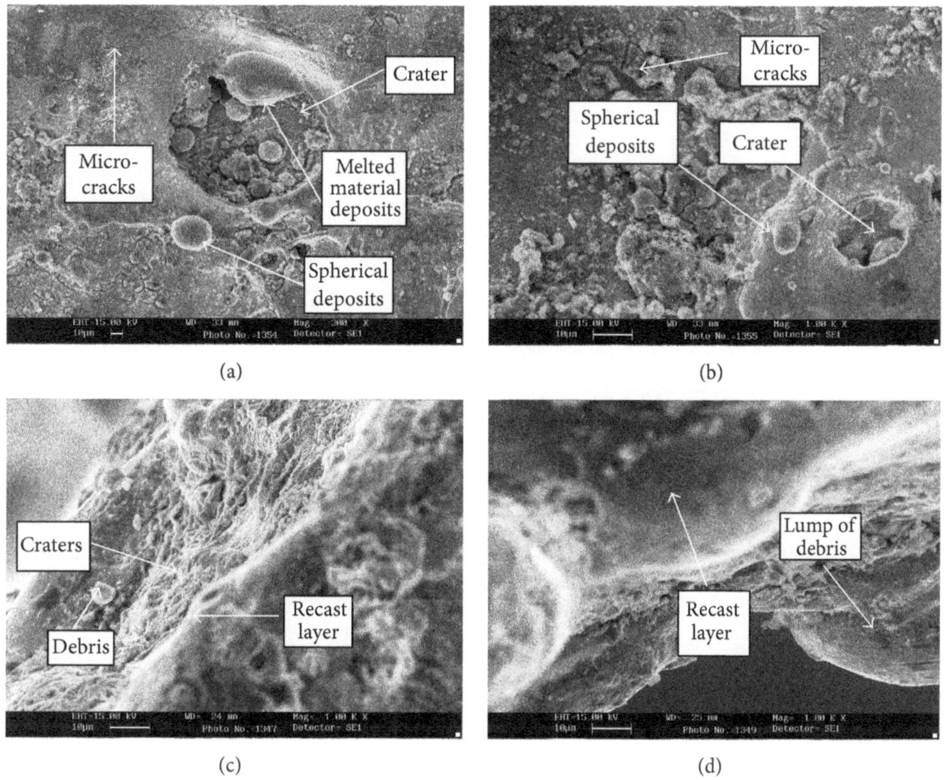

FIGURE 5: SEM micrograph: (a) Experiment number 3, at T_{ON} = 340 μs, T_{OFF} = 16 μs, and I_P = 25 A; (b) experiment number 17, at T_{ON} = 580 μs, T_{OFF} = 28 μs, and I_P = 25 A; (c) experiment number 13, at T_{ON} = 100 μs, T_{OFF} = 4 μs, and I_P = 45 A; and (d) experiment number 12, at T_{ON} = 580 μs, T_{OFF} = 16 μs, and I_P = 5 A.

of dielectric fluid, causing more melt expulsions and larger tensile stresses. These effects resulted in poor surface finish. At higher peak current, the impact of discharge energy on the surface of workpiece becomes greater, and, thus, the resulting erosion leads to the increase in deterioration of surface roughness. The recast layer is the outer region of the heat affected zone and consists of superimposed strata derived from melted and resolidified workpiece material as seen in Figure 5. This layer is caused by the material that was melted during the discharge and then when the discharge current breakdown occurred, a part of the material has resolidified in the crater and surroundings and the rest was flushed away by the dielectric [6]. The structure of the recast layer that is formed on steels consists mainly of iron carbides in acicular or globular form, distributed within an austenite matrix, which are independent of the composition of the base material and of the type of the electrode, that is, copper or graphite. The increase in carbon content in the recast layer is intrinsically related to the pyrolysis products that follows the cracking of the dielectric and is very confined to the melted and resolidified work piece material forming the iron carbides. Due to the very high cooling gradient from the surface into the matrix material, the iron carbides are normally oriented perpendicularly to the surface.

Figure 6 reveals the existence of cracks which start on the white layer and progress in the HAZ. The fundamental cause of cracking lies in the existence of the intern stresses which are created at the time of the machining operation. The surface crack density and the depth of these cracks are directly related to the machining conditions; the more we increase the discharge energy, the more the appearance frequency of these cracks increases. These types of cracking are due to very high temperature, because cracking at high temperature is due to the phenomena of segregation to solidification, which is due to the enrichment in certain elements, as solidification progresses and the intern stresses grow. The elements which have a harmful effect on steel and which increase its tendency to cracking are silicon, nickel, sulphur, phosphorus, and carbon [14]. The segregation comes owing to the fact that the grain boundaries generally constitute preferential sites for the impurities. The heating at very high temperature, during flashing, involves the enlargement of the grain of metal which implies the migration of the grain joints; during this migration, the joints of the grains rake the impurities which are in metal, and it is possible that this causes local concentrations in impurities. The presence of these cracks is also due to the presence of the elements strongly conducting in material as copper which let penetrate the current of discharge. It results in preferential energy propagation from it through these elements which can lead to total or partial separation of certain grains.

During EDM, some particles eroding from the tool electrode stick on the machined hole surface as seen in Figures 3(a) and 4(c), leading to formation of a rough surface with

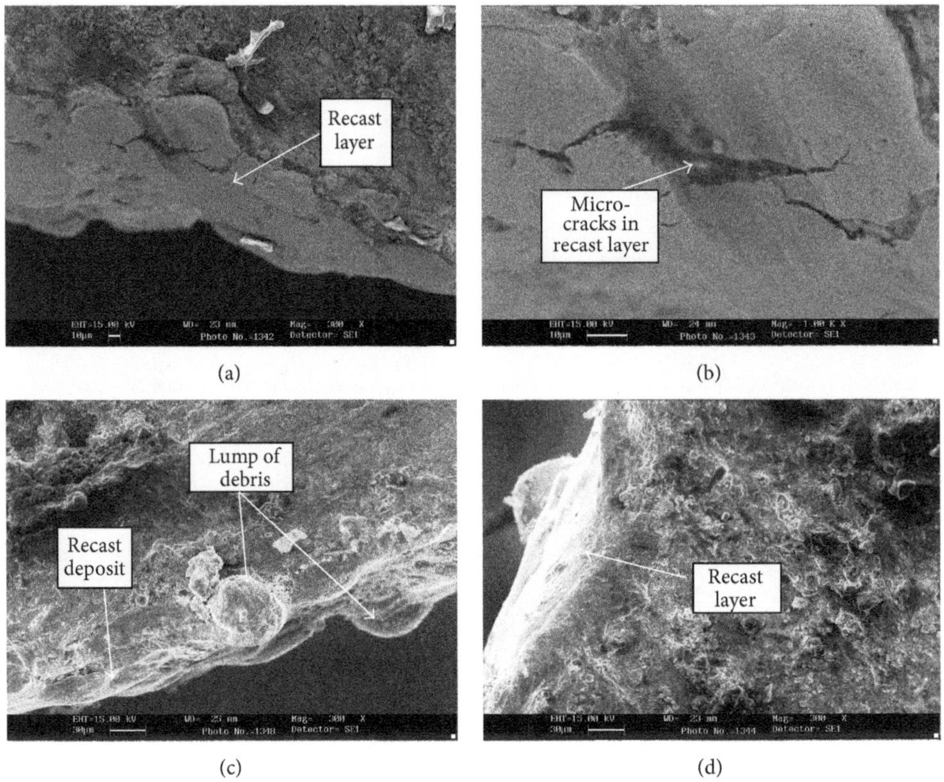

FIGURE 6: Recast layer: (a) experiment number 3, at T_{ON} = 340 μs, T_{OFF} = 16 μs, and I_P = 25 A; (b) experiment number 10, at T_{ON} = 580 μs, T_{OFF} = 16 μs, and I_P = 45 A; (c) experiment number 16, at T_{ON} = 340 μs, T_{OFF} = 4 μs, and I_P = 45 A; (d) experiment number 17, at T_{ON} = 580 μs, T_{OFF} = 28 μs, and I_P = 25 A.

irregularities. Such damages are even more intensively dispersed on the corresponding surfaces of blind holes [7]. This is due to the reason that the particles eroding from the surface being machined cannot be flushed away sufficiently. Thereby, even rougher surfaces are obtained during the production of blind holes upon solidification of debris and melted drops on the surface. It should be noted that these damages occurring during EDM process can be minimized by establishing the optimized process parameters to achieve desirable surface quality.

3.5. Multiresponse Optimization Based on Desirability. The desirability function can be used to combine multiple responses into one response called the "desirability function" by choice of value from 0 (one or more product characteristics are unacceptable) to 1 (all product characteristics are on target). The method is attractive because it is intuitive and simple [18]. The inputs are mean response estimates, target value, and upper and lower acceptability bounds. The individual desirability is combined using the geometric mean. The desirability of a product characteristic value depends on the lower and upper ranges of product specification. Improper selection of ranges can result in a very different "optimum." The basic idea of the desirability function approach is to transform a multiple response problem into a single response problem by means of mathematical transformations. The desirability function involves transformation of each estimated response variable $\hat{y}$ to a desirability value di, where $0 \le di \le 1$.

The value of di increases as the "desirability" of the corresponding response increases. The procedure followed in this work for simultaneous optimization of the three responses is a modification of the method developed by Derringer and Suich and El-Taweel.

Step 1. Calculate the individual desirability (di) for each response ($\hat{y}$).

Step 2. Combine individual desirabilities to obtain composite desirability (DG) for given weights of MRR, EWR, and surface roughness. Composite desirability is the weighted geometric mean of individual desirability for the given responses.

Step 3. Maximize the composite desirability and identify the optimal parameter combinations:

$$di = \begin{cases} 0 & \hat{y} < Ai \\ \left(\hat{y} - \dfrac{Ai}{ti} - Ai\right)^{w} & Ai \le \hat{y} \le ti \\ 1 & \hat{y} > ti. \end{cases} \tag{6}$$

TABLE 7: Constraints for desirability analysis.

Parameter	Goal	Lower limit	Upper limit	Lower weight	Upper weight	Importance
T_{ON}	Is in range	100	580	1	1	3
T_{OFF}	Is in range	4	16	1	1	3
I_P	Is in range	5	45	1	1	3
MRR	Maximum	14	29.03	1	1	3
EWR	Minimum	3.5	12.9	1	1	3
SR	Minimum	5.39	10.16	1	1	3

If the target (ti) is to minimize a response, the individual desirability (di) is calculated as

$$di = \begin{cases} 1 & \hat{y} < ti \\ \left(Bi - \dfrac{\hat{y}}{Bi} - ti\right)^{w} & ti \leq \hat{y} \leq Bi \\ 0 & \hat{y} > Bi, \end{cases} \quad (7)$$

where Ai is the lower limit value of response $\hat{y}$ and Bi is the upper limit value of response $\hat{y}$.

If the object for the response is a target value, then individual desirability (di) is calculated as

$$di = \left[\frac{(\hat{y} - Ai)}{(ti - Ai)}\right]^{w} \quad Ai \leq \hat{y} \leq ti$$

$$di = \left[\frac{(Bi - \hat{y})}{(Bi - ti)}\right]^{w} \quad ti \leq \hat{y} \leq Bi \quad (8)$$

$$di = 0 \quad \text{if } \hat{y} < Ai$$

$$di = 0 \quad \text{if } \hat{y} > Bi.$$

If the importance is same for each response, the composite desirability (DG), the geometric mean of all desirability functions, is given by

$$\mathrm{DG} = \left(d_1 \times d_2 \times \cdots \times d_n^{w_n}\right)^{1/n} = \left(\prod di\right)^{1/n}, \quad (9)$$

where n is the number of responses = 3.

It can be extensive to reflect the possible difference in the importance of different responses by giving weights. Where the weight wi satisfies $0 < wi < 1$ and $w_1 + w_2 + \cdots + w_n = 1$:

$$\mathrm{DG} = \left(d_1^{w_1} \times d_2^{w_2} \times \cdots \times d_n^{w_n}\right)^{1/n}. \quad (10)$$

Table 7 shows the constraints of input parameters and that of responses and the goal and weights assigned to each parameter. Table 8 shows the values of 36 levels of combinations of process parameters that will give a high value of composite desirability (ranged from 0.79 to 0.94). Table 8 gives the optimal input process parametric setting for multiresponse optimization.

The confirmatory experiments were performed to validate the optimal input parametric setting for MRR, EWR, and surface roughness. The observed experimental results for performance measures are material removal rate = 18.83 mm^3/min, electrode wear rate = 6.81 mm^3/min, and surface roughness = 4.86 μm. Table 9 shows the error percentage for experimental validation of the developed models for the responses with optimal parametric setting. From the analysis, it was concluded that the error between experimental and predicted values for MRR, EWR, and surface roughness are 6.98%, 5.90%, and 7.48%, respectively. Obviously, this confirms excellent reproducibility of the experimental conclusions.

3.6. Conclusion. In this study, the influence of significant EDM process parameters like peak current, pulse on time and pulse off time on response parameters like MRR, TWR, and SR while machining the EN 353 steel has been investigated. Experimentations were planed and conducted according to the response surface methods. Analysis of variance (ANOVA) was applied to study the significance on performance measures. The major conclusions drawn from this study are as follows.

(1) EDM is an adequate process to machine EN 353 steel with good MRR and TWR.

(2) The MRR obtained is ranged between 14 mm^3/min and 29.03 mm^3/min. The maximum MRR was obtained when the parameters were set at pulse on time = 580 μs, pulse off time = 16 μs, and peak current = 45 A.

(3) The minimum EWR 3.5 mm^3/min was obtained when the parameters were set at pulse on time = 100 μs, pulse off time = 28 μs, and peak current = 25 A. The interaction effect of pulse on time and pulse off time influences the most.

(4) SR while EDMing the EN 353 steel alloy is majorly influenced by all the three factors peak current, pulse on time, and pulse off time. In order to obtain better surface finish set peak current and pulse on time at low levels.

(5) The optimized values of MRR, TWR, and SR are 17.62 mm^3/min, 6.47 mm^3/min, and 4.54 μm, respectively, obtained at the optimum setting of parameters pulse on time, pulse off time, and peak current at 100.77 μs, 25.43 μs, and 45 A, respectively.

(6) The confirmation tests showed that the error between experimental and predicted values of MRR, TWR, and SR are 6.98% and 5.90%, 7.48% and 5.66%, and 2.14% and 2.87%, respectively.

TABLE 8: Favorable solutions for responses with varying weights.

Serial number	T_{ON}	T_{OFF}	I_P	MRR	EWR	SR	Desirability	Weight of		
								MRR	EWR	SR
1	579.99	11.03	5	28.03	11.99	9.18	0.87	0.8	0.1	0.1
2	503.2	4	45	26.33	8.86	9.45	0.84	0.7	0.2	0.1
3	524.61	9.49	5	26.37	11.79	8.48	0.83	0.7	0.1	0.2
4	480.40	4	45	25.41	8.50	9.18	0.83	0.6	0.3	0.1
5	448.23	4	45	24.17	7.99	8.81	0.80	0.6	0.2	0.2
6	480.14	4	45	25.41	8.49	9.18	0.83	0.6	0.1	0.3
7	442.03	4.95	45	23.90	7.8	8.75	0.82	0.5	0.4	0.1
8	413.98	4.01	45	22.91	7.45	8.43	0.80	0.5	0.3	0.2
9	433.73	11.21	5	22.42	9.60	7.41	0.79	0.5	0.2	0.3
10	385.81	5.97	5	22.57	11.37	6.82	0.80	0.5	0.1	0.4
11	383.97	6.36	45	21.85	6.89	8.13	0.82	0.4	0.5	0.1
12	357.69	7.05	45	21.01	6.48	7.86	0.81	0.4	0.4	0.2
13	100	26.88	45	17.81	7.11	4.36	0.79	0.4	0.3	0.3
14	100.05	27.58	45	17.90	7.44	4.27	0.80	0.4	0.2	0.4
15	100.03	27.53	44.89	17.88	7.40	4.27	0.82	0.4	0.1	0.5
16	301.58	8.03	45	19.42	5.61	7.30	0.84	0.3	0.6	0.1
17	275.53	9.74	45	18082	5.29	7.03	0.82	0.3	0.5	0.2
18	100.01	23.06	45	17.29	5.50	4.80	0.82	0.3	0.4	0.3
19	100	26.26	45	17.74	6.87	4.42	0.83	0.3	0.3	0.4
20	100	27.50	45	17.89	7.40	4.27	0.84	0.3	0.2	0.5
21	109.20	28	45	18.01	7.79	4.27	0.86	0.3	0.1	0.6
22	100	16.39	45	16.38	3.49	5.34	0.87	0.2	0.7	0.1
23	100	16.85	45	16.44	3.60	5.31	0.87	0.2	0.6	0.2
24	100	18.99	45	16.73	4.15	5.16	0.86	0.2	0.5	0.3
25	100	21.81	45	17.12	5.04	4.92	0.86	0.2	0.4	0.4
26	100	25.66	45	17.64	6.56	4.51	0.87	0.2	0.3	0.5
27	100.01	27.50	45	17.89	7.40	4.28	0.88	0.2	0.2	0.6
28	100.05	27.52	45	17.90	7.41	4.22	0.90	0.2	0.1	0.7
29	100	16.38	45	16.38	3.49	5.34	0.93	0.1	0.8	0.1
30	100.01	16.38	45	16.38	3.49	5.34	0.92	0.1	0.7	0.2
31	100	16.38	45	16.38	3.50	5.34	0.92	0.1	0.6	0.3
32	100	16.38	45	16.38	3.50	5.34	0.91	0.1	0.5	0.4
33	100	16.84	45	16.84	4.38	5.01	0.91	0.1	0.4	0.5
34	100	17.53	45	17.53	6.22	4.60	0.91	0.1	0.3	0.6
35	100.01	17.89	45	17.89	7.40	4.27	0.92	0.1	0.2	0.7
36	100.04	17.89	44.97	17.89	7.40	4.27	0.94	0.1	0.1	0.8

TABLE 9: Confirmatory experiments.

Serial number	T_{ON}	T_{OFF}	I_P	MRR		EWR		SR	
				Actual	Error %	Actual	Error %	Actual	Error %
1	100	25.37	45.00	18.83	6.98	6.81	5.90	4.86	7.48
2	100.94	25.58	45.00	18.64	5.66	6.68	2.14	4.65	2.87

Conflict of Interests

The authors declare that there is no conflict of interests regarding the publication of this paper.

References

[1] S. Rajesha, A. K. Sharma, and P. Kumar, "On electro discharge machining of inconel 718 with hollow tool," *Journal of Materials Engineering and Performance*, vol. 21, no. 6, pp. 882–891, 2012.

[2] S. R. Nimbhorkar and B. D. Deshmukh, "Effect of case hardening treatment on the structure and properties of automobile gears," *International Journal of Pure and Applied Research in Engineering and Technology*, vol. 1, no. 8, pp. 317–325, 2013.

[3] A. Bensely, S. Venkatesh, D. Mohan Lal, G. Nagarajan, A. Rajadurai, and J. Boland, "Determination of poisson's ratio of cryogenically treated case carburized steel (En 353) by using ultrasonic technique," in *Proceedings of the International Conference on Manufacturing Science and Technology*, pp. 361–365, 2006.

[4] S. Murugesan and K. Balamurugan, "Optimization by Grey relational analysis of EDM parameters in machining Al-15% SiC MMC using multihole electrode," *Journal of Applied Sciences*, vol. 12, no. 10, pp. 963–970, 2012.

[5] S. Singh and M. Verma, "A parametric optimization of electric discharge drill machine using taguchi approach," *Journal of Engineering, Computers & Applied Sciences*, vol. 1, no. 3, pp. 39–48, 2012.

[6] F. L. Amorim and W. L. Weingaertner, "Die-sinking electrical discharge machining of a high-strength copper-based alloy for injection molds," *Journal of the Brazilian Society of Mechanical Sciences and Engineering*, vol. 26, no. 2, pp. 137–144, 2004.

[7] O. Yilmaz, A. T. Bozdana, M. A. Okka, and I. H. Filiz, "A comparative investigation of the effects of single and multichannel electrodes in EDM fast hole drilling of aerospace alloys," in *Proceedings of the 5th International Conference and Exhibition on Design and Production of Machines and Dies/Molds*, Kusadasi, Turkey, June 2009.

[8] R. Manikandan and R. Venkatesan, "Optimizing the machining parameters of micro-EDM for Inconel 718," *Journal of Applied Sciences*, vol. 12, no. 10, pp. 971–977, 2012.

[9] N. Sharma, R. Khanna, and R. D. Gupta, "Parametric optimization on WEDM for HSLA by integrated RSM and genetic algorithm," in *AEMDS*, 2014.

[10] G. Kibria, B. R. Sarkar, B. B. Pradhan, and B. Bhattacharyya, "Comparative study of different dielectrics for micro-EDM performance during microhole machining of Ti-6Al-4V alloy," *International Journal of Advanced Manufacturing Technology*, vol. 48, no. 5-8, pp. 557–570, 2010.

[11] M. S. Azad and A. B. Puri, "Simultaneous optimisation of multiple performance characteristics in micro-EDM drilling of titanium alloy," *International Journal of Advanced Manufacturing Technology*, vol. 61, no. 9-12, pp. 1231–1239, 2012.

[12] A. D. Patel, M. C. Parekh, B. B. Patel, and B. B. Patel, "Multi-objective optimisation of die sinking electro discharge machining process using Taguchi," *International Journal of Engineering Research and Applications*, vol. 2, no. 6, pp. 1367–1371, 2012.

[13] P. Janmanee and A. Muttamara, "Optimization of electrical discharge machining of composite 90WC-10Co base on taguchi approach," *European Journal of Scientific Research*, vol. 64, no. 3, pp. 426–436, 2011.

[14] M. Boujelbene, E. Bayraktar, W. Tebni, and S. Ben Salem, "Influence of machining parameters on the surface integrity in electrical discharge machining," *International Scientific Journal*, vol. 37, no. 2, 2009.

[15] A. K. Khanra, S. Patra, and M. M. Godkhindi, "Electrical discharge machining studies on reactive sintered FeAl," *Bulletin of Materials Science*, vol. 29, no. 3, pp. 277–280, 2006.

[16] H. K. Kansal, S. Singh, and P. Kumar, "Parametric optimization of powder mixed electrical discharge machining by response surface methodology," *Journal of Materials Processing Technology*, vol. 169, no. 3, pp. 427–436, 2005.

[17] T. Y. Tai and S. J. Lu, "Improving the fatigue life of electro-discharge-machined SDK11 tool steel via the suppression of surface cracks," *International Journal of Fatigue*, vol. 31, no. 3, pp. 433–438, 2009.

[18] A. Kumar, V. Kumar, and J. Kumar, "Investigation of machining parameters and surface integrity in wire electric discharge machining of pure titanium," *Proceedings of the Institution of Mechanical Engineers B: Journal of Engineering Manufacture*, vol. 227, no. 7, pp. 972–992, 2013.

[19] A. T. Bozdana, O. Yilmaz, M. A. Okka, and İ. H. Filiz, "A comparative experimental study on fast hole EDM of Inconel 718 and Ti-6Al-4V," in *Proceedings of the 5th International Conference and Exhibition on Design and Production of MACHINES and DIES/MOLDS*, 2009.

[20] U. Çaydaş and A. Hasçalik, "Modeling and analysis of electrode wear and white layer thickness in die-sinking EDM process through response surface methodology," *International Journal of Advanced Manufacturing Technology*, vol. 38, no. 11-12, pp. 1148–1156, 2008.

[21] N. Sharma, R. Khanna, and R. Gupta, "Multi quality characteristics of WEDM process parameters with RSM," *Procedia Engineering*, vol. 64, pp. 710–719, 2013.

[22] N. Sharma, R. Khanna, R. D. Gupta, and R. Sharma, "Modeling and multiresponse optimization on WEDM for HSLA by RSM," *International Journal of Advanced Manufacturing Technology*, vol. 67, no. 9-12, pp. 2269–2281, 2013.

[23] P. Gupta, R. Khanna, R. D. Gupta, and N. Sharma, "Effect of process parameters on kerf width in WEDM for HSLA using response surface methodology," *Journal of Engineering Technology*, vol. 2, no. 1, pp. 1–5, 2012.

[24] A. Kumar, V. Kumar, and J. Kumar, "Multi-response optimization of process parameters based on response surface methodology for pure titanium using WEDM process," *International Journal of Advanced Manufacturing Technology*, vol. 68, no. 9–12, pp. 2645–2668, 2013.

[25] D. C. Whittinghill, "A note on the robustness of Box-BEHnken designs to the unavailability of data," *Metrika*, vol. 48, no. 1, pp. 49–52, 1998.

[26] K. Ojha, R. K. Garg, and K. K. Singh, "MRR improvement in sinking electrical discharge machining: a review," *Journal of Minerals & Materials Characterization & Engineering*, vol. 9, no. 8, pp. 709–739, 2010.

[27] N. Kumar, L. Kumar, H. Tewatia, and R. Yadav, "Comparative study for MRR on die-sinking EDM using electrode of copper & graphite," *International Journal of Advanced Technology & Engineering Research*, vol. 2, no. 2, pp. 170–174, 2012.

[28] A. Kumar, V. Kumar, and J. Kumar, "Investigation of microstructure and element migration for rough cut surface of pure titanium after WEDM," *International Journal of Microstructure and Materials Properties*, vol. 8, no. 4-5, pp. 343–356, 2013.

21

Optimal Designing of Variables Chain Sampling Plan by Minimizing the Average Sample Number

S. Balamurali and M. Usha

Department of Mathematics, Kalasalingam University, Krishnankoil, Tamil Nadu. 626 126, India

Correspondence should be addressed to S. Balamurali; sbmurali@rediffmail.com

Academic Editors: G.T.S. Ho, T. R. Kurfess, and I. Zeid

We investigate the optimal designing of chain sampling plan for the application of normally distributed quality characteristics. The chain sampling plan is one of the conditional sampling procedures and this plan under variables inspection will be useful when testing is costly and destructive. The advantages of this proposed variables plan over variables single sampling plan and variables double sampling plan are discussed. Tables are also constructed for the selection of optimal parameters of known and unknown standard deviation variables chain sampling plan for specified two points on the operating characteristic curve, namely, the acceptable quality level and the limiting quality level, along with the producer's and consumer's risks. The optimization problem is formulated as a nonlinear programming where the objective function to be minimized is the average sample number and the constraints are related to lot acceptance probabilities at acceptable quality level and limiting quality level under the operating characteristic curve.

1. Introduction

Although in recent years process control techniques and off-line quality control methods have taken important roles, acceptance sampling procedures remain as a major tool for many practical quality control systems. In general, acceptance sampling is a statistical tool used to help make decisions concerning whether or not a batch or lot of products should be released for consumer use. That is, sampling plans are widely used in industries for estimating lot or process characteristics and for determining the disposition of lots. Variable sampling plans constitute one of the major areas of the theory and practice of acceptance sampling. The principle prerequisite for variables sampling is that the quality characteristic of interest is measured on a continuous scale. The primary advantage of the variables sampling plan is that the same operating characteristic (OC) curve can be obtained with a smaller sample size than would be required by an attributes sampling plan. Thus, a variables acceptance sampling plan would require less sampling. The variables sampling provides more information than the attributes sampling, and therefore the same protection is attained with partly considerable smaller sample size. When destructive testing is employed, variables sampling is particularly useful in reducing the costs of inspection.

However, it is to be pointed out that the variables sampling plans have some demerits compared to attributes sampling plans. The variables sampling plans are, of course, little bit more tedious to apply than the attributes sampling plans and the assumptions on which they are based may not be met or may not be known to be met. It is also obvious that the cost of inspecting each item by attributes is less than by variables. The lower cost may also be a result of the simpler procedure of recording only the number of units failing a test rather than recording variables measurements. However, in modern quality control systems, these reasons are no longer valid because measurements are made by automated test equipment. Even the simplest automated test equipment usually has the capability of the simple calculations and data accumulation needed for variables sampling. In one of his papers, Collani [1] criticized the variables sampling plans and argued that the acceptance sampling by variables is inappropriate if one is interested in the fraction nonconforming in incoming batches. But, at the same time, Seidel [2] has proved in one of his subsequent papers that the sampling by variables is always optimal.

The conventional variables sampling plans have been investigated by many authors and are available in the literature of acceptance sampling. Lieberman and Resnikoff [3] developed tables for the selection of plan parameters for various acceptable quality levels (AQL) for MIL-STD 414 scheme. Owen [4] developed variables sampling plans based on normal distribution when the standard deviation of the process is unknown. Bender Jr [5] considered variables sampling plans for assuring the product quality when the quality characteristic of the product follows normal distribution with unknown standard deviation and provided a procedure for calculating the noncentral t-distribution. Hamaker [6] has given a procedure for finding the parameters of the unknown sigma variables sampling plans from the known sigma variables sampling plans. Govindaraju and Soundararajan [7] developed tables for selecting the parameters of variables single sampling plans that match with the OC curves of MIL-STD 105D schemes. Bravo and Wetherill [8] developed a method for designing double sampling plans (DSP) by variables with OC curves matching with the OC curves of the equivalent single sampling plans. Schilling [9] has written an exclusive book on acceptance sampling which also deals with the conventional variables sampling plans. Baillie [10] developed tables for variables double sampling plans when the process standard deviation is unknown. Govindaraju and Balamurali [11] extended the chain sampling for variables inspection. However, they have not provided any tables for easy implementation of the plan. Also, they have restricted their discussion only with known standard deviation. Hence, this paper attempts to develop tables for easy application of chain sampling to variables inspection for a normally distributed quality characteristic for both known and unknown standard deviation cases.

2. Chain Sampling Plan for Variables Inspection

The concept of chain sampling was first introduced by Dodge [12] for the application of attribute quality characteristics. The characteristics and the properties of the chain sampling plan were investigated by many authors (see, for example, Soundararajan [13]). The chain sampling plan belongs to the group of conditional sampling procedures like multiple dependent (deferred) sampling plans (see Vaerst [14] and Wortham and Baker [15]). In these procedures, acceptance or rejection of a lot is based not only on the sample from that lot, but also on sample results from past lots. The chain sampling plan is applicable in the case of Type B situations (i.e., sampling from a continuous process) where lots expected to be of the same quality are submitted for inspection serially in the order of production. The operating procedure and characteristics of the attributes chain sampling plan can be seen in Dodge [12].

The performance of an acceptance sampling plan can be measured by its OC curve which quantifies the risks of producer and consumers. It would be of interest to examine whether or not the shape of the variables plan is concave for good quality levels. If the variables plan possesses an unsatisfactory OC curve similar to that of the zero acceptance number attributes plan, then it is desirable to improve the shape of the curve. A detailed discussion on the shape of the OC curve of a single sampling variables plan with known standard deviation can be seen in Govindaraju and Kuralmani [16]. It is always necessary to increase the discriminating power of the OC curve. This can be achieved by increasing the sample size, but this may not be always possible particularly in costly and destructive testing. An alternative solution is to develop a conditional sampling procedure. Based on this idea, Govindaraju and Balamurali [11] extended the concept of chain sampling to variables inspection. As pointed out earlier, they have not provided any tables for the selection of parameters and also they have dealt only with known standard deviation case. This paper attempts to provide tables for the easy industrial application of this plan for both known and unknown standard deviation cases. It is also to be pointed out that Govindaraju and Balamurali [11] have given an approximate solution for finding the plans of unknown standard deviation case. But in this paper, we provide a different procedure for finding the unknown sigma chain sampling plan. The major advantage of this plan is to achieve better protection to the producer with minimum average sample number (ASN) and these aspects will be discussed in a separate section. The following assumptions should be valid for the application of the variables chain sampling plan.

(i) Lots are submitted for inspection serially in the order of production from a process that turns out a constant proportion nonconforming items.

(ii) The consumer has confidence in the supplier and there should be no reason to believe that a particular lot is poorer than the preceding lots.

In addition, the usual conditions for the application of single sampling variables plans with known or unknown standard deviation should also be valid. The operating procedures of the variables chain sampling plan are given in the next section.

3. Operating Procedures of Variables Chain Sampling Plan

3.1. Known Sigma Case. Suppose that the quality characteristic of interest has the upper specification limit U and follows a normal distribution with unknown mean μ and known standard deviation σ. Then, the following two procedures of the variables chain sampling plan are proposed.

Step 1. From each submitted lot, take a random sample of size n_σ, say $(X_1, X_2 \cdots X_{n_\sigma})$, and compute $V = (U - \overline{X})/\sigma$, where $\overline{X} = (1/n_\sigma)\sum_{i=1}^{n_\sigma} X_i$.

Step 2. Accept the lot if $V \geq k_\sigma$ and reject the lot if $V < k'_\sigma$. If $k'_\sigma \leq V < k_\sigma$, then accept the current lot provided that the preceding i_σ lots were accepted on the condition that $V \geq k_\sigma$ but reject, otherwise (note: $k_\sigma > k'_\sigma$).

In the case of lower specification limit L, the operating procedure is as follows.

Step 1. From each lot, take a random sample of size n_σ, say $(X_1, X_2 \cdots X_{n_\sigma})$, and compute $V = (\overline{X} - L)/\sigma$, where $\overline{X} = (1/n_\sigma)\sum_{i=1}^{n_\sigma} X_i$.

Step 2. Accept the lot if $V \geq k_\sigma$ and reject the lot if $V < k'_\sigma$. If $k'_\sigma \leq V < k_\sigma$, then accept the current lot provided that the preceding i_σ lots were accepted on the condition that $V \geq k_\sigma$ but reject, otherwise.

Thus, the proposed variables chain sampling plan is characterized by four parameters, namely, n_σ, i_σ, k_σ, and k'_σ. If $k_\sigma = k'_\sigma$, then the proposed plan will reduce to the variables SSP. Also, when i_σ tends to infinity, the proposed plan becomes variables SSP with parameters n_σ and k_σ. It is to be pointed out that the chain sampling plan can be applied for inspection of lots which are submitted serially in the order of production or in the order of being submitted. The decision of current lot depends on the results of preceding lots. So, when $i_\sigma \geq 2$, we need to keep the records of results of previous lots. Of course, maintaining records of preceding lots may be a drawback of the chain sampling plan over the single sampling plan; however, this can be compensated by minimizing the inspection efforts in terms of minimum ASN with desired protection.

3.2. Unknown Sigma Case. Whenever the standard deviation is unknown, we may use the sample standard deviation S instead of σ. In this case, the plan operates as follows.

Step 1. From each submitted lot, take a random sample of size n_S, say $(X_1, X_2 \cdots X_{n_S})$, and compute $V = (U - \overline{X})/S$, where $\overline{X} = (1/n_S)\sum_{i=1}^{n_S} X_i$ and $S = \sqrt{\sum (X_i - \overline{X})^2/(n_S - 1)}$.

Step 2. Accept the lot if $V \geq k_S$ and reject the lot if $V < k'_S$. If $k'_S \leq V < k_S$, then accept the current lot provided that the preceding i_S lots were accepted on the condition that $V \geq k_S$.

Thus, the proposed unknown sigma variables chain sampling plan is characterized by four parameters, namely, n_S, i_S, k_S, and k'_S. If $k_S = k'_S$, then the proposed plan will reduce to the variables single sampling plan with unknown standard deviation.

4. Designing Methodology of Variables Chain Sampling Plan

Generally, variables sampling plans are designed based on two points on the OC curve, namely, $(p_1, 1-\alpha)$ and (p_2, β), where p_1 is called the acceptable quality level (AQL), p_2 is the limiting quality level (LQL), α is the producer's risk, and β is the consumer's risk. Any well-designed sampling plan must provide at least $(1-\alpha)\%$ probability of acceptance of a lot when the process fraction nonconforming is at AQL level and the sampling plan must also provide not more than $\beta\%$ probability of acceptance if the process fraction nonconforming is at the LQL level. Thus, the acceptance sampling plan must have its OC curve passing through two designated points (AQL, $1-\alpha$) and (LQL, β). Some other strategies are also followed to design the sampling plans besides the statistical based paradigm, which include Bayesian approach and economic based approach. For further details, readers may refer Chen and Lam [17], Ferrell and Chhoker [18], Chen [19], Chen et al. [20], Balamurali and Subramani [21], Vijayaraghavan and Sakthivel [22], Balamurali et al. [23], and Fallahnezhad and Aslam [24]. In this paper, we have followed the designing methodology based on two points on the OC curve approach. The variables chain sampling plan is designed based on the two points on the OC curve in the following manner.

The fraction nonconforming in a lot is given as

$$p = 1 - \Phi\left(\frac{U-\mu}{\sigma}\right), \quad (1)$$

where $\Phi(Y)$ is the cumulative distribution function of standard normal distribution and is given by

$$\Phi(Y) = \int_{-\infty}^{Y} \frac{1}{\sqrt{2\pi}} \exp\left(\frac{-Z^2}{2}\right) dZ. \quad (2)$$

Here, the quality characteristic of interest is normally distributed with mean μ and standard deviation σ, and the unit is classified as nonconforming if it exceeds the upper specification limit U. So, the unknown mean μ can be determined if p is specified. Let us define the standardized quality characteristic corresponding to the fraction conforming as

$$Z_p = \Phi^{-1}(1-p). \quad (3)$$

Then, the OC function of the variables chain sampling plan, which gives the proportion of lots that are expected to be accepted for given product quality, p, is given by (see Govindaraju and Balamurali [11])

$$P_a(p) = \Pr(V \geq k) + \left[\Pr\left(V \geq k'\right) - \Pr(V \geq k)\right]\left[\Pr(V \geq k)\right]^i, \quad (4)$$

where $\Pr(V \geq k)$ is the probability of accepting a lot based on a single sample with parameters (n, k) and $\Pr(V < k')$ is the probability of rejecting a lot based on a single sample with parameters (n, k'). Under Type B situation (i.e., a series of lots of the same quality), forming lots of N items from a process and then drawing random sample of size n from these lots is equivalent to drawing random samples of size n directly from the process. Hence, the derivation of the OC function is straightforward.

The probability of acceptance of the chain sampling plan can also be written as

$$P_a(p) = \Phi(w_2) + \left[\Phi(w_1) - \Phi(w_2)\right]\left[\Phi(w_2)\right]^i, \quad (5)$$

where $w_1 = (Z_p - k')\sqrt{n}$ and $w_2 = (Z_p - k)\sqrt{n}$.

It is to be pointed out that the parameters of the known sigma variables chain sampling plan are denoted by

$(n_\sigma, k'_\sigma,$ and $k_\sigma)$ and for unknown sigma plan they are denoted by $(n_S, k'_S,$ and $k_S)$. The determination of parameters $(n_S, k'_S,$ and $k_S)$ of unknown sigma variables chain sampling plan is explained as follows. It is known that, for large samples, $\overline{X} \pm k_S S$ is approximately normally distributed with mean $\mu \pm k_S E(S)$ and variance $(\sigma^2/n) + k_S \operatorname{Var}(S)$ (see Duncan [25], and Balamurali and Jun [26]). That is,

$$\overline{X} + k_S S \sim N\left(\mu + k_S\sigma, \frac{\sigma^2}{n} + k_S^2\frac{\sigma^2}{2n}\right). \quad (6)$$

Therefore, the probability of accepting a lot at each repetition is given by

$$\begin{aligned} P(V \geq k_S) &= P\left\{\overline{X} \leq U - k_S S \mid p\right\} \\ &= P\left\{\overline{X} + k_S S \leq U \mid p\right\} \\ &= \Phi\left(\frac{U - k_S\sigma - \mu}{(\sigma/\sqrt{n_S})\sqrt{1 + (k_S^2/2)}}\right) \\ &= \Phi\left(\left(Z_p - k_S\right)\sqrt{\frac{n_S}{1 + (k_S^2/2)}}\right). \end{aligned} \quad (7)$$

If we let

$$w_{2S} = \left(\left(Z_p - k_S\right)\sqrt{\frac{n_S}{1 + (k_S^2/2)}}\right), \quad \text{then } \Pr\left(V \geq k_S\right) = \Phi\left(w_{2S}\right). \quad (8)$$

Similarly, if we let $w_{1S} = ((Z_p - k'_S)\sqrt{n_S/(1 + (k_S'^2/2))})$, then we have

$$\Pr\left(V < k'_S\right) = 1 - \Phi\left(w_{1S}\right). \quad (9)$$

Hence, the lot acceptance probability for sigma unknown case is given by

$$P_a(p) = \Phi\left(w_{2S}\right) + \left[\Phi\left(w_{1S}\right) - \Phi\left(w_{2S}\right)\right]\left[\Phi\left(w_{2S}\right)\right]^{i_S}, \quad (10)$$

where $w_{1S} = (Z_p - k'_S)\sqrt{n_S}$ and $w_{2S} = (Z_p - k_S)\sqrt{n_S}$.

If (AQL, $1-\alpha$) and (LQL, β) are prescribed, then the OC function can be written as

$$\begin{aligned} \Phi\left(w_{21S}\right) + \left[\Phi\left(w_{11S}\right) - \Phi\left(w_{21S}\right)\right]\left[\Phi\left(w_{21S}\right)\right]^{i_S} &= 1 - \alpha, \\ \Phi\left(w_{22S}\right) + \left[\Phi\left(w_{12S}\right) - \Phi\left(w_{22S}\right)\right]\left[\Phi\left(w_{22S}\right)\right]^{i_S} &= \beta. \end{aligned} \quad (11)$$

Here, w_{11S} is the value of w_{1S} at $p = p_1$, w_{21S} is the value of w_{2S} at $p = p_1$, w_{12S} is the value of w_{1S} at $p = p_2$, and w_{22S} is the value of w_{2S} at $p = p_2$.

We obtain w_{11S}, w_{21S}, w_{12S}, and w_{22S}, respectively, by

$$\begin{aligned} w_{11S} &= \left(\left(Z_{p1} - k'_S\right)\sqrt{\frac{n_S}{1 + (k_S'^2/2)}}\right), \\ w_{21S} &= \left(\left(Z_{p_1} - k_S\right)\sqrt{\frac{n_S}{1 + (k_S^2/2)}}\right), \\ w_{12S} &= \left(\left(Z_{p_2} - k'_S\right)\sqrt{\frac{n_S}{1 + (k_S'^2/2)}}\right), \\ w_{22S} &= \left(\left(Z_{p_2} - k_S\right)\sqrt{\frac{n_S}{1 + (k_S^2/2)}}\right), \end{aligned} \quad (12)$$

where Z_{p_1} is the value of Z_p at AQL and Z_{p_2} is the value of Z_p at LQL. For given AQL and LQL, the parametric values of the unknown sigma variables chain sampling plan, namely, n_S, i_S, k_S, and k'_S, are determined by satisfying the required producer and consumer conditions. Alternatively, we can determine the above parameters of the variables chain sampling plan to minimize the ASN at AQL, which is analogous to minimizing the average sample number in the variables repetitive group sampling plans and multiple dependent state sampling plan (see Balamurali et al. [27] and Balamurali and Jun [26]). Some of the authors have investigated the designing of sampling plans by using some other optimization techniques which are available in the literature (see, for example, Feldmann and Krumbholz [28], Krumbholz and Rohr [29, 30], Krumbholz et al. [31], and Duarte and Sariava [32, 33]. The ASN for the chain sampling plan is the sample size only. Therefore, the following optimization problem is considered to determine those parameters.

$$\begin{aligned} &\text{Minimize} \quad \operatorname{ASN}\left(p_1\right) = n_S \\ &\qquad\qquad n_S,\ i_S,\ k_S,\ k'_S \\ &\text{Subject to} \quad P_a\left(p_1\right) \geq 1 - \alpha \\ &\qquad\qquad P_a\left(p_2\right) \leq \beta \\ &\qquad\qquad n_S \geq 2,\ i_S \geq 1,\ k_S + \delta \geq k'_S + \delta \geq 0, \\ &\qquad\qquad \delta \geq 0,\ n_S \in N,\ i_S \in N. \end{aligned} \quad (13)$$

In the case of known sigma variable chain sampling plan, the OC function under the specified AQL and LQL conditions can be written as

$$\begin{aligned} \Phi\left(w_{21\sigma}\right) + \left[\Phi\left(w_{11\sigma}\right) - \Phi\left(w_{21\sigma}\right)\right]\left[\Phi\left(w_{21\sigma}\right)\right]^{i_\sigma} &= 1 - \alpha, \\ \Phi\left(w_{22\sigma}\right) + \left[\Phi\left(w_{12\sigma}\right) - \Phi\left(w_{22\sigma}\right)\right]\left[\Phi\left(w_{22\sigma}\right)\right]^{i_\sigma} &= \beta. \end{aligned} \quad (14)$$

Here, $w_{11\sigma}$ is the value of $w_{1\sigma}$ at $p = p_1$, $w_{21\sigma}$ is the value of $w_{2\sigma}$ at $p = p_1$, $w_{12\sigma}$ is the value of $w_{1\sigma}$ at $p = p_2$, and $w_{22\sigma}$ is the value of $w_{2\sigma}$ at $p = p_2$. That is,

$$\begin{aligned} w_{11\sigma} &= \left(Z_{p_1} - k'_\sigma\right)\sqrt{n_\sigma}, \qquad & w_{21\sigma} &= \left(Z_{p_1} - k_\sigma\right)\sqrt{n_\sigma}, \\ w_{12\sigma} &= \left(Z_{p_2} - k'_\sigma\right)\sqrt{n_\sigma}, \qquad & w_{22\sigma} &= \left(Z_{p_2} - k_\sigma\right)\sqrt{n_\sigma}, \end{aligned} \quad (15)$$

where Z_{p_1} is the value of Z_p at AQL and Z_{p_2} is the value of Z_p at LQL. The following optimization problem is considered to determine the optimal parameters of known sigma variables sampling plan such as n_σ, i_σ, k_σ, and k'_σ.

$$\begin{aligned} &\text{Minimize} \quad \text{ASN}(p_1) = n_\sigma \\ &\qquad n_\sigma, i_\sigma, k_\sigma, k'_\sigma \\ &\text{Subject to} \quad P_a(p_1) \geq 1-\alpha \\ &\qquad P_a(p_2) \leq \beta \\ &\qquad n_\sigma \geq 2,\ i_\sigma \geq 1,\ k_\sigma + \delta \geq k'_\sigma + \delta \geq 0, \\ &\qquad \delta \geq 0,\ n_\sigma \in N,\ i_\sigma \in N. \end{aligned} \tag{16}$$

We may determine the parameters of the unknown sigma chain sampling plan and the known sigma chain sampling plan by solving the nonlinear equation given in (13) and (16), respectively. There may exist multiple solutions since there are four unknowns with only two equations. Generally, a sampling would be desirable if the required number of samples is small. So, in this paper, we consider the ASN as the objective function to be minimized with the probability of acceptance along with the corresponding producer's and consumer's risks as constraints. To solve the above nonlinear optimization problems given in (13) and (16), the sequential quadratic programming (SQP) proposed by Nocedal and Wright [34] can be used. Here, it is to be pointed out that the optimization problems given in (13) and (16) include continuous values of k'_σ (or k'_S) and k_σ (or k_S) and integer values of n_σ (or n_S) and i_σ(or i_S). This kind of formalization leads to a problem of mixed integer programming (MINLP) type. The SQP method does not allow directly solving MINLP problems but only the nonlinear programming (NLP) structures. The strategy we have followed is to relax the discrete variables first and solve the NLP problem and then fix them to the upper integers and resolve the problem again now just to find k'_σ (or k'_S) and k_σ (or k_S). In such a way, we have solved the NLP problem given above. The SQP is implemented in Matlab software using the routine "fmincon." By solving the NLP based on the procedure described above, the optimal parameters (n_σ, i_σ, k_σ, and k'_σ) for known sigma plan and the parameters (n_S, i_S, k_S, and k'_S) for unknown sigma plan are determined and these values are tabulated in Table 1.

5. Designing Examples

5.1. Selection of Known Sigma Variables Chain Sampling Plan for Specified AQL and LQL. Table 1 is used to determine the parameters of the known σ variables chain sampling plan for specified values of AQL and LQL when $\alpha = 5\%$ and $\beta = 10\%$. For example, if $p_1 = 2\%$, $p_2 = 7\%$, $\alpha = 5\%$, and $\beta = 10\%$, Table 1 gives the parameters as $n_\sigma = 18$, $i_\sigma = 3$, $k'_\sigma = 1.544$, and $k_\sigma = 1.779$.

For the above example, the plan is operated as follows.

From each submitted lot, take a random sample of size 18 and compute $V = (U - \overline{X})/\sigma$, where $\overline{X} = (1/18)\sum_{i=1}^{18} X_i$.

Accept the lot if $V \geq 1.779$ and reject the lot if $V < 1.544$. If $1.544 \leq V < 1.779$, then accept the current lot provided that the preceding 3 lots were accepted on the condition that $V \geq 1.779$ with the sample size of 18.

5.2. Selection of Unknown Sigma Variables Chain Sampling Plan for Specified AQL and LQL. As mentioned earlier, the unknown sigma variables chain sampling plan is operated as a known sigma variables chain sampling plan, but the parameters n_S, i_S, k_S, and k'_S are used in the place of n_σ, i_σ, k_σ, and k'_σ, respectively. Table 1 can also be used for the selection of the parameters of the unknown σ variables chain sampling plan for given values of AQL and LQL. Suppose that AQL = 0.0075, LQL = 0.035%, $\alpha = 5\%$, and $\beta = 10\%$. From Table 1, the parameters of the variables chain sampling plan can be determined as $n_S = 47$, $i_S = 1$, $k'_S = 1.925$, and $k_S = 2.190$.

5.3. Illustrative Example. To illustrate the implementation of the proposed sampling plan for the example given in Section 5.2, we consider case study data on STN-LCD manufacturing process given in Wu and Pearn [35]. The upper specification limit is given as 0.77 mm. Here, we consider only 47 values randomly taken from the original data given in Wu and Pearn [35]. The data are shown below:

$$\begin{array}{cccccccccc} 0.717 & 0.698 & 0.726 & 0.684 & 0.727 & 0.688 & 0.708 & 0.703 & 0.694 & 0.713 \\ 0.730 & 0.699 & 0.710 & 0.688 & 0.665 & 0.704 & 0.725 & 0.729 & 0.716 & 0.685 \\ 0.712 & 0.716 & 0.712 & 0.733 & 0.709 & 0.703 & 0.730 & 0.716 & 0.688 & 0.688 \\ 0.712 & 0.702 & 0.726 & 0.669 & 0.718 & 0.714 & 0.726 & 0.683 & 0.713 & 0.737 \\ 0.740 & 0.706 & 0.726 & 0.688 & 0.715 & 0.704 & 0.724 & & & \end{array} \tag{17}$$

The implementation of the plan is shown below.

Step 1. Take a random sample of size 47. The data are given above.

Step 2. For this data, we calculate $\overline{X} = (1/n)\sum_{i=1}^{n_S} X_i = 0.708915$ and $S = \sqrt{\sum (X_i - \overline{X})^2/(n-1)} = 0.017583$.

Step 3. Calculate $V = (U - \overline{X})/S = (0.77 - 0.708915)/0.017583 = 3.4741$.

Step 4. Since $V = 3.4741 > k_S = 2.190$, we accept the current lot without considering the result of past lots.

Just for the sake of discussion, let us assume that the V value (for different data sets) is calculated as 2.15. In this case,

TABLE 1: Variables chain sampling plans indexed by AQL and LQL for $\alpha = 5\%$ and $\beta = 10\%$.

p_1	p_2	Known sigma				Unknown sigma			
		n_σ	i_σ	k'_σ	k_σ	n_S	i_S	k'_S	k_S
0.001	0.0020	119	1	2.883	3.013	650	1	2.888	3.013
	0.0025	68	2	2.810	2.965	361	1	2.834	2.984
	0.003	46	1	2.776	2.961	242	1	2.774	2.964
	0.004	31	2	2.740	2.885	147	1	2.716	2.921
	0.005	21	2	2.610	2.860	103	1	2.637	2.897
	0.006	19	4	2.597	2.807	84	2	2.599	2.829
	0.007	15	3	2.534	2.789	72	3	2.532	2.792
	0.008	13	3	2.495	2.765	62	3	2.518	2.768
	0.009	12	3	2.521	2.736	51	2	2.470	2.765
	0.010	12	6	2.517	2.697	45	2	2.425	2.750
	0.012	9	4	2.331	2.686	40	3	2.414	2.694
	0.015	7	3	2.237	2.657	34	4	2.387	2.637
	0.020	7	6	2.414	2.539	27	5	2.335	2.565
	0.025	6	6	2.370	2.485	20	3	2.197	2.552
	0.030	5	6	2.264	2.454	19	5	2.231	2.476
0.0025	0.004	230	2	2.658	2.738	1131	2	2.676	2.736
	0.005	105	2	2.598	2.703	474	2	2.585	2.705
	0.006	64	2	2.530	2.675	282	1	2.533	2.698
	0.0075	39	2	2.441	2.641	171	1	2.464	2.669
	0.010	26	3	2.379	2.579	106	1	2.430	2.610
	0.012	19	2	2.335	2.555	76	1	2.321	2.601
	0.015	16	4	2.296	2.491	55	1	2.237	2.572
	0.020	12	4	2.260	2.425	38	1	2.139	2.529
	0.025	10	6	2.206	2.366	32	2	2.163	2.413
	0.030	8	5	2.114	2.334	29	5	2.107	2.342
	0.035	7	6	2.307	2.297	25	5	2.092	2.302
	0.040	6	6	1.884	2.274	21	5	1.952	2.282
0.005	0.0075	259	1	2.434	2.524	1113	2	2.442	2.512
	0.010	86	1	2.326	2.486	347	1	2.348	2.483
	0.012	53	1	2.277	2.457	211	1	2.299	2.454
	0.015	33	1	2.211	2.421	128	1	2.233	2.418
	0.020	24	4	2.161	2.316	75	2	2.100	2.345
	0.025	16	3	2.046	2.281	55	2	2.083	2.293
	0.030	14	6	1.989	2.224	43	3	1.974	2.249
	0.035	12	6	2.002	2.182	36	3	1.979	2.209
	0.040	10	4	1.976	2.156	31	3	1.964	2.174
	0.050	8	5	1.893	2.098	25	4	1.925	2.105
0.0075	0.010	505	2	2.339	2.384	1817	1	2.329	2.394
	0.012	173	1	2.259	2.369	690	2	2.278	2.353
	0.015	78	1	2.185	2.335	286	1	2.182	2.337
	0.020	43	2	2.146	2.251	133	1	2.083	2.293
	0.025	29	5	2.008	2.198	83	1	2.003	2.258
	0.030	19	2	1.968	2.178	61	1	1.984	2.214
	0.035	17	5	1.888	2.123	47	1	1.925	2.190
	0.040	13	3	1.837	2.107	41	3	1.851	2.111
	0.050	10	3	1.792	2.052	28	2	1.756	2.081
	0.060	9	5	1.792	1.982	26	4	1.815	1.990

Table 1: Continued.

p_1	p_2	Known sigma n_σ	i_σ	k'_σ	k_σ	Unknown sigma n_S	i_S	k'_S	k_S
0.010	0.015	248	1	2.212	2.257	786	1	2.187	2.267
	0.020	74	1	2.090	2.220	247	1	2.071	2.226
	0.025	42	1	2.027	2.177	135	1	2.011	2.186
	0.030	27	1	1.906	2.161	93	2	1.954	2.124
	0.035	21	2	1.846	2.096	66	1	1.900	2.125
	0.040	18	3	1.804	2.054	53	2	1.846	2.066
	0.045	15	3	1.762	2.027	46	3	1.811	2.026
	0.050	13	3	1.731	2.001	39	2	1.824	2.004
	0.060	11	5	1.662	2.942	30	3	1.730	1.955
	0.070	9	5	1.573	1.903	26	4	1.728	1.898
	0.080	7	3	1.486	1.891	21	5	1.579	1.869
0.02	0.030	187	2	1.896	1.976	528	1	1.886	1.991
	0.035	96	2	1.835	1.945	272	2	1.831	1.946
	0.040	59	1	1.779	1.939	169	2	1.770	1.920
	0.045	45	2	1.749	1.889	117	1	1.734	1.919
	0.050	34	2	1.693	1.868	88	1	1.683	1.903
	0.060	22	2	1.563	1.833	63	2	1.670	1.820
	0.070	18	3	1.544	1.779	44	2	1.577	1.792
	0.080	14	3	1.434	1.749	36	3	1.525	1.745
	0.090	13	4	1.502	1.697	27	2	1.423	1.737
	0.10	10	2	1.436	1.691	22	1	1.386	1.766
	0.11	10	4	1.457	1.632	23	4	1.476	1.636
	0.12	7	1	1.235	1.725	17	1	1.379	1.699
	0.13	8	5	1.360	1.580	17	4	1.356	1.596
	0.15	6	4	1.110	1.560	14	5	1.279	1.544
0.030	0.040	341	2	1.771	1.821	1027	5	1.771	1.816
	0.045	157	1	1.711	1.811	416	2	1.704	1.799
	0.050	96	1	1.650	1.795	253	1	1.679	1.789
	0.060	52	2	1.566	1.736	126	1	1.585	1.760
	0.070	33	1	1.508	1.728	81	1	1.539	1.724
	0.080	26	3	1.427	1.657	62	2	1.514	1.659
	0.090	23	5	1.449	1.609	44	2	1.401	1.641
	0.100	17	3	1.378	1.593	36	2	1.384	1.609
	0.110	13	2	1.268	1.588	29	2	1.305	1.590
	0.120	11	1	1.270	1.605	26	3	1.274	1.549
	0.130	11	4	1.204	1.514	23	3	1.289	1.519
	0.150	10	6	1.273	1.443	18	3	1.230	1.475
	0.200	6	5	0.950	1.365	12	5	1.134	1.354
0.04	0.060	145	2	1.568	1.663	339	2	1.564	1.664
	0.070	73	1	1.513	1.643	166	1	1.507	1.647
	0.080	50	2	1.478	1.588	102	1	1.425	1.625
	0.090	34	2	1.394	1.564	71	1	1.374	1.600
	0.100	27	3	1.324	1.529	53	1	1.315	1.580
	0.110	24	5	1.309	1.489	42	1	1.294	1.554
	0.120	18	3	1.233	1.478	34	1	1.234	1.539
	0.130	14	2	1.130	1.475	31	2	1.271	1.461
	0.140	14	4	1.158	1.423	27	3	1.193	1.433
	0.150	11	2	1.108	1.428	23	3	1.130	1.415
	0.170	10	4	1.070	1.360	17	2	1.061	1.396
	0.200	8	5	0.990	1.295	15	4	1.120	1.290

TABLE 1: Continued.

p_1	p_2	Known sigma n_σ	i_σ	k'_σ	k_σ	Unknown sigma n_S	i_S	k'_S	k_S
0.050	0.070	188	1	1.487	1.582	427	2	1.485	1.570
	0.080	93	1	1.408	1.558	203	1	1.422	1.557
	0.090	60	2	1.359	1.509	127	2	1.365	1.510
	0.100	46	3	1.331	1.471	86	1	1.331	1.506
	0.110	33	2	1.288	1.453	63	1	1.261	1.491
	0.120	26	1	1.270	1.450	49	1	1.213	1.473
	0.130	20	1	1.157	1.452	45	3	1.235	1.395
	0.140	19	3	1.160	1.375	34	2	1.148	1.393
	0.150	17	4	1.108	1.348	30	2	1.165	1.365
	0.160	13	2	1.011	1.356	25	1	1.134	1.384
	0.170	12	2	1.045	1.330	24	3	1.112	1.312
	0.200	11	6	1.059	1.229	17	3	1.023	1.258

TABLE 2: Variables chain sampling plans indexed by AQL and LQL for $\alpha = 1\%$ and $\beta = 1\%$.

p_1	p_2	Variables chain sampling plan (known sigma) n_σ	i_σ	k'_σ	k_σ	$P_a(p_1)$	$P_a(p_2)$	ASN
0.001	0.005	79	1	2.82801	2.83801	0.99007	0.00991	79
0.0025	0.01	91	1	2.56203	2.57203	0.99024	0.00956	91
0.005	0.02	77	1	2.31004	2.32004	0.99012	0.00974	77
0.0075	0.03	69	1	2.15205	2.16205	0.99003	0.00976	69
0.01	0.04	64	1	2.03504	2.04504	0.99007	0.00929	64
0.02	0.05	124	1	1.84403	1.85403	0.99018	0.00997	124
0.03	0.06	193	1	1.71300	1.73200	0.99007	0.00977	193
0.04	0.08	172	1	1.57299	1.58299	0.9906	0.00986	172
0.05	0.10	156	1	1.45798	1.46798	0.99017	0.00999	156

we can accept the current lot provided previous lot must have been accepted with the condition that $V > k'_S = 1.925$. Otherwise, the current lot is rejected.

Further, it is to be pointed out that the proposed variables chain sampling plan is more efficient in terms of minimum ASN than the variables SSP for low values of producer's risk (α) and consumer's risk (β). In order to prove this, we provide two tables. Table 2 gives the optimal parameters of variables chain sampling plan for some selected combinations of AQL and LQL and for $\alpha = 1\%$ and $\beta = 1\%$ and Table 3 gives the parameters of variables SSP under the same set of conditions. By comparing these two tables, one can easily observe that variable chain sampling plan involves minimum ASN compared to the variables SSP.

6. Advantages of the Variables Chain Sampling Plan

This section describes the advantages of the variables chain sampling plan over the conventional variables single sampling plan. Two acceptance sampling plans will be called equivalent when they possess nearly identical OC curves. A customary procedure for achieving such equivalency consists of constructing the sampling plans so that their OC curves coincide in two suitably chosen points, namely, $(p_1, 1-\alpha)$ and (p_2, β). Suppose that, for given values of $p_1 = 0.5\%$, $\alpha = 5\%$, $p_2 = 1.5\%$, and $\beta = 10\%$, one can find the parameters of the known sigma variables chain sampling plan from Table 1 as

(i) $n_\sigma = 33$, $i_\sigma = 1$, $k'_\sigma = 2.211$, and $k_\sigma = 2.421$; ASN = 33.

For the same values of the AQL and LQL, we can determine the parameters of the single and double sampling variables plan (from Sommers [36]) as

(ii) $n_\sigma = 53$ and $k_\sigma = 2.35$; ASN = 53,

(iii) $n_\sigma = 39$, $k_{a\sigma} = 2.41$, and $k_{r\sigma} = 2.31$; ASN = 43.

It can be observed that variables chain sampling plan achieves a reduction of over 38% in sample size than the variables single sampling plan and a reduction of 23% over double sampling plan with the same AQL and LQL conditions. This indicates that the variables chain sampling plan achieves the same OC curve with minimum sample size compared to the variables single and double sampling plans. Further, Figure 1 shows the OC curves of the variables chain sampling plans with parameters $n_\sigma = 18$, $k'_\sigma = 1.544$, and $k_\sigma = 1.779$ for different values of i_σ. This figure apparently shows that the variables chain sampling plan increases the probability of acceptance in the region of principal interest, that is, for good quality levels, and maintains the consumer's

TABLE 3: Variables single sampling plans indexed by AQL and LQL for $\alpha = 1\%$ and $\beta = 1\%$.

p_1	p_2	Variables single sampling plan (known sigma) n_σ	k_σ	$P_a(p_1)$	$P_a(p_2)$	ASN
0.001	0.005	82	2.83305	0.99006	0.00992	82
0.0025	0.01	94	2.56703	0.99001	0.00980	94
0.005	0.02	80	2.31505	0.99016	0.00970	80
0.0075	0.03	72	2.15806	0.99004	0.00932	72
0.01	0.04	66	2.03907	0.99015	0.00957	66
0.02	0.05	130	1.84907	0.99018	0.00996	130
0.03	0.06	205	1.71806	0.99009	0.00971	205
0.04	0.08	182	1.57805	0.99007	0.00982	182
0.05	0.10	165	1.46305	0.99025	0.00984	165

TABLE 4: ASN values of the variables SSP, DSP and variables chain sampling plans.

p_1	p_2	Average sample number: Known sigma			Unknown sigma		
		SSP	DSP	ChSP	SSP	DSP	ChSP
0.001	0.003	74	59.4	46	381	302.4	242
0.0025	0.0075	62	50.1	39	267	214.2	171
0.005	0.015	53	43.0	33	196	157.6	128
0.0075	0.025	39	31.3	29	129	102.1	83
0.01	0.05	19	14.9	13	54	41.5	39
0.02	0.08	21	16.9	14	50	39.6	36
0.03	0.09	30	24.0	23	66	52.5	44
0.04	0.10	39	31.3	27	82	65.1	53
0.05	0.12	39	32.0	26	76	60.9	49

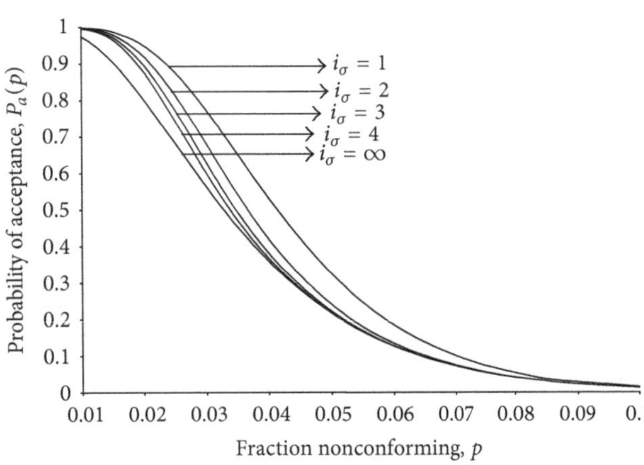

FIGURE 1: OC curves of a variables chain sampling plan for different i_σ values.

risk at poor quality levels. This is also an important feature of the variables chain sampling plan.

Further, it is also to be noted that the variables chain sampling plan is economically superior to the double sampling plans in terms of average sample number (ASN). Obviously, a sampling plan having smaller ASN would be more desirable. The variables double or multiple sampling plans are not practically very useful. Variables sampling standards avoid presenting such plans due to increased complexity involved in operating them, but the variables chain sampling plan has no such complexity. Table 4 shows the ASNs for variables single sampling plan, variables double sampling plan along with variables chain sampling plan for some arbitrarily selected combinations of AQL and LQL. These ASN values are calculated at the producer's quality level for the known sigma plans. The ASN of the variables single and double sampling plans can be found in Sommers [36].

From this table, one can easily understand that the variables chain sampling plan will have minimum ASN when compared to the single and double sampling plans. Similar reduction in the ASN can be achieved for any combination of AQL and LQL values. This implies that variables chain sampling plan will give desired protection with minimum inspection so that the cost of inspection will greatly be reduced. Thus, the variables chain sampling plan provides better protection than the variables single and double sampling plans.

7. Average Run Length of Variables Chain Sampling Plan

Schilling [37] has pointed out that average run length (ARL) is a missing and meaningful measure for characterizing and evaluating the sampling plans under Type B situations as in the process control procedures. The ARL gives an indication

TABLE 5: ARLs of variables chain sampling plan for different i_σ values.

p	Average run length				
	$i_\sigma = 1$	$i_\sigma = 2$	$i_\sigma = 3$	$i_\sigma = 4$	$i_\sigma = \infty$
0	∞	∞	∞	∞	∞
0.005	88768.34	71697.5	60133.4	51781.5	607.80
0.01	868.88	554.66	410.18	327.19	38.29
0.015	81.89	50.25	37.21	30.13	10.64
0.02	20.07	12.66	9.77	8.25	5.06
0.025	8.12	5.42	4.41	3.92	3.15
0.03	4.41	3.14	2.70	2.50	2.29
0.035	2.88	2.19	1.98	1.89	1.83
0.04	2.14	1.73	1.62	1.58	1.56
0.045	1.73	1.47	1.41	1.39	1.39
0.05	1.48	1.32	1.28	1.28	1.28
0.055	1.33	1.22	1.20	1.20	1.20
0.06	1.23	1.15	1.15	1.14	1.14
0.065	1.16	1.11	1.11	1.11	1.11
0.07	1.11	1.08	1.08	1.08	1.08
0.075	1.08	1.06	1.06	1.06	1.06
0.08	1.06	1.04	1.04	1.04	1.04
0.085	1.04	1.03	1.03	1.03	1.03
0.09	1.03	1.02	1.02	1.02	1.02
0.10	1.02	1.01	1.01	1.01	1.01

of the expected number of samples until a decision is made. The ARL can be easily calculated once the probability of acceptance ($P_a(p)$) of the plan is known for any process fraction nonconforming, p. It is clear that the distribution of the run length, L, follows the geometric distribution with probability mass function

$$f_G(L) = [P_a(p)]^{L-1}[1 - P_a(p)]. \quad (18)$$

Its mean and variance are, respectively, given by

$$\text{ARL} = E(L) = \frac{1}{(1 - P_a(p))},$$
$$\text{Var}(L) = \frac{P_a(p)}{(1 - P_a(p))^2}, \quad (19)$$

where $P_a(p)$ is given in (5).

Table 5 gives the values of ARL of the chain sampling plan with $n_\sigma = 16$, $k'_\sigma = 1.501$, and $k_\sigma = 1.841$ for different values of i_σ. This table apparently shows that when the process fraction nonconforming is small, the ARL is high and for the increased values of fraction nonconforming, the ARL is low. By comparing the ARL values for different i_σ values, when i_σ increases, the ARL values reduce even for the lower fraction nonconforming. Also, it is clear from the table that 95% of the lots will be accepted at fraction nonconforming 2% by the variables chain sampling plan ($i_\sigma = 1$) at an average rate of 20 inspections, whereas with the single sampling plan ($i_\sigma = \infty$), at 2% nonconforming, only 80% of lots will be accepted at an average rate of 5 inspections. Also, 90% of the lots will be rejected at the fraction nonconforming 7% by the variables chain sampling plan at an average rate of 1.11 inspections, and at the fraction nonconforming, 93% of the lots will be rejected by the single sampling plan at the rate of 1.08 inspections.

8. Conclusions

The purpose of this paper is to develop conditional sampling procedures for the inspection of normally distributed quality characteristics. Variables sampling plans generally require a smaller sample size than do attributes plans. If the OC curve of the variables plan is unsatisfactory, then its shape can be improved by chaining the past lot results. The proposed variables chain sampling plan is one of the variables conditional sampling plans which also ensure the protection against the consumer point of view. This plan is also simple to apply rather than double and multiple sampling variables plans. Also, this plan provides better protection than the conventional single and double sampling variables plans with minimum sample size. Such a variables chain sampling plan will be effective and useful for compliance testing. However, it is also to be pointed out that the variables chain sampling plan developed in this paper is based on the assumption that the quality characteristic of interest follows a normal distribution. Whenever the normality assumption is not true or invalid, using this variables chain sampling plan can be quite misleading. Hence, this study can be further extended for nonnormal distributions as future research.

Nomenclature

P: Submitted lot quality; $q = 1-p$
p_1: Acceptable quality level (AQL)
p_2: Limiting quality level (LQL)
α: Producer's risk
β: Consumer's risk
$P_a(p)$: The probability of lot acceptance when the fraction nonconforming is p
n_σ: Sample size for known sigma plan
k_σ: Acceptance criterion for known sigma plan
k'_σ: Rejection criterion for known sigma plan
n_S: Sample size for unknown sigma plan
k_S: Acceptance criterion for unknown sigma plan
k'_S: Rejection criterion for unknown sigma plan
i_σ: Number of preceding lots considered for accepting current lot for known sigma plan
i_S: Number of preceding lots considered for accepting current lot for unknown sigma plan
V: The value which is to be compared with acceptance criterion for making decision
$\overline{X}$: The sample mean
S: The sample standard deviation
σ: The population standard deviation
$\Phi(\cdot)$: The cumulative distribution function of standard normal distribution
ASN: The average sample number
ARL: The average run length
SSP: Single sampling plan
DSP: Double sampling plan.

Acknowledgment

The authors would like to thank the anonymous reviewers and the editor for their valuable comments and suggestions which significantly improved the presentation of this paper.

References

[1] E. V. Collani, "A note on acceptance sampling for variables," *Metrika*, vol. 38, pp. 19–36, 1990.

[2] W. Seidel, "Is sampling by variables worse than sampling by attributes? A decision theoretic analysis and a new mixed strategy for inspecting individual lots," *Sankhya B*, vol. 59, pp. 96–107, 1997.

[3] G. J. Lieberman and G. J. Resnikoff, "Sampling plans for inspection by variables," *Journal of the American Statistical Association*, vol. 50, pp. 72–75, 1955.

[4] D. B. Owen, "Variables sampling plans based on the normal distribution," *Technometrics*, vol. 9, no. 3, pp. 417–423, 1967.

[5] A. Bender Jr., "Sampling by variables to control the fraction defective: part II," *Journal of Quality Technology*, vol. 7, no. 3, pp. 139–143, 1975.

[6] H. C. Hamaker, "Acceptance sampling for percent defective by variables and by attributes," *Journal of Quality Technology*, vol. 11, no. 3, pp. 139–148, 1979.

[7] K. Govindaraju and V. Soundararajan, "Selection of single sampling plans for variables matching the MIL-STD 105 scheme," *Journal of Quality Technology*, vol. 18, pp. 234–238, 1986.

[8] P. C. Bravo and G. B. Wetherill, "The matching of sampling plans and the design of double sampling plans," *Journal of the Royal Statistical Society A*, vol. 143, pp. 49–67, 1980.

[9] E. G. Schilling, *Acceptance Sampling in Quality Control*, Marcel Dekker, New York, NY, USA, 1982.

[10] D. H. Baillie, "Double sampling plans for inspection by variables when the process standard deviation is unknown," *International Journal of Quality and Reliability Management*, vol. 9, pp. 59–70, 1992.

[11] K. Govindaraju and S. Balamurali, "Chain sampling plan for variables inspection," *Journal of Applied Statistics*, vol. 25, no. 1, pp. 103–109, 1998.

[12] H. Dodge, "Chain sampling inspection plan," *Industrial Quality Control*, vol. 11, pp. 10–13, 1955.

[13] V. Soundararajan, "Procedures and tables for construction and selection of chain sampling plans (ChSP-1)—part I," *Journal of Quality Technology*, vol. 10, no. 2, pp. 56–60, 1978.

[14] R. Vaerst, "A procedure to construct multiple deferred state sampling plan," *Methods of Operations Research*, vol. 37, pp. 477–485, 1982.

[15] A. W. Wortham and R. C. Baker, "Multiple deferred state sampling inspection," *International Journal of Production Research*, vol. 14, no. 6, pp. 719–731, 1976.

[16] K. Govindaraju and V. Kuralmani, "A note on the operating characteristic curve of the known sigma single sampling variables plan," *Communications in Statistics: Theory and Methods*, vol. 21, pp. 2339–2347, 1992.

[17] J. Chen and Y. Lam, "Bayesian variable sampling plan for the Weibull distribution with type I censoring," *Acta Mathematicae Applicatae Sinica*, vol. 15, no. 3, pp. 269–280, 1999.

[18] W. G. Ferrell Jr. and A. Chhoker, "Design of economically optimal acceptance sampling plans with inspection error," *Computers and Operations Research*, vol. 29, no. 10, pp. 1283–1300, 2002.

[19] C. H. Chen, "Economic design of Dodge-Romig AOQL single sampling plans by variables with the quadratic loss function," *Tamkang Journal of Science and Engineering*, vol. 8, no. 4, pp. 313–318, 2005.

[20] J. Chen, K. Li, and Y. Lam, "Bayesian single and double variable sampling plans for the Weibull distribution with censoring," *European Journal of Operational Research*, vol. 177, no. 2, pp. 1062–1073, 2007.

[21] S. Balamurali and J. Subramani, "Economic design of SkSP-3 skip-lot sampling plans," *International Journal of Mathematics and Statistics*, vol. 6, no. 10, pp. 23–39, 2010.

[22] R. Vijayaraghavan and K. M. Sakthivel, "Chain sampling inspection plans based on Bayesian methodology," *Economic Quality Control*, vol. 26, no. 2, pp. 173–187, 2011.

[23] S. Balamurali, M. Aslam, C. H. Jun, and M. Ahmad, "Bayesian double sampling plan under gamma-poisson distribution," *Research Journal of Applied Sciences, Engineering and Technology*, vol. 4, no. 8, pp. 949–956, 2012.

[24] M. S. Fallahnezhad and M. Aslam, "A new economical design of acceptance sampling models using Bayesian inference," *Accreditation and Quality Assurance*, vol. 18, no. 3, pp. 187–195, 2013.

[25] A. Duncan, *Quality Control and Industrial Statistics*, Richard D. Irwin, Homewood, Ill, USA, 5th edition, 1986.

[26] S. Balamurali and C. Jun, "Multiple dependent state sampling plans for lot acceptance based on measurement data," *European Journal of Operational Research*, vol. 180, no. 3, pp. 1221–1230, 2007.

[27] S. Balamurali, H. Park, C. H. Jun, K. J. Kim, and J. Lee, "Designing of variables repetitive group sampling plan involving minimum average sample number," *Communications in Statistics: Simulation and Computation*, vol. 34, no. 3, pp. 799–809, 2005.

[28] B. Feldmann and W. Krumbholz, "ASN-minimax double sampling plans for variables," *Statistical Papers*, vol. 43, no. 3, pp. 361–377, 2002.

[29] W. Krumbholz and A. Rohr, "The operating characteristic of double sampling plans by variables when the standard deviation is unknown," *Allgemeines Statistisches Archiv*, vol. 90, no. 2, pp. 233–251, 2006.

[30] W. Krumbholz and A. Rohr, "Double ASN Minimax sampling plans by variables when the standard deviation is unknown," *AStA Advances in Statistical Analysis*, vol. 93, no. 3, pp. 281–294, 2009.

[31] W. Krumbholz, A. Rohr, and E. Vangjeli, "Minimax versions of the two-stage t test," *Statistical Papers*, vol. 53, no. 2, pp. 311–321, 2012.

[32] B. P. M. Duarte and P. M. Saraiva, "An optimization-based framework for designing acceptance sampling plans by variables for non-conforming proportions," *International Journal of Quality and Reliability Management*, vol. 27, no. 7, pp. 794–814, 2010.

[33] B. Duarte and P. Saraiva, "Optimal design of acceptance sampling plans by variables for non-conforming proportions when the standard deviation is unknown," *Communications in Statistics: Simulation and Computation*, vol. 42, pp. 1318–1342, 2013.

[34] J. Nocedal and S. J. Wright, *Numerical Optimization*, Springer, New York, NY, USA, 1999.

[35] C. Wu and W. L. Pearn, "A variables sampling plan based on Cpmk for product acceptance determination," *European Journal of Operational Research*, vol. 184, no. 2, pp. 549–560, 2008.

[36] D. J. Sommers, "Two-point double variables sampling plans," *Journal of Quality Technology*, vol. 13, pp. 25–30, 1981.

[37] E. G. Schilling, "Average run length and the OC curve of sampling plans," *Quality Engineering*, vol. 17, no. 3, pp. 399–404, 2005.

22

Optimized Injection Molding of Unfilled and Glass Filled PA6 Gears

Nik Mizamzul Mehat,[1,2] Shahrul Kamaruddin,[2] and Abdul Rahim Othman[2]

[1] *Department of Mould Technology, Kolej Kemahiran Tinggi MARA, Balik Pulau, Genting, 11000 Balik Pulau, Penang, Malaysia*
[2] *School of Mechanical Engineering, Universiti Sains Malaysia, Engineering Campus, 14300 Nibong Tebal, Penang, Malaysia*

Correspondence should be addressed to Nik Mizamzul Mehat; nik_miza78@yahoo.com

Academic Editors: G. T. S. Ho and T. R. Kurfess

Shrinkage behavior is a crucial problem in manufacturing plastic molded gear. This is because it negatively affects the dimensional stability and accuracy of the involute profile, as well as the concentricity, roundness, tooth spacing uniformity, and size of the gear. By integrating the Taguchi robust design, Grey relational analysis, and principal component analysis, we investigated the dimensional stability related to the shrinkage of tooth thickness, addendum circle, and dedendum circle of molded gear via the optimization of processing parameters and glass fiber reinforcement. The results revealed that the optimal combination of the processing parameters of the molded gear to achieve minimum shrinkage is melt temperature of 260°C, packing pressure of 60%, packing time of 5 s, and cooling time of 30 s. The melt temperature showed the highest comparability sequence among the four key process parameters examined, followed by packing pressure, cooling time, and packing time. Meanwhile, the presence of glass fibers induced higher deviations of tooth thickness, addendum circle, and dedendum circle than those of the unfilled polyamide 6 gears.

1. Introduction

The evolution of plastic gears in power and motion transmission application is manifested in diverse applications in consumer electronic items, including copiers, printers, scanners, and washing machines. Manufacturers of automotive components, such as windshield wipers and power car seats and windows, have also used plastic gearing to produce lightweight and cost-effective drive-train designs. Compared with metallic gears that suffer from chemical corrosion and lubrication-related failures, as well as high operation and maintenance costs, plastic gears are lighter, generate less noise, have low friction, and are more economical to produce in large quantities with complex shapes using the injection molding process. In addition, the molded plastic gears do not require lubricant and additional finishing operation before being mounted in the transmission system.

Although plastic gears are popular, they have certain drawbacks. Plastic gears can fail by divergent mechanisms. For example, Senthilvelan and Gnanamoorthy [1] observed different types of failures on the Nylon 66 spur gears, including gear tooth wear, cracking at the tooth surface, tooth root cracking, and severe shape deformation. Mao [2] studied acetal gear failure related to fatigue. Li et al. [3] investigated the wear behavior of dissimilar plastic gears and found that higher contact force during the operation causes the driven gear tip and the driver's root to wear fast. Focusing on the same plastic gear failure mode, Mao et al. [4, 5] in their work reported that the wear rate in acetal gear increased dramatically when the load reaches a critical value for a specific geometry and running speed. The gear surface showed a slowly wear by means of a low specific wear rate if the gear is loaded below the critical value.

Numerous studies have been conducted to improve the plastic gear limitations by material reinforcement. For example, in a series of studies by Kurokawa et al. [6–8] the effect of the presence of carbon fibers (CFs) as reinforcement on the performance of the plastic gear has been investigated; their results showed that the wear properties of the composite gear filled with CFs differed significantly depending on the composition of CFs at various load capability levels. In the work of Senthilvelan and Gnanamoorthy [9], by incorporating 20% short glass and 20% carbon fiber to the Nylon 6/6 gear, the damping capacity of the composite gear has been decreased, thereby influencing the noise and heat generation during service. Hirogaki et al. [10] used cotton fiber plain woven filled phenolics resin laminate as the gear material to study the tooth root stress resulting from bending movement under running conditions; the results revealed that the strength of a tooth with a filled fiber angle of 45° is at the lowest when the load is applied to the tooth tip. In addition, Mendi et al. [11] found that reinforcing the polypropylene gear tooth with metallic wires, subsequently generating a residual compression stress in the structure of the gear tooth, increases the number of cycles of fatigue failure and enhances fatigue strength.

Despite the quality of these studies, the issue related to the dimensional stability in plastic gear during the early stage of gear manufacturing has not been thoroughly described elsewhere. An investigation of the literature related to plastic gears study showed that so much attention is given on plastic gear reinforcement in order to overcome the limitations of plastic gears usage. The dimensional stability of a plastic gear is indispensable to the perfect functioning of the gear when it is running, which should not be underrated of its impact on the final quality of plastic gears produced. During the injection molding process, the dimensions of the molded plastic gear change as the parts cool due to shrinkage. The irreversible changes in dimensions, especially in the involute profile of the plastic gear, as a result of uncontrolled shrinkage due to complexity and inappropriate process parameters setting negatively influence concentricity, roundness, tooth spacing uniformity, and the size of the gear produced. Consequently, the quality of the end molded gear, noise, and vibration as well as the service life of the gear are affected by damage mechanisms such as tooth fatigue, creep, excessive wear, and plastic deformation. In fact, the optimum properties of the plastic material with the most innovative gear and mold design cannot be attained, making them irrelevant without optimal processing parameters during gear manufacturing.

Thus, the main objective of the current study is to explore the effects of optimized injection molding processing parameters on the dimensional stability related to shrinkage behavior in the involute profile of a molded plastic gear. A systematic study on the optimization of processing parameters is conducted via a hybrid integration of the Taguchi parameter design, statistical Grey relational analysis (GRA), and principal component analysis (PCA) by selecting three quality characteristics, namely, tooth thickness and the addendum and dedendum circles. A side-by-side comparison between the unfilled and glass fiber-filled polyamide 6 (PA6) gears was also carried out with regard to the dimensional stability related to shrinkage in the involute profile after the optimization process.

2. Experimental Procedure

2.1. Materials. The unfilled PA6 under the grade name of ZISAMIDE TP4210 (Zig Sheng Industrial Co.) as well as 15% and 30% glass fiber-filled PA6 denoted as Vitamide BR13 and Vitamide BY16, respectively (Perrite), were specified in this study for the gear materials. The study used the spur gear design in compliance with the American Gear Manufacturers Association (AGMA) standards. The geometry and specifications of the gear are shown in Figure 1. The final mold design is prepared after the gear part design has been specified and all requirements affecting the design of the mold including number of cavities, type of mold, types of ejection, and types of venting as well as injection molding machine consideration have been clarified. In this case, the preliminary filling analysis has been previously conducted on the gear model before optimizing the gating system for the gear part. The diaphragm gating system is selected to be incorporated with two plate mold, single gear cavity based on the uniformity of the transient progression of the polymer flow front in the cavity as shown in the analysis.

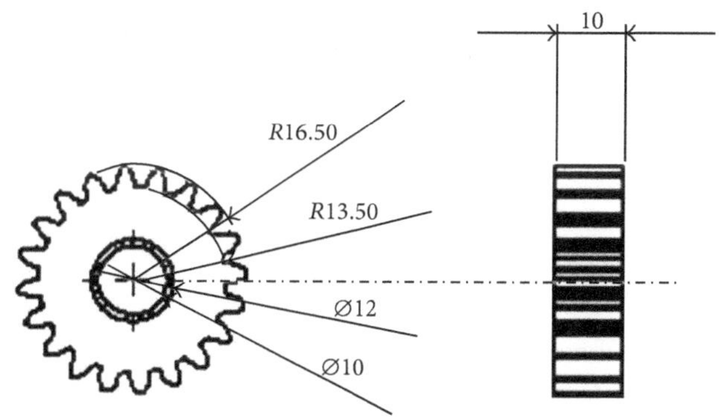

FIGURE 1: Geometry and specifications of spur gear with module = 1.5; pressure angle = 20°; number of teeth = 20; face width = 10 mm.

2.2. Gear Preparation and Fabrication. The PA6 gears were injected using a Battenfeld TM750/210 injection molding machine. The experiment was conducted with four controllable three-level processing parameters, namely, melt temperature, packing pressure, packing time, and cooling time. Other processing parameters, such as mold temperature (60°C), injection pressure (60 bar), and stroke distance (60 mm), were kept constant during the experiment. In addition, the Taguchi method based on orthogonal arrays (OA) was used to design the experiments. As discussed by Mehat and Kamaruddin [12], Taguchi's OA are highly fractional orthogonal designs that can be employed to study an entire parameter space with a small number of experiments. Table 1 lists the design of the experiment based on the L_9 (3^4) OA in producing nine trials of unfilled PA6 gears with five repetitions.

2.3. Shrinkage Measurement. All nine trials of the unfilled PA6 gears with five repetitions of each trial were left for 48 hours at room temperature before the measurement of shrinkage (ASTM D955-89). A Rax Vision DC 3000 Mitutoyo profile projector was used to inspect the accuracy of the specific profile of the involute gear teeth following the molding process. The projector was used in measuring the two-dimensional (2D) addendum circle, dedendum circle, and tooth thickness using the coordinates of selected points along the gear profile. The gear used in this study was designed

TABLE 1: Orthogonal array L_9 (3^4) of the experimental runs.

Trial no.	Processing parameters/			
	Melt temperature (°C)	Packing pressure (%)	Packing time (s)	Cooling time (s)
1	260	60	5	30
2	260	80	10	40
3	260	100	15	50
4	280	60	10	50
5	280	80	15	30
6	280	100	5	40
7	300	60	15	40
8	300	80	5	50
9	210	100	10	30

FIGURE 2: Two-dimensional (2D) addendum and dedendum circles measurement.

with 20 teeth, and therefore in the process of addendum and dedendum circle measurement, 20 edge points of each tooth gear were set in the profile projector (Figure 2). Five injection molded gears from the same batch were measured to determine the repeatability of the part geometry. The relative changes in dimensions due to the shrinkage of the selected quality characteristics were calculated based on the following equation:

$$S = \frac{D_m - D}{D_m} \times 100\%. \quad (1)$$

In this equation, S represents the relative changes in dimensions due to shrinkage, D is the reading of a corresponding dimension measurement using a profile projector, and D_m is the mold cavity dimension.

3. Implementation of Hybrid Optimization

After obtaining all the shrinkage data of the molded gears, the sequential steps described below were adopted to determine the optimal combination of the injection molding process parameters based on the proposed Taguchi/GRA/PCA hybrid method.

Step 1: Signal-to-Noise (S/N) Analysis. For constant processing parameters, we calculated the average value of five repeated results for relative changes in dimensions due to shrinkage in tooth thickness, as well as in the addendum and dedendum circles for the final result. Subsequently, the results of relative changes in dimensions in tooth thickness, addendum circle, and dedendum circle were transformed into the S/N ratio. The S/N ratio is a measure of performance of the development of products or processes that are insensitive to noise factors in a controlled manner. The noise factors refer to uncontrollable factors, such as humidity and weather, whose influence on the product or process is unknown and inevitable. Depending on the objective, three different methods of calculating the S/N ratio in the Taguchi method are commonly used, namely, smaller-the-better, bigger-the-better, and nominal-the-better quality characteristics. The conversion of the results into the S/N ratio involves a series of calculations of the mean squared deviation (MSD).

For this study, the smaller-the-better category was used to characterize the relative changes in the dimensions of tooth thickness, addendum circle, and dedendum circle given as follows:

$$\frac{S}{N} = -10\mathrm{Log}\left(\frac{1}{N}\sum_{i=1}^{n} y_{ij}^2\right), \quad (2)$$

where y_{ij} is the value of the quality characteristics for the ith trials and N is the number of repetitions. The final measured results and the S/N ratios for the three quality characteristics (i.e., tooth thickness, addendum circle, and dedendum circle) are shown in Table 2.

Step 2: Grey Generation of Raw Data. The S/N ratios were normalized by adopting the GRA, a method to measure the correlation degree between factors based on their similarities or differences. GRA is characterized by less data and multifactor analysis. For the current study, the normalization of the S/N ratios involved the transfer of the original sequence to a comparable sequence. Thus, the smaller-the-better characteristics of the data sequence were selected to investigate the shrinkage behavior in molded gear and were calculated as follows:

$$x_i^*(k) = \frac{\max x_i^{(O)}(k) - x_i^{(O)}(k)}{\max x_i^{(O)}(k) - \min x_i^{(O)}(k)}. \quad (3)$$

TABLE 2: Average changes in dimension and S/N ratios for each trial.

Trials	Average changes in dimension			S/N ratio		
	Tooth thickness (mm)	Addendum circle (mm)	Dedendum circle (mm)	Tooth thickness (dB)	Addendum circle (dB)	Dedendum circle (dB)
1	0.0928	0.0225	0.0195	20.4439	32.9303	34.1914
2	0.0947	0.0173	0.0150	20.3710	35.1167	36.1893
3	0.0615	0.0159	0.0124	23.9734	35.9816	38.1000
4	0.0397	0.0156	0.0102	27.3704	36.1456	39.8327
5	0.0356	0.0167	0.0118	28.3401	35.5302	38.5394
6	0.0318	0.0144	0.0102	26.9833	36.8383	39.7901
7	0.0811	0.0162	0.0116	21.2971	35.7943	38.6992
8	0.0652	0.0159	0.0121	23.5700	35.9580	38.3414
9	0.0688	0.0153	0.0123	22.9145	36.2715	38.1526

TABLE 3: The sequences after data preprocessing (Grey generation).

Trials	Tooth thickness	Addendum circle	Dedendum circle
Reference sequence	1.0000	1.0000	1.0000
Comparability sequence			
1	0.9909	1.0000	1.0000
2	1.0000	0.4405	0.6458
3	0.5480	0.2192	0.3072
4	0.1217	0.1773	0.0000
5	0.0000	0.3347	0.2293
6	0.1703	0.0000	0.0076
7	0.8838	0.2672	0.2009
8	0.5986	0.2253	0.2644
9	0.6808	0.1450	0.2978

The values of the tooth thickness, addendum circle, and dedendum circle were set as the reference sequence $x_0^{(O)}(k)$, $k = 1\text{–}3$, and the comparability sequences $x_i^{(O)}(k)$, $i = 1, 2, 3, \ldots, 9$, $k = 1\text{–}3$. Table 3 reports the sequences after data preprocessing.

According to Table 3, the deviation sequences $\Delta_{01}(k)$ can be calculated as follows:

$$\Delta_{01}(1) = \left|x_0^*(1) - x_1^*(1)\right| = |1.0000 - 0.9909| = 0.0091,$$
$$\Delta_{01}(2) = \left|x_0^*(2) - x_1^*(2)\right| = |1.0000 - 1.0000| = 0.0000,$$
$$\Delta_{01}(3) = \left|x_0^*(3) - x_1^*(3)\right| = |1.0000 - 1.0000| = 0.0000. \quad (4)$$

Therefore, $\Delta_{01} = (0.0091, 0.0000, 0.0000)$.

The same calculation method was performed for $i = 1\text{–}9$. The results of all Δ_{0i} for $i = 1\text{–}9$ are listed in Table 4. Based on the data presented in Table 6, $\Delta_{\max}(k)$ and $\Delta_{\min}(k)$ can be calculated as follows:

$$\Delta_{\max} = \Delta_{02}(1) = \Delta_{01}(2) = \Delta_{01}(3) = 1.0000,$$
$$\Delta_{\min} = \Delta_{05}(1) = \Delta_{06}(2) = \Delta_{04}(3) = 0.0000. \quad (5)$$

Step 3: Computation of Grey Relational Coefficients of Response Variables. The corresponding Grey relational coefficients (GRC) were calculated using (6). Given that all the process parameters had equal weight, the value of ζ in this study was defined as 0.5 in the following equation:

$$\varepsilon_i(k) = \frac{\Delta_{\min} + \zeta\Delta_{\max}}{\Delta_{0i}(k) + \zeta\Delta_{\max}}. \quad (6)$$

Examples of the GRC $\varepsilon_{1(k)}$ are as follows:

$$\varepsilon_{1(1)} = \frac{0.0000 + (0.5)(1.0000)}{0.0091 + (0.5)(1.0000)} = 0.9820,$$
$$\varepsilon_{1(2)} = \frac{0.0000 + (0.5)(1.0000)}{0.0000 + (0.5)(1.0000)} = 1.0000, \quad (7)$$
$$\varepsilon_{1(3)} = \frac{0.0000 + (0.5)(1.0000)}{0.0000 + (0.5)(1.0000)} = 1.0000.$$

Thus, $\varepsilon_{1(k)} = (0.9820, 1.0000, 1.0000)$, $k = 1\text{–}3$. A similar procedure was applied for $i = 1\text{–}9$. Table 5 lists the GRC for each trial of the L_9 OA.

Step 4: Computation of the Contribution of the Respective Quality Characteristics Using PCA. An engineering judgment or subjective estimation is required to determine the weighted values for each quality characteristic in the process of optimizing a problem related to multiple quality characteristics or

TABLE 4: Deviation sequences.

Deviation sequences	Δ_{01}	Δ_{02}	Δ_{03}
Trial 1	0.0091	0.0000	0.0000
Trial 2	0.0000	0.5595	0.3542
Trial 3	0.4520	0.7808	0.6928
Trial 4	0.8783	0.8227	1.0000
Trial 5	1.0000	0.6653	0.7707
Trial 6	0.8297	1.0000	0.9924
Trial 7	0.1162	0.7328	0.7991
Trial 8	0.4014	0.7747	0.7356
Trial 9	0.3192	0.8550	0.7022

TABLE 5: Calculated GRC for nine comparability sequences.

Trials	Grey relational coefficient (GRC)		
	Tooth thickness (mm)	Addendum circle (mm)	Dedendum circle (mm)
1	0.9820	1.0000	1.0000
2	1.0000	0.4719	0.5854
3	0.5252	0.3904	0.4192
4	0.3628	0.3780	0.3333
5	0.3333	0.4291	0.3935
6	0.3760	0.3333	0.3350
7	0.8114	0.4056	0.3849
8	0.5547	0.3922	0.4046
9	0.6104	0.3690	0.4159

TABLE 6: Eigenvalues and explained variations for principal components.

Principal component	Eigenvalue	Explained variation (%)
First	2.562	85.398
Second	0.424	14.139
Third	0.014	0.463

performances. Adopting the conventional method in determining the values for each quality characteristic depends mainly on experience and trial-and-error, which can result in uncertainties during the decision making process. The PCA was utilized in the current study to quantitatively reveal the relative importance of each quality characteristic in the GRA. Specifically, the PCA was adopted to determine the corresponding weighted values for each quality characteristic as described below.

The GRCs of all quality characteristics presented in Table 4 were used to evaluate the correlation coefficient matrix and to determine the corresponding eigenvalues and eigenvectors as in the following equation:

$$(R - \lambda_k I_m) V_{ik} = 0. \quad (8)$$

In this equation, λ_k are the eigenvalues; $\sum_{k=1}^{n} \lambda_k = n$, $k = 1, 2, \ldots, n$; $V_{ik} = [a_{k1} a_{k2} \cdots\cdots a_{kn}]^T$ are the eigenvectors corresponding to the eigenvalue λ_k. The corresponding eigenvalues and eigenvectors are listed in Tables 6 and 7.

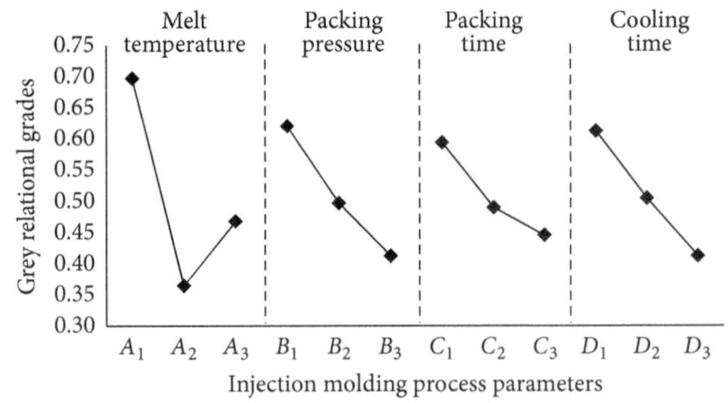

FIGURE 3: Effect of injection molding parameter levels on the dimensional stability related to shrinkage behavior in plastic gear.

The squares of the respective eigenvectors were then selected as the weighted values of the related quality characteristics. The contributions of shrinkage behavior to tooth thickness, addendum circle, and dedendum circle of the molded PA6 gear are reported in Table 8.

The variance contribution for the first principal component characterizing the three quality characteristics is as high as 85.398%. Therefore, the coefficients of w_1, w_2, and w_3 in (9) were set as 0.3488, 0.3773, and 0.2739, respectively.

Step 5: Computation of Grey Relational Grades. After determining the corresponding weighted values, w_k for each quality characteristic to reflect its relative importance in the GRA by using PCA, the Grey relational grades (GRG) were calculated as follows:

$$\gamma_i = \sum_{k=1}^{n} w_k \cdot \varepsilon_i(k), \quad (9)$$

where w_k represents the normalized weighted value of factor k. The results are listed in Table 9.

4. Discussion

4.1. Optimal Combination of Injection Molding Processing Parameters and Their Levels. To determine the optimal combination of injection molding processing parameters for the case of dimensional stability related to shrinkage behavior in tooth thickness, addendum circle, and dedendum circle of the molded gear, the average GRG for each parameter level was calculated by employing the main effects analysis of the Taguchi method. The calculation was done by classifying the GRG corresponding to the levels of the parameters in each column of the orthogonal array and taking the average of those with the same level. The main effects analysis was constructed as shown in Table 10. The results were plotted in Figure 3.

Given that the GRG represents the level of correlation between the reference and the comparability sequences, the larger GRG indicates a stronger correlation between those sequences. Basically, a higher GRG indicates better multiple quality characteristics. Figure 3 shows that the relative changes in dimensions resulting from shrinkage behavior in

TABLE 7: Eigenvectors for principal components.

Quality characteristic	Eigenvector		
	First principal component	Second principal component	Third principal component
Tooth thickness	0.591	−0.487	−0.643
Addendum circle	0.614	−0.245	0.750
Dedendum circle	0.523	0.838	−0.154

TABLE 8: Contribution of each individual quality characteristic to the principal component.

Quality characteristic	Contribution
Tooth thickness	0.3488
Addendum circle	0.3773
Dedendum circle	0.2739

TABLE 9: Grey relational grade after computation.

Trial	Grey relational grade
1	0.9951
2	0.6594
3	0.4382
4	0.3570
5	0.3894
6	0.3457
7	0.5089
8	0.4414
9	0.4528

the tooth thickness, addendum circle, and dedendum circle of the unfilled PA6 molded gear are influenced significantly by the adjustments of the processing parameters. The increment of melt temperature initially decreased the GRG, but then the value slightly increased at level 3. On the contrary, the GRG is lower when the packing pressure and packing time increased, resulting in more pronounced shrinkage behavior in tooth thickness, addendum circle, and dedendum circle. On the other hand, the GRG value also decreased as the cooling time increased. The more time applied during the cooling phase, the greater of shrinkage behavior in tooth thickness, addendum circle, and dedendum circle of the gear.

As in this case, the best combination of processing parameters and levels could be obtained easily from the main effect analysis by selecting the level of each parameter with the highest GRG. As shown in Figure 3, A_1, B_1, C_1, and D_1 provide the highest GRG for factors A, B, C, and D, respectively. Consequently, the optimal parameter setting, which statistically results in minimum shrinkage behavior of the tooth thickness, addendum circle, and dedendum circle of the PA6 molded gear, was predicted to be A_1, B_1, C_1, D_1, that is, melt temperature of 260°C, packing pressure of 60%, packing time of 5 s, and cooling time of 30 s.

Once the combination of optimal process parameters was identified, the experimental verification test was carried out on five repetitions of gear samples. The purpose of conducting the verification test is to assess the accuracy and to verify the improvement of relative changes in dimension related to those shrinkage behaviors by adopting the hybrid Taguchi/GRA/PCA optimization method. Table 11 lists the results of the verification test by means of the optimal process parameters A_1, B_1, C_1, D_1, for the unfilled PA6 molded gear.

Using the optimal process parameter setting of melt temperature of 260°C, 60% of packing pressure, packing time of 5 s, and cooling time of 30 s, the final unfilled PA6 molded gear exhibited the average of relative changes in dimension related to shrinkage behavior in tooth thickness, addendum circle, and dedendum circle as 9.28%, 2.25%, and 1.95%, respectively. Compared with the cavity dimensions of tooth thickness, addendum circle, and dedendum circle, the unfilled PA6 gear demonstrate the optimum average dimensions of 2.1774, 32.9013, and 27.0038 mm, respectively. The results are generally better than those prior to optimization.

4.2. Significance of Process Parameters. To examine the extent to which injection molding parameters significantly influence the performance of molded gear, analysis of variance (ANOVA) of the Taguchi method was performed on the GRG for nine comparability sequences (Table 9). The computed quantities of degree of freedom (DOF), sum of squares (S), variance (V), and percentage contribution (P) are presented in Table 12.

In this case, the percentage contribution of each processing parameter was directly calculated from S because of the nonvalue in the DOF for the error term. The significance of each processing parameter on the dimensional stability can be determined by the percentage contribution. Roy [13] suggested an alternative by using the 10% rule, which considers the insignificant parameter when its effect is less than 10% of the highest parameter influence. From the results of ANOVA in Table 12, the melt temperature appears to be the most decisive processing parameter in reducing the shrinkage behavior in tooth thickness, addendum circle, and dedendum circle of the molded gear, with the highest percentage contribution of 52.07%, outweighing the other process variables. The analysis also demonstrates that packing pressure, packing time, and cooling time are significant because their percentages are over 10%, more than the highest parameter influence (5.21%). The packing pressure, cooling time, and packing time are recorded as 19.58%, 17.95%, and 10.04%, respectively. As a broad rule in the effect of shrinkage, it must be anticipated in the design of the mold and requires expert knowledge. Accurate and specific treatment of this phenomenon is a result of years of experience in building molds for products particularly gears. As indicated above the final size of a molded gear can be influenced by the processing parameters.

Table 10: Main effects table for Grey relational grades.

Symbol	Process parameters	Level 1	Level 2	Level 3
A	Melt temperature (°C)	0.6975	0.3640	0.4677
B	Packing pressure (%)	0.6203	0.4967	0.4122
C	Packing time (s)	0.5941	0.4897	0.4455
D	Cooling time (s)	0.6124	0.5047	0.4122

Table 11: Results of verification test.

	Tooth thickness	Addendum circle	Dedendum circle
Mold cavity dimension (mm)	2.4000	33.6600	27.5400
Average dimension after optimization (mm)	2.1774	32.9013	27.0038
Relative changes in dimension (%)	9.28	2.25	1.95

From the above, it becomes obvious that with the same mold by changing molding parameters, parts of different sizes can be produced [14, 15]. In addition the form of the gear teeth itself altered as a result of shrinkage, irrespective of it shrinking away from the mold. The outcome will be impacted on the injected gear where it will be too thin at the top and too thick at the base. Therefore, the pressure angle will amplified, resulting in the possibility of binding as well as greater wear and eventually reduce the gear performance.

4.3. Influence of Glass Fiber Reinforcement on the Dimensional Stability of PA6 Molded Gear. To further understand the effects of glass fiber reinforcement on the dimensional stability related to shrinkage behavior in tooth thickness, addendum circle, and dedendum circle, gears made from 15% and 30% glass fiber-filled PA6 were also produced using similar optimal processing parameters obtained in the case of the unfilled PA6. The purpose of using the similar optimal processing parameters is to control the changeability in process setting; therefore, the influence of glass fiber reinforcement on the dimensional stability of the molded gears can be directly investigated. Table 13 shows the variation of dimensional stability related to shrinkage behavior in the tooth thickness, addendum circle, and dedendum circle of glass fiber-filled PA6 produced by melt temperature of 260°C, packing pressure of 60%, packing time of 5 s, and cooling time of 30 s.

As observed in Table 13, the relative changes in dimension related to shrinkage behavior in tooth thickness, addendum circle, and dedendum circle decrease greatly with the addition of glass fiber as reinforcement. At the optimal process parameter settings of melt temperature, packing pressure, packing time, and cooling time, the relative changes in dimension due to shrinkage behavior for PA6 + 15% GF result in 5.32%, 0.92%, and 0.78% for tooth thickness, addendum circle, and dedendum circle, respectively. Compared with those of the unfilled PA6, the addition of 30% glass fiber as reinforcement significantly decreases the relative changes in dimension due to shrinkage behavior in tooth thickness, addendum circle, and dedendum circle to 4.83%, 0.58%, and 0.49%, correspondingly. The deviation in the involute profile, especially in the addendum and dedendum circles, is also more than those of the unfilled gears due to the presence of hard glass fiber in the molded surface of the gear. However, less total profile deviation is observed along the gear tooth thickness due to restricted shrinkage resulting from the alignment of glass fibers. The presence of fibers across the tooth section also has an effect on material homogeneity, thus accelerating the deviation compared to the unfilled PA6 gears. Similar circumstances have been reported by Senthilvelan and Gnanamoorthy [16].

5. Conclusion

In relating to plastic gear proper sizing have an effect on the overall performance of the gear injected. A metal gear is commonly evaluated by referring to the load data, however, plastics have different properties than metals and are receptive to changing operating conditions. A plastic gear, therefore, must be designed and produced by taking into consideration the load data, environmental conditions, and material properties. In relation to the work carried out, the mold gear was designed with single cavity, two plate mold, and diaphragm gate system in order to provide the uniformity in flow pattern of polymer inside the cavity of the gear part during the filling process. Several trials runs are carried out by using Moldflow Plastic Insight 6.1 (MPI) software in previous work to forecast and visualize the filling pattern or the transient progression of the polymer flow front within the feed system and mold cavity. By established diaphragm gate system and no changing in cavity size, the main focus of the present study is to investigate the effects of the optimized injection molding processing parameters on the dimensional stability related to shrinkage behavior in the involute profile of a molded plastic gear.

Through a systematic Taguchi parameter design/GRA/PCA hybrid method in the optimization of process parameters, the variation in the dimensional stability due to shrinkage behavior of the thickness, addendum circle, and dedendum circle in PA6 molded gears has been minimized. In addition, through a series of analyses and optimizations of the selected multiple quality characteristics, we found that the optimal combination of processing parameters for the molded gear to achieve minimum relative changes in

Table 12: ANOVA results for the Grey relational grade for nine comparability sequences.

Column	Processing parameters	DOF	S	V	P
A	Melt temperature (°C)	2	0.1748	0.0874	52.07
B	Packing pressure (%)	2	0.0657	0.0329	19.58
C	Packing time (s)	2	0.0349	0.0175	10.40
D	Cooling time (s)	2	0.0603	0.0301	17.95
All others/error		0	0.0000	0.0000	0.00
Total		8	0.3357		100.00

Table 13: The effect of GF addition on the involute profile of PA6 molded gear.

	Relative changes in dimension (%)		
	Tooth thickness	Addendum circle	Dedendum circle
Unfilled PA6	9.28	2.25	1.95
PA6 + 15% GF	5.32	0.92	0.78
PA6 + 30% GF	4.83	0.58	0.49

dimension related to shrinkage behavior are A_1, B_1, C_1, and D_1. From the results of ANOVA, the melt temperature appears to be the most decisive processing parameter in reducing the shrinkage behavior of tooth thickness, addendum circle, and dedendum circle of the molded gear with the highest percentage contribution of 52.07%, which outweighs the other process variables. In the case of filled PA6 gears, the presence of glass fibers has induced greater deviation of tooth thickness, addendum circle, and dedendum circle than the nonfilled PA6 gears.

Conflict of Interests

The authors would like to declare that all financial and material support for the conducting this research and preparation of this paper is clearly termed in the Acknowledgments and there will be no conflict of interests in relation to the financial gain subject to the publication of this paper.

Acknowledgment

The authors acknowledge the Short-Term Research Grant provided by the University Sains Malaysia, Pulau Pinang, for funding the study that resulted in this paper.

References

[1] S. Senthilvelan and R. Gnanamoorthy, "Damage mechanisms in injection molded unreinforced, glass and carbon reinforced nylon 66 spur gears," *Applied Composite Materials*, vol. 11, no. 6, pp. 377–397, 2004.

[2] K. Mao, "A new approach for polymer composite gear design," *Wear*, vol. 262, no. 3-4, pp. 432–441, 2007.

[3] W. Li, A. Wood, R. Weidig, and K. Mao, "An investigation on the wear behaviour of dissimilar polymer gear engagements," *Wear*, vol. 271, no. 9-10, pp. 2176–2183, 2011.

[4] K. Mao, W. Li, C. J. Hooke, and D. Walton, "Friction and wear behaviour of acetal and nylon gears," *Wear*, vol. 267, no. 1-4, pp. 639–645, 2009.

[5] K. Mao, W. Li, C. J. Hooke, and D. Walton, "Polymer gear surface thermal wear and its performance prediction," *Tribology International*, vol. 43, no. 1-2, pp. 433–439, 2010.

[6] M. Kurokawa, Y. Uchiyama, and S. Nagai, "Performance of plastic gear made of carbon fiber reinforced poly-ether-ether-ketone," *Tribology International*, vol. 32, no. 9, pp. 491–497, 1999.

[7] M. Kurokawa, Y. Uchiyama, and S. Nagai, "Performance of plastic gear made of carbon fiber reinforced poly-ether-ether-ketone: part 2," *Tribology International*, vol. 33, no. 10, pp. 715–721, 2000.

[8] M. Kurokawa, Y. Uchiyama, T. Iwai, and S. Nagai, "Performance of plastic gear made of carbon fiber reinforced polyamide 12," *Wear*, vol. 254, no. 5-6, pp. 468–473, 2003.

[9] S. Senthilvelan and R. Gnanamoorthy, "Damping characteristics of unreinforced, glass and carbon fiber reinforced nylon 6/6 spur gears," *Polymer Testing*, vol. 25, no. 1, pp. 56–62, 2006.

[10] T. Hirogaki, E. Aoyama, T. Katayama, S. Iwasaki, Y. Yagura, and K. Sugimura, "Design systems for gear elements made of cotton fiber-reinforced plastics," *Composite Structures*, vol. 66, no. 1-4, pp. 47–52, 2004.

[11] F. Mendi, H. Can, and M. K. Külekci, "Fatigue properties of polypropylene involute rack gear reinforced with metallic springs," *Materials and Design*, vol. 27, no. 5, pp. 427–433, 2006.

[12] N. M. Mehat and S. Kamaruddin, "Optimization of mechanical properties of recycled plastic products via optimal processing parameters using the Taguchi method," *Journal of Materials Processing Technology*, vol. 211, no. 12, pp. 1989–1994, 2011.

[13] R. K. Roy, *Design of Experiment Using the Taguchi Approach: 16 Steps to Product and Process Improvement*, John Wiley & Sons, New York, NY, USA, 2001.

[14] T. C. Chang and E. Faison III, "Shrinkage behavior and optimization of injection molded parts studied by the Taguchi method," *Polymer Engineering and Science*, vol. 41, no. 5, pp. 703–710, 2001.

[15] S. J. Liao, D. Y. Chang, H. J. Chen et al., "Optimal process conditions of shrinkage and warpage of thin-wall parts," *Polymer Engineering and Science*, vol. 44, no. 5, pp. 917–928, 2004.

[16] S. Senthilvelan and R. Gnanamoorthy, "Influence of reinforcement on composite gear metrology," *Mechanism and Machine Theory*, vol. 43, no. 9, pp. 1198–1209, 2008.

23

Mechanistic Force Modeling for Broaching Process

Raghavendra Kamath Cholpadi[1] and Appu Kuttan[2]

[1] *Mechanical & Manufacturing Engineering Department, MIT, Manipal, Karnataka 576104, India*
[2] *Mechanical Engineering Department, NITK, Surathkal, Karnataka 575025, India*

Correspondence should be addressed to Raghavendra Kamath Cholpadi; crkamath77@gmail.com

Academic Editor: Tuğrul Özel

There is a demand for mechanistic force model that can predict and simulate the broaching process. In this paper, an attempt has been made for mechanistic force modeling of the broaching operation and experimental corroboration with the simulated result. The stiffness and damping coefficients for dynamic model are computed from the natural frequency of the broaching system and actual damped natural frequency has been obtained experimentally. Experimental work has been carried to compute the dynamic model parameters such as mass, stiffness, and damping coefficients. The simulated dynamic forces are illustrated graphically and are closely in agreement with the results obtained through manual broaching process.

1. Introduction

The broaching process may be used to generate irregular internal and external part features, and therefore has many potential industrial applications. Slotting, spline cutting, and production of internal helical gears are the major applications in which the broaching process has been used [1, 2]. One of the advantages of broaching over other competitive processes is its higher productivity. However, since the material removal rates are relatively high for broaching, the cutting forces are also high. As a result there will be large deflection which gives rise to high surface errors. Therefore, careful attention must be directed towards the design of the broach geometry and the selection of process conditions [3–5].

Broach teeth usually are divided into three separate sections along the length of the tool. These include roughing teeth, semifinishing teeth, and finishing teeth. The first roughing tooth is proportionately the smallest tooth on the tool. The subsequent teeth progressively increase in size up to and including the first finishing tooth. The difference in height between each tooth, or tooth rise, usually is greater along the roughing section, less along the semifinishing section. All finishing teeth of a broach are almost the same size with microns difference [3–5].

Historically, broaches have been designed based on experience or through trial and error. As more emphasis is placed on part accuracy and precision, it becomes less likely that satisfactory broach geometry can be designed based on experience [6–8]. Additionally, as manufacturers struggle to reduce costs and nonproductive time, it becomes clear that trial and error approaches to tooling design will not be satisfactory. To design a broach early in the life cycle of a product, a model for how a broach will perform during the operation would be extremely advantageous, for a given raw part and broach geometries. The mechanistic modeling of cutting forces in broaching process is important for the design of broach tool.

Maximum force on an internal pull broach is a function of minimum cross section of the tool and yield point of tool material [9, 10]. The allowable pulling force used to be determined empirically in earlier days [3, 4]. Russian scientists have given empirical relation to determine the maximum allowable broaching force in broach tool which can withstand without damage. They specified a broaching constant whose value is a function of workpiece material. The cutting force is dependent on width of cut, depth, or rise per tooth, and so forth [5, 11]. Many mechanistic models for systems in machining are described in the literature; these models are however used in turning and boring operation [5, 6]. Models have been developed for multipoint tool process such as end milling and face milling [12–14].

In this paper an attempt has been made to investigate mechanistic model for the cutting force in rectangular slotting operation; the static and dynamic forces are modeled during the broaching operation. For static model the relevant equations are formulated to compute the forces considering chip thickness area and proportionality constant. The proportionality constant is determined using a specific energy constant obtained experimentally. Dynamic forces are due to the variations of chip thickness during broaching process. Dynamic equations are formulated and parameters of the dynamic equations are computed using theoretically obtained natural frequency and damped natural frequency obtained through experiment. Static and dynamic forces obtained through the model are verified experimentally. To compute the chip load area, the broach coordinates are measured using toolmakers microscope and broach profile configuration is obtained using AutoCAD. Axial, radial, and tangential forces acting on the broach are simulated and presented graphically.

2. Mechanistic Modeling

2.1. Specific Cutting Energy Constant Computation. In the mechanistic modeling, for any machining process the basic equations that relates the F_x, F_y, and F_z to the chip cross sectional area are given by [15, 16]

$$F_x = K_x A_c,$$
$$F_y = K_y A_c, \quad (1)$$
$$F_z = K_z A_c,$$

where F_x, F_y, and F_z are the three dimensional forces acting on the tool tip.

The specific cutting energy constant depends on t_c, v_c, and γ_a of the cutting tool. Mathematically [12],

$$\log K_x = a_0 + a_1 \log t_c + a_2 \log v_c + a_3 \log t_c \log v_c + a_4 \log \gamma_a,$$
$$\log K_y = b_0 + b_1 \log t_c + b_2 \log v_c + b_3 \log t_c \log v_c + b_4 \log \gamma_a$$
$$\log K_z = c_0 + c_1 \log t_c + c_2 \log v_c + c_3 \log t_c \log v_c + c_4 \log \gamma_a. \quad (2)$$

The coefficients a_i's, b_i's, and c_i's ($i = 1, 2, 3, 4$) depend upon the tool and work piece material and range of cutting speed and chip thickness and they are independent of the machining process. Usually these constants are determined from calibration test for a given tool and work piece combination and given range of cutting conditions [12]. Keeping the rake angle and the velocity of the tool movement as constant corresponding to the broach, (2) reduces to

$$\log K_x = a_0 + a_1 \log t_c,$$
$$\log K_y = b_0 + b_1 \log t_c, \quad (3)$$
$$\log K_z = c_0 + c_1 \log t_c.$$

The values of specific cutting energy constants can be determined using a simple calibration experiment [17–19]. The experiments were conducted for shaping operation and specific cutting energy constants are determined.

2.2. Experimental Investigation. Experiments were conducted on three work materials namely mild steel, aluminium, and cast iron and was repeated three times for a cutter with varying different depth of work material. Initially, three experiments were conducted on each set of processes. The first set of experiments was used to ascertain which tool and cut geometry's variables affect the proportionality constants. The second set of experiments were used to develop adequate model for proportionality constant based on the important tool and cut geometry variables, determined from first set of experiments. Third experiment was used to evaluate the specific cutting energy constants and hence the proportionality constants for mechanistic modeling.

Chip thickness area for the calibration purpose is measured from the chip curl. The volume of the chip is measured using water displacement method. After that, the chip curl is heated and elongated. The width of the chip b and length ℓ are measured using micrometer and vernier caliper, respectively. Then, the actual chip thickness t_c is obtained by volume divided by the product of length and width; that is, $t_c = v/b\ell$ where "v" is the volume of chip curl. Knowing the chip thickness "t_c" and width "b" of the cut, the chip load area can be computed as the product of t_c and b.

2.3. Results and Discussions. Figures 1(a), 1(b), and 1(c) illustrate the cutting force versus chip thickness during shaping operation for the mild steel material. In the plot, x-axis indicates the logarithm value of chip thickness t_c and y-axis is the logarithm of the cutting forces along x, y, and z directions, marked as F_x, F_y, and F_z, respectively in Newton. The chip thickness is corresponding to depth of cut which varied from 0.04 mm to 0.16 mm. The calibration test is performed for small depth of cut to avoid the error due to impulsive cutting forces coming on the workpiece at higher depth of cut. A linear curve fitting is made using Matlab software to determine the specific energy constants. Tool geometries were considered the same as the broaching tool. Hence, the variation in specific cutting energy constants is not considered for the rake angle variation in the tool.

Negative slope linear curves are obtained and the coefficients of the linear curve fitting give the specific cutting energy constants. The proportionality constants K_x, K_y, and K_z are determined using (3). Tables 1 and 2 give the values of specific cutting energy constants and proportionality constants for materials mild steel, aluminium, and cast iron, respectively.

3. Static Model

Accurate modeling of cutting force is required to predict the vibration, surface quality, and stability of the machining process [20]. In the static force model, axial cutting force, F_x during broaching is acting on the tooth as the product

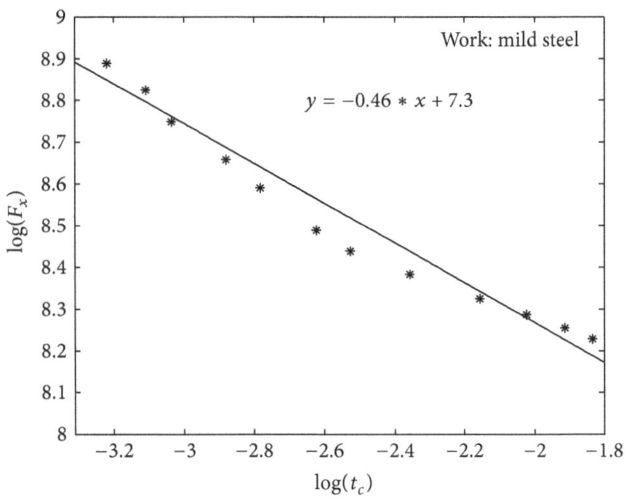

(a) Graph for determining a_0 and a_1

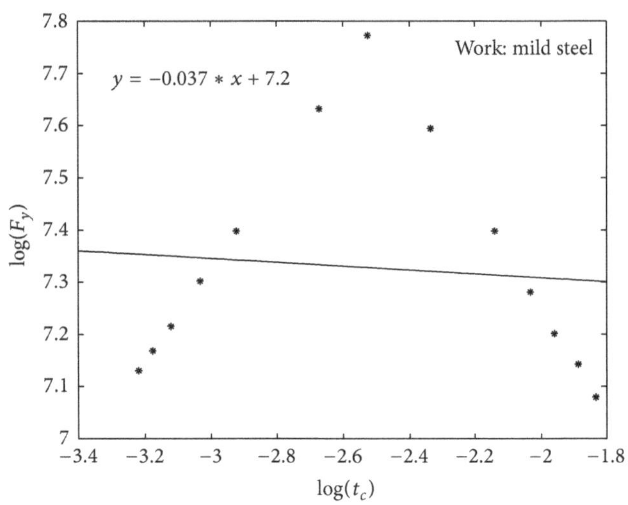

(b) Graph for determining b_0 and b_1

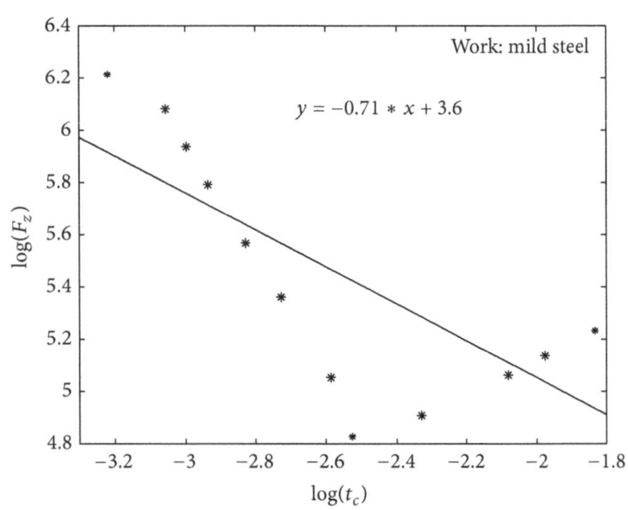

(c) Graph for determining c_0 and c_1

FIGURE 1: Specific cutting energy constants for mild steel obtained during shaping operation.

TABLE 1: Specific cutting energy constants during shaping process.

Material	a_0	a_1	b_0	b_1	c_0	c_1
Mild steel	7.3	−0.46	7.2	−0.037	3.6	−0.71
Aluminium	7.4	−0.39	5.0	−0.68	2.1	−1.2
Cast iron	7.3	−0.29	5.9	−0.31	2.3	−1.0

TABLE 2: Proportionality constants for broaching process.

Material	K_x (N/mm^2)	K_y (N/mm^2)	K_z (N/mm^2)
Mild steel	5732.3	1654.0	307.1
Aluminium	5370.0	1088.0	279.4
Cast iron	3562.0	924.4	199.5

of chip cross section area A_c and a proportionality constant, K_x. Similarly normal force F_y acting along the cutting edge is obtained by multiplying the chip cross section area with proportionality constant K_y and lateral force F_z is similarly obtained by multiplying chip cross section area by proportionality constant, K_z. The values of K_x, K_y, and K_z are determined by simple calibration test using the shaping operation.

A static analysis calculates the effects of steady loading conditions on a structure, while ignoring inertia and damping effects, such as those caused by time-varying loads. A static analysis can, however, include steady inertia loads (such as gravity and rotational velocity) and time-varying loads that can be approximated as static equivalent loads [21, 22].

3.1. Chip Load Area Measurement. Tool maker's microscope is used to get the coordinates of tooth profile. AutoCAD drawing of the broaching tool is drawn to get the dimensions of the tool profile and to get chip load area. Figure 2 shows the profile of one tooth and Figure 3 shows the AutoCAD drawing of the complete broaching tool.

From the AutoCAD drawing, pitch = 6.98 mm, rake angle = 3°, rise per tooth = 0.05 mm, width = 8 mm. Hence chip load area = width × rise per tooth = 0.4 mm^2.

3.2. Simulation of Static Cutting Force Model Using Matlab Program. A MatLab program has been written to simulate the static force model. The simulation results have been plotted as shown in Figures 4(a) and 4(b) for different depth of the workpiece. When the workpiece depth is very small it gives a periodic variation in the cutting force in axial direction. The cutting forces in radial and transverse direction are negligibly small. As the depth of the workpiece increases the cutting force progressively increases until reaching a steady state force when all the teeth are engaged in the workpiece. Then, suddenly the cutting force falls as the broach disengages with the workpiece. Experiment was repeated for different workpieces of depths 5 mm and 40 mm and simulated pattern of the cutting force was obtained as in Figures 4(a) and 4(b). Experiment was conducted for mild steel workpiece. Axial cutting force proportionally increases as the depth of the material increases. When all the cutting

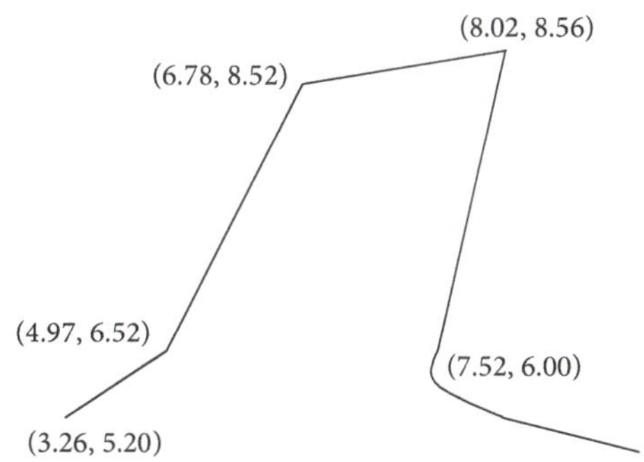

FIGURE 2: Broach tooth profile obtained from Toolmaker's microscope and AutoCAD drawing.

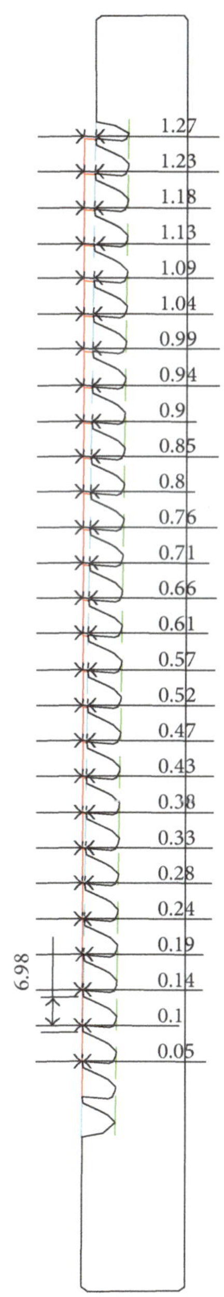

FIGURE 3: AutoCAD drawing of broach tool.

tips are engaged in the workpiece the cutting force required is stable at a steady state cutting conditions.

For higher depth of the workpiece, initial increase in the cutting forces, there is a step by step increase in the cutting forces. This clearly indicates the engagement of the cutting tooth as the broach goes inside the workpiece. When the depth of the workpiece is less than 5 mm, the nature of the cutting force is periodic and frequency of the spectrum obtained from the graph shows maximum cutting force at 40 Hz. This indicates that cutting force variation mainly depends upon the chip thickness variation since there is no peak values observed at higher frequency. The feed rate of the broach is 10 mm/second.

4. Development of Dynamic Force Model

The broaching operation uses only two dimensional models and cutting force in lateral direction is negligibly small and not relevant. This is because axial force is more predominant and lateral force is very small in comparison. Figure 5 shows the cutting force direction on a broaching tool.

Broaching is analyzed as orthogonal cutting because the chip velocity is orthogonal to the cutting edge in the slotting operation. Even though all the cutting processes are oblique cutting, in the case of broaching, the cutting happens to be in depth-wise, hence, the assumption of orthogonal cutting is valid [23–25].

For determining the dynamic cutting forces, the two degrees of freedom vibratory system is considered and the model parameters of the machine tool structural system have been determined. By assuming that worktable and spindle head of the machine tool are rigid bodies, the vibratory system of the workpiece and tool in a one directional reference system was modeled as two degrees of freedom system as shown in Figure 6. The parameters of this system were determined from the theoretical natural frequencies and damped natural frequency corresponding to frequency obtained from the actual cutting system.

Assuming tool and workpiece as a rigid body, the system can be modeled as

$$[M]\ddot{X} + [C]\dot{X} + [K]X = [F], \tag{4}$$

where $X = \{x_1\ x_2\ y_1\ y_2\}^T$ is the relative displacement vector between tool and workpiece and $F = \{F_x\ 0\ F_y\ 0\}^T$ is the force vector; $[M]$ is the mass matrix which depends on tool and workpiece and $[C]$ is the damping coefficient which depends on cutting feed and speed between the tool and workpiece and $[K]$ is the stiffness matrix connecting the tool and workpiece.

The values for $[M]$, $[C]$, and $[K]$ are:

$$[M] = \begin{bmatrix} M_1 & 0 & 0 & 0 \\ 0 & M_2 & 0 & 0 \\ 0 & 0 & M_1 & 0 \\ 0 & 0 & 0 & M_2 \end{bmatrix},$$

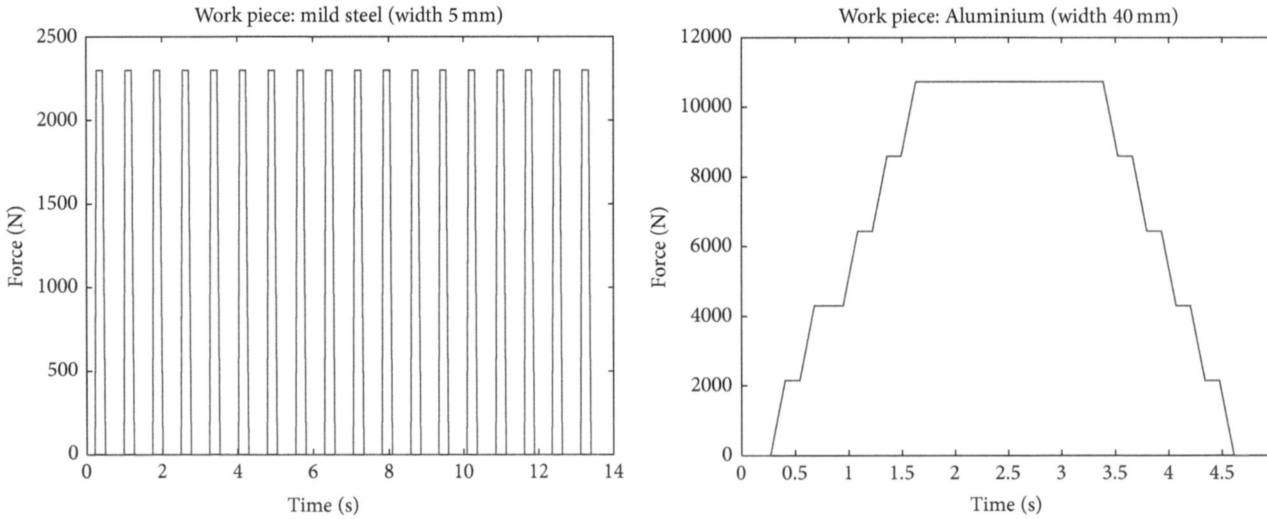

(a) Static axial cutting force on mild steel (depth 5 mm)

(b) Static axial cutting force on mild steel (depth 40 mm)

FIGURE 4: Static force variation for mild steel workpiece of depth from 5 mm to 40 mm.

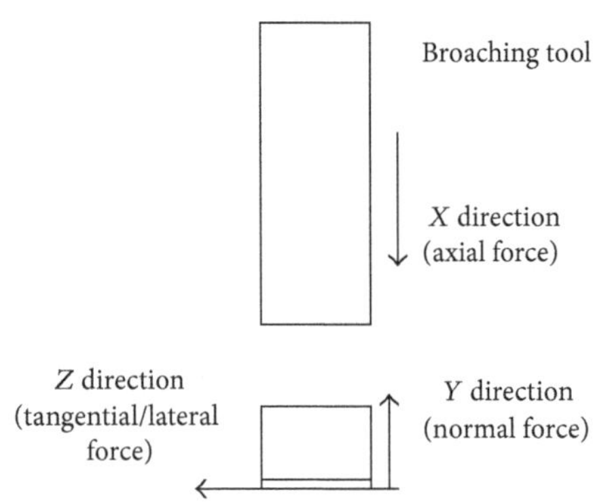

FIGURE 5: Schematic diagram showing direction of the cutting forces acting on a broach.

$$[C] = \begin{bmatrix} C_{11}+C_{12} & -C_{12} & 0 & 0 \\ -C_{12} & C_{12} & 0 & 0 \\ 0 & 0 & C_{21}+C_{22} & -C_{22} \\ 0 & 0 & -C_{22} & C_{22} \end{bmatrix},$$
$$[K] = \begin{bmatrix} K_{11}+K_{12} & -K_{12} & 0 & 0 \\ -K_{12} & K_{12} & 0 & 0 \\ 0 & 0 & K_{21}+K_{22} & -K_{22} \\ 0 & 0 & -K_{22} & K_{22} \end{bmatrix}. \tag{5}$$

Using (4), deflections $X(t)$ have been computed using Wilson theta method [12, 24]. To solve (4), for the first iteration static forces F_x and F_y are used in the force vector F. Deflections $X(t)$ were then computed and dynamic forces dF_x have then solved. In the next iteration, the computed dynamic forces were used to solve (4). The procedure was continued until it converged.

In Wilson Theta method, basic assumption is made such that the acceleration varies linearly not only with time for a period but also over an extended time period or step. The instantaneous dynamic force component $dF_x(t, \varphi)$ caused by total chip thickness variation can be expressed as

$$dF_x(t,\phi) = K_x U(t,\phi) D_i(t,\phi), \tag{6}$$

where $U(t, \phi)$ is the resultant chip thickness variation and it is given by the equation

$$U(t,\phi) = U_x(t,\phi) = [X_0(t,\phi) - X_I(t,\phi)], \tag{7}$$

where

$$X_1(t,\phi) = X_1(t,\phi) - X_2(t,\phi). \tag{8}$$

X_1 and X_2 represent the displacements of the tool and work piece corresponding to broach rotational angle, ϕ, respectively. Instantaneous outer modulation $X_0(t, \phi)$ can be obtained by delaying the inner modulation by the time lag of one tooth of the broach as

$$X_0(t,\phi) = X_I\left[t, \left(\phi - \frac{2\pi N}{60 Z_n}\right)\right]. \tag{9}$$

Since keyway broach is used in the slotting operation, ϕ is taken as zero and $D_i(t, \phi)$ is the instantaneous depth of cut.

4.1. Computation of Dynamic Force Model Parameters. To determine the dynamic force, using (1), mass matrix $[M]$, damping coefficient matrix $[C]$, and stiffness matrix $[K]$ have to be determined. The parameters for the dynamic system were determined form experimentally obtained force spectrum and natural frequency of the cutter. Two directional damped natural frequency ω_d has been obtained experimentally during actual cutting conditions. Knowing the natural frequency and damped natural frequency, damping ratio was computed. Knowing the mass and stiffness of the broach tool, damping coefficient C is computed and the computed values are shown in Table 3.

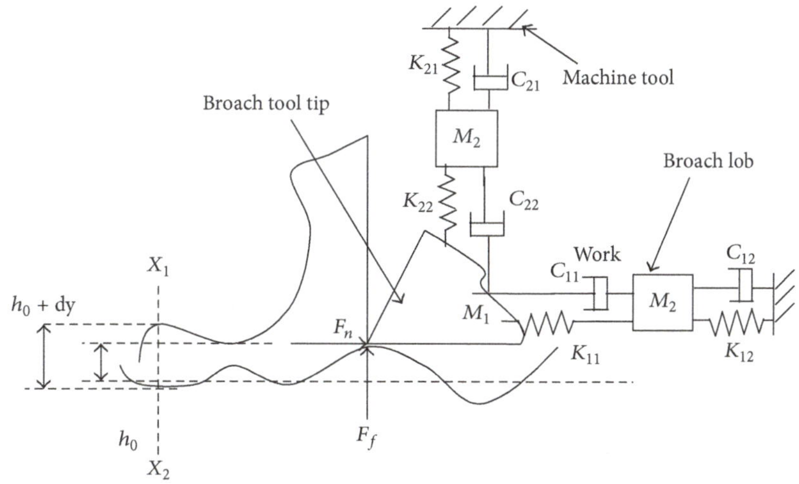

FIGURE 6: Dynamic force model for broaching.

TABLE 3: Damping coefficients and stiffness parameters for dynamic model.

Damping coefficients	Value (N/mm/sec)	Stiffness parameters	Value (N/mm)
c_{11} c_{21}	140	k_{11} k_{12}	7.0×10^6
c_{12} c_{22}	60	k_{21} k_{22}	6.0×10^6

FIGURE 7: Photograph of a Push type broaching machine.

The computed mass, stiffness, and damping coefficient matrix are

$$[M] = \begin{bmatrix} 0.35 & 0 & 0 & 0 \\ 0 & 0.116 & 0 & 0 \\ 0 & 0 & 0.35 & 0 \\ 0 & 0 & 0 & 0.116 \end{bmatrix},$$

$$[K] = 10^6 \begin{bmatrix} 23 & -9 & 0 & 0 \\ -9 & 23 & 0 & 0 \\ 0 & 0 & 17 & -6 \\ 0 & 0 & -6 & 15 \end{bmatrix},$$

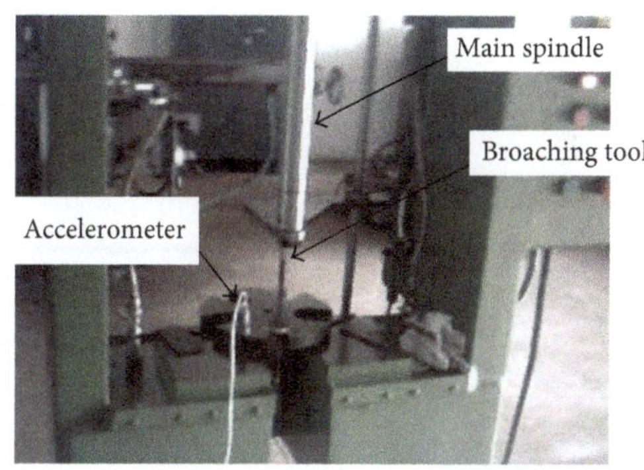

FIGURE 8: Photograph indicating the position of an accelerometer.

$$[C] = \begin{bmatrix} 200 & -60 & 0 & 0 \\ -60 & 150 & 0 & 0 \\ 0 & 0 & 200 & -50 \\ 0 & 0 & -50 & 50 \end{bmatrix}. \quad (10)$$

5. Experimental Setup

The experimental setup includes a push type broaching machine as shown in Figure 7 which was used for measuring the cutting force in three directions, keyway broach, Kistler accelerometer, and data acquisition system. The data acquisition system consists of hardware unit and an application programming interface. The real-time analog signals obtained from the accelerometer are sent to the hardware unit. The hardware unit performs the conditioning like amplification and filtering of the signals to increase the signal-to-noise ratio. This conditioned analog signal is sent through the proper cables to a data acquisition card which has electronic integrated circuit to convert the analog signal into the digital counterpart and sends the signal to the computer for further processing where it can store 1000 characters per second data in the computer. The program in the personal computer (PC) written using the Application Programming Interface (API) software performs the relevant operations on the input signal

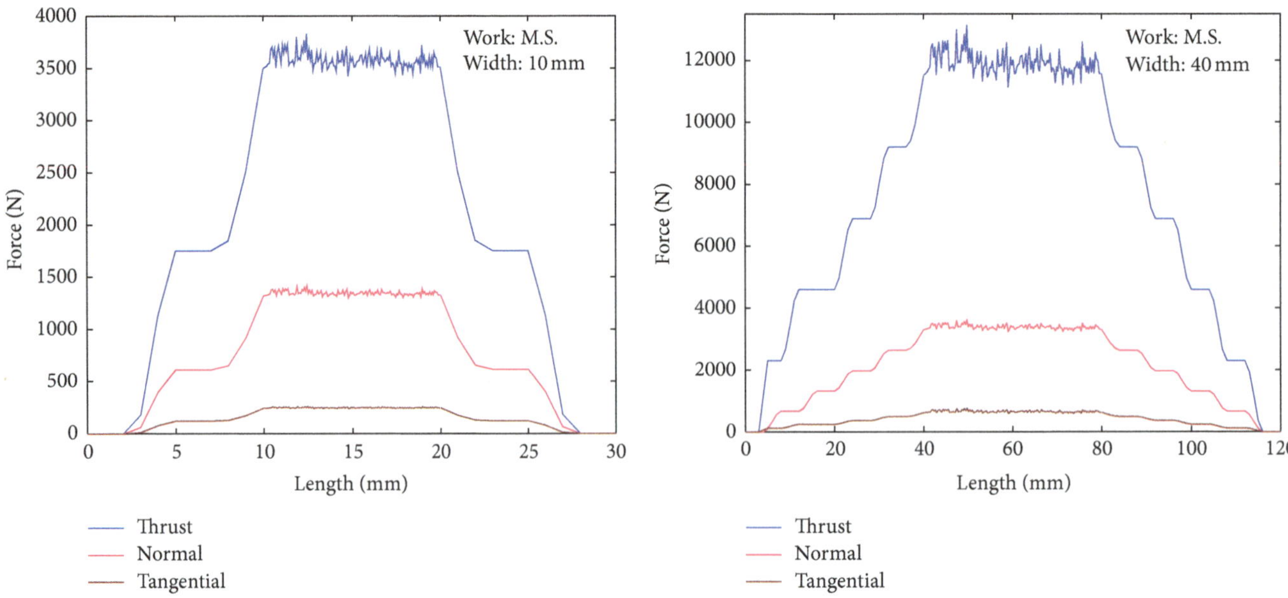

(a) Cutting forces for 10 mm depth

(b) Cutting forces for 40 mm depth

FIGURE 9: Measured force values for mild steel work material of 10 mm and 40 mm depth.

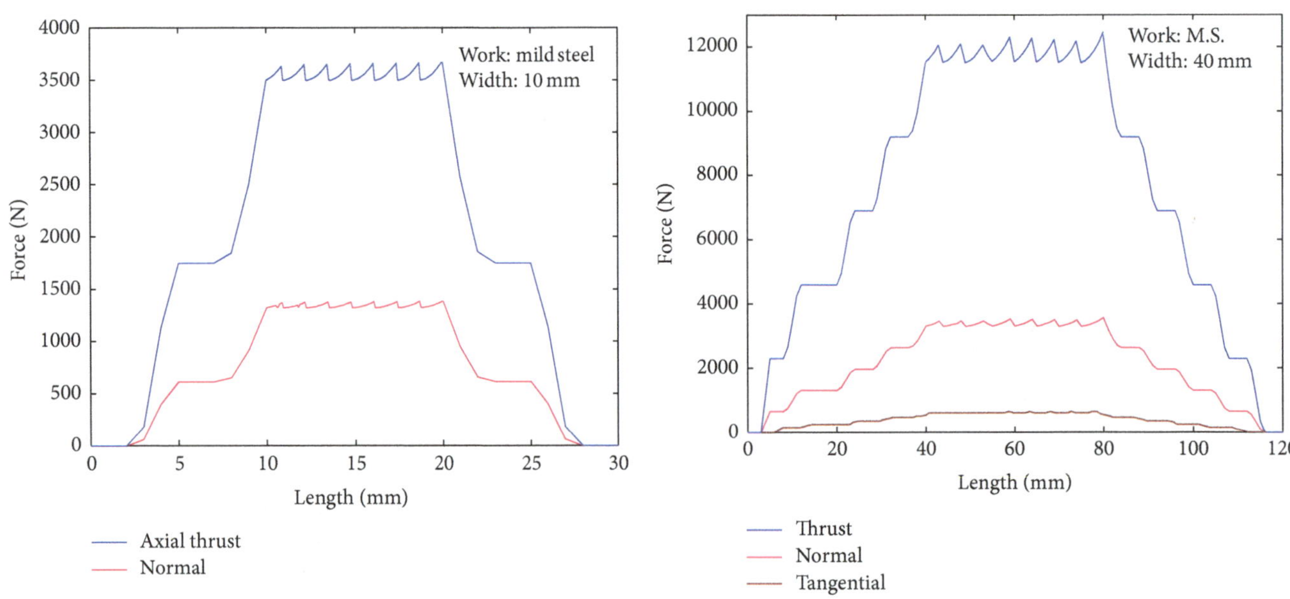

(a) Simulated cutting forces for 10 mm depth of cut

(b) Simulated cutting forces for 40 mm depth of cut

FIGURE 10: Simulated force values for mild steel work material of 10 mm and 40 mm depth.

and displays the output in the form of graphs or as digital displays and also has the provision for writing it to a text file. The data acquisition system includes National Instrument (NI)'s SCXI hardware unit, NI4472 card, and the API used is NI's LabVIEW 7.0 software. The number of specimens was prepared for different depth to perform the experiment. The experiments were conducted for different materials such as mild steel, aluminium and cast iron. The stored data was later processed to get graphical representation of cutting forces.

5.1. Technical Specifications of the Machine Tool. The specification of the machine tool is given in Table 4. It is a low capacity broaching machine with push type feed.

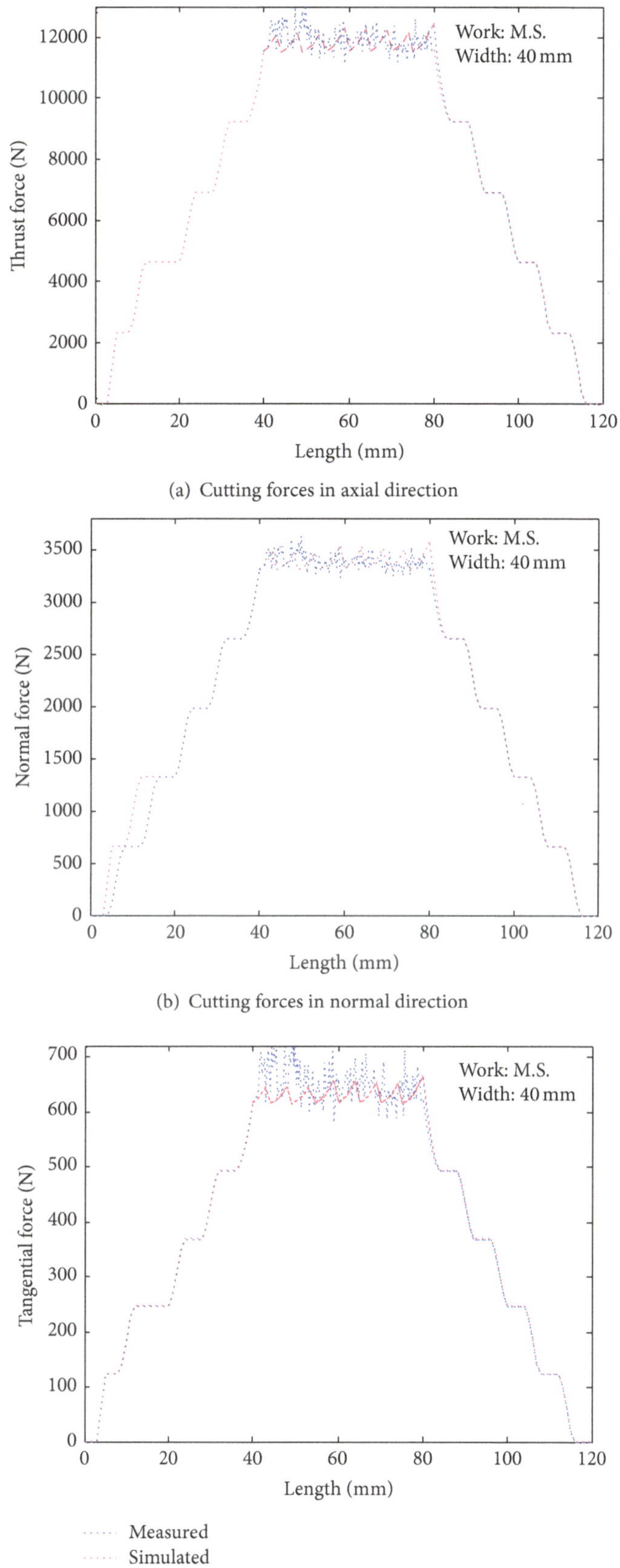

(a) Cutting forces in axial direction

(b) Cutting forces in normal direction

(c) Cutting forces in lateral direction

FIGURE 11: Comparison of measured and simulated force values during broaching operation on mild steel work.

TABLE 4: Technical specifications.

Maximum push load capacity	5000 kg
Maximum stroke	500 mm
Face plate dimension	405 × 210 mm
Slot dimension in face plate	100 mm
Broaching speed	1.5 m/min
Return speed	3 m/min
Power of electric motor	1.5 kW/2HP/1500 RPM

The photograph of the accelerometer position on the machine tool is shown in Figure 8.

6. Results and Discussions

Figures 9(a) and 9(b) show the measured values of cutting forces acting on mild steel workpiece of 10 mm and 40 mm depth, respectively. The pattern of experimental result is closely in agreement with mechanistic model results in Figures 10(a) and 10(b). The percentage of error between experimental results and mechanistic model results is within 2%.

Figures 11(a), 11(b), and 11(c) show the comparison of experimental results with simulated results for cutting force variation for a mild steel workpiece of 40 mm depth in axial, normal, and lateral directions. As height of the work piece increases, forces are also increasing and repeatedly in sinusoidal manner and come to zero as broach disengages with the work piece.

The experiment and simulated results shown in Figures 11(a), 11(b), and 11(c) confirm that axial force is very much higher than normal force for both 40 mm and 10 mm depth of mild steel work material. The force in the tangential direction is even smaller in comparison to the axial force.

7. Conclusions

This paper reports the dynamic force modeling of broaching process and validation of output with experimental results. A software has been developed to compute the dynamic forces using Matlab. Experiment has been carried out using a vertical broaching machine for the mild steel material of different depths. Experimental validations of the cutting forces are required to establish the prediction from the model so that the model is acceptable to the manufacturing industry. The work piece is the most complicated link of the machining system. The second validation deals with a case where a machine tool spindle is the weakest part of the system. The fixture of the work piece for the broaching was mounted such that feed direction becomes very compliant. Slots were provided on the fixture so that cutter can perform axis symmetry cutting. The position of the cutter also permits a proper pressing action on the fixture so that cutting process has effect on the overall stability of the cutting process.

The parameters which affect the cutting process are depth of cut, feed rate and cutter geometry. The feed rate, determines the axial chip load. Overall stability increases as the feed rate is increased. This is due to the fact that specific cutting energy constant decreases as a function of chip load. Stability in general is accountable for dynamics in the axial direction for broaching because the work piece and machine tool dynamics in axial direction are usually of order of magnitude stiffer than in the normal direction. The occurrence of the chatter predominant during the experiment is largely dependent on the material specific cutting energy constants since they determine the magnitude of the forces which present during machining. During broaching, the work holding fixture has an important role to play in controlling the force acting on the tool. Major force acting on the broaching tool is in axial direction.

All the factors discussed above that affect stability and dynamics of the structural system are influential. A good understanding of how the dynamics affect the overall stability can be helpful to assess the quality of the product. The dynamics variables include stiffness, damping, natural frequency, and mode cycles. Special care has to be taken to ensure that experimentally collected data is clear and accountable. The simulated and experimental results are presented graphically and are in close agreement. It is concluded that mechanistic model is suitable for determining cutting forces for broaching process.

Nomenclature

A_c: Chip cross-section area
a_i, b_i, c_i: Specific cutting energy constants
F_a: Axial cutting force
F_t: Tangential cutting force
F_n: Normal cutting force
K_x, K_y, K_z: Proportionality constants
t_c: Chip thickness
v_c: Cutting velocity
γ_a: Rake angle.

Conflict of Interests

The authors declare that there is no conflict of interests regarding the publication of this paper.

Acknowledgments

The authors would like to express their gratitude and sincere thanks to Professor Dr. Appu Kuttan, Director, MANIT, Bhopal (MP), for guidance, valuable help, enthusiastic attitude, and suggestions throughout the period and National Institute of Technology Karnataka (NITK) Surathkal, for providing necessary facilities to conduct experiments.

References

[1] E. Linsley Horace, *Broaching Tooling and Practice*, Industrial Press, 1961.

[2] C. Monday, *Broaching*, Machinery Publication, London, UK, 1960.

[3] J. W. Sutherland, E. J. Salisbury, and F. W. Hoge, "A model for the cutting force system in the gear broaching process," *International Journal of Machine Tools and Manufacture*, vol. 37, no. 10, pp. 1409–1421, 1997.

[4] V. S. Belov and S. M. Ivanov, "Factors affecting broaching condition and broach life," *Journal of Stanki Instruments*, vol. 45, pp. 31–33, 1974.

[5] E. Kuljanic, "Cutting force and surface roughness in broaching," *Annual CIRP*, vol. 24, no. 1, pp. 77–82, 1975.

[6] M. E. Merchant, "Mechanics of the metal cutting process. I. Orthogonal cutting and a type 2 chip," *Journal of Applied Physics*, vol. 16, no. 5, pp. 267–275, 1945.

[7] M. E. . Merchant, "Basic mechanics of the metal cutting process," *Journal of Applied Mechanics*, vol. 168, pp. 175–178, 1954.

[8] T. Tyan and W. H. Yang, "Analysis of orthogonal metal cutting processes," *International Journal for Numerical Methods in Engineering*, vol. 34, no. 1, pp. 365–389, 1992.

[9] V. Sajeev, L. Vijayraghavan, and U. R. K. Rao, "An analysis of the effect of burnishing in internal broaching," *International Journal of Mechanical Engineering*, vol. 28, pp. 163–173, 2000.

[10] D. A. Axinte and N. Gindy, "Tool condition monitoring in broaching," *Journal of Wear*, vol. 254, no. 3-4, pp. 370–382, 2003.

[11] B. Daniel Dallas, *Tool and Manufacturing Engineers Handbook*, 3rd edition, 2002.

[12] H. S. Kim and K. F. Ehmann, "A cutting force model for face milling operations," *International Journal of Machine Tools and Manufacture*, vol. 33, no. 5, pp. 651–673, 1993.

[13] H. J. Fu, R. E. DeVor, and S. G. Kapoor, "A mechanistic model for the prediction of the force system in face milling operations," *Journal of Engineering for Industry, Transactions of the ASME*, vol. 106, no. 1, pp. 81–88, 1984.

[14] E. Felder, P. Gillormini, L. Tronchet, and F. Leroy, "A comparative analysis of three machining process: broaching, tapping and slotting," *etude DGRST Materiaux*, vol. 80, pp. 512–513, 1982.

[15] S. G. Kapoor, R. E. DeVor, R. Zhu, R. Gajjela, G. Parakkal, and D. Smithey, "Development of mechanistic models for the prediction of machining performance: model building methodology," *Machining Science and Technology*, vol. 2, no. 2, pp. 213–238, 1998.

[16] S. Jayaram, S. G. Kapoor, and R. E. Devor, "Estimation of the specific cutting pressures for mechanistic cutting force models," *International Journal of Machine Tools and Manufacture*, vol. 41, no. 2, pp. 265–281, 2001.

[17] E. Usui, A. Hirota, and M. Masuko, "Analytical predictions of three dimensional cutting process—part I: basic cutting model an energy approach," *Journal of Engineering for Industry, Transactions of the ASME*, vol. 100, no. 2, pp. 222–228, 1978.

[18] W. J. Endres, R. E. DeVor, and S. G. Kapoor, "A dual mechanism approach to the prediction of machining forces," *ASME Journal of Engineering For Industry*, vol. 117, no. 4, pp. 526–533, 1999.

[19] J. A. Kir, D. K. Anand, and C. Mahindra, "Matrix representation and prediction of three dimensional cutting forces," *ASME Journal of Engineering For Industry*, vol. 99, pp. 822–828, 1978.

[20] F. Koenigsberger and J. Tlusty, *Machine Tool Structures-Vol. 1: Stability against Chatter*, Pergamon Press, 1967.

[21] P. Albrecht, "Dynamics of the metal cutting process," *ASME Journal of Engineering for Industry*, vol. 87, pp. 429–441, 1965.

[22] J. D. Smith and S. A. Tobias, "The dynamic cutting of metals," *International Journal of Machine Tool Design and Research*, vol. 1, no. 4, pp. 283–292, 1961.

[23] D. A. Stephenson and S. M. Wu, "Computer models for the mechanics of three dimensional cutting process part 1: theory and numerical methods," *ASME Journal of Engineering for Industry*, vol. 110, no. 1, pp. 32–37, 1988.

[24] H. Q. Zheng, X. P. Li, Y. S. Wong, and A. Y. C. Nee, "Theoretical modelling and simulation of cutting forces in face milling with cutter runout," *International Journal of Machine Tools and Manufacture*, vol. 39, no. 12, pp. 2003–2018, 1999.

[25] J. A. Yang, V. Jaganathan, and R. Du, "A new dynamic model for drilling and reaming processes," *International Journal of Machine Tools and Manufacture*, vol. 42, no. 2, pp. 299–311, 2002.

24

Cutting Forces in Milling of Carbon Fibre Reinforced Plastics

Luca Sorrentino and Sandro Turchetta

Department of Civil and Mechanical Engineering, University of Cassino and Southern Lazio, 03043 Cassino, Italy

Correspondence should be addressed to Luca Sorrentino; sorrentino@unicas.it

Academic Editor: Thomas R. Kurfess

The machining of fibre reinforced composites is an important activity for optimal application of these advanced materials into engineering fields. During machining any excessive cutting forces have to be avoided in order to prevent any waste product in the last stages of production cycle. Therefore, the ability to predict the cutting forces is essential to select process parameters necessary for an optimal machining. In this paper the effect of cutting conditions during milling machining on cutting force and surface roughness has been investigated. In particular the cutting force components have been analysed in function of the principal process parameters and of the contact angle. This work proposes experimental models for the determination of cutting force components for CFRP milling.

1. Introduction

Composite materials milling is a rather complex task owing to its heterogeneity and some problems such as surface delamination appearing during the machining process, associated with the characteristics of the material and cutting parameters.

Milling is the machining operation most frequently used in manufacturing of fibre reinforced plastics parts as a corrective operation to produce well-defined and high-quality surfaces that often require the removal of excess material to control tolerances [1]. The machinability of fibre reinforced plastics is strongly influenced by the type of fibre embedded in the composite and by its properties. Mechanical and thermal properties have an extreme importance in machining FRP. The fibre used in the composites has a great influence in the selection of cutting tools (cutting edge material and geometry) and machining parameters. It is fundamental to ensure that the tool selected is suitable for the material. The knowledge of cutting mechanisms is necessary to optimize the cutting mechanics and machinability in milling [1, 2]. Composite materials such as carbon fibre reinforced plastics (CFRPs) made by using carbon fibres for reinforcing plastic resin matrices, such as epoxy, are characterised by excellent properties as light weight, high strength, and high stiffness. These properties make them especially attractive for aerospace applications [2]. Surface roughness is a parameter that has a great influence on dimensional precision, on performance of mechanical pieces, and on production costs. For these reasons, research developments have been carried out with the purpose of optimising the cutting conditions to reach a specific surface roughness [3, 4]. The required quality of the machined surface depends on the mechanisms of material removal and the kinetics of machining processes affecting the performance of the cutting tools [5]. The works of a number of authors [6–12], when reporting on milling of FRP, show that the type and orientation of the fibre, cutting parameters, and tool geometry have an essential influence on the machinability. Everstine and Rogers [6] have presented the first theoretical work on the machining of FRPs in 1971, since then the research carried out in this area has been based on experimental investigations. Koplev et al. [7], Kaneeda [8], and Puw and Hocheng [9] established that the principal cutting mechanisms are strongly correlated to fibre arrangement and tool geometry. Santhanakrishnan et al. [10] and Ramulu et al. [11] carried out a study on machining of polymeric composites and concluded that an increasing of the cutting speed leads to a better surface finish. Hocheng et al.

[12] studied the effect of the fibre orientation on the cut quality, cutting forces, and tool wear on the machinability. Hintze Wolfgang et al. [13] investigated the case of delamination of the top layers during the machining of CFRP tape milling and have showed that delamination depends highly on the fibre orientation and the tool sharpness.

Liu et al. [14] summarized an up-to-date progress in mechanical drilling of composite laminates reported in the literature; they cover drilling operations (including conventional drilling, grinding drilling, vibration-assisted twist drilling, and high speed drilling), drill bit geometry and materials, drilling-induced delamination and its suppressing approaches, thrust force, and tool wear.

Enemuoh et al. [15] realized that, with the application of the technique of Taguchi and a multiobjective optimization criterion, it is possible to achieve cutting parameters that allow the absence of damage in drilling of fibre reinforced plastics. Davim et al. [16] studied the cutting parameters (cutting velocity and feed rate) under specific cutting pressure, thrust force, damage, and surface roughness in drilling glass fibre reinforced plastics (GFRPs). A plan of experiments, based on the techniques of Taguchi, was established considering drilling with prefixed cutting parameters in a hand lay-up GFRP material. Sheikh-Ahmad et al. [17] studied the comprehensive model for orthogonal milling of unidirectional composites at various fibre orientations. Kalla et al. [18] studied the mechanistic modelling techniques for simulating the cutting of carbon fibre reinforced polymers (CFRPs) with a helical end mill. A methodology was developed for predicting the cutting forces by transforming specific cutting energies from orthogonal cutting to oblique cutting.

Yashiro et al. [19] confirmed that the measurement of the cutting temperature is important when dealing with CFRP: temperatures higher than the glass-transition temperature of the matrix resin are not favourable as they damage the laminate. In Liu et al. [20] a heat transfer model is developed to investigate the temperature distribution of CFRP workpiece in helical milling process. The relationship between cutting speed and temperature of processing is pointed out. In summary, it can be noticed that the works carried out on the machinability of FRP are basically related to the wear of cutting tools and the quality of the surfaces, as a function of the cutting conditions, the distribution, and angle of inclination of fibres in the polymeric matrix.

The aim of this work is to value cutting forces of carbon fibre reinforced plastics during the milling machining. This work aims to investigate the relationship among the cutting force and the surface finishing of machined laminate as a function of relevant cutting parameters, such as the cutting speed, axial depth of cut, and the feed rate. In particular, the most interesting machining conditions from an industrial point of view have been investigated in order to measure the cutting force components by a dynamometer. The obtained values of cutting force have been put into relationship with the angular position of the cutter. In the following the models of the relationship between cutting force, chip thickness, and process parameters have been presented.

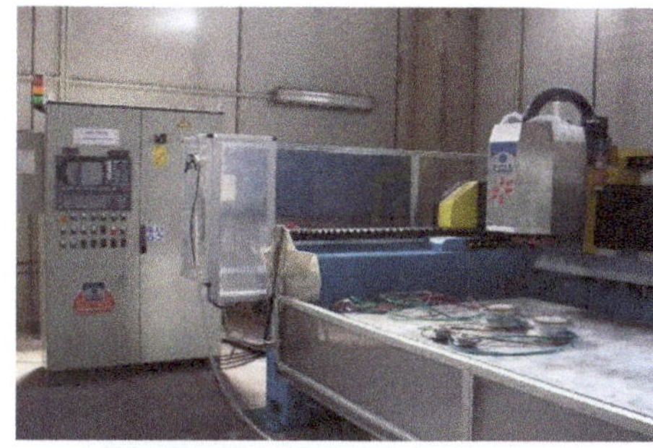

Figure 1: CNC milling machine.

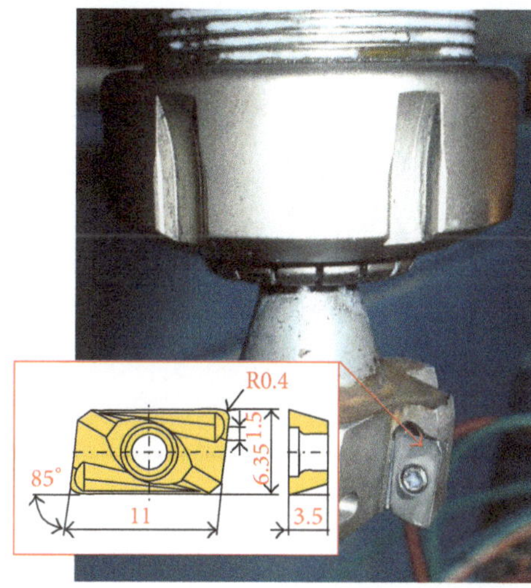

Figure 2: End mill with a single cutting tooth (MITSUBISHI).

2. Experimental Setup

The experiments have been undertaken on a CNC milling machine with 15 kW spindle power where maximum spindle speed of 15000 rpm has been used to perform the experiments; see Figure 1. An end mill commonly used for the machining of CFRP has been used, whose diameter is 40 mm, with a single cutting tooth (insert APMT1135PDER-H1 UTi20T of MITSUBISHI); see Figure 2. The composite material used in the tests (epoxy matrix reinforced with 50% of carbon fibre) has been produced by autoclave with a fibre orientation of 0/90°. The experiments have been carried out in a laminate plate, made up with 40 alternating layers of fibres with 13 mm of thickness; the tests have been carried out by up-milling, in the absence of cooling fluid. In Figure 3 the surface machined can be observed: machining (a) has been performed with the aim of realizing the correct tool access for the following machining: (b), (c), and (d).

Three cutting speed values, four axial cutting depth values, and two feeds per tooth per revolution have been taken into account; they have been chosen in order to reproduce the commonly used industrial range of process variables. Each cut has been replicated three times, yielding a total of 72 measured forces. The plan of experiments is shown in Table 1. The cutting conditions have been represented by the angular position of the cutter (θ). The experimental cuts have been

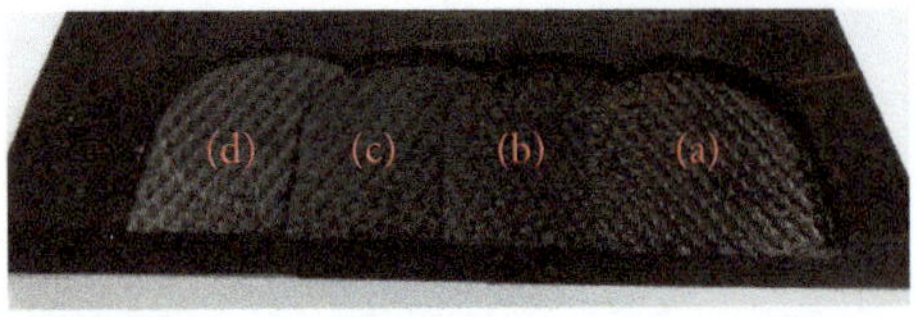

FIGURE 3: Sample after milling machining.

FIGURE 4: Gripping system/dynamometer/sample.

TABLE 1: Experimental plan.

Factors	Levels [#]	Value
p_a, axial cutting depth [mm]	4	1.0-1.5-2.0-3.0
p_r, radial cutting depth [mm]	1	25
V_t, cutting speed [m/min]	3	100-200-300
f_t, feed per tooth [mm]	2	0.022-0.044
Replications	3	
Total cuts	72	

performed in a random sequence, in order to reduce the effect of any possible systematic error. The cutting forces F_x and F_y have been measured by a Kistler piezoelectric platform dynamometer (type 9257 BA); see Figure 4. The signals acquired by the dynamometer have been sampled at different frequencies and for different time intervals in order to set the acquisition parameters giving the whole information about force signal with the minimum time waste. The signal along X and Y axes is periodic and 16384 acquisition points seem to be enough to keep the whole signal information. Therefore, the output of the dynamometer has been fed into an A/D converter and sampled at 10 kHz by a PC. Each observation consisted of about 1.6384 s time signal [21, 22].

The typical force signals acquired by dynamometer along X and Y axes, respectively, have been reported in Figures 5, 6, and 7. In Figure 8 the signal acquired by dynamometer for 8 tool's round is shown. The acquired signal is time periodic; in Figure 9 only one period is shown.

The tangential F_T and radial F_R components of the cutting force have been calculated according to (1) (see Figure 10):

$$F_T = F_y \cos\theta - F_x \sin\theta,$$
$$F_R = F_x \cos\theta + F_y \sin\theta. \quad (1)$$

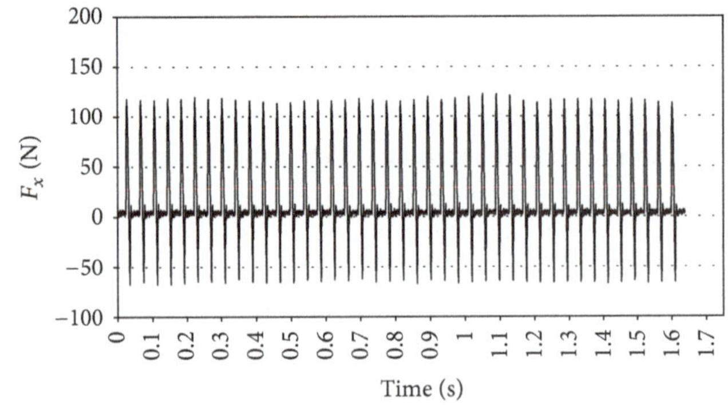

FIGURE 5: Example of time domain signal monitored in X direction.

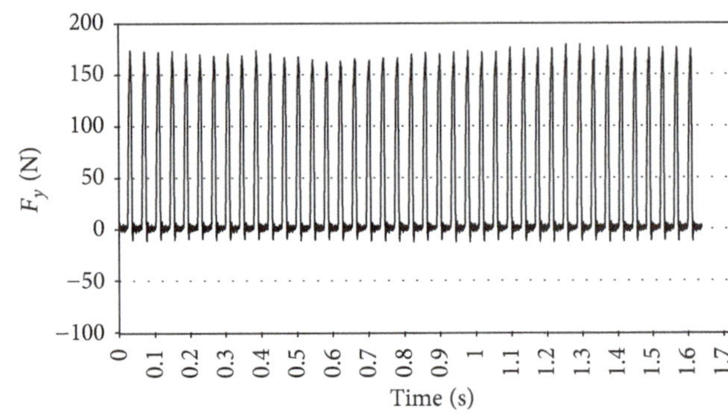

FIGURE 6: Example of time domain signal monitored in Y direction.

3. Results and Discussion: Forces Analysis

Some preliminary tests have been carried out with fixed process parameters in order to verify the testing system stability and the repeatability of the test itself.

Cutting force components in tangential and normal direction have been calculated by acquiring force signal for each test. The signals have been analyzed in time domain as function of contact angle between 0° and 90°. Since a similar trend of the forces for every tool round has been observed only the results for first round of the tool have been reported. The trends of tangential cutting force components as function of contact angle and cut depth have been reported in Figures 11, 12, and 13. Figure 11 reports tangential cutting force component as function of contact angle, obtained with a cutting speed of 100 m/min and a tooth feed of 0.022 mm; as it can be seen from the graph the cutting force tangential component increases with cut depth and with contact angle. In particular, with thin contact angles there is a fast force increase, while with angles near 90° it is practically constant. Similar trends have been obtained with cutting speed of 200 m/min and 300 m/min. Figure 12 reports tangential cutting force component versus contact angle obtained with a cutting speed of 200 m/min and tooth feed of 0.022 mm. Figure 13 reports tangential cutting force component versus contact angle obtained with a cutting speed of 300 m/min and tooth feed of 0.022 mm. The trends of radial cutting force component as function of contact angle and cut depth have

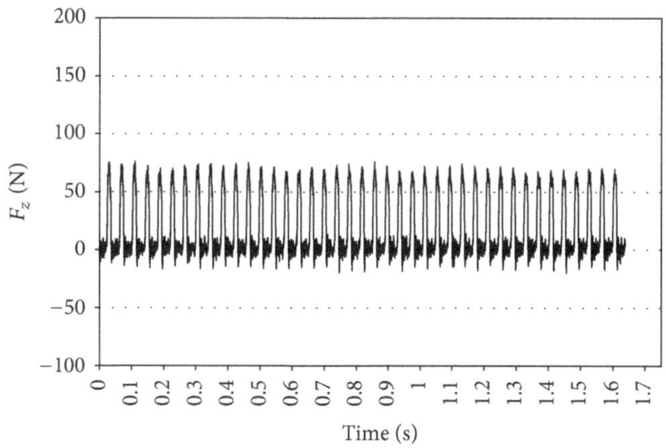

FIGURE 7: Example of time domain signal monitored in Z direction.

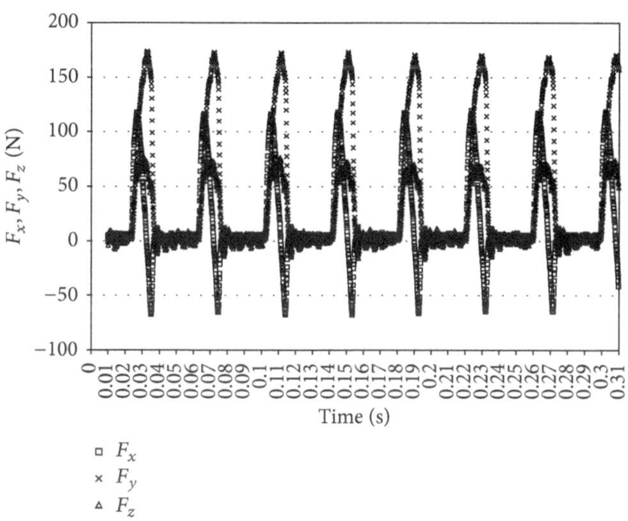

FIGURE 8: Time domain signal monitored in X, Y, and Z directions for eight tool's round.

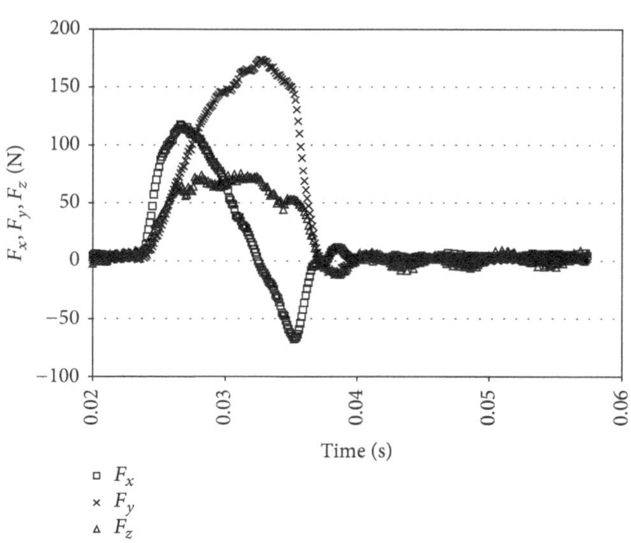

FIGURE 9: Time domain signal monitored in X, Y, and Z directions for one tool's round.

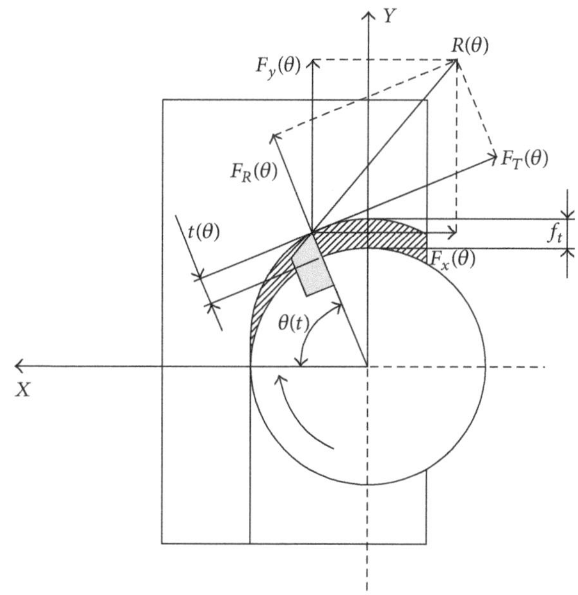

FIGURE 10: Scheme of F_T and F_R force components.

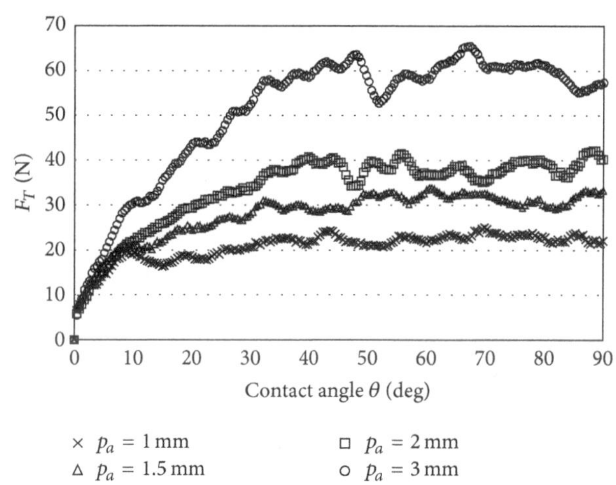

FIGURE 11: F_T component versus contact angle (cutting speed = 100 m/min; feed per tooth = 0.022 mm).

been reported in Figures 14, 15, and 16. Figure 14 reports radial cutting force component versus contact angle obtained with a cutting speed of 100 m/min and a tooth feed of 0.022 mm. The cutting force radial component as the tangential cutting force component increases with cut depth and with contact angle. In particular, with thin contact angles there is a fast force increase, while with angles in range of 40°–90° it is practically constant. Similar trends have been obtained with cutting speed of 200 m/min and 300 m/min.

Figure 15 reports radial cutting force component versus contact angle obtained with cutting speed of 200 m/min and tooth feed of 0.022 mm. Figure 16 reports radial cutting force component versus contact angle obtained with cutting speed of 300 m/min and tooth feed of 0.022 mm. Cutting tests carried out with a tooth feed of 0.044 mm have shown an increase of cutting force components and a trend, as

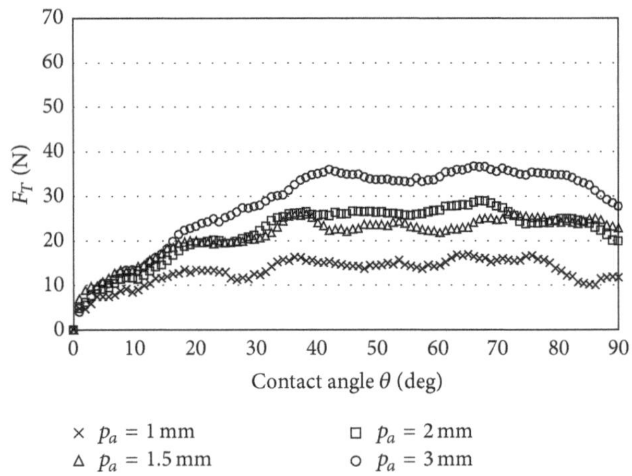

FIGURE 12: F_T component versus contact angle (cutting speed = 200 m/min; feed per tooth = 0.022 mm).

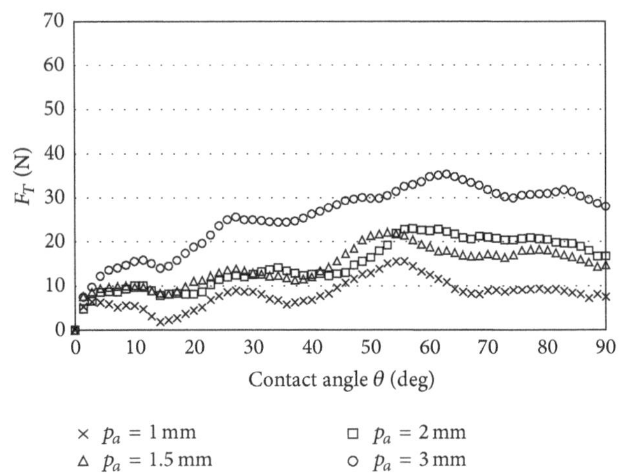

FIGURE 13: F_T component versus contact angle (cutting speed = 300 m/min; feed per tooth = 0.022 mm).

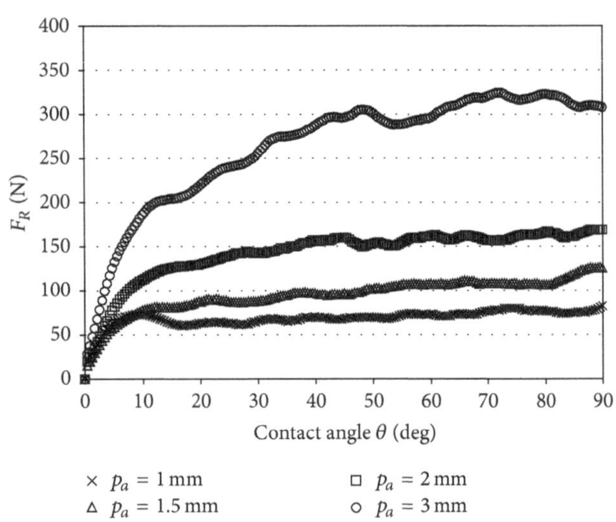

FIGURE 14: F_R component versus contact angle (cutting speed = 100 m/min; feed per tooth = 0.022 mm).

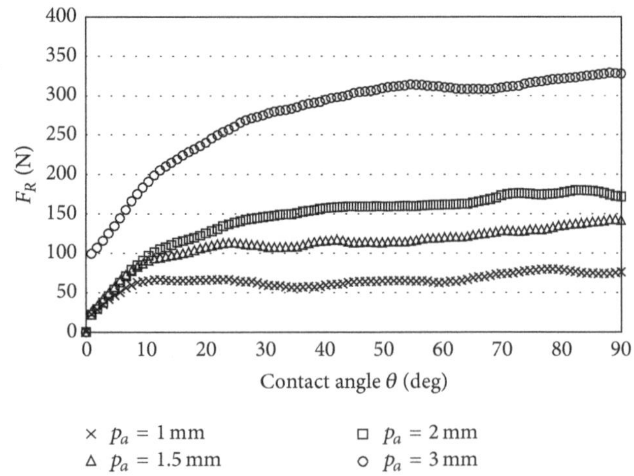

FIGURE 15: F_R component versus contact angle (cutting speed = 200 m/min; feed per tooth = 0.022 mm).

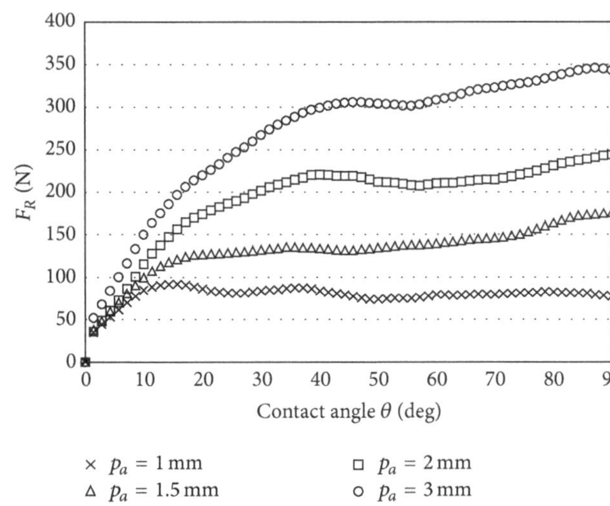

FIGURE 16: F_R component versus contact angle (cutting speed = 300 m/min; feed per tooth = 0.022 mm).

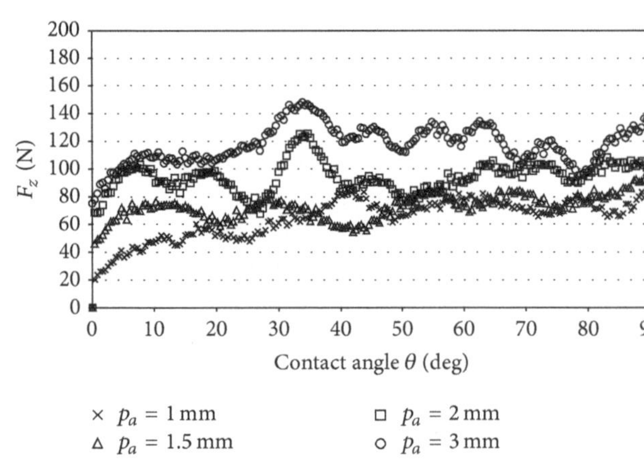

FIGURE 17: F_z component versus contact angle (cutting speed = 100 m/min; feed per tooth = 0.044 mm).

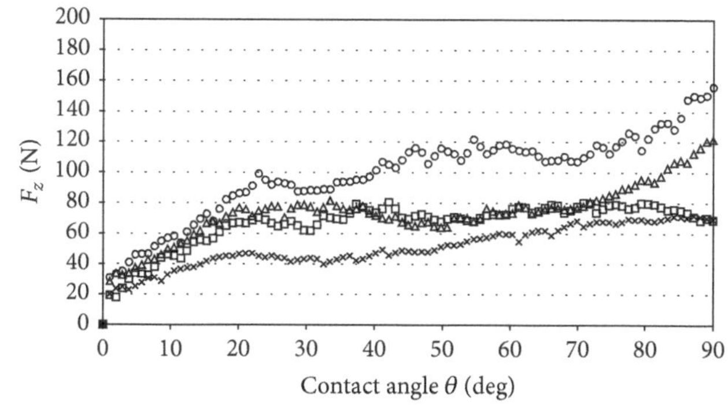

FIGURE 18: F_z component versus contact angle (cutting speed = 200 m/min; feed per tooth = 0.044 mm).

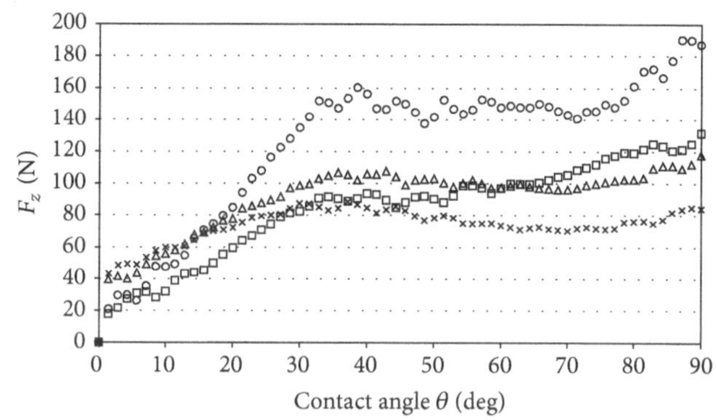

FIGURE 19: F_z component versus contact angle (cutting speed = 300 m/min; feed per tooth = 0.044 mm).

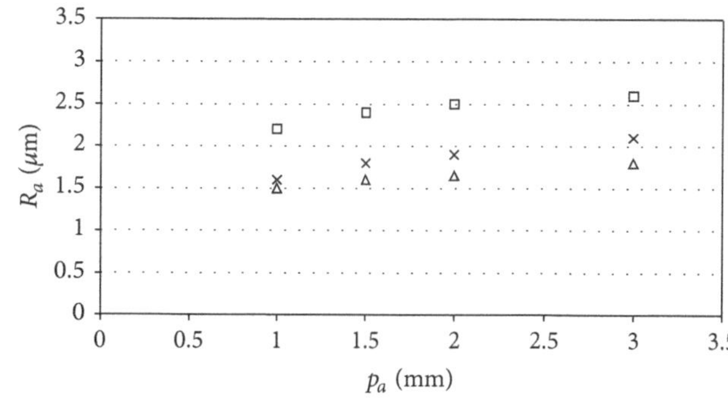

FIGURE 20: Surface roughness versus p_a (feed per tooth = 0.022).

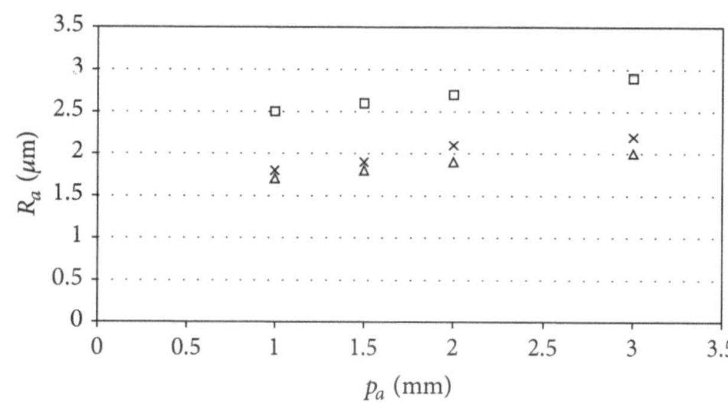

FIGURE 21: Surface roughness versus p_a (feed per tooth = 0.044).

TABLE 2: Constant values of the model.

K_T	υ_T	α_T	β_T
2455.00	0.333	0.777	−0.705
K_R	υ_R	α_R	β_R
83.17	0.262	1.190	0.191

a function of contact angle and cut deep, similar to that one found with tooth feed of 0.022 mm.

With particular regard to component F_T of the cutting force, it decreases with increasing of the cutting speed; in fact, the maximum value of "F_T" is 65 N, in correspondence of V_t = 100 m/min and p_a = 3 mm, and it is 38 N, in correspondence of V_t = 300 m/min and p_a = 3 mm. This reduction does not result for force radial component "F_R" which remains substantially constant as a function of cutting speed.

In Figures 17–19 the trends of component F_z of the cutting force have been reported as a function of the process parameters considered. In particular, Figure 17 shows the trend of component F_z in function of axial cutting depth "p_a" relatively to the cutting speed of 100 m/min. From the graph it is possible to note that component F_z increases rapidly with the increase of the contact angle to then assume almost a constant value. The maximum value of this force increases with the increase of axial cutting depth from about 60 N to 140 N. Figures 18-19 show the trend of component F_z in function of "p_a" relatively to the cutting speed of 200 m/min and 300 m/min; the graphs show how by increasing V_t the values of component F_z do not significantly change but the initial transient extends for a greater contact angle.

Regression analysis has allowed determining the model for evaluation of cutting force components shown in

$$F_T = K_T \cdot t_\theta^{\upsilon_T} \cdot p_a^{\alpha_T} \cdot V_T^{\beta_T},$$
$$F_R = K_R \cdot t_\theta^{\upsilon_R} \cdot p_a^{\alpha_R} \cdot V_T^{\beta_R} \quad (2)$$

with $t_\theta = f_t \sin\theta$.

The values of regression model constants are reported in Table 2. The coefficients of determination are higher than 95%, while the hypotheses (normality and homogeneity of variance) about the residuals are satisfied.

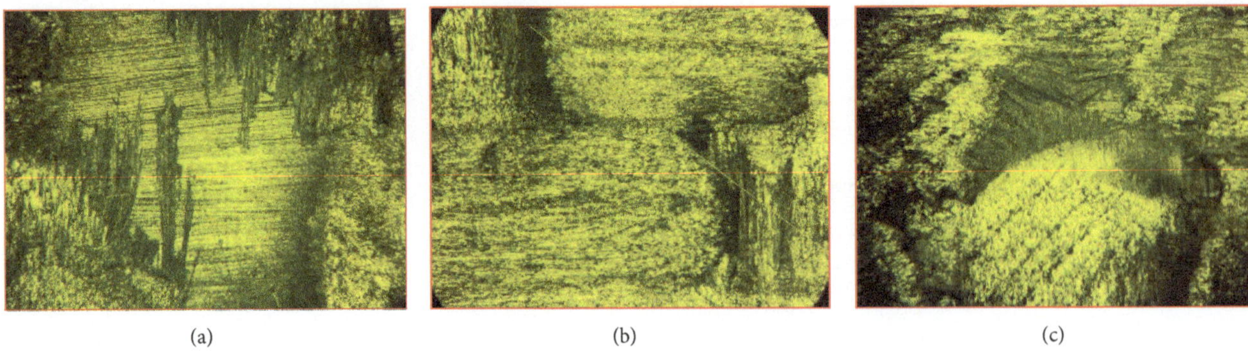

FIGURE 22: Images by microscope related to surface machined for a cutting depth of (a) 3 mm, (b) 2 mm, and (c) 1 mm.

4. Results and Discussion: Surface Analysis

The surface quality has been analyzed by measurements of surface roughness and optical microscopic observations. Roughness measurements have been carried out according to the direction perpendicular to the feed direction; 5 measures of roughness for each machined surface have been acquired. Figures 20-21 show the average roughness measured on each surface according to the axial cutting depth for the three levels of the cut speed and for two levels of feed tooth (0.022 mm and 0.044 mm); the maximum standard deviation is equal to 7%. Figure 20, relating to a feed tooth of 0.022 mm, shows that the roughness increases with the increase of axial cutting depth and decreases with the increase of the cutting speed in agreement with what has been shown for the components of the cutting force. Figure 21, relating to a feed tooth of 0.044 mm, shows an increased roughness compared to that obtained for the advancement of 0.022 mm.

Preliminary analysis of the surface by LAICA microscope (magnification 50x) shows the presence of defects highlighted as fiber breakage. Figures 22(a), 22(b), and 22(c) show the machined surface related to tests carried out, respectively, with a cutting depth of 3 mm, 2 mm, and 1 mm, but cutting speed of 100 m/min and feed of 0.022 mm; in Figure 22(a) it is possible to note the presence of surface defects greater than those reported in Figures 22(b) and 22(c).

5. Conclusion

In the present work the cutting force components and the surface roughness in CFRP milling have been analyzed. In particular the cutting force radial and tangential components in relation to contact angle and to principal process parameters have been determined. The analysis of results has underlined that the feed speed, depth of cut, and chip thickness significantly influence force components F_T and F_R. An increase of both the cut depth and the chip thickness causes an increase of the force components. Regression analysis has allowed defining the cutting models to predict the cutting force components as function of principal process parameters.

As regards surface roughness it has been possible to value the trend of parameter Ra; the optimal condition has been obtained in correspondence with low axial cutting depth and high cutting speed.

Nomenclature

p_a: Axial cutting depth [mm]
p_r: Radial cutting depth [mm]
t_θ: Chip thickness [mm]
θ: Contact angle [degree]
v_t: Cutting speed [m/min]
f_t: Feed per tooth [mm]
F_X: Cutting force component along X direction [N]
F_Y: Cutting force component along Y direction [N]
F_Z: Cutting force component along Z direction [N]
F_T: Tangential cutting force [N]
F_R: Radial cutting force [N].

Conflict of Interests

The authors declare that there is no conflict of interests regarding the publication of this paper.

Acknowledgments

The authors acknowledge AgustaWestland, Anagni Plant, for providing the CFRP composite material used in the experimental tests. Special thanks are due to Tiziana S.

References

[1] S. Jahanmir, M. Ramulu, and P. Koshy, *Machining of Ceramics and Composites*, Marcel Dekker, New York, NY, USA, 2000.

[2] W. F. Smith, *Principles of Materials Science and Engineering*, McGraw-Hill, New York, NY, USA, 1990.

[3] M. Ramulu, C. W. Wern, and J. L. Garbini, "Effect of fibre direction on surface roughness measurements of machined graphite/epoxy composite," *Composites Manufacturing*, vol. 4, no. 1, pp. 39–51, 1993.

[4] E. Eriksen, "Influence from production parameters on the surface roughness of a machined short fibre reinforced thermoplastic," *International Journal of Machine Tools and Manufacture*, vol. 39, no. 10, pp. 1611–1618, 1999.

[5] P. S. Sreejith, R. Krishnamurthy, S. K. Malhotra, and K. Narayanasamy, "Evaluation of PCD tool performance during machining of carbon/phenolic ablative composites," *Journal of Materials Processing Technology*, vol. 104, no. 1, pp. 53–58, 2000.

[6] G. C. Everstine and T. G. Rogers, "A theory of machining of reinforced materials," *Journal of Composite Materials*, vol. 5, pp. 94–106, 1971.

[7] A. Koplev, A. Lystrup, and T. Vorm, "The cutting process, chips, and cutting forces in machining CFRP," *Composites*, vol. 14, no. 4, pp. 371–376, 1983.

[8] T. Kaneeda, "CFRP cutting mechanism," in *Proceedings of the 16th North American Manufacturing Research Conference*, pp. 216–221, 1989.

[9] H. Y. Puw and H. Hocheng, "Anisotropic chip formation models of cutting of FRP," in *Proceedings of the ASME Symposium on Material Removal and Surface Modification Issues in Machining Processes*, New York, NY, USA, 1995.

[10] G. Santhanakrishnan, R. Krishnamurthy, and S. K. Malhotra, "Machinability characteristics of fibre reinforced plastics composites," *Journal of Mechanical Working Technology*, vol. 17, pp. 195–204, 1988.

[11] M. Ramulu, D. Arola, and K. Colligan, "Preliminary investigation of effects on the surface integrity of fiber reinforced plastics, PD-Vol-64-2," *Engineering Systems Design and Analysis, ASME*, vol. 2, pp. 93–101, 1994.

[12] H. Hocheng, H. Y. Puw, and Y. Huang, "Preliminary study on milling of unidirectional carbon fibre-reinforced plastics," *Composites Manufacturing*, vol. 4, no. 2, pp. 103–108, 1993.

[13] W. Hintze Wolfgang, D. Hartmann, and C. Schütte, "Occurrence and propagation of delamination during the machining of carbon fibre reinforced plastics (CFRPs)—an experimental study," *Composites Science and Technology*, vol. 71, no. 15, pp. 1719–1726, 2011.

[14] D. Liu, Y. Tang, and W. L. Cong, "A review of mechanical drilling for composite laminates," *Composite Structures*, vol. 94, no. 4, pp. 1265–1279, 2012.

[15] E. U. Enemuoh, A. S. El-Gizawy, and A. C. Okafor, "An approach for development of damage-free drilling of carbon fiber reinforced thermosets," *International Journal of Machine Tools and Manufacture*, vol. 41, no. 12, pp. 1795–1814, 2001.

[16] J. P. Davim, P. Reis, and C. C. António, "Experimental study of drilling glass fiber reinforced plastics (GFRP) manufactured by hand lay-up," *Composites Science and Technology*, vol. 64, no. 2, pp. 289–297, 2004.

[17] J. Sheikh-Ahmad, J. Twomey, D. Kalla, and P. Lodhia, "Multiple regression and committee neural network force prediction models in milling FRP," *Machining Science and Technology*, vol. 11, no. 3, pp. 391–412, 2007.

[18] D. Kalla, J. Sheikh-Ahmad, and J. Twomey, "Prediction of cutting forces in helical end milling fiber reinforced polymers," *International Journal of Machine Tools and Manufacture*, vol. 50, no. 10, pp. 882–891, 2010.

[19] T. Yashiro, T. Ogawa, and H. Sasahara, "Temperature measurement of cutting tool and machined surface layer in milling of CFRP," *International Journal of Machine Tools & Manufacture*, vol. 70, pp. 63–69, 2013.

[20] J. Liu, G. Chen, C. Ji, X. Qin, H. Li, and C. Ren, "An investigation of workpiece temperature variation of helical milling for carbon fiber reinforced plastics (CFRP)," *International Journal of Machine Tools and Manufacture*, vol. 86, pp. 89–103, 2014.

[21] S. Turchetta, "Cutting force on a diamond grit in stone machining," *International Journal of Advanced Manufacturing Technology*, vol. 44, no. 9-10, pp. 854–861, 2009.

[22] S. Turchetta, "Cutting force and diamond tool wear in stone machining," *International Journal of Advanced Manufacturing Technology*, vol. 61, no. 5–8, pp. 441–448, 2012.

25

Flexural Strength Analysis of Starch Based Biodegradable Composite Using Areca Frond Fibre Reinforcement

Srinivas Shenoy Heckadka,[1] Manjeshwar Vijaya Kini,[1] Raghuvir Pai Ballambat,[1] Satish Shenoy Beloor,[2] Sathish Rao Udupi,[1] and Ullal Achutha Kini[1]

[1] *Department of Mechanical and Manufacturing Engineering, Manipal Institute of Technology, Manipal University, Manipal 576 104, India*
[2] *Department of Aeronautical and Automobile Engineering, Manipal Institute of Technology, Manipal University, Manipal 576 104, India*

Correspondence should be addressed to Srinivas Shenoy Heckadka; shenoysrinivas@rediffmail.com

Academic Editor: Archie Lockamy

Natural fibres and biodegradable matrices are being considered nowadays as substitutes to synthetic fibre reinforced polymer composites mainly in sectors where high load carrying capacity and high strength are not prerequisites. Present study utilizes biodegradable matrix composite prepared by varying the weight of the base material (95–170 g), binder (5–10 g), and plasticizer (5–20 g) with treated areca frond fibres as reinforcement. Contents are transferred to a pneumatic press, compacted, and subjected to curing. Taguchi method with L8 orthogonal array was used to reduce the number of experiments. Specimens for the flexural tests are cut out from the prepared laminates and tests are performed using UTM. Maximum flexural strength of 16.97 MPa was obtained with a combination of base (170 g), binder (10 g), and plasticizer (5 g). Analysis of the results indicated that plasticizer has the maximum effect on flexural strength of the biodegradable composites.

1. Introduction

1.1. Composites. Composites are novel engineering materials made from two or more constituents that remain separate and distinct on a macroscopic level while forming a single component. There are two categories of constituent materials: matrix and reinforcement. At least one portion of each type is required. The matrix material surrounds and supports the reinforcement by maintaining their relative positions. The reinforcement imparts special physical (mechanical and electrical) properties to enhance the matrix properties [1]. A synergism produces material properties unavailable from naturally occurring materials. Composites provide the designer, fabricator, equipment manufacturer, and consumer with sufficient flexibility to meet the demands presented by different environments and special requirements. Thus composites, due to their heterogeneous composition, provide unique flexibility in design along with other attributes like superior directional properties, high specific strength, and stiffness properties. Manufacturing of composites with complex shapes especially moulding with polymer composites, reparability, corrosion resistance, durability, adaptability, and cost effectiveness has attracted their use in several engineering and other applications [2].

1.2. Biodegradable Composites. Environmental concerns and stricter government regulations on recycling materials pressed scientists to develop new materials mainly from renewable resources. Using natural fibres in biodegradable matrices can give numerous advantages with regard to fibres traditionally used in composites. The attractive features of these fibres are their abundant availability, low cost, light weight, and high specific modulus in contrast to the synthetic fibres. Biodegradable composites are developed with a perspective of sustainability. The concept is to maximize the performance of the blend and deliver the optimal functional

FIGURE 1: Areca frond fibre extraction and treatment: (a) extracted fibres; (b) treated fibres.

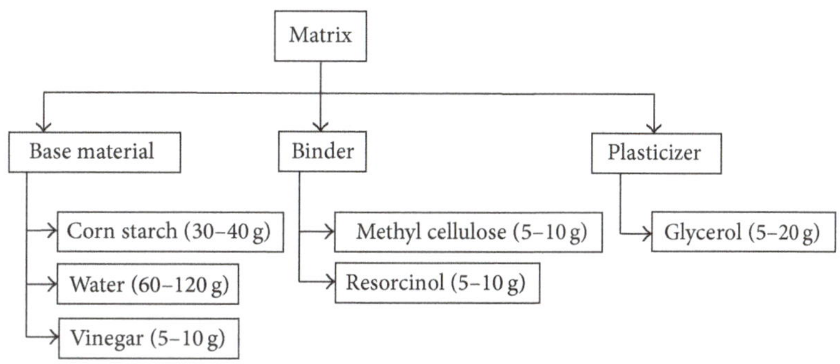

FIGURE 2: Constituents of biodegradable matrix.

requirements. The hydrophilic nature of the natural polymers together with reinforcing strength of the natural fibres results in environment friendly composite materials. Biodegradable composite are formulated by using a suitable matrix (biodegradable resin) and a reinforcement of natural fibres. Natural fibres such as hemp, sisal, kenaf, bamboo, bagasse, areca, and jute along with biodegradable polymers such as polylactic acid (PLA) [3, 4] and polybutylene succinate (PBS) are extensively used as green composites [5].

Normally, natural fibre polymer composites are fabricated by using traditional manufacturing techniques which are designed for conventional fibre reinforced polymer composites and thermoplastics. These techniques include resin transfer moulding, vacuum infusion, compression moulding, direct extrusion, compounding, and injection moulding [6]. However techniques such as compression moulding and hand layup methods have been successfully used for producing biodegradable composites with controllable quality. Venkata Reddy et al. fabricated alkali treated kapok/sisal reinforced polyester composites by hand layup technique [7]. Srinivasababu et al. carried out a study on okra, sisal, and banana fibre reinforced polyester composites using hand layup technique. The variations in mechanical properties and thermal characteristics of these composites are evaluated [8]. Anuar and Zuraida evaluated the improvement in mechanical properties of kenaf bast fibre reinforced thermoplastic elastomer composite using compression moulding technique [9].

1.3. Taguchi Concept. Taguchi method is one of the effective techniques to reduce number of experiments while retaining quality of data collection. The first step in designing the experiments is the proper selection of factors and levels. In the present study, three main factors base material (A), binder (B), and plasticizer (C) in two levels (Level 1 and Level 2) are considered.

Thian et al. have investigated the sintered properties such as sintering temperature, heating rate, holding time, and cooling rate of Ti-6Al-4V/HA tensile bars using Taguchi method using L_9 3-level orthogonal array [1]. Majdzadeh-Ardakani and Nazari fabricated thermoplastic nanocomposites using melt extrusion technique. The effects of clay cation, water, polyvinyl alcohol, clay contents on clay intercalation, and mechanical properties are also investigated according to the Taguchi experimental design method [10]. Palanikumar studied the influence of drilling parameters such as spindle speed and feed rate on the thrust force, surface roughness, and delamination characteristics of the composites using Taguchi's L_{16}, 4-level orthogonal array [11]. Satapathy et al. have investigated the processing and characterization of epoxy matrix composites reinforced with short flakes obtained from the scales of a typical fresh water fish and also reported erosion wear characteristics. Analysis of variance (ANOVA) is performed and signal to noise (S/N) ratios are determined. Experimental results revealed that the rate of erosion by impact of solid erodent is greatly influenced by various control factors [12]. Pal and Gauri studied the effectiveness of five performance metrics that are used for optimization of multiple response problems. Multiple regression-based weighted signal-to-noise ratio as

(a) (b)

FIGURE 3: Preparation of biodegradable matrix composites: (a) mixing and heating of matrix constituents; (b) compaction.

FIGURE 4: Flexural specimen testing.

TABLE 1: Selected factors and their respective levels (coded units).

Factors	Level 1	Level 2
Base material (A)	−1	+1
Binder (B)	−1	+1
Plasticizer (C)	−1	+1

a performance metric is the most effective in finding an optimal solution for multiple response problems [13].

2. Experimental Details

2.1. Fibre Extraction and Treatment. Areca fronds are collected from locally available areca plantations. Fronds thus obtained are soaked in water for 4 days. The soaking process loosens the fibres and can be extracted out easily [14]. The Areca frond fibres are separated from the fronds using a wire brush. Finally, the fibers are washed again with fresh water and dried at room temperature. The dried fibres are "untreated fibres." The extracted areca frond fibres are treated

TABLE 2: Taguchi L_8 orthogonal array of designed experiments (coded units).

Experiment number	Factors		
	Base material, A	Binder, B	Plasticizer, C
1	−1	−1	−1
2	−1	−1	+1
3	−1	+1	−1
4	−1	+1	+1
5	+1	−1	−1
6	+1	−1	+1
7	+1	+1	−1
8	+1	+1	+1

TABLE 3: Selected factors and their respective levels (uncoded units).

Factors	Level 1	Level 2
Base material, A (g)	95	170
Binder, B (g)	5	10
Plasticizer, C (g)	5	20

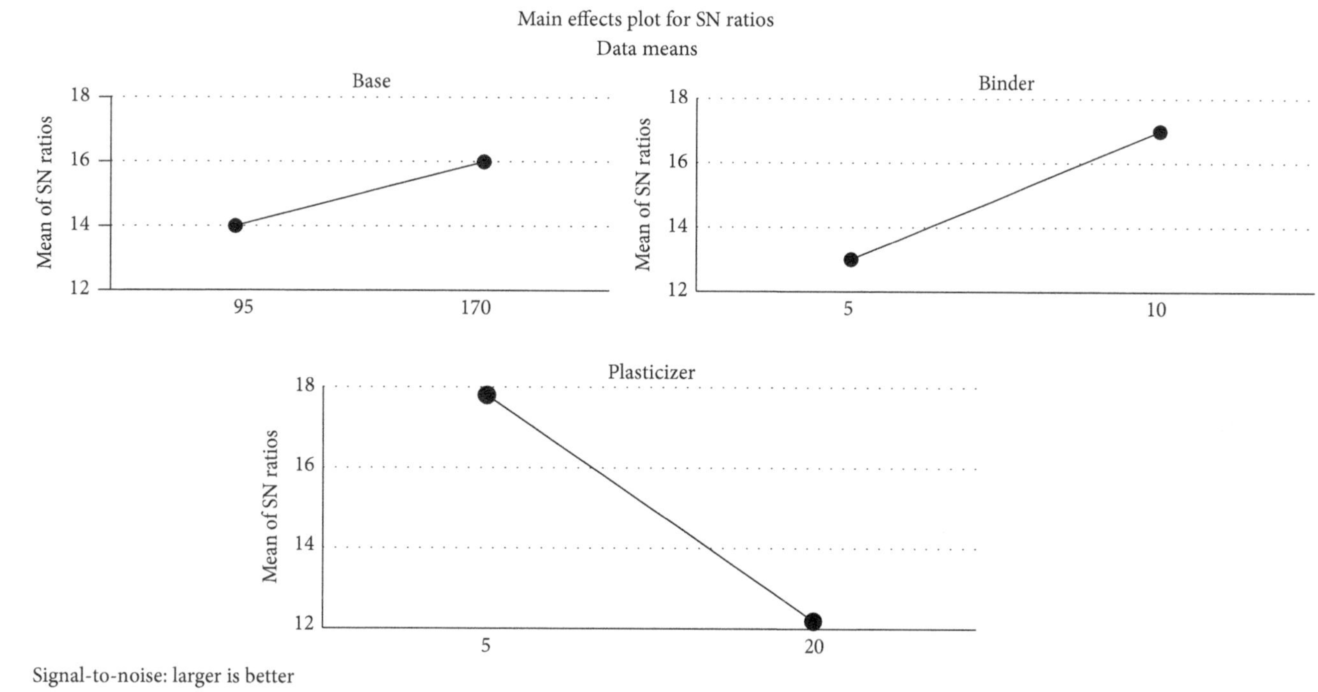

FIGURE 5: Response plot for flexural strength.

TABLE 4: Taguchi L_8 orthogonal array of designed experiments (uncoded units).

Experiment number	Combination of factors and levels	Factors		
		Base material, A (g)	Binder, B (g)	Plasticizer, C (g)
1	A1B1C1	95	5	5
2	A1B1C2	95	5	20
3	A1B2C1	95	10	5
4	A1B2C2	95	10	20
5	A2B1C1	170	5	5
6	A2B1C2	170	5	20
7	A2B2C1	170	10	5
8	A2B2C2	170	10	20

TABLE 5: S/N ratios for different characteristics.

S/N ratio	Description	Equation
Larger is better	Maximize the response	$S/N_i = -10\log\left(\sum(1/Y_u^2)/N_i\right)$
Smaller is better	Minimize the response	$S/N_i = -10\log\left(\sum(Y_u^2)/N_i\right)$

Note: i = experiment number; u = trial number; N_i = number of trials for experiment i.

with 1 N solution of Na_2CO_3 where the total volume of solution was 10 times the weight of the fibres. The fibers are kept in this alkaline solution for 12 h at room temperature. Finally fibres are dried at room temperature for 48 h to get the "alkali treated fibres" [15]. Figure 1 shows the extracted and treated areca fibres. Chemical composition of areca frond fibres is Hemi cellulose (35–64.8%), Lignin (13–24.8%), and ash (4–4.4%) [16].

2.2. Matrix Preparation. Figure 2 shows the basic ingredients of the biodegradable matrix.

The preparation of biodegradable matrix starts by weighing the constituents of the Base material (corn starch powder, water, and vinegar). The weighed ingredients are then transferred into a mixing vessel and evenly mixed by using a magnetic stirrer. Binders (methyl cellulose and resorcinol) and plasticizer (glycerol) are added to the mix and stirred for about 15 minutes. Further the mixture is transferred to a steel vessel and heated at 140°C until it turns into a semisolid paste [14]. As the matrix mixture turns into a semisolid paste the fibres are added to the mix and the contents are transferred to a mould (300 mm × 300 mm) and spread evenly to obtain a homogenous distribution using hand layup

Table 6: Flexural strength results.

Experiment number	Combination of factors and levels	Factors Base A (Wt. grams)	Binder B (Wt. grams)	Plasticizer C (Wt. grams)	Response average maximum flexural strength (MPa)
1	A1B1C1	95	95	95	10.48
2	A1B1C2	95	95	170	4.59
3	A1B2C1	95	170	95	6.49
4	A1B2C2	95	170	170	3.38
5	A2B1C1	170	95	95	4.13
6	A2B1C2	170	95	170	2.81
7	A2B2C1	170	170	95	16.97
8	A2B2C2	170	170	170	6.64

Table 7: Response table for signal-to-noise ratios.

Level	Base (A)	Larger is better Binder (B)	Plasticizer (C)
1	14.42	13.04	17.70
2	15.58	16.97	12.31
Delta	1.16	3.93	5.39
Rank	3	2	1

technique. Figure 3 shows the preparation of biodegradable matrix composites. The mix is compacted to form sheets using pneumatic press. The inner walls of the mould are coated with a thin layer of wax as a release agent. Further composites are cured in a hot chamber (steam heating) at 85°C for 24 h.

2.3. Designing the Experiment. In this study, main factors considered are base material (A), binder (B), and plasticizer (C) in two levels (Table 1). The factors and their levels have been selected based on past literature, practical aspects, and results of screening tests. For Taguchi method with three factors and two levels for each factor, a standard L_8 orthogonal array is employed as shown in Table 2. The uncoded units of Taguchi L_8 orthogonal array are provided in Tables 3 and 4. Flexural strength is used for response calculations.

2.4. Signal-to-Noise (S/N) Ratio. The Taguchi method is commonly used to verify the robustness of products. The method helps to identify control factors that reduce variability by minimizing the effects of uncontrollable factors. Taguchi method often uses a 2-step optimization process. In step 1 use the *S/N* ratio to identify those control factors that reduce variability. In step 2, identify control factors that bring the mean to target and have little or no effect on the *S/N* ratio. The signal-to-noise (*S/N*) ratio measures how the response varies relative to the nominal or target value under different noise conditions. Table 5 shows the *S/N* ratios for different characteristics. For the present study, "larger is better aspect" is considered.

2.5. Flexural Testing. Composite specimens are tested for flexural strength according to ASTM D 790 [17, 18]. Experiments are performed using Universal Testing Machine and Instron (Model 3366). Test specimens with dimensions 64 × 12 × 5 mm and cross head speed of 2 mm/min are used. Figure 4 shows the UTM test rig.

3. Results and Discussion

Results of the flexural strength tests performed on the fabricated biodegradable matrix composite are given in Table 6.

The maximum value of flexural strength is 16.97 MPa and minimum value is 2.81 MPa.

Based on ranking of the factors using Taguchi analysis, plasticizer has the maximum effect (Table 7). Ranking obtained by Taguchi method points out that plasticizer is ranked 1 followed by binder (rank 2) and base material (rank 3).

Referring to Figure 5 the graph for Base material is very close to being parallel to mean line, indicating that it has a very low effect on the flexural strength. Both Binder and Plasticizer have steep slopes indicating that they are important factors contributing to increase in flexural strength.

4. Conclusions

The conclusions are as follows.

(i) Optimal combination of the factors and levels for the biodegradable matrix composite is base material (170 g), binder (10 g), and plasticizer (5 g).

(ii) The flexural strength of the laminate obtained suggests that the laminate could be used to fabricate products like pen stand, packaging boxes, and similar products.

Conflict of Interests

The authors declare that there is no conflict of interests regarding the publication of this paper.

References

[1] E. S. Thian, N. H. Loh, K. A. Khor, and S. B. Tor, "Microstructures and mechanical properties of powder injection molded Ti-6Al-4V/HA powder," *Biomaterials*, vol. 23, no. 14, pp. 2927–2938, 2002.

[2] J. P. Davim, "Study of drilling metal-matrix composites based on the Taguchi techniques," *Journal of Materials Processing Technology*, vol. 132, no. 1–3, pp. 250–254, 2003.

[3] B. Bax and J. Müssig, "Impact and tensile properties of PLA/Cordenka and PLA/flax composites," *Composites Science and Technology*, vol. 68, no. 7-8, pp. 1601–1607, 2008.

[4] S. Ochi, "Mechanical properties of kenaf fibers and kenaf/PLA composites," *Mechanics of Materials*, vol. 40, no. 4-5, pp. 446–452, 2008.

[5] T. H. Nam, S. Ogihara, N. H. Tung, and S. Kobayashi, "Effect of alkali treatment on interfacial and mechanical properties of coir fiber reinforced poly(butylene succinate) biodegradable composites," *Composites B: Engineering*, vol. 42, no. 6, pp. 1648–1656, 2011.

[6] M.-P. Ho, H. Wang, J.-H. Lee et al., "Critical factors on manufacturing processes of natural fibre composites," *Composites Part B: Engineering*, vol. 43, no. 8, pp. 3549–3562, 2012.

[7] G. Venkata Reddy, S. Venkata Naidu, T. Shobha Rani, and M. C. S. Subha, "Compressive, chemical resistance, and thermal studies on kapok/sisal fabrics polyester composites," *Journal of Reinforced Plastics and Composites*, vol. 28, no. 12, pp. 1485–1494, 2009.

[8] N. Srinivasababu, K. Murali Mohan Rao, and J. Suresh Kumar, "Experimental determination of tensile properties of okra, sisal and banana fiber reinforced polyester composites," *Indian Journal of Science and Technology*, vol. 2, no. 7, pp. 35–38, 2009.

[9] H. Anuar and A. Zuraida, "Improvement in mechanical properties of reinforced thermoplastic elastomer composite with kenaf bast fibre," *Composites Part B: Engineering*, vol. 42, no. 3, pp. 462–465, 2011.

[10] K. Majdzadeh-Ardakani and B. Nazari, "Improving the mechanical properties of thermoplastic starch/poly (vinyl alcohol)/clay nanocomposites," *Composites Science and Technology*, vol. 70, no. 10, pp. 1557–1563, 2010.

[11] K. Palanikumar, "Experimental investigation and optimisation in drilling of GFRP composites," *Measurement*, vol. 44, no. 10, pp. 2138–2148, 2011.

[12] A. Satapathy, A. Patnaik, and M. K. Pradhan, "A study on processing, characterization and erosion behavior of fish (*Labeo-rohita*) scale filled epoxy matrix composites," *Materials and Design*, vol. 30, no. 7, pp. 2359–2371, 2009.

[13] S. Pal and S. K. Gauri, "Assessing effectiveness of the various performance metrics for multi-response optimization using multiple regression," *Computers and Industrial Engineering*, vol. 59, no. 4, pp. 976–985, 2010.

[14] R. P. Swamy, G. C. M. Kumar, Y. Vrushabhendrappa, and V. Joseph, "Study of areca-reinforced phenol formaldehyde composites," *Journal of Reinforced Plastics and Composites*, vol. 23, no. 13, pp. 1373–1382, 2004.

[15] N. H. Padmaraj, M. Vijaya Kini, B. Raghuvir Pai, B. Sathish Shenoy, and Krishnamoorthy, "Effect of different treatment media on the breaking strength of long Areca nut fibers," in *Proceedings of the 4th National Conference on Trends in Mechanical Engineering (TIME '10)*, pp. 30–33, 2010.

[16] C. V. Srinivasa, A. Arifulla, N. Goutham et al., "Static bending and impact behaviour of areca fibers composites," *Materials and Design*, vol. 32, no. 4, pp. 2469–2475, 2011.

[17] B.-H. Lee, H.-J. Kim, and W.-R. Yu, "Fabrication of long and discontinuous natural fiber reinforced polypropylene biocomposites and their mechanical properties," *Fibers and Polymers*, vol. 10, no. 1, pp. 83–90, 2009.

[18] M. M. Kabir, H. Wang, K. T. Lau, F. Cardona, and T. Aravinthan, "Mechanical properties of chemically-treated hemp fibre reinforced sandwich composites," *Composites Part B: Engineering*, vol. 43, no. 2, pp. 159–169, 2012.

Permissions

All chapters in this book were first published in IJME, by Hindawi Publishing Corporation; hereby published with permission under the Creative Commons Attribution License or equivalent. Every chapter published in this book has been scrutinized by our experts. Their significance has been extensively debated. The topics covered herein carry significant findings which will fuel the growth of the discipline. They may even be implemented as practical applications or may be referred to as a beginning point for another development.

The contributors of this book come from diverse backgrounds, making this book a truly international effort. This book will bring forth new frontiers with its revolutionizing research information and detailed analysis of the nascent developments around the world.

We would like to thank all the contributing authors for lending their expertise to make the book truly unique. They have played a crucial role in the development of this book. Without their invaluable contributions this book wouldn't have been possible. They have made vital efforts to compile up to date information on the varied aspects of this subject to make this book a valuable addition to the collection of many professionals and students.

This book was conceptualized with the vision of imparting up-to-date information and advanced data in this field. To ensure the same, a matchless editorial board was set up. Every individual on the board went through rigorous rounds of assessment to prove their worth. After which they invested a large part of their time researching and compiling the most relevant data for our readers.

The editorial board has been involved in producing this book since its inception. They have spent rigorous hours researching and exploring the diverse topics which have resulted in the successful publishing of this book. They have passed on their knowledge of decades through this book. To expedite this challenging task, the publisher supported the team at every step. A small team of assistant editors was also appointed to further simplify the editing procedure and attain best results for the readers.

Apart from the editorial board, the designing team has also invested a significant amount of their time in understanding the subject and creating the most relevant covers. They scrutinized every image to scout for the most suitable representation of the subject and create an appropriate cover for the book.

The publishing team has been an ardent support to the editorial, designing and production team. Their endless efforts to recruit the best for this project, has resulted in the accomplishment of this book. They are a veteran in the field of academics and their pool of knowledge is as vast as their experience in printing. Their expertise and guidance has proved useful at every step. Their uncompromising quality standards have made this book an exceptional effort. Their encouragement from time to time has been an inspiration for everyone.

The publisher and the editorial board hope that this book will prove to be a valuable piece of knowledge for researchers, students, practitioners and scholars across the globe.

List of Contributors

ChaBum Lee
Department of Mechanical Engineering, University of South Carolina, 300 Main Street, Columbia, SC 29208, USA

Afef Bougharriou,Mohieddine Jeridi,Mohamed Hdiji, Anoir Boughrira and Kacem Sa
UGPMM, National Engineering School of Sfax, Route Soukra Km 2.5, BP 1173, Sfax, Tunisia

Sagil James and MuraliM. Sundaram
Department of Mechanical and Materials Engineering, University of Cincinnati, Cincinnati, OH 45221, USA

Thongchai Arunchai, Kawin Sonthipermpoon and Phisut Apichayakul
Department of Industrial Engineering, Faculty of Engineering, Naresuan University, Phitsanulok 65000,Thailand

Kreangsak Tamee
Department of Computer Science and Information Technology, Faculty of Science, Naresuan University, Phitsanulok 65000,Thailand

Godfrey C. Onwubolu and Farzad Rayegani
School of Mechanical and Electrical Engineering & Technology, Faculty of Applied Science and Technology, Sheridan Institute of Technology and Advanced Learning, Brampton, ON, Canada L6Y 5H9

Ravi Koirala, Sailesh Chitrakar, Amod Panthee, Hari Prasad Neopane and Bhola Thapa
Turbine Testing Lab, Kathmandu University, Dhulikhel 45200, Nepal

Ameen Topa and Qasim H. Shah
Department of Mechanical Engineering, International Islamic University Malaysia, Jalan Gombak, 53100 Kuala Lumpur, Malaysia
Nicolas Gardan
CAD/CAE Research Centre DINCCS of MICADO, 08 005 Charleville-Méziéres, France

Deepak Doreswamy, Devineni Anjaiah and N. Yagnesh Sharma
Department of Mechanical and Manufacturing Engineering, Manipal Institute of Technology, Manipal University, Manipal 576104, India

Basavanna Shivamurthy
School of Engineering & IT, Department of Mechanical Engineering, Manipal University Dubai Campus, Dubai International Academic City, P.O. Box 345050, Dubai, UAE

F. Javier Ramírez
EXPAL Systems, Avenida Parten´on 16, 28042Madrid, Spain
School of Industrial Engineering, University of Castilla-La Mancha, Albacete, Spain

Rosario Domingo
School of Industrial Engineering, UNED, 28042 Madrid, Spain

Michael S. Packianather
School of Engineering, Cardiff University, Cardiff, UK

Miguel A. Sebastian
School of Industrial Engineering, UNED, 28042 Madrid, Spain

I. Vicario, I. Erauskin, L. M. Plaza, I. Crespo and P. Caballero
Tecnalia Research & Innovation, 20009 San Sebasti´an, Spain

J. K. Idoiaga
Industrias Lebario, 48213 Izurza, Spain

E. Arratibel
Fundiciones Inyectadas Alavesas S.A., 01230 Nanclares de la Oca, Spain

Siva Prasad Kondapalli
Department of Mechanical Engineering, Anil Neerukonda Institute of Technology & Sciences, Visakhapatnam 530 003, India

Srinivasa Rao Chalamalasetti
Department of Mechanical Engineering, AU College of Engineering, Andhra University, Visakhapatnam 531 162, India

Nageswara Rao Damera
Centurion University of Technology&Management, Odisha 761 211, India

Raju Bangalore Singe Gowda
Department of Mechanical Engineering, REVA ITM, Bangalore, Karnataka 560037, India

Chandra Sekhar Udayagiri
Institution of Engineers (India), Gachibowli, Hyderabad 500032, India

Drakshayani Doulat Narendra
Department of Mechanical Engineering, Sir MVIT, Bangalore, Karnataka 562157, India

J. J. Liou and T. I. El-Wardany
United Technologies Research Center, 411 Silver Lane, East Hartford, CT 06108, USA

Rastee D. Koyee, Uwe Heisel and Rocco Eisseler
Institute for Machine Tools, University of Stuttgart, Holzgartenstraße 17, 70174 Stuttgart, Germany

Siegfried Schmauder
IMWF, University of Stuttgart, Pfaffenwaldring 32, 70569 Stuttgart, Germany

Deok-Hyun Yoon
College of Civil Engineering, Nanjing Forestry University, Jiangsu, Nanjing 210037, China
Samin Information Co., Ltd., Busan 612-020, Republic of Korea

Hee Jin Kang and Jin Choi
Advanced Ship Research Division, Korea Research Institute of Ships & Ocean Engineering, Daejeon 305-343, Republic of Korea

Deok Eun Kim
Samin Information Co., Ltd., Busan 612-020, Republic of Korea

Xin Huang
College of Civil Engineering, Nanjing Forestry University, Jiangsu, Nanjing 210037, China

Nripen Mondal, Biswajit Sing Sardar, Ranendra Nath Halder and Santanu Das
Department of Mechanical Engineering, Kalyani Government Engineering College, Kalyani,West Bengal 741235, India

Vasim Shaikh
Department of Applied Engineering, Safety and Technology, Millersville University of Pennsylvania, Millersville, PA 17551, USA

Nourredine Boubekri
Department of Engineering Technology, University of North Texas, Denton, TX 76207, USA

ThomasW. Scharf
Department of Materials Science and Engineering, University of North Texas, Denton, TX 76207, USA

Karuppuchamy Annadurai
Department of Mathematics, Government Arts College, Udumalpet, Tirupur District, Tamil Nadu 642126, India

Teepu Sultan, Anish Kumar and Rahul Dev Gupta
Department of Mechanical Engineering, M. M. University, Mullana, Ambala, Haryana 133203, India

S. Balamurali and M. Usha
Department of Mathematics, Kalasalingam University, Krishnankoil, Tamil Nadu. 626 126, India

NikMizamzulMehat
Department of Mould Technology, Kolej Kemahiran Tinggi MARA, Balik Pulau, Genting, 11000 Balik Pulau, Penang, Malaysia

School of Mechanical Engineering, Universiti Sains Malaysia, Engineering Campus, 14300 Nibong Tebal, Penang, Malaysia

Shahrul Kamaruddin and Abdul Rahim Othman
School of Mechanical Engineering, Universiti Sains Malaysia, Engineering Campus, 14300 Nibong Tebal, Penang, Malaysia

Raghavendra Kamath Cholpadi
Mechanical & Manufacturing Engineering Department, MIT, Manipal, Karnataka 576104, India

Appu Kuttan
Mechanical Engineering Department, NITK, Surathkal, Karnataka 575025, India

Luca Sorrentino and Sandro Turchetta
Department of Civil and Mechanical Engineering, University of Cassino and Southern Lazio, 03043 Cassino, Italy

Srinivas Shenoy Heckadka, Manjeshwar Vijaya Kini, Raghuvir Pai Ballambat, Sathish Rao Udupi and Ullal Achutha Kini
Department of Mechanical and Manufacturing Engineering, Manipal Institute of Technology, Manipal University, Manipal 576 104, India

Satish Shenoy Beloor
Department of Aeronautical and Automobile Engineering, Manipal Institute of Technology, Manipal University, Manipal 576 104, Ind